# Medicinal Agroecology

## Reviews, Case Studies, and Research Methodologies

Edited by
Immo Norman Fiebrig

**CRC Press**
Taylor & Francis Group
Boca Raton  London  New York

CRC Press is an imprint of the
Taylor & Francis Group, an **informa** business

First edition published 2023
by CRC Press
6000 Broken Sound Parkway NW, Suite 300, Boca Raton, FL 33487-2742

and by CRC Press
4 Park Square, Milton Park, Abingdon, Oxon, OX14 4RN

*CRC Press is an imprint of Taylor & Francis Group, LLC*

© 2023 Taylor & Francis Group, LLC

*Library of Congress Cataloging-in-Publication Data*
Names: Fiebrig, Immo, editor.
Title: Medicinal agroecology : reviews, case studies and research methodologies / edited by Immo Fiebrig.
Description: First edition. | Boca Raton, FL : CRC Press, 2023. |
Includes bibliographical references and index. |
Summary: "Medicinal Agroecology : Reviews, Case Studies and Research Methodologies presents information on applications of 'green therapies' in restoration towards global sustainability. These practices connect the world of medicinal plants with ecologic farming practice creating a compassionate socio-political worldview and heartfelt scientific research towards food sovereignty and a healthier future on planet Earth"– Provided by publisher.
Identifiers: LCCN 2022040882 (print) | LCCN 2022040883 (ebook) | ISBN 9780367705565 (hbk) |
ISBN 9780367702977 (pbk) | ISBN 9781003146902 (ebk)
Subjects: LCSH: Medicinal plants. | Agricultural ecology. | Materia medica, Vegetable.
Classification: LCC SB293 .M423 2023 (print) |
LCC SB293 (ebook) | DDC 633.8/8–dc23/eng/20221219
LC record available at https://lccn.loc.gov/2022040882
LC ebook record available at https://lccn.loc.gov/2022040883

ISBN: 9780367705565 (hbk)
ISBN: 9780367702977 (pbk)
ISBN: 9781003146902 (ebk)

DOI: 10.1201/9781003146902

Typeset in Times New Roman
by Newgen Publishing UK

*Dedicated to all life forms on Planet Earth*

*Book cover:*
The Three Sisters, including their shaman sibling.
*The mystic artwork represents the intercropping agricultural system of corn* (Zea mays *L.*), beans (legumes), *and squash* (Curcubita *ssp. L.*), *more generally known in sustainable agriculture as* Three Sisters. *The planting combination is attributable to the Native Americans and is still used in modern-day Mayan culture (e.g., Mexico);* Three Sisters *in* Medicinal Agroecology *stands for soil regeneration, genetic diversity, and a balanced diet, including a leguminous medicinal plant* (Mucuna pruriens (L.) DC.
*Lucero Sarabia Salgado: visualisation & paintings*
*Vinicius Cruz: artwork & graphic design*
*MEDICINAL AGROECOLOGY*
*Imprint*

# Contents

## SECTION I   POLICIES and FRAMEWORKS

## SECTION II   INSIGHTS and OVERVIEWS

## SECTION III    CASE STUDIES and RESEARCH METHODS

Contents ix

# The Editor

**Immo Norman Fiebrig**
Honorary Associate Professor
National Centre for Macromolecular Hydrodynamics (NCMH), School of Biosciences
University of Nottingham, UK

Immo N. Fiebrig received his bachelor's degree in pharmaceutical sciences at Munich University LMU in 1991, to become a licensed pharmacist shortly afterwards. This was followed by a PhD in physical biochemistry at the University of Nottingham, UK. His work experience outside academia comprised a position as deputy manager of a German chemist and organic food shop, as monitor of clinical studies in the pharmaceutical industry, head of department of a pharmaceutical database publisher, and healthcare project manager in an advertising agency. For the 12 years that followed, Immo worked as a freelance service provider for the healthcare publishing industries and as a project manager in pharmaceutical development. He published his first book, *Edible Cities*, on urban permaculture following a career change and returned to academia to carry out research on permaculture and agroecological systems in commercial settings at the Centre of Agroecology, Water and Resilience (CAWR) of Coventry University, UK. Since 2019, he has returned to Munich and works as an independent researcher, lecturer (University of Nottingham and Hohenheim University), author, and book editor. This book, *Medicinal Agroecology*, represents the joint effort of many researchers and the synthesis of Immo's career experience.

# Contributors

**Anna Elisabeth Bach**
Foundation Sowa Rigpa – The Buddhist Science of Healing, Boppard, Germany

As a teacher of Buddhist Medicine, Anna has been at home in two cultures since 1998: the western culture in Germany and the eastern one in Nepal. She is recognised as a natural health professional in Germany and has studied Tibetan Buddhism, as well as Sowa Rigpa (Tibetan Buddhist Medicine). During the lifetime of Lopön Tsechu Rinpoche, and for a total of six years, she lived in Sangye Chö Ling monastery close to Swoyambhu Stupa in Kimdol/Kathmandu. Under the guidance of Prof. Dr. Pasang Yonten Arya, she completed a four-year study programme of Tibetan Medicine in Switzerland. Anna is an *Amchi* (Tibetan physician) as well as a practitioner of Tibetan Buddhist Medicine; she received the Sowa-Rigpa-Transmissions from Dr. Kunsang Dorje in Kathmandu and is fluent in the Tibetan language.

**Subramani Paranthaman Balasubramani**
Department of Natural Sciences, Albany State University, Albany, GA, USA

Being a first-generation graduate from an agriculture-based family, my interest in Ayurveda was elicited when I joined FRLHT, in Bangalore, India, during 2008. Discussions with colleagues at FRLHT fuelled several curiosity-based questions that eventually led to my doctoral research on the concept of 'Rasayana' (Ayurvedic nutraceuticals). I studied the role of Dadima (pomegranate) Rasayana in the regulation of iron metabolism and healthy ageing. Understanding the necessity of laboratory models for studying Ayurvedic concepts, I developed yeast cell and drosophila tests for studying the Ayurvedic properties of herbs with the help of national and international collaborators at FRLHT-TDU. I moved to the United States in 2018 as a Research Associate with Florida Agricultural and Mechanical University's Center for Viticulture and Small Fruit Research and established a small fruit-based nutraceutical testing facility. Currently, I work as an Assistant Professor of Biology in the Department of Natural Sciences, Albany State University, GA, USA. My research interest is in metabolic homeostasis through medicinal herbs and agro-nutraceuticals.

**Larissa Bombardi**
Department of Geography, FFLCH-USP, University of São Paulo, Brazil

Larissa Bombardi is an Associate Professor at the Department of Geography – University of São Paulo – Brazil, currently living in Brussels. She has been a specialist on the subject of pesticide use for the last 12 years, with dozens of lectures, several published articles, and more than 100 interviews given on the matter, both nationally (in Brazil) and internationally. She is the author of the atlases, *A Geography of Agrotoxins Use in Brazil and its Relations to the European Union*, launched in 2019 in its English edition in Europe (Scotland and Germany) and *Geography of Asymmetry: Circle of Poison and Molecular Colonialism in the Commercial Relationship Between Mercosur and the European Union*, launched in 2021 at the European Parliament. She is also a member of the National Forum to Combat the Impacts of Pesticides (Brazil) and a Board Member of the international organisation 'Justice Pesticides'.

**Immo Norman Fiebrig**
National Centre for Macromolecular Hydrodynamics (NCMH), School of Biosciences
University of Nottingham, Sutton Bonington, UK

Having been born in Frankfurt, Germany, I spent many formative years in São Paulo, Brazil and Mexico City. Back in Germany, I graduated with a bachelor's degree in pharmaceutical sciences, followed by a PhD in biochemistry. Starting to work as a community pharmacist in 1996, I quickly realised that patients' health remained compromised despite the medication I would dutifully dispense to them. At the time, I had a striking experience with a seriously ill customer ordering a rare herbal medicine. This fortuity set the stage for Chapter 3 in this book. After more than 20 years in healthcare, I was forced to change to a new sector for health issues of my own. I went back to academic research. This time, my investigations were related to permaculture as a sustainability concept, with agroecology as its underlying scientific discipline. One of the results of my work is the present edition of *Medicinal Agroecology*, aimed at interconnecting 'restoration of health' and 'regeneration of the environment'. Since 2019, I have been working as an independent researcher and author in medicinal agroecology, whilst advocating for overall 'greener' therapies.

### Francisco Galindo Maldonado
Department of Ethology, Wildlife and Laboratory Animals, Faculty of Veterinary Medicine and Animal Science, National Autonomous University of Mexico, Ciudad Universitaria, Mexico City, Mexico

Profesor Galindo received his bachelor's degree at the Faculty of Veterinary Medicine and Zootechnics, FMVZ-UNAM followed by a doctorate in Ethology and Animal Welfare (PhD), University of Cambridge, UK (1996). He became a full Professor in the Department of Ethology and Wildlife, FMVZ-UNAM and was Head of the Department of Ethology and Wildlife, FMVZ-UNAM (from its foundation in 1993 to 2002).

Francisco founded the Mexican Society of Veterinary Ethology, AC (SOMEV), presiding over it between 1996 and 1998, and since 2001, he has been the secretary for Latin America of the International Society of Applied Ethology (ISAE). He organised the first national congresses of SOMEV and the first Latin American congress in the area, achieved a fruitful meeting of professionals and academics from different professional sectors interested in the applications of the study of animal behaviour and promoted, nationally and internationally, teaching and research in applied ethology.

### Pedro Geraldo González-Pech
Autonomous University of Yucatán, Mérida, Yucatán, México

Professor González is a lecturer and researcher at the Faculty of Veterinarian Science in Yucatán, Mexico.
He graduated from the FMVZ/UADY, México, gained an MSc in Agriculture and Environment (ENSAIA de Nancy, France) and a PhD in Agricultural Sciences (AgroParisTech-INRA, France), with experience in the feeding behaviour of small ruminants in free-grazing of heterogeneous vegetation in France, Mexico, and Brazil. Current work is centred on the use of a tropical deciduous forest for sheep and goats, and self-medication (control of gastrointestinal nematodes infection) through the selection and consumption of secondary compounds contained in native plants.

### Anja Greinwald
University of Applied Forest Sciences, Schadenweilerhof, Rottenburg am Neckar, Germany.

Dr. Greinwald is a PharmaPlants Project Associate and obtained her PhD on the topic of "Wild collection of medicinal plants as an ecosystem service supporting the conservation of high nature value grasslands".]

### Marion Johnson
Listen Limited, New Zealand

Growing up between Africa, New Zealand, and the UK and spending all my spare time in the bush or on farms, I have always believed in the power of plants. Whether they are healing the landscape or its inhabitants they are always teaching us – if only we will listen. I have degrees in Agriculture and Environmental Biology and explored parasites in my postgraduate education. My focus since then has been on the connections between plants, animal health, and landscapes, marrying agriculture and environmental science. I have been fortunate to work with many farmers and healers on agroecological projects, combining a cultural knowledge of plants with management objectives to reinstate healthy landscapes and to provide livestock with choices, facilitating healthy and contented lives.

### Bettina Klocke

Julius Kühn Institute – Federal Research Centre for Cultivated Plants,
Institute for Strategies and Technology Assessment, Kleinmachnow, Germany

I was born in Paderborn and became interested in natural sciences during my school years. In 1992, I started studying horticultural sciences at Leibniz University Hannover, followed by a PhD in agricultural sciences at Martin Luther University Halle-Wittenberg. My focus still lies in the field of phytopathology. The spread of fungal pathogens and their control has been central to my work since I joined the Institute for Strategies and Technology Assessment in 2012. Sustainable plant protection and the transfer of knowledge to agricultural practice are of great importance here, in order to secure yields in the long term and still act in an environmentally friendly way.

### Andrea Krähmer

Julius Kühn Institute – Federal Research Centre for Cultivated Plants,
Institute for Ecological Chemistry, Plant Analysis and Stored Product Protection, Berlin, Germany

Born in Berlin, I became interested in chemistry and plants at an early age through observing nature in my parents' garden and through the chemical textbooks of my grandmother, a chemical engineer by training. My school education, which therefore focused on the natural sciences, was followed by a degree in chemistry in Jena from 1999. After completing my diploma in the field of organic synthesis of nature-like complex compounds, I switched to analytical chemistry for my doctoral thesis. The focus here was set on the analysis of plant leaf surfaces as the first physical and chemical barrier against pathogens and insects. After my doctorate, plants remained the central object of research, which I continued from 2009 as a scientist in the Department of Phytochemistry at the Julius Kühn Institute. In addition to chemical phenotyping and analysis accompanying breeding, current projects focus on the identification of active plant constituents as bio-based alternatives to synthetic pesticides. Becoming Head of the Institute for Ecological Chemistry, Plant Analysis and Stored Product Protection in spring 2020, I now support the area of residue and by-product utilisation of plant materials in a more co-ordinating and administrative role.

### Sibylle Kümmritz

Julius Kühn Institute – Federal Research Centre for Cultivated Plants,
Institute for Ecological Chemistry, Plant Analysis and Stored Product Protection, Berlin, Germany

Since my childhood in Potsdam, I have been interested in nature and focussed on medicinal and aromatic plants. My mother, a pharmacist, who is also a role model for me, might have influenced this. However, she did not advise me to study pharmacy. So, I decided to study food chemistry to stay healthy, because 'you are what you eat'. Therefore, I went to Halle/S. in 2004 and came back in 2008 to finish my studies with a diploma and a thesis on aroma substances in wine at the German Federal Institute for Risk Assessment in Berlin. Then, I continued plant research with various projects on the production of secondary metabolites with plant cells and tissue cultures in the field of bioprocess engineering at the TU Dresden and completed this with my PhD on triterpenic acids from sage cell cultures. Since 2019, I have been investigating residues from steam distillation for the extraction of essential oils, especially hydrolates, as a research scientist at the Julius Kühn Institute for Ecological Chemistry, Plant Analysis and Stored Product Protection. It is amazing what (plant) nature can do!

**Juan Carlos Ku-Vera**
Autonomous University of Yucatán, Mérida, Yucatán, México

Professor Ku-Vera studied Veterinary Medicine and Zootechnics at the University of Yucatan (1977–1982) and completed a master's degree in animal nutrition (1982–1983), as well as a doctorate in nutrition from ruminants (1985–1988) at the University of Aberdeen, Great Britain with fellowships from The British Council and CONACYT, respectively. He did a postdoctoral degree (1990–1991) in the metabolism study of ruminants at the National Institute of Animal Science in Tsukuba, Japan with a fellowship from the Science and Technology Agency of that country. He has served as a teacher of ruminant feeding at the Autonomous Universities of Chapingo, Tamaulipas and Yucatan.

He is currently developing research using local tree and shrub species to mitigate methane emissions in livestock systems.

**Subrahmanya Kumar Kukkupuni**
University of Trans-Disciplinary Health Sciences and Technology, Jarakabande Kaval, Bangalore, India

I am an Ayurveda physician with an M.D. and a PhD in Ayurvedic pharmacology. I am working on quality standards of natural resources used in Ayurvedic medicines, including research on substitutes for rare and endangered medicinal plants with an aim of the conservation and sustainable use of plant drugs. Effective use of rejuvenator nutraceuticals of Ayurveda is another area of my interest. I also teach pharmacognosy and pharmacology of plant drugs used in Indian traditional medicine.

**Rainer Luick**
Hochschule Rottenburg, Nature and Enviromental Protection

Prof. Dr. Rainer Luick teaches and researches at the Rottenburg University of Applied Forest Sciences. He studied biology (focus on geobotany and plant physiology) and ethnology at the Albert-Ludwigs-University Freiburg and evolutionary biology at the University of Michigan, Ann Arbor, USA. Rainer obtained his Dr. sc. agr. from University of Hohenheim. Many years of work in private water management and landscape planning followed. Since 1999 he is Professor for Nature Conservation and Landscape Management at the Rottenburg University of Applied Forest Sciences. Main research interests: natural processes in rural areas, agricultural, nature conservation and regional policy, extensive land use systems, technology assessments for the energy transition and commitment to protect the last European primeval forests. ]

**Georgina McAllister**
Centre for Agroecology, Water and Resilience, Ryton Organic Gardens, Coventry University, Coventry, UK

I am currently Assistant Professor in Stabilisation Agriculture at the Centre for Agroecology, Water and Resilience (CAWR), where my work considers the causes of fragility, including the exploitation of the natural world and concentration of wealth and power, that drive social division and increase inequalities and vulnerabilities to hazards. My interests include the transformative potential of agroecological processes and relationships to negotiate social-ecological change by re-forging more-than-human relationships and social networks based on principles of reciprocity and trust. Having spent 30 years working with non-governmental and civil society organisations on agroecology, before more recently transitioning into academia, my work has involved the defence of indigenous cosmologies and the plantlore of traditional healers and small-scale farmers. As an action researcher, I am interested in social mobilisation, and influenced by political ecology to better understand how historical and contemporary struggles for land, culture, power and nature continue to shape our sense of meaning and identity.

**Nina Moeller**
Centre for Agroecology, Water and Resilience, Ryton Organic Gardens, Coventry University, Coventry, UK

Currently Associate Professor of Political Ecology and People's Knowledge at CAWR, I have a mixed academic background in philosophy, sociology and anthropology. My research interests comprise the dynamics of sustainability transitions, including their unintended socio-ecological effects; diversity of knowledge and value systems; and more-than-human relations. My interest in plant medicine, traditional health and food systems goes beyond research and has been shaped in significant ways through friendships and exchanges with indigenous Amazonians and subsistence farmers across the world. I have worked in Latin America and Europe – as academic as well as consultant to indigenous federations, NGOs and the UN's Food and Agriculture Organization.

**M N Balakrishnan Nair**
Jarakabande Kaval, Post Attur via Yelahanka Bengaluru, India

Professor emeritus and Head of the Centre for Ethno-Veterinary Sciences and Practice at the University of Trans-Disciplinary Health Sciences and Technology (TDU), Nair holds an MSc. and PhD in botany. He has 39 years of teaching and research experience, over 45 publications and has contributed to different books. His research interests are trans-disciplinary research, One Health, documentation of ethno-veterinary practices, validation and promotion of them through mainstreaming in veterinary curricula to reduce antimicrobial's use in dairy production, and residues in animal products.

**Florin Păcurar**
Department of Grassland Management and Forage Crops, Faculty of Agriculture, University of Agricultural Sciences and Veterinary Medicine Cluj-Napoca, Cluj-Napoca, Romania

Prof. Dr. Florin Păcurar is full professor at the University of Agricultural Sciences and Veterinary Medicine of Cluj-Napoca/Romania, and Head of the Department of Grassland Management and Forage Crops. He has been working for many years on the management and maintenance of grassland, with particular interest in the biodiversity of grassland in mountainous areas. His research topics deal with the traditional management of oligotrophic grassland with HNV, the impact of climate change on grassland succession, and the effects of different fertilisation methods and quantities on floristic composition and on yield. His research group seeks to also take into account other ecosystem services beyond traditional grassland functions and to integrate them into the sustainable use of agricultural resources.

He has had experience in internationally funded interdisciplinary research projects since 2000 and established an intensive co-operation with the University of Freiburg (in particular), the University of Hohenheim and the University of Applied Sciences Rottenburg. The results of these studies were implemented in practice with the participation of various groups of stakeholders. Consequently, one can say that USAMV is one of the first institutions in Romania working, from the beginning, in project implementation with the participation of local, regional, and national stakeholders.

**Martin Pedersen**
Centre for Agroecology, Water and Resilience, Ryton Organic Gardens, Coventry University, Coventry, UK

As an Honorary Research Fellow at CAWR, I am currently exploring the idea of magic, religion and science, particularly on the intersection of soil, community, individual and planetary health, as three primary and complementary frameworks of meaning that have shaped human history. I studied at the

School of Independent Studies, University of Lancaster, composing my own degrees, with a focus on globalisation, technology, the evolution of consciousness, and the history of ideas; and wrote a transdisciplinary PhD titled 'Property, Commoning and the Politics of Free Software', which was published in The Commoner. I have keen interests in plants, food and medicine that began taking shape with a Permaculture Design Course and while working (since 2005) with indigenous people in the Amazon. For five years, I ran an agroecological guest house catering for people with special dietary needs and remain interested in many subjects from astrophysics to zoology, as well as current affairs.

**Marina Cristina Campos Peralta**
Department of Geography, Faculty of Philosophy, Literature and Human Sciences, University of São Paulo, São Paulo, SP, Brazil

Marina holds a degree in geography from the Department of Geography of the Faculty of Philosophy, Literature and Human Sciences of the University of São Paulo. She is currently a Master's student in the Graduate Program in Human Geography (FFLCH-USP) at the University of São Paulo and has a FAPESP scholarship. Since her undergraduate course, she has dedicated herself to deepening her knowledge about agroecology, agrarian geography, and rural popular university projects, having participated in teaching and agroecological projects in rural settlements and public schools in São Paulo.

**Natesan Punniamurthy**
Jarakabande Kaval, Post Attur via Yelahanka Bengaluru, India

Professor emeritus at the University of Trans-Disciplinary Health Sciences and Technology (TDU) Bangalore, India, Punniamurthy holds a BVSc, MVSc and a PhD from Madras Veterinary College, India. As professor of Pharmacology and Toxicology, he taught veterinary students (UG, PG) for over 12 years and has overall 41 years of experience in teaching, research, and extension at the Tamil Nadu Veterinary and Animal Sciences University (TANUVAS) and TDU. He has worked in neuropharmacology of feeding behaviour in poultry whilst having 20 years of research and development experience in documenting and validating veterinary herbal medicine clinically based on *Siddhayur* folk traditions. N. Punniamurthy currently trains veterinarians and farmers across India and is mainstreaming EVP in veterinary curricula.

**Jose Prieto Garcia**
School of Pharmacy and Biomolecular Sciences, Liverpool John Moores University, UK

Dr Jose M Prieto read pharmacy (1993) and obtained a PhD in pharmacology (2001) at the University of Valencia (Valencia, Spain) in the field of topical inflammation. His post-doctoral research activities include the EU-funded projects "Insect Chemical Ecology" (Department of Bioorganic Chemistry, Universita Degli Studi di Pisa, Italy) (2001-2004) and "Medicinal Cannabis" (Department of Pharmaceutical and Biological Chemistry, School of Pharmacy, University of London, UK) (2005–2006). He was then appointed as Lecturer in Pharmacognosy (UCL School of Pharmacy, 2006–2020) and Senior Lecturer in Natural Products and Phytochemistry (LJMU School of Pharmacy and Biomolecular Sciences, present). His research focuses on elucidating the effects of medicinal plants and natural products on skin conditions (inflammation and cancer), herb-drug interactions and the application of advanced techniques (Direct NMR, Artificial Intelligence) to the analysis and biological effects of complex natural products. He has supervised more than 10 PhDs, authored over 80 original papers and chapters in books, sits on the editorial board of several journals, acts as principal investigator of The Yeheb Project (www.yeheb.org) and is an active consultant to the herbal industry.

## Albert Reif

Prof. Dr. H.C. Chair of Nature Conservation and Landscape Ecology, Faculty of Environment and Natural Resources, Albert-Ludwigs University Freiburg, Freiburg, Germany

Born in Ansbach, Germany, I studied and graduated in 1978 in biology and chemistry in Würzburg; graduated with a PhD degree in 1982 in Bayreuth; worked there until my 'habilitation' in 1988; and was Professor for Site Classification and Vegetation Science at the University of Freiburg between 1989 and 1998. In 2007, I was appointed 'doctor honoris causa' of the University of Agricultural Sciences and Veterinary Medicine (USAMV) in Cluj-Napoca/Romania.

During my scientific life, when I was teaching, I and my students worked on forest vegetation in relation to site, forest use, and nature conservation. A second issue was grassland ecology. I had projects in Germany, southeast Europe (Romania, Greece), New Zealand, Asia (Indonesia, Philippines, China), and South America (Venezuela, Chile).

Since 1998 I have been retired, but still work part time with teaching and research at universities, and for nature conservation organisations and administration in SW Germany.

## Evelyn Ruşdea

Chair of Nature Conservation and Landscape Ecology, Faculty of Environment and Natural Resources, Albert-Ludwigs University Freiburg, Freiburg, Germany

Dr. Evelyn Ruşdea is a senior researcher working at University Freiburg since 1990. She was born in Romania, where she studied biology at the university in Cluj-Napoca. After her PhD in zoology at the University of Münster/Germany she started to work at the Institute for Landscape Management of the University Freiburg in the fields of land use and landscape development, monitoring of biotopes and species. She has wide experience in project management at national and EU level and is co-ordinating a career development programme for women. In addition, she is working as project evaluator and was appointed as Associate Professor at the University of Agricultural Sciences and Veterinary Medicine (USAMV) in Cluj-Napoca.

## Carlos A. Sandoval Castro

Autonomous University of Yucatán, Mérida, Yucatán, México

Professor Sandoval currently works at the Faculty of Veterinary Medicine and Animal Science, Universidad Autónoma de Yucatán. He carries out research in animal science, animal health and animal nutrition. His current project is herbivore-tannin-parasite relationships in small ruminants.

## Lucero Sarabia Salgado

Department of Ethology, Wildlife and Laboratory Animals, Faculty of Veterinary Medicine and Animal Science, National Autonomous University of Mexico (UNAM), Ciudad Universitaria, Mexico City, Mexico

Born in Ciudad Altamirano, Guerrero, Mexico, Lucero studied biology at the National Institute Technological. She finished her Master's degree in animal nutrition at the Autonomous University of Yucatan and took her doctorate at the Rural University of Rio de Janeiro. Her work focuses on the restoration of livestock landscapes and the mitigation of greenhouse gases. Lucero is currently doing a post-doctoral degree at the Autonomous University of Mexico.

## Francisco J. Solorio-Sánchez

Autonomous University of Yucatán, Mérida, Yucatán, México

Professor Solorio graduated from the Agricultural Technological Institute, Morelia, Mexico. He completed his Master's degree at the University of Aberdeen and a PhD at the University of

Edinburgh. For more than 30 years, he has carried out work on the restoration of livestock landscapes and the mitigation of greenhouse gases. Solorio has recently directed research to implement strategies to strengthen the health of agroecosystems, restoring soil fertility and reducing dependence on external inputs such as fertilisers and agronomics.

### Anne Stobart
Crediton, Devon, UK;
affilliated to the Centre for Rural Policy Research, University of Exeter, UK

I am an independent herbal grower, practitioner and researcher. As a child I lived in London near to the Royal Botanic Gardens and my parents encouraged me to appreciate the natural world. I well remember days out foraging for blackberries. Following an education in the sciences, I trained as a clinical herbal practitioner. This led to a teaching post in the School of Health Sciences at Middlesex University in London. I was able to pursue research in the history of herbal medicine, published as 'Household Medicine in Seventeenth-Century England' (Bloomsbury, 2020). On moving to southwest England, I discovered permaculture projects including forest gardens. These showed me how much potential there is for design for sustainable growing in three-dimensions with trees and shrubs. I explored possibilities in growing my own herbal medicines. This led to the establishment of a medicinal woodland in north Devon, based on transforming a redundant conifer plantation. The aim was to provide my own practice supplies of herbal medicines. There was much interest in the project and it is described more fully in 'The Medicinal Forest Garden Handbook '(Permanent Publications, 2020). Based on that experience, I have set up the Medicinal Forest Garden Trust to promote the sustainable cultivation and use of medicinal trees and shrubs.

### Christian Stollberg
Hochschule Wismar, Faculty of Engineering, Department of Mechanical, Process and Environmental Engineering, Malchow, Germany
Professor in Process Technology of Biogenous Raw Materials and Head of laboratory.

### Joanna Sucholas
University of Applied Forest Sciences, Schadenweilerhof, Rottenburg am Neckar, Germany.

Ms Sucholas is a PharmaPlants Project Associate at the University of Applied Forest Sciences Rottenburg and a PhD student at the Ecology and Conservation Biology Department, Institute of Plant Sciences, University of Regensburg, Germany.]

### Mark Tilzey
Centre for Agroecology, Water and Resilience, Ryton Organic Gardens, Coventry University, Coventry, UK

I am currently Associate Professor in the Political Ecology and Governance of Food Systems. I have degrees in geography and anthropology, with my PhD addressing the issue of agrarian change and environmental sustainability. While, academically, I consider myself to be principally a social scientist, I am also qualified as a conservation ecologist and spent many years working in nature conservation for what is now Natural England. I have had a lifelong passion for seeking to understand and address the politico-economic causes of our accelerating ecological (including health) crises, rather than simply documenting these crises as ecological impacts. I am concerned, therefore, to chart courses out of our current crises through politico-economic change towards (agro)ecologically sustainable and socially equitable social systems, and here the 'peasant way' seems to offer a vitally important escape route. Having grown up and lived mainly in the UK, I have also spent considerable time, and undertaken research in, Australia, Europe, Latin America, North America, and South Asia. I am the author of 'Political Ecology, Food Regimes, and Food Sovereignty: Crisis,

Resistance, Resilience' (Palgrave Macmillan, 2018). My next book is entitled 'Peasants, Capitalism, and Colonialism: Revisiting the Work of Eric R Wolf in an Age of Politico-Ecological Crisis' (Routledge, forthcoming).

### Chiara Tornaghi

Centre for Agroecology, Water and Resilience; Ryton Organic Gardens, Coventry University, Coventry, UK

I have a background in politics, sociology and planning, and I am currently Associate Professor in Urban Food Sovereignty and Resilience at the Centre for Agroecology, Water and Resilience (CAWR), Coventry University, UK. My main research interests include grassroots contestation and re-appropriation of public space, politics of urban land, political pedagogies for urban agroecology, indigenous cosmologies and knowledge of plants as food and medicine, feminist political ecology and the conceptualisation of an agroecological urbanism. I inherited an interest in medicinal plants from the women in my family, particularly my grandmother who, as part of a sharecropper family in her childhood, had been relying on foraging to complement her own diet in the peri-urban area of Milan. Becoming a migrant and a mother has been a journey that led me to understand not only the role and quality of the urban environment in (dis)abling our ability to reproduce food and medicinal knowledge through generations and across borders, but also the important role of women in preserving and sharing that knowledge, and building alternative urbanisms. My latest co-edited book (with Michiel Dehaene) is titled 'Resourcing an Agroecological Urbanism: Political, transformational and territorial dimensions', Routledge (2021).

### Felipe Torres Acosta

Autonomous University of Yucatán, Mérida, Yucatán, México

Professor Torres holds a Bachelor's degree (1989) from the Faculty of Veterinary Medicine and Zootechnics at the Autonomous University of Yucatan, Mexico, as well as a Master in Science degree (1991): Tropical Animal Production and Health, Centre for Tropical Veterinary Medicine, University of Edinburgh, Scotland, and a PhD (1999) on Internal Parasitism of Goats. Royal Veterinary College. University of London, England. His current project is on herbivore-tannin-parasite relationships in small ruminants.

### Mariya Ukhanova

University of Applied Forest Sciences, Schadenweilerhof, Rottenburg am Neckar, Germany.

PharmaPlants Project Associate.

### Monique van de Vijver

Innovation Manager Health; Citizen Scientist in Parkinson's disease caretaking;
Solidaridad Network Foundation Secretariat, Utrecht, The Netherlands

My personal experience has given me insights into how health, agriculture, trade and consumption interconnect and in fact, form a closed circle or rather, a spiral, that moves either upwards or downwards. This experience came from my role as day-to-day caretaker of a Parkinson's patient and my resulting involvement in a project group on Mucuna pruriens with co-authors Immo Fiebrig and Corrie van Kan, in addition to my lifelong professional endeavours in fair trade, sustainable agriculture and supply chain transformation to the benefit of smallholder farmers. It is my hope and belief that medicinal plants as part of an integrated approach can help us to both prevent the incidence of pesticide-related Parkinson's disease altogether and improve the health and wellbeing of those unfortunately suffering from the condition already.

**Corrie van Kan**
Amsterdam, The Netherlands

As an independent researcher and practitioner in the broad field of natural and functional wellbeing sciences, Corrie has a particular interest in the central and peripheral nervous system and how it is influenced by internal and external environments.

In 2000, she took up the study into the cause and cure of Parkinson's disease and as a volunteer she actively participated in the Workgroup Complementary Therapies of the Dutch Parkinson Association (Parkinson Vereniging, Bunnik) for some years. This led to a specific investigation, together with Immo Fiebrig and Monique van de Vijver, into the role Mucuna pruriens could have for Parkinson's.

Given the globally increasing interest in natural medicine as an option to meet current and future needs, Corrie looks forward to the establishment of those regenerative ecologies where due field research in agroecology plus clinical research with Mucuna pruriens can be done in the interest of the many.

**Padma Venkatasubramanian**
SRM Institute of Science and Technology, Kattankulathur, Greater Chennai, India

Dr. Padma Venkatasubramanian is Dean, School of Public Health (SPH) at SRM Institute of Science & Technology (SRMIST), Chennai, India. She received the prestigious Cambridge-Nehru Merit Scholarship for her doctoral work at Cambridge University, UK. Her MMgmt (Health Leadership) degree was from McGill University, Canada. A life scientist by training, she has worked to provide appropriate solutions for public health. She believes that traditional medicines have the potential to provide affordable and sustainable healthcare solutions for the masses. Her research into medicinal plants and traditional medicine has provided insights into the duelling epistemologies and brought out several products for wellness, nutrition and clean drinking water. She coined the term *'Reverse Pharmacognosy'*, a traditional knowledge-guided approach to scientific standardisation of the quality of medicinal plants and products. At SRMIST, she rolled out an innovative pedagogic framework called the University Wellness Program (UWP) to promote wellness and campus sustainability. She is a member of several national and international advisory boards, including the National AYUSH R&D Working Group for solutions against Covid-19 infection.

**Anja Vieweger**
Research Institute of Organic Agriculture (FiBL), Switzerland

I have been working at FiBL in Switzerland since November 2020, where I am leading the group on vegetable and herb production. I am a trained vegetable grower, with a degree in horticulture and a Master's in biodiversity. Between 2011 and 2020, I was a senior researcher at the Organic Research Centre in England, working closely with farmers and growers to develop their own health concepts for organic agriculture, and healthy and sustainable food systems.

**Christel Weins**
Agency for Effect-Directed Analysis in Environment and Food, Saarbrücken, Germany

Christel Weins was born in Düsseldorf (Rhineland, Germany). From 1971-1978 she studied biology at the university of Cologne and biochemistry at Eberhard Karls University Tübingen (Baden-Württemberg, Germany). Until 2002, alongside her family of three children, she managed fundamental research projects in the medical and environmental fields. In 1992, she established the combination of analytical chemistry and biological and biochemical detection methods resulting in effect-directed analysis (EDA) at Saarland University in Saarbruecken, followed by a PhD in chemistry at the University of Basel (Switzerland) in 2006. As an expert in the field of EDA and risk analysis she is a consultant for public authorities and NGOs. Since 2021, she has been a head of a working group for the 'Determination of the estrogenic components in water and waste water: Planar Yeast Estrogen Screen (p-YES)' at the DIN (German Institute of Standardization). In 2022, she became president of the HPTLC Association (Germany Chapter).

**Sally Westaway**
Royal Agricultural University, Cirencester, UK

I was previously a senior researcher in the agroforestry team at the Organic Research Centre working on practical applied agroecological research focused on bringing trees into farming systems. I am now completing a part-time PhD with the Royal Agricultural University carrying out spatial modelling of the impact of increased tree cover scenarios on ecosystem service provision at a farm and landscape scale,whilst also running a community-supported agriculture project in Cornwall. I am interested in the connection between people and the landscapes that nourish us. I have worked in land management for more than 20 years, initially as an ecologist and more recently, in agroecological research and local food production, and am motivated by the need to create resilient landscapes and local farming systems.

**Lindsay Whistance**
Organic Research Centre, Cirencester, UK

Following a PhD in dairy cow behaviour and welfare, I have worked in various aspects of farm animal welfare research, alongside promoting the importance of understanding and respecting the nature of animals (these modern-day representatives of their wild counterparts) in our managed landscapes, their home.

My conscious journey on this road began at 14 years of age when I fell under the spell of cows and my first job – on a small, grass-based dairy farm – fixed a life-long desire to understand and promote farming systems that delivered care and respect according to – what I now know to be – agroecological principles.

Alongside other forms of resilience (e.g., climate, economic, welfare), there is much potential to build medicinal resilience into farming systems by incorporating appropriate plants, either for human harvesting or animal foraging. Care can be offered in many ways and one of the most sustainable forms is the provision of landscapes in which animals can be empowered to help themselves, according to their own wants and needs. Our role then, as landscape managers and animal carers, becomes one of learning to gain detailed insight and knowledge of plant properties and animal-plant relationships.

# Foreword One

Being part of this important book and the novel concept of medicinal agroecology introduced and promoted within it, fills me with joy. It feels like a crossroads where people from very different backgrounds and disciplines meet and join forces towards meaningful nature-based solutions for the health and wellbeing of humanity and the planet.

For more than 15 years, my life has been inspired and guided by a small, but powerful, medicinal bean from the subtropics called *Mucuna pruriens* (MP). The journey around it started when my husband, Bert Beekman, was diagnosed with Parkinson's disease, early in 2003. He had been battling with fatigue, pain in his right shoulder and burnout-like symptoms, as well as a tremor in his right hand. A short examination by a neurologist, including some typical tests – like walking up and down to check on arm swinging – ended ten minutes later with a medical prescription and a lifelong sentence. We could hardly – a blessing at the time – imagine what this diagnosis would mean for his livelihood, for mine and the life of our two toddlers aged one and nearly three.

I started researching about Parkinson's to relieve my initial ignorance. I read about the disease, its possible causes, how it evolves and about the rather unsatisfactory perspectives for treatment. I concluded that in my husband's case, the cause of Parkinson's was most likely linked to continued pesticide exposure during a time when he had worked and lived in Honduras, Central America. Here, he had spent a lot of time with farmers in the fields. In those days, small planes – so he had told me – would fly over and spray pesticides on the fields, whilst paying little attention neither to adults nor children living and working there, and being exposed during spraying and afterwards. The local inhabitants, in turn, were largely oblivious to the harm such pesticides could cause. They would often not even bother to go inside their homes to protect themselves.

Another story Bert told me was about how he would drive with his motorcycle to visit a dairy farm and buy fresh milk every weekend. After driving back on the rather bumpy road, once home, he would enjoy eating the cream that had formed on top of the milk by the spoonful. It has always been one of his big treats. In those days, early 1970s, the insecticide DDT (dichlorodiphenyltrichloroethane), whilst having been banned in Europe and North America, was still widely used in Latin America. Presumably, cows left to eat the young sugar cane sprouts in post-harvest sugar cane fields ingested DDT due to its environmental persistence. Being lipophilic and known to concentrate in the body fat of mammals, he might have been exposed to DDT by ingesting it indirectly. Stemming from two long-living parents with no signs of PD whatsoever, the 'pesticide cocktail exposure' at the time has most likely – and sadly – been the major cause of Bert's illness, developing and manifesting itself some 20 to 25 years later as an illness that is increasingly known to be caused by hard-to-prove, long-term side effects of 'Green Revolution' agriculture methods.

Bert had just turned 50 when diagnosed. The prospects of more, or rather less effective, treatment with conventional medication for possibly another 15 years seemed grim. Thus, I started looking for alternative treatments with better outcomes. My attention was caught by studies on *Mucuna pruriens* bean preparations, a medicine known from the ancient Indian Ayurvedic medicine system. The seed of this subtropical plant, with a naturally high content of levodopa, seemed to provide additional benefits and fewer to no side-effects to patients, as opposed to what I then read about conventional, synthetic levodopa. We will elaborate on this in our contribution in Chapter 3. Besides, my Bert had decided initially not to take the prescribed medication – dopamine agonists – anyway. He preferred to wait and see how things would develop. He visited a retired neurologist instead, who practised acupuncture for neurological disorders. He felt it was helpful, at least for a while. After a couple of years, when symptoms gradually worsened, we looked for a neurologist who was a bit more communicative and open to alternative or 'complementary' treatments. I also decided to respond to a call for new members from the Working Group on Complementary Therapies in the magazine of the Dutch Parkinson's Association. The group was initiated by people who, like me, were looking beyond treatment with the standard prescription medication. Their aim was to improve the health and wellbeing of Parkinson's patients who would not respond well, or no longer well enough, to standard treatment. I contacted the working group, claiming my specific research interest: *Mucuna pruriens*.

This is how I then got in touch with Corrie van Kan, co-author of Chapter 3, who had just left the group, but who had indicated that if anyone would come by who was interested in researching *Mucuna p.*, she would be happy to step in and provide support. And this is how our project started, way back in 2006, when we met up for the first time, in a funny and misplaced Chinese-style highway restaurant along the A2, near a small municipality called Breukelen, just north of Utrecht. While we talked, my children were happily exploring and playing, unaware that we were initiating strides to create a better outlook for their father and many others like him.

One of the main problems back then was the availability of a reliable *Mucuna* preparation, and one of my internet searches brought me to a local German pharmacy, in a – like Breukelen – small town, called Grabenstätt. After contacting the pharmacy, I was referred to Dr. Immo Fiebrig, who had done all the required research on *Mucuna p.* on behalf of patients, had organised for an Ayurvedic preparation to be imported from India and had developed the informative web pages on the medicinal bean I had run into.

Bert was the next one in a row of patients who experimented with, and clearly benefited from, supplementing his medication with *Mucuna p.* It led to an improved sleeping pattern, better mood and fewer to no side-effects than the synthetic levodopa he started taking as his basic therapeutic agent, replacing any other medication he was taking, as that had proven to do him more harm than good. This benefit became clearly visible when we started replacing part of his – meanwhile quite high – dose of synthetic levodopa by first *Mucuna p.* 'powder', then by a *Mucuna p.* 'extract'. Even after almost 15 years, his shaking (tremors) has remained practically absent and the abnormal movements (dyskinesia) under synthetic levodopa never came back. However, under the logic of prevailing Western medicine, these experienced utterly important and striking benefits are not perceived. And they are not perceived nor recognised because the benefit has not been proven to exist, statistically… And this again is the case because studies to demonstrate these effects scientifically and statistically do not exist. There are just not enough studies. Those that do exist are not showing significant benefits from the perspective of standard motoric symptom control, unable to prove other 'so far unmeasured' or 'hard to measure' benefits; or studies are claimed 'not to meet Western standards'.

What is it that is needed for science to become more open and to integrate a more holistic perspective in clinical research? Why not include the patient's perspective and interests more consistently? Why not include other types of symptoms that are just as relevant to the patient's wellbeing? Most likely, commercial interests, and the conventional mind-sets on which our entire pharmaceutical system is based, still stand in the way. For scientists born and raised within, it must certainly not be easy to move away from trying to understand exactly how every component of a herbal medicine behaves – towards accepting that we can simply not fully understand nature's complexity. For the relief of patients, an important first step would be to recognise their individual differences and needs and to use their needs and experiences to guide additional research.

But now, how do these reflections connect with the other areas of research of medicinal agroecology? In 1987, Bert's connection with agriculture and small-holder farmers in Central America had led him to become the first director of the Max Havelaar fair-trade label for coffee, front running the global fair-trade labelling and wider sustainability certification movement. The label had been developed by the Dutch NGO Solidaridad[1], my employer since 2004. Bert and I met

---

1    Nowadays, Solidaridad is an international network organisation for development cooperation aimed at improving the livelihoods of small-holder farmers and workers. Solidaridad has offices in five continents and develops projects in more than 40 countries (www.solidaridadnetwork.org).

through work in 1989 and both dedicated our careers to improving the fate of small-holder farmers worldwide.

As Bert's utterly disabling disease progressed, he soon had to stop working, in spite of medication. Even if we were lucky to receive a pension from the national disability scheme and enjoy the services and protection of a good healthcare system, our income went down considerably, while additional costs for care and other things went up. I forcedly had to work part time, accepting a job I was over-qualified for, back then only suspecting the sizable care tasks for him and our children that were awaiting me, and hardly aware of how all his contributions to the household, administrative tasks and other chores would gradually land on my plate in the years to follow. I could not stop thinking about what would have happened to us as a family if we had had to face these challenges in less fortunate circumstances, which we had both witnessed from close by. International research shows that poor people in developing countries hit by chronic disease fall back into extreme poverty exactly because of this mechanism. Their earning capacity is severely affected or lost and health services and medication are out of reach or unaffordable. In the meantime, evidence had piled up that exposure to certain pesticides was linked to the occurrence of Parkinson's among people in rural areas. The thought of the aeroplanes spraying pesticides on poor farmer families and imagining what this disease would do to them was unbearable. More than ever, I saw how fortunate we had been to face our misfortune in a developed country. And for the first time in my career, I clearly realised how health is a precondition to sustainable development and that raising people's income is not enough to ensure this. I also realised that preventing Parkinson's disease with those who are most exposed to pesticides is at least as important, or even more, as access to healthcare and medication.

While researching *Mucuna p.* for private reasons, I was struck at an early stage by the multiple uses of this apparently simple bean and the socio-economic and ecological potential of this 'weed' for small-holder farmers and their farm ecology. I ran into uses of this leguminous plant as soil fertiliser, cover crop, animal feed and even food. What if small-holder farmers could grow this medicinal plant as a diversification crop and sell it to the high-value natural health and pharma markets? This could provide an interesting additional source of income for them and form a natural part of their farming set-up. It could offer new income and entrepreneurship opportunities for women and youth in farming communities, who could be introduced to growing and processing this and other medicinal plants. If well connected to markets, it could even become a driver for farmers to transition to more ecological farming practices, as they would have to deliver a 'clean' product, free from harmful pesticide residues. As such, introducing *Mucuna* in their farms could indeed be a means to improve their incomes, but it would also greatly contribute to the availability of beans for those committed to making *Mucuna p.* safely and affordably available to patients needing it, no matter which part of the world. More than ten years after setting up the Project Group on *Mucuna pruriens,* I started sharing my ideas with colleagues in Solidaridad and developed the 'Herbal Medicine and Health' discussion paper, which – for the purpose of this book – I published as a white paper on Researchgate (van de Vijver 2022). The managing director of Solidaridad Asia, located in India, needed only a few words from my side to embrace the agenda and encouraged me to develop it further.

In the meantime, Dr. Fiebrig, a pharmacist by education and disappointed in the industry and how it underserved patients, decided to shift careers and explore a very different path: that of permaculture and agroecology. We started discussing all these different aspects of *Mucuna p.* and our – at the outset rather separate – professional paths started aligning.

In early June 2017, Dr. Fiebrig received a call for contributions for a symposium being organised by Prof. Dr. Ing. Christian Stollberg at Wismar University, North Germany, Faculty of Engineering. The title was 'Symposium on medicinal plants in the context of globally sustainable land use and bioeconomy'. For the first time, I saw someone actively bringing together the different aspects of sustainability which I had integrated into the envisioned multi-impact development strategy based on medicinal plants: sustainable cultivation, income generation, value creation and market linkages, backed up by science. We decided to give it a try and present an abstract of the work done on *Mucuna pruriens* to the organisers. We gave it the title 'Torn between hope, desperation and remedy: Parkinson's disease, patient perspectives and *Mucuna pruriens*'. To our joy, the abstract was accepted and Dr. Fiebrig was invited to present. We consolidated the work carried out, our own experiences and systemised patient experiences. This is how we connected to Prof. Christian Stollberg. He is an expert in the engineering of biogenous raw materials and machine building for the same and has extensively researched medicinal properties of plants in support of their use by the pharmaceutical industry. He would research the origin of the plants too and came to an important conclusion: if those producing the raw materials are not properly enabled and rewarded for producing high-quality ingredients in a sustainable way, no proper business case can be built from it by any player in the supply chain. And this is where his path aligned with that of Dr. Fiebrig and mine. He will tell you about it in his own foreword which follows.

I have been working on the development of this medicinal agenda for Solidaridad as 'Innovation Manager Health' since May 2019, mainly together with my colleagues from Solidaridad in India. Considering its long and lasting history in traditional medicine, India is the most idoneous country in which to start. Several of my Indian colleagues appeared to have personal connections with traditional medicine or its practitioners and embraced this promising challenge. The programme was launched during a stakeholder meeting in Bhopal in September 2019, where I had the opportunity to share my ideas and learn from experts, like Dr. Gurupal Singh Jarial, a leading expert in medicinal plants cultivation and protection and now Senior Advisor with Solidaridad, and Mr. Rajaram Tripathi and his wife, who are leading by example with their entrepreneurship, putting Dr. Gurupal Singh Jarial's lessons into practice. By now, Mr. Tripathi has managed to build a perfect business case with his company, Maa Danteshwari Herbal Group, in which over 20,000 farmers participate. As part of the Medicinal Plants Programme, Solidaridad is collaborating with different governmental departments and institutions to build sustainable and transparent supply chains for medicinal plants. At community level, not only the cultivation of medicinal plants is promoted, but also their use as a low-cost treatment for the basic ailments from which people in the communities are suffering. The Solidaridad book, 'Reclaiming Sustainability in Medicinal Plants and Herbal Medicine Sector', written by Dr. Gurupal Singh Jarial, Dr. Suresh Motwani and Mr. Himanshu Bais outlines the potential, issues at stake and strategy of the programme and serves as a manual for the cultivation of 33 carefully selected medicinal plants (Jarial et al. 2022). You can imagine my joy when I saw that 'Kewanch' – known by the scientific name of *Mucuna pruriens* – was included in the book as 'one of the important medicinal plants of India'.

As an answer to the self-posed question 'Why bother with herbals?', Dr. Fiebrig for his presentation during the symposium took two quotes from a 1996 World Bank Technical Paper: "Medicinal plants are those that are commonly used in treating and preventing specific ailments and diseases, and that are generally considered to play a beneficial role in health care." and: "Medicinal plants are viewed as a possible bridge between sustainable economic development, affordable health care and conservation of vital biodiversity." Since then, many people in different countries all over the world have made all kinds of efforts, varying from research in the related fields, to policy influencing, the development of national pharmacopoeias, processing technologies and business development, learning all kinds of lessons. In addition, research on the causes of Parkinson's has advanced. The book, *The Parkinson Pandemic*, by the visionary Prof. Dr. Bas Bloem and his team, brought me back to where it all started: Parkinson's caused by exposure to harmful chemicals, among which are pesticides (Bloem et al. 2021). The book is an urgent call to action we cannot ignore.

In 2011, Prof. Bloem and his team were very close to a multi-centre, multi-country double-blind clinical patient trial with *Mucuna p.* for which our project group also provided input from the work done. Everything was lined up, but the most crucial part – the research preparation – could not be made available, as the pharmaceutical company that was to produce it in the end decided not to invest. Was it because of the fact that the intellectual property of herbal medication – the basis of the pharmaceutical business model and R&D agenda – is hard to protect? Or was it because of the complexity of *Mucuna p.* in terms of matrix components and the supposed necessity of science to research, understand and control all the chemical constituents with pharmacological activity to be able to use it? If one thing is clear to me, it is that any of these arguments stand in the way of investments in much-needed research and the development of research methodologies that are more appropriate for herbals. They are the reason that longer-term, well-designed trials with herbal medicine are not taking place, which by the way is analogous to what is happening in the agrochemical industry with respect to green pest control solutions. As a consequence, the required research and evidence base of herbals cannot easily be built, which in turn provides arguments to not invest in further research and development. Or to say: "let us better stick to something we know works, even if deficient and even if only good for some."

The important contributions of many authors to this publication confirm our own findings and ideas. I am sure that with our Chapter 3 on *Mucuna pruriens*, based on the thorough literature research by Dr. Fiebrig, the input of a significant group of daring patients and caregivers, and the important work of exceptional clinical researchers like Dr. Roberto Cilia in Ghana, we will be able to contribute to the relief of Parkinson's in a non-conventional and effective way. I am sure that jointly we can put the pieces of the puzzle together to start building a new, comprehensive and scalable programme for health, wellbeing and sustainability based on plants and their amazing properties – in this case, a strategy aimed at preventing the incidence of Parkinson's caused by exposure to pesticides, while making adequate treatment available to those affected by it in different ways. I am convinced that the concept of medicinal agroecology, starring *Mucuna pruriens*, will guide us there and thank Dr. Fiebrig for leading the way.

**Monique van de Vijver**, *Rhenen, NL, Spring 2022*

## LITERATURE

Bloem, Bastiaan R, Jorrit Hoff, and Ray Dorsey. 2021. *De Parkinsonpandemie: Een Recept Voor Actie*. Koog Aan De Zaan: Poiesz Uitgevers.

Jarial, Gurpal Singh, Suresh Motwani and Himanshu Bais. 2022. *Reclaiming Sustainability in Medicinal Plants and Herbal Medicine Sector*. New Delhi, India: Blackspine Publishing Private Limited. https://www.researchgate.net/profile/Suresh-Motwani/publication/360310043_Reclaiming_Sustainability_in_Medicinal_Plants_and_Herbal_Medicine_Sector/links/626f31e7b277c02187dc6ee4/Reclaiming-Sustainability-in-Medicinal-Plants-and-Herbal-Medicine-Sector.pdf?origin=publication_detail.

Vijver, Monique van de. 2022. *Exploring the Potential of Medicinal Plants in Sustainable Development*. ResearchGate. https://www.researchgate.net/profile/Monique-Vijver/publication/361173853_EXPLORING_THE_POTENTIAL_OF_MEDICINAL_PLANTS_FOR_SUSTAINABLE_DEVELOPMENT/links/62a0c87055273755ebdd5b97/EXPLORING-THE-POTENTIAL-OF-MEDICINAL-PLANTS-FOR-SUSTAINABLE-DEVELOPMENT.pdf?origin=publication_detail. Accessed 9/6/2022.

# Foreword Two

Deep in the heart of the Indian state of Punjab, also known as the 'breadbasket of the Indian subcontinent', lies the ashram of Kirpal Sagar – a mere stone's throw from the town of Nawanshahr and the hamlet of Dariapur. Surrounded by small fields and plots of poplar or moringa and protected by a sun-bleached wall, this oasis ascends unobtrusively from the landscape and gives the visitor a feeling of peaceful coexistence. A coexistence of mankind and nature as well as mankind with spirituality.

This side view lies in stark contrast with many other places, especially in Punjab, which if managed more appropriately, could provide so much more quality of life to its people, especially considering the land's remarkable natural wealth. Most of Punjab's population, however, lives in moderate to extreme poverty, squashed under a heavy burden of financial debt amid vast monoculture farmed land. Looking back only a few decades, the average peasant farmer grew 19 different crops within a richly structured landscape.

Today, however, seemingly never-ending arable fields stretch out from one horizon to the other. Hybrid wheat, hybrid rice and hybrid corn appear to be the new magic crops. Whatever these crops may mean to the (often) illiterate rural population, they have, in fact, created an unhealthy, if not deadly, dependency on costly pesticides and special fertilisers. Such agrochemicals represent a dubious "sacred medicine" aimed at increasing and maintaining yields – in the short term. The side effects of this "medicine", however, have led to a rise in human cancer rates – in Punjab alone by more than 800% within just a few years.

Punjab and its modern agricultural development holds many a ghastly story of human drama, for example, the undiluted application of pesticide concentrate on ecologically certified areas where Ayurvedic medicinal plants are grown and harvested, whilst small children are building sandcastles in the ditch next to the plantation row. There are also stories about debt slavery and the unfulfilled wish of the enslaved boys to be allowed to go to school. Instead, 16 youths between 10 and 16 years old share a room of about 12 m$^2$ in size as accommodation, surrounded by red hot fields, without a toilet or running water, after having worked from sunrise to sunset in the field at two flatbreads a day. The look of fear in their feverish eyes – fear of punishment if they fail to get enough work done due to illness or injury or simply exhaustion – has etched itself deeply into my memory. Parents my age or younger, on the other hand, live with the agony that the next unborn child might be a girl and, therefore, become an untenable liability.

Kirpal Sagar is the place where my journey to India begins. Every time, it seems. Here, in 2014, together with three German scientists, one farmer, and the people of the ashram, we started the sustainable and thus ecologically compatible cultivation and processing of *Moringa oleifera* – a plant widely applied and not only in Ayurvedic Medicine.

A rocky road, as so often, was the path of achieving cultivation, processing, packaging and certification. And yet, distribution is succeeding in Germany, Austria and, above all, in Punjab itself. For me, this achievement means not only joy but also insight – a look far beyond my horizon as a chemical engineer, whose heart had so often beaten solely for the extraction of highly purified plant ingredients. For me, my gaze had primarily been turned towards processes such as 'extraction, preparative chromatography or crystallization', which, according to the Western view, are a prerequisite to produce medicinally active ingredients.

However, from now on I could no longer get past the obvious connection between healthy organic cultivation followed by responsible processing, ultimately leading to a high quality of the product. Each of the former aspects cannot be achieved without sustainability in mind and in collaboration with the local people.

This very insight and the hope it raised for both a better quality and a fairer agriculture had led me to invite our worldwide partners to the first 'Symposium on Medicinal Plants in the Context of Globally Sustainable Land Use and Bioeconomy' in 2017. The symposium took place on *Insel Poel*, a small island in the western Baltic Sea, where our University Institute for Biotechnology, Chemistry and Process Engineering of Biogenic Raw Materials is located. The aim of this event was to set an example in a region dominated by industrial agriculture, whilst bringing together all players who are part of the value chain of medicinal plants to find a shared language.

And they came, be it the Indonesian farmer, the Armenian ethnobotanist, the European certifier, importer or manufacturers of herbal products, the development aid worker or the Ayurvedic scholar, including the German chemist or pharmacologist specialised on natural products. They came to talk to one another, to discuss, but also to listen and learn from each other. As I learnt only recently, it was this very symposium that had set off the inspiration for this book project... How wonderful!

*Medicinal Agroecology* - I truly hope - will bring about many more joyful images and stories as it can contribute to creating hope and positive outcomes for future generations. I am sure that *Medicinal Agroecology* supports the walk on a path towards sustainable land use with plant-based, holistic medicine beyond corporate intellectual property rights – with a medicine being regarded as medicinal in the widest sense: positive for human health, animal health, plant health and soil health. It should be based on ethnobotany, be supported by exploratory data analysis and backed by modern science, whilst guiding regulatory reforms in tune with future developmental programmes.

By now, the Darjeeling Himalayan Railway (DHR) starts to move panting and whistling. This takes me up a mountain on my long trip to Sikkim – my next promising research visit. Sikkim is the second smallest Indian state and as such it seems a bit trapped between West Bengal, Nepal, Tibet, and Bhutan. And yet, Sikkim shines far beyond its borders and above the sky-high mountains it finds itself surrounded by. As the first federal state in our ever-more industrialising world, Sikkim has committed itself entirely to organic farming.

That too can be - and is - India and this gives me hope. Hope for a healthier, fairer world, where humans co-exist with other humans and generally with Nature, spiritually guided whilst respectful of all beings as 'equally worthy creatures'.

This is my sincerest hope. And with *Medicinal Agroecology*, I feel this hope is nourished.

**Christian Stollberg**
*Professor for Process Engineering of Biogenous Raw Materials*
*Wismar University of Applied Sciences, Technology, Business and Design*
*Malchow/Poel Island, Germany*
*Autumn, 2022*

# Introduction to *Medicinal Agroecology*

*Medicinal Agroecology* as you are holding it in your hand – or as you are reading it electronically – started early in 2020 as an editorial project. While it was being encouraged and supported by Alice Oven, Julia Tanner, Randy Brehm, and Tom Connelly from the publisher, CRC Press/Routledge Taylor & Francis Group, the world was being hit by CoViD-19. The first lockdown forced me to stick to my flat, re-organise daily routines, and incubate the conceptual details of the book. Interestingly, none of the big internet search engines showed any hit on 'medicinal AND agroecology' at the time, motivating me further to set a stake where I felt there was a gap of consilience in current sustainability research agendas.

The idea of medicinal agroecology (MAE) as a field of research is to view plant material as medicinal, not only within human medicine but together with veterinary medicine, phyto-medicine (i.e., medicine for plants) and as a medicine for the soil (which one might want to call 'edapho-medicine'). The aim is to unify these different areas under the umbrella of 'health and healthiness to the benefit of all beings'. However, MAE is not only concerned with all sorts of medicines from the natural environment. More than that, it is to maintain a balanced – and at least a sustainable, if not a regenerative – ecosystem that is also freed from petrochemical substances with ecocidal effect, such as environmentally toxic agrochemicals. Alongside this idea, ecosystems should just as well be freed from anthropogenic non-compostable waste, like (micro)plastics. It may sound like something that agroecology embraces. Old wine in new – now medicinal – bottles?

The medicinality in the present book, *Medicinal Agroecology*, is a matter of emphasis and point of view, whereby:

- Conceptually, plants as part of our healthy diet, in particular those defined as having medicinal properties, constitute our 'medicine' or may be defined as 'nutraceutical'. Secondary plant metabolites may also function as an everyday nutraceutical to animals, including domesticated ones like farm animals and pets. Such secondary metabolites can carry properties that are also useful to other plants, including farmed fruits and vegetables, and that can be functional in pest management. Particular plant substances may promote the livelihood of (beneficial) insects (e.g., pollinators) as well as fungi (e.g., rizobiaceae) and the general fauna of the soil ('soil food web'). They can also attract insects (e.g., floral scent from essential oils) or protect from UV light (e.g., photoprotection via flavonoids). However, plants and their constituents can also be rather toxic to all life forms; secondary metabolites are the plants' natural mechanism of defence against predation (e.g., from insects, vertebrates, or fungi). MAE also embraces the symbiotic and pathogenic interactions in complex and dynamic systems, extending into environmental (pollutant) phyto-remediation and bio-transformation (Rane et al. 2022; Singer et al. 2003; Berenbaum and Rosenthal 1992).
- In agroecological production systems, medicinal plants may play a functional role in animal health (e.g., parasite control), plant health (e.g., pest control), and soil health (e.g., soil fertility and microbiome health). However, they may also constitute an additional, high-value crop for both local and global healthcare provision – whilst making a contribution towards avoiding the extinction of medicinal species that have so far mainly been collected from wild sources and have thus been over-exploited.
- Secondary plant metabolites that are important to humans are not limited to vitamins in food. They comprise a world of bioactive compounds of its own and only a few selected components will be exemplarily dealt with in this book (see also Harborne et al. 1999 for wider reference).

Beyond the distinct molecules used in conventional drug development and herbal remedies, as well as the nutraceutical components utilised for the production of food supplements, secondary plant components have been used as flavours and fragrances (e.g., essential oils like rose oil from *Rosa damascena*), dyes and pigments (e.g., indigo from *Indigofera tinctoria*), pesticides (e.g., azadirachtin from *Azadirachta indica*), or food additives (e.g., sodium alginate [E 401] from brown algae [*Phaeophyceae*]) (Hussain et al. 2012). MAE, as I propose it, should see these substances as foremostly embedded in a live system with its respective metabolism. Research should make knowledge and practice around these bioactive systems open source, as being of public interest and thus available to anyone who is in need of it. What MAE is NOT meant to be is, as a matter of stance, to primarily see a bioactive component as a singular molecule that can be isolated and exploited by corporations. This exclusivity – via the acquisition of intellectual property rights – ignores people and the environment being affected by such actions, whilst at the same time depriving living beings of the insights or products from Nature.

- Finally, consilience, a convergence of evidence between research findings of the Natural Sciences and the Social Sciences, is something I intend to promote with *Medicinal Agroecology* and the research presented herein – be it from literature or from the field. Respect and empathy towards all life forms should go alongside evidence from the landscape level down to the molecular dimension towards a unity of knowledge as proposed by E. O. Wilson (2007), calling – in my view – not only for inter-disciplinarity but also trans-disciplinarity, including citizen science approaches.

With these ideas in mind, and thanks to a large number of scientists and researchers engaged in restoring planetary health, *Medicinal Agroecology* is structured into the following sections:

- *Forewords:* The two forewords are about how personal experience has led to positive visions and how such visions guided the work of professionals and scientists in the development sector.
  - *Foreword One* relates to the challenges of people affected by chronic diseases such as Parkinson's Disease, caused by harmful production methods, to sustainability solutions based on medicinal plants and with outcomes that are fair to farmers, consumers, and the environment alike.
  - *Foreword Two* is about how personal experience leads to insights and inspires scientific research, in the lab and on the ground, in the form of holistic projects that safeguard the availability of precious medicinal plants on which many existing and potential health solutions rely.
- *Policies and Frameworks* gives us an overview of what is meant by MAE.
- *Overviews and Insights* is focused on literature reviews often supported by the authors' own research data.
- *Case Studies and Research Methods* gives practical information and examples to those readers planning to set up MAE projects of their own. At the end, the trilogy of Weins' (Chapters 12–14) introduces an innovation in chemical analysis enabling scientists, decision makers in the private sector, and public institutions, as well as policy makers, to become aware of the risks (harmful substances) and opportunities (beneficial bioactive compounds) in Nature. Details of the methods are supported by numerous and expressive images.
- *Epilogue One* opens up a glimpse into the principle of Molecular Colonialism through corporate biocides; and finally
- *Epilogue Two* adds a spiritual dimension to MAE with a view towards Buddhism.

Enjoy your reading! Sulden/Soldano, It., 30 April 2022

**Immo Norman Fiebrig**, Editor

## LITERATURE

Berenbaum, M. and Gerald A. Rosenthal. 1992. *Herbivores, Their Interaction with Secondary Plant Metabolites*. San Diego; London: Academic Press.

Harborne, J B, Herbert Baxter, and Gerard P Moss. 1999. *Phytochemical Dictionary: A Handbook of Bioactive Compounds from Plants*. London; Philadelphia: Taylor & Francis.

Hussain, Md. Sarfaraj, Sheeba Fareed, Saba Ansari, Md. Akhlaquer Rahman, Iffat mkZareen Ahmad, and Mohd. Saeed. 2012. "Current Approaches toward Production of Secondary Plant Metabolites." *Journal of Pharmacy and Bioallied Sciences* 4 (1): 10. https://doi.org/10.4103/0975-7406.92725.

Rane, Niraj R., Savita Tapase, Aakansha Kanojia, Anuprita Watharkar, El-Sayed Salama, Min Jang, Krishna Kumar Yadav, et al. 2022. "Molecular Insights into Plant–Microbe Interactions for Sustainable Remediation of Contaminated Environment." *Bioresource Technology* 344 (January): 126246. https://doi.org/10.1016/j.biortech.2021.126246.

Singer, Andrew C, David E Crowley, and Ian P Thompson. 2003. "Secondary Plant Metabolites in Phytoremediation and Biotransformation." *Trends in Biotechnology* 21 (3): 123–30. https://doi.org/10.1016/s0167-7799(02)00041-0.

Wilson, Edward O. 2007. *Consilience: The Unity of Knowledge*. Cambridge: International Society for Science and Religion.

# Section I

## POLICIES and FRAMEWORKS

# 1 Building Medicinal Agroecology
## *Conceptual grounding for healing of rifts*

*Chiara Tornaghi, Georgina McAllister, Nina Moeller and Martin Pedersen*

## 1. INTRODUCTION

"All things are connected. Whatever befalls the earth befalls the children of the earth." (attributed to Chief Seattle, c. 1786–1866[1]; a widely shared sentiment)

It is common to talk about the extinction of 'beautiful others', like snow leopards, or 'necessary others' like buzzing bees that pollinate our food crops (Small 2011); but *Homo sapiens* is caught up in the same web of life, resting on the same foundation of complex inter-species relations from the ground up: fungi, bacteria, nematodes, worms, insects, plants, trees, birds and so on. The word human and the word humus, for soil, have the same root.[2] All things are indeed connected and agroecological wisdom sprouts from this awareness.

As political agroecologists, our approach to medicinal agroecology is inspired by and aligns with the struggles for collective and individuals' rights by local, indigenous and small-scale producers engaged in protecting their way of life. We therefore begin by acknowledging that diverse medicinal agroecologies already exist, while emphasising that a fundamental epistemic shift is required to appreciate them.

As such, our chapter proposes that medicinal agroecology refers to understandings of food and nutrition *as* medicine, and heals the rifts which have opened up beneath and between us, and which separate us from the natural world and each other. This is a political proposition that seeks not only to resist the objectification of nature, but to embrace a culture of care. This requires that issues of power, control and governance be addressed, and that social and political action for systemic change be centred (Anderson et al, 2021). It simultaneously entails: a) nurturing agroecology by empowering grassroots, place-based and people-led processes, and b) deconstructing the existing food regime that disables agroecology.

To advance a conceptual and philosophical grounding for a medicinal agroecology, we react to three provocations that, we propose, are intensifying the erasure of medicinal agroecologies:

- Cheap 'food from nowhere' as solution to feed the poorly resourced – the contention that the industrialised food system is justified through its provision of cheap food commodities, the origin of which is unknown and irrelevant;

---

1   Chief Seattle was a Susquamish leader who lived on the islands of the Puget Sound. He is thought to have delivered a speech in 1894 that included the cited words, though their veracity is contested (e.g. Gifford 1998).
2   Both human and humus come from the Proto-Indo-European root *dhghem – meaning 'earth'.

DOI: 10.1201/9781003146902-2

- Approaches to medicine that tackle symptoms, disconnected from food and diet – the de facto approach of the modern healthcare system and its technologies, in contradistinction with other systems, most notably Ayurveda and Chinese Traditional Medicine, which see health as related to an ecosystem of practices, and understand food as medicine;
- An acceptance of urban lifestyles as fundamentally extractive – the acquiescence to, and participation in, a socio-economic paradigm that is based on continuous resource extraction from mostly 'rural' areas to enable contemporary 'urban' lives of consumption.

At the centre of our argument lies a contention that multiple rifts have been radically accelerated by industrial capitalism and the technocratic responses it produces (Schneider and McMichael, 2010; Arora et al., 2020). Fundamentally extractive in their commodification of labour, nature and knowledge, these altered relations have cemented deep metabolic, epistemic and social rifts, the roots of which reach back to at least the invention of the plough (Hyams 1952; Montgomery 2007; Scott 2017). The implications continue to shape our food system and health outcomes today.

Critical of industrialised food and healthcare provision, the chapter departs from tracing – within fragmented contemporary debates – how the pursuit of profit has led to a quest for high yields and genetic uniformity, which has increased the toxic load and reduced the diversity and nutrient density of available foods, as well as how the promotion of inappropriate diets and growing disease burdens are compounded by technocratic responses that further embed and reproduce corporate power structures. In reflecting on the implications of stacking health, biodiversity and climate crises, we engage with the concepts of 'syndemic'[3] and 'planetary health'.

At its most transformative, agroecology is concerned with the intersection of soil, community, individual and planetary health and justice, thus constituting a complex approach that involves systems thinking and biosocial and biocultural lenses of inquiry in combination. The syndemic perspective, we argue, intersects with agroecology in a number of ways, especially when the medicinal aspect of agroecology is emphasised.

The chapter is structured as follows: after this introduction, section 2 grounds our argument through a historical glance at the idea that humans, and their health, are connected to all other things in the web of life. This is our point of departure.

In section 3 we trace the roots of the epistemic rift and unpack some of its consequences on agricultural practices and food system design, such as soil destruction, the erasure of diverse medicinal knowledge ecologies and the establishment of unhealthy and unjust diets.

In section 4 we delve deeper into some of the consequences of the rift, particularly on human and planetary health through a look at some recent figures regarding the global burden of disease.

In section 5, we warn against emerging co-opting discussions around planetary health through industrial-friendly dishing out of planetary diet discourses.

We conclude the chapter in section 6 with a summary of the main points raised in this chapter and an offering for nourishing research agendas with regard to medicinal agroecology.

## 2.   ALL THINGS ARE CONNECTED: HUMANS IN THE WEB OF LIFE

Chief Seattle's quote, with which we begin this chapter, expresses a truism among most, if not all, indigenous people and many peasant cultures: that *all things are connected*. Since the Buddha realised the "diversity and interconnectedness of the biocommunity" (Bilimoria 2017), Asian and European thinkers of different kinds have spoken about how we are bound up with one another and our habitat. Many have taken action to defend life by defending habitats against degradation and contamination on that note. The idea of a complex web of life has been central to animism across cultures for thousands of years and expresses a relationship between people and their surroundings

---

3    As expanded below, 'syndemic' refers to a synergistic epidemic and was coined in the 1990s (Singer and Snipes 1992).

that rests on an understanding of the complexities that has slowly dawned on modern science in the 20th century, notably in ecology. That we are one with the earth and that we live in a complex web of life that sustains our livelihoods is no longer seriously disputed by any commentators, but, of course, continues to be disregarded or circumnavigated to suit creative accounting.

In the introduction to the 50th anniversary edition of Rachel Carson's *Silent Spring* (2002 [1962]) Linda Lear writes that by 1957, as a result of her research on DDT and related petrochemical pesticides, Carson "believed that these chemicals were potentially harmful to the long-term health of the whole biota" (Lear in Carson 2002: xv) and "strongly" stated that they should be called "biocides" (ibid: xvi). She also argued that "public health and the environment, human and natural, are inseparable" (ibid: xx), thus reflecting themes that had emerged already in the 1940s in the work of Eve Balfour, whose pioneering studies into soil and health took place during the Haughley Experiment. The first ever comparative study on, respectively, organic and petrochemical farming (on two farms in Suffolk, England), which began in 1939, led to Balfour's milestone publication *The Living Soil* (1943). Here, Balfour presented evidence for her argument that the "interrelationship between soil vitality and the health of plants, animals, and man [sic], is of so important a nature, and of such far-reaching implication, that it is high time the general public were given an opportunity to study it".

Shortly after that, Albert Howard published *The Soil and Health: A Study of Organic Agriculture* (1945), which shared this understanding. In 1952 Edward Hyams's *Soil and Civilization* followed, which ruminates on the collapse of civilizations as a function of soil degradation. In the 1960s and 70s, James Lovelock, Lynn Margulis and others worked on similar themes that have become known as Gaia Theory (Lovelock and Margulis 1974; Lovelock 1979), thus establishing a foundation for these transdisciplinary perspectives. More recently, it has been noted by Deem et al. (2019) that: "The idea that human, animal, and environmental health are connected has been around, in various renderings, for many years" (Deem et al. 2019: 9). The authors list various versions: One Medicine from the 1960s, Conservation Medicine from the 1990s, EcoHealth, Ecosystem Health, One Health and Planetary Health.

Yet, despite these seminal contributions, the rifts between people and nature have only deepened with intensifying impacts. Central to our argument is that the marginalisation of traditional medicinal and nutritional knowledge and agroecological food growing practices is due to the deep-rooted establishment of what Schneider and McMichael (2010) have called the 'epistemic rift'. Understanding and healing the rift are, we consider, necessary steps of agroecological transitions. They are also important to illustrate and fulfil the epistemic ambitions of this chapter, to lay the conceptual ground for an authentic medicinal agroecology and, ultimately, to understand the present and future battlegrounds of this emerging field.

## 3.   CAPITALISM, THE EPISTEMIC RIFT AND THE EROSION OF PEOPLE'S MEDICINAL KNOWLEDGE

The concept of epistemic rift emerges from a rich debate in rural sociology, agrarian studies and Marxist scholarship on the idea of a metabolic rift – or the rift that opens when the place where plants are grown (for food and medicine), or animals are bred, does not coincide with the place where organic matter and food waste (including human 'manure') are disposed of. This rift carries a consequence of soil depletion. The metabolic rift has been connected to the division of labour between town and country, the expansion of urbanisation and later, the mainstreaming of green revolution technologies (see for example the work of Jason Moore 2000 and John Bellamy Foster 1999) that have resulted in a reliance on chemical fertilisers. The growth of towns and cities that follows the expansion of capitalist industrialisation is associated with new practices of waste management (of human manure and household waste) that once used to be regarded as precious nutrients and returned to the compost to later feed the soil. Except for a few timid attempts in France and Belgium (Kohlbrenner 2014) to build an infrastructure for those nutrients to be returned to farmland for production, modern urban settlements were equipped with landfills, incineration and sewage systems that ignored centuries of wisdom in the management of natural resources and human settlements. These systems have instead contributed to the depletion of agricultural soils – through loss of minerals and organic matter – by breaking the closed loops of resource cycles that preceded them.

The rise of consumerism, population boom and the establishment of industrial agriculture have exercised further and ongoing pressure on agricultural soils and the ecosystem more generally. While capitalist urbanisation has played a crucial role in this story, the debate on the rift has also highlighted that extractive and exploitative practices preceded such urbanisation dynamics. There is indeed a lively debate on the origins of humans' exploitation of soils, with some authors tracing these back to the rise of agrarian capitalism in the XV century; and others to land enclosures and the slave trade (see for example Moore in Schneider and McMichael 2010). Others place important points of departure in the destruction of soils and medicinal wisdom to even earlier times, such as the invention of the plough (Hyams 1952; Montgomery 2007) or the growing power of mercantile interests and religious factions that obstructed or even outlawed the reproduction of medicinal practices (Buhner 1988). A shared point in these debates is the focus on the displacement of people – whether people displaced from common lands where they used to pasture their animals, cultivate and forage were forced to migrate to industrial centres to find a job or enslaved and deported overseas and forced to work in plantations on the 'new' continent. These debates shed light on the role of a knowledge rift in the mistreatment of both people and nature, and the subsequent erasure of food/medicinal knowledge. If knowledge can only be kept alive and reproduced so far as it is used, then we understand that forced displacement, be it through the enclosure of land and/or slavery, was a contributing factor to the loss of traditional ecological agricultural and medicinal practices. The metabolic rift has therefore always been enmeshed with the knowledge rift: people's bodies and people's knowledge were systematically disempowered by mercantilist ambitions and the changing geopolitical relations these ambitions forged.

The long-term establishment of industrialisation, capitalist urbanisation and colonialism (today in the form of land grabs and neo-colonialism) have fuelled the merging and consolidation of metabolic and knowledge rifts, in what has been termed the 'epistemic rift' between society and nature – a rift so deep that 'we have forgotten that we have forgotten'. The epistemic rift is reflected in the dystopic urban-rural, town-country divide, which has normalised the idea that agriculture and food do not belong to cities, but to the countryside; the idea that society and nature are two different entities, rather than the former being a part of the latter. The practical eradication of communities from the land, the proliferation of urbanised lifestyles alienated from a daily contact with soils, plants and the wider ecology, have removed the possibility to use and reproduce food, medicinal and agroecological knowledges that were shared and built through practical and spiritual practices by previous generations over centuries.

As cities continue to grow and struggle to address their crises of pollution and overcrowding, processes of planetary urbanisation have rendered many rural areas hinterlands of resource extraction and toxic residue, making the human habitat hazardous to health. As traditional food ways are eroded and replaced by an industrial diet, and as the quality of food as commodities decreases and the use of additives increases, people's health is further gravely undermined. Urbanised lifestyles that, in the context of a planetary urbanisation (Brenner 2014), are pervasive in both rural and urban contexts (Dehaene et al 2016), represent what has been called the imperial mode of living (Brand and Wissen 2021). This way of living is based on the extraction of resources from the land and from the people; on speculative mindsets; on the oppression and imperialist domination of the south by the north; and on profit over people. Urbanisms are both ideologies and ways of life; or ways of life co-shaped by ideas on how to organise them. An urbanism based on the centrality of soil stewardship and the equity-based value systems of agroecology – an alternative to the capitalist urbanism we know today – is needed to reverse the epistemic rift (see for example the work on an agroecological urbanism by Deh-Tor 2017 and 2021 and by Tornaghi and Dehane 2020 and 2021).

The epistemic rift did not only occur as a happenstance of more urbanised ways of life, but was consciously fed by ideas of progress, practices of knowledge validation and the supremacy of technological innovations omnipresent in mediated communication, education institutions and international governance institutions that have labelled traditional knowledge as 'superstition', 'folklore', 'unproven', 'dangerous' or placebo.

The 'multiplexity' of knowledges that have historically co-evolved in different locations world-wide, through layered encounters of botanical, therapeutic and ecological knowledges, has been, to a great extent, co-opted, exported and codified as science (Augusto, 2007). Yet, despite underpinning modern allopathic healthcare, these rich ecologies of knowledge are under constant threat of erasure.

The development of the epistemic rift, and the establishment of modern medicine and agriculture as norm and convention, had an amplifying effect on the consolidation and expansion of detrimental practices, the full scale of which is yet to be felt.

The day-to-day workings of the epistemic rift, and the separation between humans and nature, effectively switches off humans' natural ability to connect with and understand their own bodies, as well as plants and other living beings in their territory. The deep connections between us and the climate, the ecology, the bacteria in our guts and the medicinal metabolites of the plants we consume have progressively been severed and denied. The neglect of these faculties is a reflection of an attitude to tame, dominate, domesticate and artificially reproduce nature, rather than to cherish a world we are constitutive of and co-evolved with. The loss of such knowledge means not only the lack of ability to cater for our own health (by selecting food and plants, forage, etc.), but sometimes for our own survival. Research from Grivetti (1978, 1979, 2006) and Grivetti et al. (1987), for example, has observed the comparative impact of drought on different indigenous communities in Africa. Those who have actively retained the knowledge of edible and medicinal wild, drought-resistant plants (often called the 'weeds of agriculture') were able to resist famines caused by extreme droughts, while those who did not, died in their millions. Traditional knowledge of medicinal plants, and food as medicine, has been developed by humans across the world. The Indian Ayurveda, the Chinese and the Tibetan medicines are just a few renowned examples of traditional knowledge, validated by centuries, or millennia of practice and observation, and which, together with many other community-based knowledges rooted in their lands, biomes and cultures, have been discredited by imperial science and a European sense of superiority. The exclusive consumption of cultivated plants has become hegemonic, alongside the vilification of foraging and the eating of wild foods on the basis of the constructed tale that hunter-gatherer communities had poor diets, a fact disproven by several researchers who studied the intake of forager communities living in extreme environmental conditions (Codding and Kramer, 2016; Reyes-García and Pyhälä, 2017; Sahlins, 2009).

An argument persistently levelled against calls for better food and farming standards is that 'people need cheap food'. This is augmented, and even weaponised, by the assertion that 'more food is needed' to feed our ever-growing cities and global population. Seen together, these arguments justify efficiency drives that further consolidate control of the food system in the hands of a few corporate conglomerates. Indeed, anyone seeking to challenge this consolidation, and to resist the structural inequalities it exacerbates, is regularly accused of being 'anti-science', a counter-narrative with a long historical tradition. The results of the 'cheap food' narratives, perpetuated over centuries of slow violence by landed and corporate interests, can be seen in declining food and farming standards, environmental destruction and a series of intersecting public health crises. Alongside industrially produced food, "[i]ndustrial society has given us effective medical treatments, but it's also making us sick" (De Decker, 2021).

The establishment of what we now call 'conventional' agriculture (what a splendid example of the epistemic rift!), that is: the mainstreaming of fossil fuel-intensive agriculture and food from nowhere, has impacted on, among other things, the reduced number of crops and plant varieties available today as food. A recent study, for example, compares the diversity of seeds and breeds of peasant systems to that of commercial food systems (8774 versus 100), and notes that not only the diversity of species is extremely reduced, but also that the genetic diversity *within* species has been severely eroded (since the 1960s this diversity has been reduced by 75%). The same study also observes that the nutritional qualities of chain-bred varieties have declined by up to 40% (ETC 2017). This productivist quest for genetic uniformity is increasingly cited as the root of pest epidemics, infectious diseases and of novel animal-human transmissions. (ETC 2017; Wallace 2020).

The reliance on only a few selected plant (and animal) species as food, and the selective breeding of plants for characteristics suitable for commercialisation (i.e. colour, sweetness, etc.) has considerably reduced the number of medicinal compounds in plants, as well as their nutritional value, and impoverished peoples' diets with systematic and devastating health consequences. Emerging research on plants' medicinal compounds (secondary metabolites) notes that commercial breeding directed to please consumers with familiar sweet tastes have progressively removed those compounds responsible for bitter tastes, which often have important medicinal properties (Ku et al. 2020; Clemensen et al. 2020).

The discrediting of human metabolic intelligence and people's knowledge, and the impoverished medicinal and nutritional value of commercial plants come together in a powerful and catastrophic third main consequence of the epistemic rift: the mainstreaming of biologically and ecologically inappropriate and unhealthy diets. The reduced and impoverished food choices, and the ongoing loss of traditional knowledge with the advance of urbanised ways of life, are further amplified by people's diaspora, displacement, colonisation or migration (Owen 2006; Grivettti et al 1987) communities exposed to fragility, and therefore in vulnerable financial conditions, are often prone to toxic 'cheap food', and without access to culturally or biologically appropriate diets. The global burden of disease is enormous, and food is a significant contributor to this.

## 4.   PEOPLE'S HEALTH & (IN)APPROPRIATE DIETS

According to the Global Burden of Disease Study (Afshin et al 2019) – based on data from 204 countries and territories, 369 diseases and 87 risk factors – the deaths of 11 million people in 2017 "were attributable to dietary risk factors". Not too little food – the annual figure for death from starvation is around 9 million (TheWorldCounts 2021) – but the *wrong* food. There might be some overlaps between these estimates, meaning that they cannot necessarily be added together. While this is beyond the scope of this chapter, what is certain is that millions of people die every year from either eating unhealthy food or having inadequate access to food. For reference, the current total annual number of deaths (of all causes) is around 60 million (UN 2022).

The United Nations' Food and Agriculture Organisation (FAO 2021) underlines that "[m]alnutrition, in all its forms, includes undernutrition (wasting, stunting, underweight), inadequate vitamins or minerals, overweight, obesity, and resulting diet-related noncommunicable diseases". With 1.9 billion adults and 38.9 million children overweight or obese and 462 million adults and 194 million children suffering from undernutrition, the global burden of malnutrition has serious and lasting impacts on societies across the world (Ibid.).

Globally, non-communicable diseases are said to cause the death of an estimated 41 million people annually (Johns Hopkins 2021a. That means that approximately 70% of all deaths worldwide are caused by what we can call the problems of a modern lifestyle, much of which turns on our dinner plate – wrongly filled or empty, as the case may be – and is compounded by poor labour conditions and increasing toxic environmental loads.

Humanity is increasingly eating the same diet across the world – industrially produced starches, vegetable oils, refined sugars and processed meat (Khoury et al. 2014). We may hence assume that the US figures from Johns Hopkins on non-communicable diseases – often attributed to diet and lifestyle – could serve as a useful indicator for the future: 35% of the US population is obese (115 million), 20% have cardiovascular diseases (66 million), 15% have Type 2 diabetes mellitus

(50 million), 4% have some form of cancer (13 million) and 3% have an autoimmune disease (10 million) (Johns Hopkins 2021b). This is over three quarters of the population.

These staggering figures ought to give pause, especially when seen in the light of data from the current pandemic. According to data from Johns Hopkins, 18 months into the pandemic first recorded in Wuhan in December 2019, approximately 3.8 million people have died following infection with SARS-CoV-2 (2021) – and studies strongly correlate pre-existing conditions – especially obesity (Peters et al. 2020; Wang et al. 2021; Poly et al. 2021; Wu et al. 2021) and diabetes (Corona et al. 2021), which in turn are diet-related conditions – with increased severity of the course of the disease known as COVID-19.

There are many ways of dissecting these numbers and we have here presented merely a cursory overview for indicative purposes to undergird the notion that food matters when it comes to health. All of these numbers will weigh heavily on health systems that are geared to curative and palliative, rather than preventive medicine.

With the ongoing drive for genetic uniformity and intensification of food production systems in some areas, and land sparing for biodiversity in others, health vulnerabilities associated with extraction are exported, creating new points of rupture and leaving some more exposed than others. Yet, as 2020 aptly demonstrated, novel pandemics will not be readily contained by such technocratic practices and imaginaries, travelling fast along supply chains of ever-cheaper goods and labour.

To many who had sounded alarm about zoonotic spillover due to increasing pressures on ecosystems combined with growing industrial farming of animals (e.g. Wallace 2016), the COVID-19 pandemic did not come as a surprise: it is a foreseeable product of an economic system based on over-extraction.[4] Wallace et al. (2020) put it this way:

> "the cause of COVID-19 and other such pathogens is not found just in the object of any one infectious agent or its clinical course, but also in the field of ecosystemic relations that capital and other structural causes have pinned back to their own advantage".

What is more, pandemic effects are exacerbated by other aspects of the same structures. For this reason, the current pandemic has been described as part of a *syndemic* – a 'synergistic epidemic' (Horton 2020a).

The concept of syndemics developed in the context of working with impoverished inner-city populations in the United States in the 1990s (Singer and Snipes 1992). It is a systems approach to address public and community health in which complementary elements in the course of any disease are considered (Singer 1994, 1996, 2009). From the syndemic perspective, ill health is viewed as the coming together of multiple factors, both physiological as well as social. On the one hand, pre-existing health conditions and concurrent diseases of whatever kind are considered in conjunction; while on the other, the course of a disease assemblage is combined with a critical perspective on the existing social, cultural, economic and political backdrop, leading to an appreciation of disease as a complex whole. 'Syndemic', then, is a conception of (ill) health as a biosocial phenomenon (Singer et al. 2017). Through an embrace of multiple dimensions, a syndemic understanding permits transcending conceptualisations of health that consider solely the individual, bound organism with at best comorbidity and multimorbidity factors (ibid: 941–942; Mendenhall et al. 2017: 951).

The syndemic framework enables more socially conscious approaches to medicine and public health by insisting on the importance of social and ecological systems and relationships. The insights gained from approaching health through a syndemic framework raise crucial questions with regard to narrowly conceived public health strategies as responses to COVID-19 or any other health issue.

---

4  However, despite the imminent threat of deforestation and industrial agriculture as a source of zoonosis, at the time of writing, the origins of SARS-CoV-2 have not been established; the debate on origin was recently reignited by Wade, 2021.

If we are living a syndemic crisis, then our responses need to be equally systemic and address the socio-ecological context that fans the syndemic flames.

It is our contention that food – its production, distribution and consumption – plays a central role in this crisis. The consequences of the epistemic rift we discussed above, especially the dwindling of opportunities to choose biologically appropriate diets – for reasons of ignorance, corporate greed, and/or economic hardship – as well as the ominous decline in nutritional content of crops, are crucial factors in malnutrition. Industrial food production is intricately implicated in and exacerbates the collapse of biodiverse ecosystems and habitats as well as human health through carbon emissions, agricultural toxicity, but also deforestation and the expansion of the agricultural frontier. Food storage and transport contribute equally by relying heavily on energy-intensive modalities (including digitalisation), ever-increasing asphaltation and long-distance, just-in-time logistics. In conjunction with austerity-entrenched poverty and anguish, historical disinvestments in public health, deteriorating infrastructures, including housing that is hazardous to health (damp, mouldy, over-crowded), the global food system is an explosive part of the syndemic mix.

Understanding growing disease burdens, and how we respond to them from a systems perspective also calls on us to acknowledge the importance of structural inequalities that inflict 'slow violence' on the most vulnerable whose life-support systems are being eroded and polluted (Nixon, 2011).

## 5.  FROM PLANETARY HEALTH TO A PLANETARY DIET?

In light of the many dysfunctions and disconnections regarding health and food, one might surmise that the old dictum 'all things are connected' is gaining renewed traction given that even the Lancet – one of the world's leading, establishment medical journals – and the Rockefeller Foundation have embraced the idea of planetary health (Horton and Lo 2015; Demaio and Rockström 2015; Clark 2015; Whitmee et al. 2015). At first sight there are similarities between The Rockefeller Foundation-Lancet Commission's take on planetary health and the idea of medicinal agroecology we are developing here. From both these perspectives human health should not be – indeed, cannot be – understood in isolation from the health of our habitat. Both agree: the planet's health is people's health.

A closer look, however, reveals fundamental differences. While we would welcome that such a prominent journal and philanthropic power players promote a holistic health framework, their track record of industrial solutions stands in contrast to the agroecological conception we promote here, which seeks to build health and equity from the ground up, literally, providing access to soil to improve health by way of food sovereignty and the building of interspecies alliances.

In order to realise planetary health, the landmark EAT-Lancet report (Willett et al 2019) and resulting 'Planetary Diet' has been strongly criticised for being highly reductive and deeply embedded in global power (e.g. Shiva 2019). Highlighting some of the contradictions and vested interests, investigative journalist Joanna Blythman writes:

"[EAT] has a partnership with Fresh, a body made up of 40 of the world's most powerful corporations, a roll call of the big names in pharmaceuticals, pesticides, GM, and ultra-processed food. They include Bayer, which now owns Monsanto and its infamous Round-Up (glyphosate) pesticide, Big Sugar (PepsiCo), Big Grain (Cargill), palm oil companies, and leading manufacturers of food additives and processing aids" (Blythman 2019).

While a full critique and detailed analysis of the Planetary Diet is beyond the scope of this chapter, we focus here on one further point of contention. The report's approach to taxation indicates how the Planetary Diet falls short of offering pathways to establishing and enhancing planetary health in ways that address the systemic causes of mounting and interconnected crises. In their aim to make food prices "fully reflect the true cost of food", including through "[e]xperimentation with sugar or soft drink taxes", the report recommends a critical review of subsidies for fertilisers, water, fields, electricity, and pesticides, and gestures towards the possibility of removing these entirely (Willett et al. 2019, 479). However, as such taxation measures will drive up food prices, the recommendation continues that "where appropriate, social protection or safety nets (e.g., increasing income through cash transfers) can be established to protect vulnerable populations, particularly children and women, while keeping trade open" (ibid.).

What does that mean? It means that the funding flows that subsidise the petrochemical food industry should be reorganised. Yet reducing the direct funding of petrochemicals will increase production cost in a system reliant upon petrochemical inputs. In order to keep trade open – as the report elegantly puts the widely accepted imperative that profit margins must continue to grow – food prices will rise. As the number of people who cannot afford to feed themselves will increase with a price hike, the report proposes to remedy this problem by subsidising the *purchase* of petrochemically grown food, rather than its *production*. Instead of feeding the petrochemical and fossil fuel industry directly, the financial flows should be funnelled through consumers, who are thus made more dependent on state benefits and aid. And all the while, the petrochemical trade stays open and the food continues to contain harmful residues in the context of a food industry strongly geared to run on seeds that need fertilisers and pesticides.

It must be noted that the Rockefeller Foundation, which "as the unparalleled 20th century health philanthropy heavyweight, both profoundly shaped WHO and maintained long and complex relations with it, even as both institutions changed over time" (Birn 2014, 129), is no newcomer to public health, nor to agriculture. Although now supplanted by the Bill and Melinda Gates Foundation as the primary agenda setter in global agriculture, health governance, drug research and development, the Rockefeller Foundation, together with other philanthrocapitalists such as the Ford Foundation, were important funders of the Green Revolution, which aggressively expanded petrochemical agriculture across the globe (Kohler 2007). These players continue to play a powerful role in the shaping of contemporary food and health systems. Shaped in their image, the industrial food system has been designed, rather than arisen by happenstance – and the EAT approach to planetary health is merely an updated version of the same. What the Rockefeller Foundation, EAT, the Lancet et al. are promoting is an effort to deliver on the demands of the dictum that 'everything must change so that everything can stay the same'.

Effectively, then, rather than a contribution to the medicinal agroecology debate, the EAT-Lancet's version of a planetary diet runs counter to it, going hand in hand with a wide range of corporate responses to the climate crisis – such as precision farming, robotics, genome editing and other technology-heavy innovations. These 4th Industrial revolution (4IR) responses – also referred to by the World Economic Forum as 'The Great Reset' – are about change *within* the industrial mode of production, thus maintaining current social and economic inequalities. Medicinal agroecology, on the other hand, embraces the demands from a growing movement of peasants and civil society: the crises humanity faces cannot be remedied by more of the same, and the industrial mode of food production must be discontinued. Directly addressing an array of nexuses – including but not limited to environmental justice, soil and gut health, the nutritional value of food, prevention of zoonotic spill-over, regeneration of biodiversity – the agroecological pathway to planetary health gets to the roots of causes rather than merely suppressing their symptoms.

We must question the implications of forever chasing the symptoms with new technologies that are likely to lead to the emergence of new problems, or to temporarily obscure or suppress others, including the rising environmental footprint of increasingly high-tech healthcare (De Decker, 2021). Seen together, the combined ecological footprint of 4IR in agriculture and technological responses to ill-health applied within wealthy nations are likely to accelerate climate change – pushing against the limits of planetary boundaries.

## 6. CONCLUSIONS

Amidst the mounting consensus that the dominant food regime is failing against its own metrics, including rates of hunger and malnutrition, there is acute concern that it is not only contributing to intersecting biodiversity and climate crises (HLPE 2019; IPCC, 2019, Swinburn et al, 2019; Herren and Haerlin, 2020), but causal to the epidemic of chronic health conditions (FAO, 2021; Johns Hopkins, 2021a and 2021b). Despite recognition of the need to transition to agroecological food and farming systems that produce nutrient-dense foods that are more resistant to climate stress (IAASTD, 2009; UNCTAD; 2013; Anderson et al, 2019; Leippert et al, 2020), technofixes for this failing food system and the ill health it produces, both in human and planetary terms, continue the deceit that it emerges from the logics of relentless progressivism. These are the same logics that now drive us inexorably towards the 4IR – with new modes of production being proposed to further industrialise our food systems, this time through a process of de-naturing and de-labouring the food system entirely. And so, it seems, we are forever chasing our symptomatic tails, and that "Each new phase of world capitalist development, is accompanied by a new form and scale of rupture in socio-ecological relations" (Schneider and McMichael 2010:465).

In this chapter we have argued that food – its production, distribution and consumption – plays a central role in perpetuating or transforming present crises. Medicinal agroecology offers a pathway to planetary health by directly addressing key nexi of these crises (soil health, gut health, nutritional value of food, prevention of zoonotic spillover, regeneration of biodiversity, amongst other), that is, it addresses root causes rather than symptoms and constitutes a planetary medicine.

In a medicinal agroecology framework, human and more-than-human relations matter. Whom you are connected to and in which ways, the conditions in which you live, what you eat, drink, breathe and the ways in which you are able to dispose of your waste matter; always and in all ways matter, but especially to health. Rather than pursuing techno- or pharmacological interventions to mask symptoms, the agroecological framework brings complexities to the fore and considers holistic approaches to healing people and planet.

We want to conclude this chapter by offering a few questions to nourish a research agenda for medicinal agroecologies to thrive.

What are the opportunities to move away from our present trajectory – of medicating populations against the passively accepted symptoms of modernity, and against otherwise entirely natural processes? And what are the challenges of this trajectory in the context of planetary urbanisation and ongoing urbanising societies? What might the implications be for reconceptualising planetary health if discussions were instead embedded in autonomous modes of production and reproduction? If we reconnected and remembered our rich ecologies of knowledge, how then might knowledge travel and connect up heterogeneously to optimise exchange, adoption and adaptation? How might we, in support of medicinal agroecologies, shift away from productivist approaches in favour of practices that care for and cherish the proliferation of healthy and happy wild plants and recognise human and planetary health's dependency on wildlife and wild landscapes?

These questions raise a series of tensions – not least about how and when agroecology embraces some technologies but not others. As political agroecologists with an interest in questions of power, control and social agency, we recognise that the path towards nourishing research and practice for medicinal agroecologies remains embattled and fragmented. Walking it requires an unrelenting engagement in healing the (ever deepening) epistemic rift. Addressing public health issues by applying agroecology as planetary medicine is now more relevant and important than ever.

## REFERENCES

Afshin, Ashkan, Patrick John Sur, Kairsten A. Fay, Leslie Cornaby, Giannina Ferrara, Joseph S. Salama, Erin C. Mullany, et al. 2019. "Health Effects of Dietary Risks in 195 Countries, 1990–2017: A Systematic

Analysis for the Global Burden of Disease Study 2017". *The Lancet* 393 (10184): 1958–72. https://doi. org/10.1016/S0140-6736(19)30041-8.

Anderson, Colin Ray, Janneke Bruil, Michael Jahi Chappell, Csilla Kiss, and Michel Patrick Pimbert. 2019. "From transition to domains of transformation: Getting to sustainable and just food systems through agroecology." *Sustainability* 11, no. 19: 5272.

Anderson, Colin Ray, Janneke Bruil, Michael Jahi Chappell, Csilla Kiss, and Michel Patrick Pimbert. (2021). *Agroecology now!: Transformations towards more just and sustainable food systems* (p. 199). Springer Nature

Arora, Saurabh, Barbara Van Dyck, Divya Sharma, and Andy Stirling. 2020. "Control, care, and conviviality in the politics of technology for sustainability." *Sustainability: Science, Practice and Policy* 16, no. 1: 247–262.

Augusto, Geri. 2007. "Knowledge free and 'unfree': Epistemic tensions in plant knowledge at the Cape in the 17th and 18th centuries." *International Journal of African Renaissance Studies* 2, no. 2: 136–182.

Balfour, Lady Evelyn Barbara. 1943. *The Living Soil: Evidence of the Importance to Human Health of Soil Vitality, with Special Reference to Post-War Planning*. Faber.

Bilimoria, Purushottama. 2017. "Buddha Fifth Century BCE." In *Key Thinkers on The Environment*, edited by Joy A. Palmer Cooper and David E. Cooper, 6–11. Routledge.

Birn, A.-E. 2014. "Backstage: The Relationship between the Rockefeller Foundation and the World Health Organization, Part I: 1940s-1960s." *Public Health* 128 (2): 129–40. https://doi.org/10.1016/ j.puhe.2013.11.010.

Blythman, Joanna. 2019. "Scrutinise the Small Print of Eat-Lancet." In *Wicked Leeks* (blog). 29 January 2019. https://wickedleeks.riverford.co.uk/opinion/veganism-meat/scrutinise-small-print-eat-lancet.

Brand, Ulrich, and Markus Wissen. 2021. *The Imperial Mode of Living: Everyday Life and the Ecological Crisis of Capitalism*. Verso.

Brenner, Neil. 2014 *Implosions/Explosions Towards a Study of Planetary Urbanization*. Berlin: JOVIS Verlag.

Buhner, Stephen Harrod. 1998. *Sacred and herbal healing beers: the secrets of ancient fermentation*. Brewers Publications.

Carson, Rachel. 2002. *Silent Spring*. Houghton Mifflin Harcourt.

Clark, Helen. 2015. 'Governance for Planetary Health and Sustainable Development'. In *The Lancet* 386 (10007): e39–41. https://doi.org/10.1016/S0140-6736(15)61205-3.

Clemensen, Andrea K., Frederick D. Provenza, John R. Hendrickson, and Michael A. Grusak. "Ecological Implications of Plant Secondary Metabolites-Phytochemical Diversity Can Enhance Agricultural Sustainability." *Frontiers in Sustainable Food Systems* 4 (2020): 233.

Codding, Brian F. and Karen L. Kramer. 2016. *Why Forage? Hunters and Gatherers in the Twenty-First Century*. University of New Mexico Press.

Corona, Giovanni, Alessandro Pizzocaro, Walter Vena, Giulia Rastrelli, Federico Semeraro, Andrea M. Isidori, Rosario Pivonello, Andrea Salonia, Alessandra Sforza, and Mario Maggi. 2021. 'Diabetes Is Most Important Cause for Mortality in COVID-19 Hospitalized Patients: Systematic Review and Meta-Analysis'. In *Reviews in Endocrine and Metabolic Disorders* 22 (2): 275–96. https://doi.org/10.1007/ s11154-021-09630-8.

De Decker, Kris. 2021. "How Sustainable is High-tech Health Care? Can we make modern health care carbon-neutral and maintain the levels of care, pain relief, and longevity that we have come to take for granted?" *Low<High Tech Magazine*.

Deem, Sharon L., Kelly E. Lane-deGraaf, and Elizabeth A. Rayhel. 2019. *Introduction to One Health: An Interdisciplinary Approach to Planetary Health*. John Wiley & Sons.

Deh-Tor, C. M. 2021. "Food as an urban question, and the foundations of a reproductive, agroecological urbanism." In *Resourcing an Agroecological Urbanism*, pp. 12–33. Routledge.

Deh-Tor, C.M. 2017. "From agriculture in the city to an agroecological urbanism: the transformative pathway of urban (political) agroecology." *Urban Agriculture Magazine* 33: 8–10.

Dehaene, Michiel, Chiara Tornaghi, and Colin Sage. 2016. "Mending the metabolic rift: Placing the 'urban' in urban agriculture." *Urban Agriculture Europe*, pp. 174–177. Jovis Verlag.

Demaio, Alessandro R, and Johan Rockström. 2015. "Human and Planetary Health: Towards a Common Language." *The Lancet* 386 (10007): e36–37. https://doi.org/10.1016/S0140-6736(15)61044-3.

ETC. 2017. 'Who Will Feed Us? The Industrial Food Chain vs. The Peasant Food Web'. 3rd edition. ETC group. http://www.etcgroup.org/sites/www.etcgroup.org/files/files/etc-whowillfeedus-english-webshare.pdf.

FAO. 2021. *Fact Sheets – Malnutrition*. https://www.who.int/news-room/fact-sheets/detail/malnutrition.

Foster, John Bellamy JB. 1999. "Marx's Theory of Metabolic Rift: Classical Foundations for Environmental Sociology." *American Journal of Sociology*, 105 (2), pp. 366–405

Gifford, E. 1998. *The Many Speeches of Seathl: The Manipulation of the Record on Behalf of Religious, Political and Environmental Causes*. Sonoma State University.

Grivetti, L. E., C. J. Frentzel, K. E. Ginsberg, K. L. Howell, and B. M. Ogle. 1987. "Bush foods and edible weeds of agriculture: perspectives on dietary use of wild plants in Africa, their role in maintaining human nutritional status and implications for agricultural development." in *Health and Disease in Tropical Africa: Geographical and Medical Viewpoints*. London: Harwood Academic Publishers: 51–81.

Grivetti, Louis E. 2006. "Edible wild plants as food and as medicine: reflections on thirty years of fieldwork." *Eating and healing: Traditional food as medicine*: 11–39.

Grivetti, Louis E. 1978. "Nutritional success in a semi-arid land: examination of Tswana agro-pastoralists of the eastern Kalahari, Botswana." *The American Journal of Clinical Nutrition* 31, no. 7: 1204–1220.

Grivetti, Louis Evan. 1979. "Kalahari agro-pastoral-hunter-gatherers: the Tswana example." *Ecology of Food and Nutrition* 7, no. 4: 235–256.

Herren, H.R. and Haerlin, B. (eds). 2020. *Transformation of our Food Systems: The Making of a Paradigm Shift. Reflections from IAASTD +10*.

HLPE. 2019. *Agroecological and other innovative approaches for sustainable agriculture and food systems that enhance food security and nutrition*. A report by the High-Level Panel of Experts on Food Security and Nutrition of the Committee on World Food Security, Rome.

Horton, Richard, and Selina Lo. 2015. 'Planetary Health: A New Science for Exceptional Action'. *The Lancet* 386 (10007): 1921–22. https://doi.org/10.1016/S0140-6736(15)61038-8.

Horton, Richard. 2020. 'Offline: COVID-19 Is Not a Pandemic'. In *The Lancet* 396 (10255): 874. https://doi.org/10.1016/S0140-6736(20)32000-6.

Hyams, Edward. 1952. *Soil and Civilization*. London: Thames and Hudson.

IAASTD. 2009. *Agriculture at a Crossroads*. Global report of the International Assessment of Agricultural Knowledge, Science, and Technology, Washington, DC: Island Press.

Johns Hopkins. 2021a. "Non-Communicable Diseases – Johns Hopkins CCP." *Johns Hopkins Center for Communication Programs* (blog). 2021. https://ccp.jhu.edu/what-we-do/focus-areas/non-communicable-diseases/.

Johns Hopkins. 2021b. "Prevalence of Autoimmune Diseases." *Johns Hopkins Pathology* (blog). 2021. https://pathology.jhu.edu/autoimmune/prevalence/.

Khoury, Colin K., Anne D. Bjorkman, Hannes Dempewolf, Julian Ramirez-Villegas, Luigi Guarino, Andy Jarvis, Loren H. Rieseberg, and Paul C. Struik. 2014. "Increasing Homogeneity in Global Food Supplies and the Implications for Food Security." *Proceedings of the National Academy of Sciences* 111 (11): 4001–6. https://doi.org/10.1073/pnas.1313490111.

Kohlbrenner, Ananda. 2014. "From fertiliser to waste, land to river: a history of excrement in Brussels." In *Brussels Studies*. La revue scientifique pour les recherches sur Bruxelles/Het wetenschappelijk tijdschrift voor onderzoek over Brussel/The Journal of Research on Brussels.

Kohler, Scott. 2007. "The Green Revolution: Rockefeller Foundation, 1943." Case 20. Durham, N.C: The Green Revolution: Rockefeller Foundation, 1943. Center for Strategic Philanthropy and Civil Society (Duke).

Ku, Yee-Shan, Carolina A. Contador, Ming-Sin Ng, Jeongjun Yu, Gyuhwa Chung, and Hon-Ming Lam. 2020. "The effects of domestication on secondary metabolite composition in legumes." *Frontiers in Genetics* 11.

Leippert, Fabio, Maryline Darmaun, Martial Bernoux, and Molefi Mpheshea. 2020. *The potential of agroecology to build climate-resilient livelihoods and food systems*. Rome. FAO and Biovision. https://doi.org/10.4060/cb0438en

Lovelock, James E., and Lynn Margulis. 1974. "Atmospheric Homeostasis by and for the Biosphere: The Gaia Hypothesis." *Tellus* 26 (1–2): 2–10.

Lovelock, James. 1979. *Gaia: A New Look at Life on Earth*. Oxford University Press.

Mendenhall, Emily, Brandon A. Kohrt, Shane A. Norris, David Ndetei, and Dorairaj Prabhakaran. 2017. "Non-Communicable Disease Syndemics: Poverty, Depression, and Diabetes among Low-Income Populations." *The Lancet* 389 (10072): 951–63. https://doi.org/10.1016/S0140-6736(17)30402-6.

Montgomery, David R. 2007. *Dirt: The Erosion of Civilizations*. 1st edition. Berkeley: University of California Press.

Moore, Jason W. 2000. "Environmental crises and the metabolic rift in world-historical perspective." *Organization & Environment* 13, no. 2: 123–157.

Nixon, Rob. 2011. *Slow Violence and the Environmentalism of the Poor*. Cambridge, Mass: Harvard University Press.

Owen, Patrick L. 2006. "Tibetan foods and medicines: antioxidants as mediators of high-altitude nutritional physiology." *Eating and Healing: Traditional Food as Medicine*. Food Products Press New York, London and Oxford.

Peters, Sanne A. E., Stephen MacMahon, and Mark Woodward. 2021. "Obesity as a Risk Factor for COVID-19 Mortality in Women and Men in the UK Biobank: Comparisons with Influenza/Pneumonia and Coronary Heart Disease." *Diabetes, Obesity and Metabolism* 23 (1): 258–62. https://doi.org/10.1111/dom.14199.

Poly, Tahmina Nasrin, Md Mohaimenul Islam, Hsuan Chia Yang, Ming Chin Lin, Wen-Shan Jian, Min-Huei Hsu, and Yu-Chuan Jack Li. 2021. 'Obesity and Mortality Among Patients Diagnosed With COVID-19: A Systematic Review and Meta-Analysis'. *Frontiers in Medicine*. https://doi.org/10.3389/fmed.2021.620044.

Reyes-García, Victoria, and Aili Pyhälä, eds. 2017. *Hunter-Gatherers in a Changing World*. Cham: Springer International Publishing.

Sahlins, Marshall. 2009. "Hunter-Gatherers: Insights from a Golden Affluent Age." *Pacific Ecologist* 18: 3–9.

Schneider, Mindi, and Philip McMichael. 2010. "Deepening, and repairing, the metabolic rift." *The Journal of Peasant Studies* 37, no. 3: 461–484.

Scott, James C. 2017. *Against the Grain: A Deep History of the Earliest States*. 1st edition. New Haven: Yale University Press.

Shiva, Vandana. 2019. "A New Report Sustains Unsustainable Food Systems." *Seed Freedom* (blog). 20 January 2019. https://seedfreedom.info/poison-cartel-toxic-food-eat-report/.

Singer, M. 1994. "AIDS and the Health Crisis of the U.S. Urban Poor; the Perspective of Critical Medical Anthropology." *Social Science & Medicine* 39 (7): 931–48. https://doi.org/10.1016/0277-9536(94)90205-4.

Singer, Merrill, and Charlene Snipes. 1992. "Generations of Suffering: Experiences of a Treatment Program for Substance Abuse During Pregnancy." *Journal of Health Care for the Poor and Underserved* 3 (1): 222–34. https://doi.org/10.1353/hpu.2010.0180.

Singer, Merrill, Nicola Bulled, Bayla Ostrach, and Emily Mendenhall. 2017. "Syndemics and the Biosocial Conception of Health.". *The Lancet* 389 (10072): 941–50. https://doi.org/10.1016/S0140-6736(17)30003-X.

Singer, Merrill. 1996. "A Dose of Drugs, a Touch of Violence, a Case of AIDS: Conceptualizing the Sava Syndemic." *Free Inquiry in Creative Sociology* 24 (2): 99–110.

Singer, Merrill. 2009. *Introduction to Syndemics: A Critical Systems Approach to Public and Community Health*. John Wiley & Sons.

Small, E. 2011. "The new Noah's Ark: beautiful and useful species only. Part 1. Biodiversity conservation issues and priorities." *Biodiversity*, 12(4), 232–247.

Swinburn, Boyd A., Vivica I. Kraak, Steven Allender, Vincent J. Atkins, Phillip I. Baker, Jessica R. Bogard, Hannah Brinsden et al. 2019. "The global syndemic of obesity, undernutrition, and climate change: the Lancet Commission report." *The Lancet* 393, no. 10173: 791–846.

TheWorldCounts. 2021. *How many people die from hunger each year? 2021*. https://www.theworldcounts.com/challenges/people-and-poverty/hunger-and-obesity/how-many-people-die-from-hunger-each-year/story.

Tornaghi, Chiara, and Michiel Dehaene. 2021. *Resourcing an Agroecological Urbanism: Political, Transformational and Territorial Dimensions*. London: Routledge.

Tornaghi, Chiara, and Michiel Dehaene. 2020. "The prefigurative power of urban political agroecology: rethinking the urbanisms of agroecological transitions for food system transformation." *Agroecology and Sustainable Food Systems* 44, no. 5: 594–610.

UN. 2022. World Population Prospects 2022. UN Department of Economic and Social Affairs, Population Division. https://population.un.org/wpp/

UNCTAD. 2013. *Wake up before it's too late: make agriculture truly sustainable now for food security in a changing climate*. Report by the United Nations Conference on Trade and Development. New York and Geneva.

Wade, Nicholas. 2021. "Origin of Covid — Following the Clues." *Medium*. May 28. https://nicholaswade.med ium.com/origin-of-covid-following-the-clues-6f03564c038.

Wallace, Rob. 2016. *Big Farms Make Big Flu: Dispatches on Influenza, Agribusiness, and the Nature of Science*. New York: Monthly Review Press.

Wallace, Rob. 2020. "Agriculture, capital, and infectious diseases." Herren, HR and Haerlin, B. (eds) *Transformation of our Food Systems: The Making of a Paradigm Shift. Reflections from IAASTD +10*.

Wallace, Rob, Alex Liebman, Luis Fermando Chaves, and Rodrick Wallace. 2020. "COVID-19 and Circuits of Capital." *Monthly Review*, 1 May 2020. https://monthlyreview.org/2020/05/01/covid-19-and-circuits-of-capital/.

Wang, Jingzhou, Toshiro Sato, and Atsushi Sakuraba. 2021. "Coronavirus Disease 2019 (COVID-19) Meets Obesity: Strong Association between the Global Overweight Population and COVID-19 Mortality." *The Journal of Nutrition* 151 (1): 9–10. https://doi.org/10.1093/jn/nxaa375.

Whitmee, Sarah, Andy Haines, Chris Beyrer, Frederick Boltz, Anthony G Capon, Braulio Ferreira de Souza Dias, Alex Ezeh, et al. 2015. "Safeguarding Human Health in the Anthropocene Epoch: Report of The Rockefeller Foundation–Lancet Commission on Planetary Health." *The Lancet* 386 (10007): 1973–2028. https://doi.org/10.1016/S0140-6736(15)60901-1.

Willett, Walter, Johan Rockström, Brent Loken, Marco Springmann, Tim Lang, Sonja Vermeulen, Tara Garnett, et al. 2019. "Food in the Anthropocene: The EAT–Lancet Commission on Healthy Diets from Sustainable Food Systems." *The Lancet* 393 (10170): 447–92. https://doi.org/10.1016/S0140-6736(18)31788-4.

Wu, Zeng-hong, Yun Tang, and Qing Cheng. 2021. "Diabetes Increases the Mortality of Patients with COVID-19: A Meta-Analysis." *Acta Diabetologica* 58 (2): 139–44. https://doi.org/10.1007/s00592-020-01546-0.

# 2 Achieving health sovereignty with medicinal plants on an agroecological farm – from theory to practice

*Joanna Sucholas, Anja Greinwald, Mariya Ukhanova and Rainer Luick*

## 1. INTRODUCTION

Farming based on 'agroecology' has gained prominence in scientific, agricultural and political discourse in recent years (HLPE 2019). Across these domains, it is widely seen as a transformative paradigm (Anderson et al. 2021; Anderson et al. 2019; HLPE 2019; IPES-Food 2016; Gliessman 2015; Méndez et al. 2015). It can offer local, adaptive and sustainable solutions to many global challenges, including climate change, biodiversity loss and loss of traditional knowledge (Anderson et al. 2019; IPCC 2019; Wezel et al. 2018; Méndez et al. 2015).

The transformative potential of agroecology is often explored in literature in relation to food systems (HLPE 2019; Chappell and Lappé 2018; Burlingame and Dernini 2012; Lichtfouse 2011), more specifically in reference to 'the affirmation of the right to food' (Anderson et al. 2019) and achieving food sovereignty (Bezner et al. 2019; Rothschild 2017; IPES-Food 2016).

Leaning on the understanding of transformative agroecology by Anderson et al. (2021), not as a goal but as a process 'rooted in the tradition of political ecology', we strive to harness the full transformative potential of agroecology and expand it to a further domain – health sovereignty. (1) We first reflect on the application of the multi-level perspective (MLP) theoretical framework as in Anderson et al. (2021), while focusing specifically on medicinal plants (MPs) within the context of an exemplary agroecological farm design. (2) Second, we provide an overview of MP-inclusive agroecological systems of traditional societies so as to exemplify niche innovation in the transition of communities towards health sovereignty. (3) Then, recognising the central part of knowledge sharing and continuity of practices within agroecology, we provide practical information on the presence of MPs on the model agroecological farm. For this small-scale agroecological farm with diverse vegetation sites, we provide a manual on identifying and including MPs within these sites. (4) Finally, we summarise the lessons learned from theory and practice, and pave the way for agroecological innovations towards 'health sovereignty'. We emphasise the process of innovation and its central role in agroecological transition towards health sovereignty. Our methodological approach to designing an exemplary agroecological farm is guided by U-theory by Scharmer (2018). This chapter does not constitute medical advice. It is intended for informational purposes only, and is not a substitute for professional medical advice, diagnosis or treatment. We do not make any medical recommendations and do not take any responsibility for the application of plants.

DOI: 10.1201/9781003146902-3

## 2. TRANSITION TOWARDS HEALTH SOVEREIGNTY WITH MEDICINAL PLANTS

According to Article 25 of the Universal Declaration of Human Rights, "Everyone has the right to a standard of living adequate for the health and well-being of himself and of his family, including food, clothing, housing and medical care..." (UN 1948). Very importantly, "denial of self-determination over food and medicine is a repudiation of fundamental rights of autonomy as guaranteed by Article 24 Section 1 of the UN Declaration on the Rights of Indigenous Peoples" (Kassam et al. 2010).

Against this background, we understand health sovereignty as a community's "ability to choose medicines that are socio-culturally and ecologically appropriate, thereby providing practical, reliable and contextually relevant health care options" (Kickbusch 2000), as summarised in Kassam et al. (2010). Due to the transformative and process-oriented nature of agroecology (Anderson et al. 2021), we analyse the transition towards health sovereignty through the MLP (Fig.1), which is a theoretical framework initially designed to conceptualise "dynamic patterns in socio-technical transitions" (Geels 2011). MLP is frequently applied in the analysis of transitions towards sustainability (Anderson et al. 2021; Geels 2011).

According to Geels (2011), MLP looks at transitions on three levels:

- exogenous landscape where, for example, climate change, the COVID-19 pandemic or demographic changes take place;
- the dominant regime that accounts for the stability of an existing system; and
- niches, for example, new ideas, ways of thinking or practical approaches that offer advancement towards sustainability like 'medicinal agroecology'.

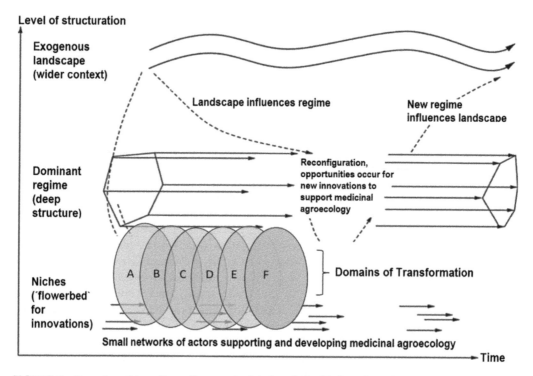

**FIGURE 1    Domains of transformation are depicted as definable interfaces between niche and regime superimposed onto a multi-level perspective figure presented in the context of medicinal agroecology.**
Source: own compilation adopted after Anderson et al. (2021) and Geels and Schot (2007).

All three levels and interactions across them are relevant for health sovereignty and in particular for MPs in the context of agroecology. However, in the case of transitions towards health sovereignty within the agroecological context, three features distinguish health sovereignty from food systems transitions.

First and most importantly, health sovereignty does not imply a complete overhaul of the existing healthcare and services regime. The aim is to provide more sustainable and, in some cases, more accessible and affordable options, but not the complete substitution of conventional medicine and services. In many cases, traditional medicine, including herbal remedies, co-exists with conventional medical services (WHO 2013).

Second, the issues of access, awareness and knowledge continuity. According to Díaz-Reviriego et al. (2016), "the adaptive capacity of local medical systems generally rests on two pillars: species diversity and a robust local knowledge system...". Moreover, the "...loss of access to plant species is detrimental to retention of knowledge about plant use and ecology" (Baumflek et al. 2015).

Third, there are stronger interactions between the niche and the landscape levels in the case of health sovereignty by MPs than with the food systems. For instance, there is a growing body of evidence that MPs are impacted by climate change (landscape level), including their quality (toxicity) and healing properties (Applequist et al. 2020; Maclean and Wilson 2011; Grabherr 2009). In addition, in many cases, ageing of the population or migration (landscape level) can lead to the lack of knowledge and access continuity in the use of MPs (niche level).

As the idea of self-determination and the concept of agency are by definition at the very core of health sovereignty, we see many overlaps with the notion of domains of transformation in the case of MPs in Table 1 (Anderson et al. 2021).

Overall, the application of the MLP shows high relevance of the concept of health sovereignty to the context of MPs in agroecology. MLP in our case revealed the relevance of all domains of transformation by Anderson et al. (2021), to the concept of health sovereignty within agroecology. Particularly relevant are the domains of Rights and Access to Nature, Knowledge and Culture, Networks and Equity. In these domains, agroecological practices involving MPs are being adopted, shared, exchanged and scaled up, which fosters co-evolution and 'cross-pollination' of these practices. Such processes are at the core of challenging the dominant regime and transitioning towards health sovereignty. These processes depend on and are shaped by the access of individuals and communities to knowledge, the networking capacity of individuals and communities. Most importantly, without rights and access to nature for communities and individuals, agroecological

## TABLE 1
**Domains of transformation (focus on MPs), whose innovations from the niche level come into conflict with the dominant regime.** Source: own compilation, based on Anderson et al. (2021). Domains A, B, D and E are most relevant for health sovereignty in agroecological context.

| Domains of transformation | Application for health sovereignty |
|---|---|
| A   **Rights and access to nature** | Access to land and biodiversity of MPs |
| B   **Knowledge and culture** | Access, awareness and knowledge continuity across generations, within and across communities |
| C   **Systems of economic exchange** | Access to market and additional income generation from MPs (commercial farming) |
| D   **Networks** | Knowledge and medicinal plant material exchange |
| E   **Equity** | Equal access to knowledge, land, biodiversity and ecosystem services |
| F   **Discourse** | Storylines and conceptualisation of the debate around health sovereignty gained with MPs |

innovations cannot occur, as agroecological knowledge is practical by nature. However, further research is needed to explore enabling and disabling factors within the domains of transformation.

To summarise in the words of Anderson et al. (2021), "agroecological transformation reflects not one grand theory of change but a recognition of a co-evolutionary and adaptive approach that involves multiple transformations". Through examples of niche innovations involving MPs in agroecological farming, in the following parts of the chapter we provide an overview of how these transformations are bringing communities and individuals towards health sovereignty.

## 3.  MEDICINAL PLANTS IN THE AGROECOLOGICAL FARMING SYSTEMS OF TRADITIONAL SOCIETIES

Since the times of early humans, plants were used for medicinal purposes (Fabricant and Farnsworth 2001; Solecki 1975). Over the centuries, every human society has developed medical systems to enhance and maintain health of the population (de Alwis 1997). Plants have formed a basis for medical systems such as Traditional Chinese Medicine (in China 4,676 plant species were identified as used medicinally [Qicheng 1980]) and Traditional Tibetan Medicine (99 species used [Kletter and Kriechbaum 2001]), Western herbal medicine (Lemonnier et al. 2017) and healing systems of traditional societies (TSs) all around the world. Research provides a number of plants used by TSs; for instance, 546 medicinal plants (MPs) are used by indigenous peoples of the Canadian boreal forest (Uprety et al. 2012), up to 700 plant species are used in traditional medicine in Sri Lanka (de Alwis 1997) and at least 1300 MPs are used by indigenous peoples in Amazonia (Schultes 1979). Obviously, TSs had to develop ways and methods of obtaining MPs for medical needs. Sourcing of plants by TSs had/has in some cases (e.g. indigenous peoples of Northern America such as the Ojibwa and Cree) a seasonally pure 'hunter-gatherer' character (Ingold 2000), but was/is often closer to 'agriculture'; for example, practices of transplanting, sowing, irrigating, burning and cultivating of plants, etc., in surrounding 'wild' nature (Anderson 2011). Ingold (2000) claimed that human land management is often a mixture of practices of obtaining resources from 'wild sites' and cultivation of domesticated species, which were developed in different proportions in various systems by human societies. Resource management and farming practices of TSs are indicated as important sources of inspiration for novel agroecological farms (Astier et al. 2017; Parmentier 2014; Gliessman 2013; Altieri 2004; Kumar and Nair 2004). Therefore, we take a closer look at agroecological systems of TSs to see if MPs are present there and how well they are integrated.

Numerous studies of TSs' agroecological farming concern systems such as agroforestry or homegardens (Kumar and Nair 2004), which are the object of our further analyses. Hoogerbrugge and Fresco (1993) defined a homegarden as "a small scale, supplementary food production system by and for household members that mimics the natural, multi-layered ecosystem". A typical homegarden in the vicinity of a dwelling is perceived as an indispensable element of the entire small-scale farming system (Kumar and Nair 2004) that is often comprised of more distant field systems, pastures, etc. Usually, homegardens are biodiverse and complex systems, intensively intercropped, mulched and irrigated, of multi-storeyed vegetation (Finerman and Sackett 2003). Worldwide, considerable research has been conducted on plant composition in general, plant usage and structure of homegardens, especially in the tropics (Wiersum 2006; Ceccolini 2002; Padoch and de Jong 1991; Rico-Gray et al. 1990).

Small-scale farming systems provide communities with various plant resources, not only food plants but also ornamental, timber, ritual, hedge, pest control and, not least, medicines. Agelet et al. (2000), recognising the great diversity, abundance and relevancy of MPs in homegardens of Catalonians, claim that the definition of homegardens by Hoogerbrugge and Fresco (1993) as 'food production systems' should be extended explicitly to production of MPs. In addition, Sunwar et al. (2006) point out that MPs are one of the most important ecosystem services of homegardens. As we are interested in the presence and integration of MPs in small-scale farming systems, we analyse practical integration of MPs in homegardens of TSs. We aim to give a few examples from broader regional, floristic and theoretical perspectives, while choosing traditional small-scale farming systems from different continents and referring to different domains of transitions towards health sovereignty. We provide an overview of scarce and therefore unique studies, which focus both on structure of the farm as well as on the ways of MPs' integration. These examples helped us in identifying the crucial farm vegetation sites and development of exemplary farm design.

### 3.1 HOMEGARDENS WITH MULTI-USE PLANTS IN TANZANIA

The Chagga people who live on the slopes of Mount Kilimanjaro in Tanzania manage their farms as agroforestry systems. They create multi-storeyed agroecological systems whose tree species are purposely interplanted by sometimes over a dozen food crop species and non-woody plants underneath. Only weeds are spontaneously growing plants. In this multicropping system, every species – from over 100 identified in the survey – had at least two out of the following uses: food, fuelwood, ornamental, hedge and medicine. The Chagga people use 30% of tree species from their homegardens as medicine (e.g. Eucalyptus species) as well as almost all of the non-woody plants and some of the weeds (Fernandes et al. 1985; O'kting'ati et al. 1984).

### 3.2 KNOWLEDGE AS A KEY ELEMENT OF MEDICINAL PLANT SPECIES' USE IN HOMEGARDENS IN CUBA

The biodiverse homegardens in Cuba are maintained as agroforestry systems usually comprised of three-layered vegetation (Wezel and Bender 2003). These homegardens used to play an enormous role in the self-sufficiency of local communities. The study of some 30 such homegardens revealed 101 different plant species that are used for food (vegetables, fruits, spices and tubers) and medicinal purposes. Some of the identified MPs were (intentionally) planted; however, the majority grow spontaneously in the homegardens. Even though almost all of the studied species have medicinal properties (according to Roig y Mesa [1945]), in the majority of cases they are identified and consumed by people as food plants while only small numbers of people had knowledge about their medicinal properties.

### 3.3 WOMEN MANAGE HIGHLY BIODIVERSE HOMEGARDENS IN ECUADOR

Finerman and Sackett (2003) analysed traditional homegardens of the indigenous Saraguro community from Ecuador. The farming system of the Saraguro people comprises three components: (1) distant field sites, multicropped, ploughed by animals; (2) pastureland located also at a distance from the house; and (3) areas of homegardens. The homegardens are usually smaller than 0.1 ha and form a complex system of intercropped and multi-storeyed plant varieties – ca. 95 varieties per garden – like fruit trees, vegetables, flowers, herbs and tolerated weeds. In one farm even, 194 plant varieties were identified; 132 varieties thereof had medicinal properties. These great numbers are not surprising if we note that in Ecuador, 3,118 plant species are reported as medicinal (La Torre et al. 2008). On each of the studied farms, the majority of the plants from all mentioned plant live forms had medicinal properties – ca. 65 per garden – and were processed and used by women to heal their family members.

### 3.4 FARMING PRACTICES FROM THE MIDDLE AGES PRESERVED IN HOMEGARDENS ON THE IBERIAN PENINSULA

Few studies focus on the presence and role of MPs in the traditional farming systems of homegardens in Europe (Agelet et al. 2000). Agelet et al. (2000) extensively studied the Catalonian mountainous homegardens of the Iberian Peninsula, which have a long tradition and are rooted in medieval times.

The homegardens supply people not only with edible plants but also with plants for other uses, like ornamental uses and, most importantly, MPs. The two last plant groups are usually planted either at garden borders near rocky walls, in pots or in other areas. Most interestingly, the structure of these gardens has not changed much since the Middle Ages. Catalonian homegardens are defined as the entire cultivated perimeter surrounding a dwelling. The study specifies a total of 250 MPs: cultivated MPs, plants with other uses than medicinal and wild growing species. The growth of these wild species is facilitated by the structure of the garden and the management by the gardener. The local people were also using plants from orchards and fields for medicinal purposes but those were not counted in the study.

## 3.5  Agroecologically managed MPs in summary

Studies focussing explicitly on how MPs are integrated into the structure of traditional agroecosystems are scarce. On the one hand, research on the structure of traditional small-scale farms is rarely focused on MPs. On the other hand, research on plant-based traditional medicinal systems tends not to focus on farms providing medicines for communities. However, MPs are indicated as one of the most important plant resources, along with food plants, that are 'produced' on traditional farms (Caballero-Serrano et al. 2019; Sunwar et al. 2006; Finerman and Sackett 2003; Agelet et al. 2000). In almost all mentioned traditional systems, the communities deliberately grow MPs on the farm for their own use, which greatly contributes to the health sovereignty of these rural communities. In all analysed homegardens from agroforestry systems, MPs belonged to tree species, shrubs or herbaceous species. They were either cultivated species, wild species introduced into cultivation or tolerated weeds. The diversity of MPs on the farm and likewise their location in the structure of the farm depends on a range of factors spanning local environmental conditions and factors relating to a specific socio-cultural context.

## 4.  PRESENCE OF MEDICINAL PLANTS ON AN AGROECOLOGICAL FARM

In this chapter, we provide practical information in the form of a manual that proves the presence of MPs on an agroecological farm and shows where and how these plant species are included in the system. Although MPs and agroecological farms occur worldwide, we chose one floristic region to provide more details. We focused on the South-Eurosiberian floristic region[1], more precisely on the Central European floristic province of this region. Countries included are Germany, Switzerland, Austria, Poland, Czech Republic, Slovakia, the Baltic countries, Southern Scandinavia, the East of France, parts of Hungary, Romania and Slovenia (Pfadenhauer and Klötzli 2020). However, many of the MPs presented are common plant species worldwide and are not restricted to the Central European floristic province. We indicate which MPs are present in various sites of the farm. We have also compiled basic pharmacognostic information about the species to show the wide range of medicinal properties plants have and the broad medicinal purposes of these plants, which can potentially be used by the community. Furthermore, we provide information about ecological conditions required by each species to specify the habitat that is suitable for growing it. The information should help to identify the best on-farm sites to maintain or cultivate certain species.

### 4.1  An example of a small-scale agroecological farm design with diverse vegetation sites

An agroecological farm strives to be heterogeneous and as biodiverse as possible, and by default should provide many habitats for various species (Anderson et al. 2021; Pimbert 2018; Altieri et al.

---

1   Floristic regions are geographical distribution areas of plant genera and species with similar distribution patterns (Pfadenhauer and Klötzli 2020)

1987). There is neither one paradigm of an agroecological farm, nor one technical manual indicating what kind of structures and habitats the farm should consist of. However, there are universal principles upon which an agroecological farming system is based, viewing and treating a farm as a natural ecosystem with plentiful ecological interactions, which do not allow the use of artificial fertilizers and pesticides, whilst banning genetically modified organisms (GMOs), as opposed to intensive 'conventional' farming (Gliessman 2015). In practice, agroecological farming systems are complemented by techniques shared by the related, more practice-based farming systems like agroforestry, silvopastorialism, polyculture, permaculture, regenerative farming, etc., and variations thereof. Nevertheless, a strong emphasis should be made on locally adapted solutions, primarily derived from local farming traditions and culture (e.g. HLPE 2019; Rothschild 2017; Altieri et al. 1987).

Based on practical farming systems mentioned above (e.g. described by Raskin and Osborn 2019; Holzer 2011; Hemenway 2009), homegardens of analysed TSs, scientific recommendations for ecological farming (Albrecht et al. 2020; Wezel et al. 2014), the agroecological farm 'Bec Hellouin' developed in France (presented during the Oxford Real Farming Conference 2021) and our ecological knowledge, we designed an exemplary small-scale agroecological farm (Fig. 2). The design process was guided by the U-theory (Scharmer 2018). Inspired by the sources mentioned, we identified the essential and most relevant vegetation site types a farm could consist of to ensure sustainability whilst being high in biodiversity. Additionally, we analysed the MPs that can be found among different plant groups on these vegetation sites. We combined both information parts and categorised nine different plant species groups of the recognised vegetation sites. However, the

**FIGURE 2   An exemplary small-scale agroecological farm of the Central European floristic province of the South-Eurosiberian floristic region.** Numbers represent the plant species groups of the different vegetation sites explained in the text.

division of the plant species into groups is not strict, as some species can be found in other groups too (growing on other vegetation sites). The plant species are assigned to a specific group to emphasise the main function of the plant (ecological or practical). Encoding by numbers (see Table 2 and Table 3) provides information about categorisation into identified plant groups for every medicinal plant species.

Plant species groups of the different vegetation sites are (Fig. 2):

1. **Culinary crop species of crop sites:** cultivated plant species of gardens and crop fields primarily used as food.
2. **Medicinal crop species of crop sites:** cultivated plant species of gardens and crop fields primarily used as medicine.
3. **Green manure species of cultivated sites where fertilization is needed:** species that naturally improve the fertility of the soil when they decompose on the cultivated area (e.g. legumes that fix nitrogen) and species producing large amounts of biomass (e.g. sunflower) (Wezel et al. 2014; Holzer 2011; Hemenway 2009). They are usually interplanted on cultivated crop sites together with cultivated plants, or sown in any other place where fertilization is needed, or they can be used as fodder for livestock.
4. **Species of forests and woody patches:** typical forest shrubs, trees and herbaceous plant species.
5. **Grassland species of open, semi-natural areas:** species of unimproved grasslands, pastures, old orchards and vineyards, wood pastures and others. These vegetation sites produce primarily hay and fodder for livestock (e.g. pigs, cattle, sheep, horses, etc.), and, depending on soil, elevation, wetness, and management, provide habitats for various grassland plant species.
6. **Species of riparian buffer strips and areas next to waterbodies:** within the farm landscape, waterbodies (e.g. ponds, streams and rivers) can occur. They play a crucial role in the water management of the farm and are used for irrigation and as drinking spots for animals (Bane 2012). Additionally, they create unique microclimates, provide habitats for amphibians (Holzer 2011) and on their edges develop vegetation tolerant of high water levels. Along rivers or creeks, in between crop fields, riparian buffer strips can be formed, which are promoted (e.g. in agroforestry) due to their ability of protecting the quality of the water, improving aquatic habitats and serving many other functions in the ecosystem (Raskin and Osborn 2019).
7. **Species of hedgerows and windbreaks:** sometimes humans deliberately pick a plant composition due to the special features of the plants. We propose here shrub and tree species growing in hedgerows as well as windbreaks that create natural physical barriers between fields or farms, catching wind, protecting soil, providing shelter and food for animals plus having numerous other functions (Raskin and Osborn 2019).
8. **Flowering species of flower strips:** species sown to create crop edges or sown in various other parts of the garden to attract pollinators and melliferous bee species, and in general to increase and support overall insect populations (Azpiazu et al. 2020; Rundlöf et al. 2018; Scheper et al. 2015).
9. **Weed species:** spontaneously growing herbs. They can occur next to pathways and crop fields, on fallow sites and generally in places that are disturbed by humans, like pathways.

**4.2  THE MEDICINAL PLANT SPECIES OF THE DIFFERENT VEGETATION SITES – A MANUAL**

In our manual, we provide information about MPs that are likely to occur on identified vegetation sites of our exemplary agroecological farm. We chose MP species occurring naturally or being easily cultivated in the European temperate climate of the Central European floristic province, excluding typical alpine species and species that are recognised as invasive in the delineated region. We analyse widely recognised MPs with well-described and proven medicinal properties. All the pharmacognostic data is based on medicinal plants and information included in the European Union herbal monographs of the EMA (available online), the European Pharmacopoeia (Ph Eur 10.0 2020) and respected herbal books (Ożarowski and Jaroniewski 1989; van Wyk and Wink 2008). All these plant species are believed to be suitable for use under domestic conditions. This means medicines can be prepared from the plant with basic home equipment and their application can be considered sufficiently safe. We excluded from our analysis medicinal plants that are a basis for medicines requiring specialist knowledge and lab conditions to prepare, as well as species that can be poisonous (depending on dosage).

In the main part of the manual, we focus on a few example species for each of the nine recognised plant groups and create pharmacognostic and ecological profiles (Table 2). For these exhaustive plant profiles, we chose species that are common worldwide, have many other uses besides medicinal ones and provide various and complementary types of medicine or species with multiple medicinal properties. Here, we aim at providing a compilation of certain plants for each of the identified groups, whose inclusion on a farm could increase health sovereignty of a given community. The profiles include medicinal properties, diseases, forms of application and possible warnings. The ecological information in the profiles can assist the detection of plants on different sites of the farm or the cultivation of plants in suitable ecological conditions. The ecological data derive from various sources, such as the database of the German Federal Office for Nature Conservation (Bundesamt für Naturschutz – BfN: www.floraweb.de), the Pladias Database of the Czech Flora and Vegetation (www.pladias.cz), the Swiss vegetation database (www.infoflora.ch) and the excursion flora of Oberdorfer (Oberdorfer and Schwabe 2001). The origin and life span of the species, the vegetation sites and plant groups where the species occurs and a short overview of the ecological needs of the species are shown. The data about ecological needs of the species are based on the Ellenberg indicator values and the indicator values of Landolt et al. (Ellenberg et al. 1992; Landolt et al. 2010). Additionally, we provide information about other uses of species by humans and other functions the species provide to an agroecosystem, based on various literature and online sources (Łuczaj 2013; Holzer 2011; Hemenway 2009; www.pfaf.org database).

Separately and additionally, we provide a long list of other MPs that might occur on an agroecological farm in European temperate climate (Table 3). However, the list includes less-detailed information compared to the plant profiles in Table 2. Table 3 provides the species name, plant family, plant organ of medicinal use, medicinal properties and where to find the plant on the farm based on the identified plant species groups.

## 5.  CONCLUSION

In the first part of this chapter, we explored the potential of agroecology in facilitating transition towards health sovereignty via the inclusion of plants with medicinal properties in agroecological systems. The application of the MLP theoretical framework as in Anderson et al. (2021) to the context of MPs on an agroecological farm had proven the high relevance of MPs to a potential transition towards health sovereignty. For instance, we found that four out of six domains of transformation as depicted in Anderson et al. (2019) and Anderson et al. (2021) are highly relevant for enabling health sovereignty of a community or an individual: Rights and Access to Nature, Knowledge and Culture, Networks and Equity. However, further exploration of the application of all six domains of transformation as well as their enabling and disabling factors in relation to envisaged health sovereignty and MPs is necessary. Such research may provide insights into identification of barriers to and opportunities for transition.

In the second part, we turned to 'niche innovations' that have potential to lead the transformation towards health sovereignty. Our analysis identified the presence of a great variety of MPs in traditional small-scale farming systems (e.g. homegardens) from different parts of the world. Most importantly, we discovered a broad spectrum of sustainable innovative practices in cultivation,

## TABLE 2
**Detailed pharmacognostic and ecological profiles of medicinal plant species possibly occurring on an agroecological farm of the Central European province.** Pharmacognostical source: EMA; European Pharmacopoeia (Ph Eur 10.0 2020); Ożarowski and Jaroniewski 1989; van Wyk and Wink 2008. Ecological source: www.floraweb.de; www.pladias.cz; www.infoflora.ch; Ellenberg et al. 1992; Landolt et al. 2010; Oberdorfer and Schwabe 2001. Abbreviations of the table headings: Ls = life span, F = plant family, Or = origin, L = light conditions, T = temperature, M = moisture, R = soil-pH, N = nitrification,

| 1 | Culinary crop species of crop sites | | | | | | | |
|---|---|---|---|---|---|---|---|---|
| Plant name | Ls | F | Or | L | T | M | R | N |
| *Allium cepa* L. | Perennial | Liliaceae | M-As | light | hot | fresh | acid/alk | rich |
| *Avena sativa* L. | Annual | Poaceae | undefined | full light | moderate | fresh | sl acidic | rich |
| *Cannabis sativa ssp. sativa* L. | Annual | Cannabaceae | E-Eu, As | light | warm | moist | sl acidic | rich |
| *Cucurbita pepo* L. | Annual | Cucurbitaceae | Am | full light | warm | fresh | acid/alk | saturated |
| *Cynara cardunculus ssp. scolymus* (L.) Berger | Perennial | Asteraceae | Eu | light | warm | fresh | no data | no data |
| *Phaseolus vulgaris* L. | Annual | Fabaceae | Am | light | warm | moist | mod alk | moderate |
| *Vitis vinifera* L. | Perennial | Vitaceae | Eu | light | warm | fresh | acid/alk | rich |

| 2 | Medicinal crop species of crop sites | | | | | | | |
|---|---|---|---|---|---|---|---|---|
| Plant name | Ls | F | Or | L | T | M | R | N |
| *Calendula officinalis* L. | Annual | Asteraceae | EU | full light | warm | fresh | acid/alk | rich |
| *Echinacea purpurea* (L.) Moench. | Perennial | Asteraceae | N-Am | light | warm | mod dry | acid/alk | moderate |
| *Matricaria recutita* L. | Annual | Asteraceae | Eu | light | warm | fresh | sl acidic | moderate |
| *Melissa officinalis* L. | Perennial | Lamiaceae | Eu, W-As | light | warm | mod dry | acid/alk | moderate |
| *Salvia officinalis* L. | Perennial | Lamiaceae | Eu | light | warm | dry | acid/alk | moderate |
| *Silybum marianum* (L.) Gaertn. | Annual, biennial | Asteraceae | Eu, S/W-As | light | warm | mod dry | acid/alk | rich |
| *Tanacetum parthenium* (L.) Sch. Bip. | Perennial | Asteraceae | Eu | light | warm | fresh | acid/alk | rich |

| 3 | Green manure species of cultivated sites where fertilization is needed | | | | | | | |
|---|---|---|---|---|---|---|---|---|
| Plant name | Ls | F | Or | L | T | M | R | N |
| *Fagopyrum esculentum* Moench | Annual | Polygonaceae | E-As | light | moderate | fresh | acid/alk | rich |
| *Galega officinalis* L. | Perennial | Fabaceae | Eu | light | warm | moist | acid/alk | saturated |
| *Linum usitatissimum* L. | Annual | Linaceae | undefined | full light | moderate | fresh | sl acidic | rich |
| *Malva sylvestris* L. | Perennial | Malvaceae | Eu, W-As | full light | warm | mod dry | acid/alk | saturated |

OU = other use. Abbreviations of the origin: Europe = Eu, Asia = As, America = Am, Africa = Af, Australia/New Zealand = Aus-New, North = N, East = E, South = S, West = W and Middle = M. Abbreviations of the ecological data: acid = acidic, alk = alkaline, acid/alk = slightly acidic/slightly alkaline, sl = slightly, mod = moderate. Abbreviations of the other uses: basketry = bas, biomass = bio, bird food = birf, culinary = cul, dye = dye, fertilizer = fer, fibre = fib, fodder = fod, melliferous = mel, oil = oil, ornamental = orn, pollinator = pol, repellent = rep, sap = sap, stabilizer = sta.

| On farm | OU | Part | Properties | Disease | How to use |
|---|---|---|---|---|---|
| 1 | Cul, dye, rep | Bulbs | Antiseptic, anti-cholesterolemic | Cold, hyper-cholesterolemia | Fresh juice |
| 1 | Bio, cul, fod | Fruits, herb | Roborant, anti-inflammatory | Exhaustion, skin irritations, | Tea and tincture for consumption, bath |
| 1 8 9 | Birf, cul, fib, mel, pol | Seeds | Anti-inflammatory, anti-cholesterolemic | Skin irritation, hyper-cholesterolemia | Oil and seeds for consumption |
| 1 | Cul, oil, pol | Seeds | Antiparasitic | Internal parasites, hypertrophy of the prostate | Dried seeds and oil for consumption |
| 1 | Cul, pol | Leaves | Digestive, hepato-protective | Hyper-cholesterolemia, liver diseases | Tea made from dried leaves |
| 1 3 | Cul, fer, pol | Fruits without seeds | Hypoglycaemic, diuretic | Diabetes, urinary complaints | Tea made from dried fruits without seeds |
| 1 | Birf, cul | Leaves | Antioxidant, astringent | Capillary fragility, varicose veins | Tea made from dried leaves |

| On farm | OU | Part | Properties | Disease | How to use |
|---|---|---|---|---|---|
| 2 8 | Mel, orn, pol, rep | Flowers | Antiseptic, vulnerary | Sore throat, skin injuries | Tea made from dried flowers, ointment (externally) |
| 2 8 | Orn, pol | Roots, herb | Immuno-stimulant, vulnerary | Cold, skin injuries | Expressed juice WARNING: use not longer than 10 days |
| 2 9 | Mel, pol, rep | Flowers | Digestive, antiseptic | Gastrointestinal spasms, throat inflammations | Tea made from dried fruits, inhalation |
| 2 | Pol, rep | Leaves | Antianxiety, sedative | Mental stress, sleep disorders | Tea made from dried leaves |
| 2 | Pol, rep | Leaves | Antiseptic, antibacterial | Sore throat, excessive sweating | Tea made from dried leaves WARNING: not during pregnancy |
| 2 | Pol | Fruits | Hepato-protective, hepato-regenerative | Severe liver damage | Powdered fruits for consumption |
| 2 6 9 | Orn, Rep | Herb | Antispasmodic, anti-inflammatory | Migraine, headache, rheumatism | Powdered herb for consumption, Tea made from dried herb |

| On farm | OU | Part | Properties | Disease | How to use |
|---|---|---|---|---|---|
| 1 3 8 | Cul, dye, fer, fod, mil, pol | Herb | Antiswelling, rich in rutin | Microcirculatory dysfunction | Tea made from dried herb |
| 3 5 6 | Fer, pol | Herb | Hypoglycaemic, lactogenic | Diabetes, insufficient lactation | Tea made from dried herb |
| 1 2 3 | Cul, fer, fib, pol, oil | Seeds | Laxative, emollient | Constipation, gastrointestinal inflammation | Decoction |
| 3 6 89 | Fer, orn, pol | Flowers | Expectorant, demulcent | Dry cough, gastrointestinal inflammation | Tea made from dried flowers |

*(continued)*

**TABLE 2 (Continued)**
**Detailed pharmacognostic and ecological profiles of medicinal plant species possibly occurring on an agroecological farm of the Central European province.** Pharmacognostical source: EMA; European Pharmacopoeia (Ph Eur 10.0 2020); Ożarowski and Jaroniewski 1989; van Wyk and Wink 2008. Ecological source: www.floraweb.de; www.pladias.cz; www.infoflora.ch; Ellenberg et al. 1992; Landolt et al. 2010; Oberdorfer and Schwabe 2001. Abbreviations of the table headings: Ls = life span, F = plant family, Or = origin, L = light conditions, T = temperature, M = moisture, R = soil-pH, N = nutrients,

| 3 | Green manure species of cultivated sites where fertilization is needed | | | | | | | |
|---|---|---|---|---|---|---|---|---|
| Plant name | Ls | F | Or | L | T | M | R | N |
| *Medicago sativa* L. | Perennial | Fabaceae | M-As | full light | warm | mod dry | acid/alk | indifferent |
| *Melilotus officinalis* (L.) Lam. | Annual | Fabaceae | Eu, W-As | full light | warm | mod dry | mod alk | poor |
| *Trifolium pratense* L. | Perennial | Fabaceae | Eu, W-As | light | indifferent | indifferent | indifferent | indifferent |

| 4 | Species of forests and woody patches | | | | | | | |
|---|---|---|---|---|---|---|---|---|
| Plant name | Ls | F | Or | L | T | M | R | N |
| *Betula pendula* Roth. | Perennial | Betulaceae | Eu | light | indifferent | indifferent | indifferent | indifferent |
| *Frangula alnus* Mill. | Perennial | Rhamnaceae | Eu | light | warm | wet | mod acidic | indifferent |
| *Humulus lupulus* L. | Perennial | Cannabaceae | Eu | light | warm | wet | sl acidic | saturated |
| *Pinus sylvestris* L. | Perennial | Pinaceae | Eu, As | light | indifferent | indifferent | indifferent | indifferent |
| *Rubus fruticosus* agg. Null | Perennial | Rosaceae | Eu, N-Am | light | moderate | fresh | sl acidic | moderate |
| *Sambucus nigra* L. | Perennial | Caprifoliaceae | Eu | light | moderate | fresh | indifferent | saturated |
| *Viburnum opulus* L. | Perennial | Caprifoliaceae | Eu, W-As | light | moderate | indifferent | acid/alk | rich |

| 5 | Grassland species of open, semi-natural areas | | | | | | | |
|---|---|---|---|---|---|---|---|---|
| Plant name | Ls | F | Or | L | T | M | R | N |
| *Achillea millefolium* L. | Perennial | Asteraceae | Eu | full light | indifferent | mod dry | indifferent | moderate |
| *Armoracia rusticana* Gaertn., B. Mey. & Scherb. | Perennial | Brassicaceae | Eu | full light | warm | fresh | indifferent | saturated |
| *Euphrasia rostcoviana* Hayne | Perennial | Orobanchaceae | Eu | full light | moderate | fresh | mod acidic | poor |
| *Hypericum perforatum* L. | Perennial | Hypericaceae | Eu, W-As | light | warm | mod dry | sl acidic | poor |
| *Potentilla erecta* (L.) Raeusch. | Perennial | Rosaceae | Eu | light | indifferent | indifferent | indifferent | depleted |
| *Primula veris* L./ *P. elatior* (L.) Hill. | Perennial | Primulaceae | Eu, Ws-As | light | indifferent | mod dry | mod alk | poor |
| *Thymus serpyllum* L. | Perennial | Lamiaceae | Eu | light | warm | dry | sl acidic | depleted |

OU = other use. Abbreviations of the origin: Europe = Eu, Asia = As, America = Am, Africa = Af, Australia/New Zealand = Aus-New, North = N, East = E, South = S, West = W and Middle = M. Abbreviations of the ecological data: acid = acidic, alk = alkaline, acid/alk = slightly acidic/slightly alkaline, sl = slightly, mod = moderate. Abbreviations of the other uses: basketry = bas, biomass = bio, bird food = birf, culinary = cul, dye = dye, fertilizer = fer, fibre = fib, fodder = fod, melliferous = mel, oil = oil, ornamental = orn, pollinator = pol, repellent = rep, sap = sap, stabilizer = sta.

| On farm | OU | Part | Properties | Disease | How to use |
|---|---|---|---|---|---|
| 3  9 | Fer, fod, pol | Herb | Roborant, anti-cholesterolemic, oestrogenic | Weight loss, hypercholesterolemia, menstrual problems | Tea made from dried herb |
| 3  5  89 | Fer, mel,pol, rep | Herb | Antiswelling, antispasmodic | Varicose veins, heaviness of legs | Tea made from dried herb |
| 3  5 | Cul, fer, fod,pol, mel | Herb | Expectorant, vulnerary | Dry cough, skin irritation | Tea made from dried herb |

| On farm | OU | Part | Properties | Disease | How to use |
|---|---|---|---|---|---|
| 4  6  7 | Bas, bio, sap | Leaves | Antirheumatic, Diuretic, | Rheumatism, kidney stones | Tea made from dried leaves |
| 4  6  7 | Birf | Bark | Laxative | Constipation | Dried bark needs to be stored 12 months, decoction |
| 4  6  9 | Cul, pol | Flowers | Antianxiety, antispasmodic | Mental stress, sleep disorder | Tea made from dried flowers; WARNING: > 12 years old |
| 4 | Bio, Cul | Young shoots | Expectorant, antiseptic | Respiratory infections | Tea made from dried shoots; syrup made from shoots |
| 1  4  7 | Birf, cul, dye, pol | Leaves | Anti-inflammatory, antidiarrhoeal | Sore throat, diarrhoea | Tea made from dried leaves |
| 4  6  79 | Birf, cul, dye, fod, pol | Flowers, fruits | Diaphoretic (flowers), mild laxative (fruits) | Cold (flowers), constipation (fruits) | Tea made from dried flowers/fruits, WARNING: not during pregnancy |
| 4  6  7 | Birf, cul, dye, pol | Bark | Antispasmodic | Menstrual pain, any spasms | Tea made from dried bark |

| On farm | OU | Part | Properties | Disease | How to use |
|---|---|---|---|---|---|
| 5  8  9 | Cul, fod, mel, pol, rep | Herb | Digestive, antispasmodic, anti-haemorrhagic | Dyspepsia, spasms, wounds | Tea made from dried herb, fresh juice on wounds |
| 1  5  9 | Cul, pol, rep | Roots | Antibacterial, antiseptic | Respiratory infections, rheumatism | Fresh roots internally and externally |
| 5 | Pol | Herb | Anti-inflammatory, astringent | Eye inflammation | Ocular compress |
| 5  8  9 | Dye, pol | Herb | Antidepressant, cholagogue | Mild depression, gastrointestinal disorder, skin inflammation | Tincture and tea for consumption, oil externally WARNING: Phototoxic tinctures, drug interactions |
| 5  9 | Dye | Roots | Antidiarrhoeal, astringent | Diarrhoea, sore throat | Tea made from dried roots |
| 5 | Orn, pol | Flowers, roots | Expectorant, anti-inflammatory | Dry cough, cold | Tea made from dried flowers and roots WARNING: Allergies against salicylates are known |
| 5 | Cul, mel, pol | Herb | Expectorant, antibacterial | Cough, cold | Tea made from dried herb |

(continued)

**TABLE 2 (Continued)**

**Detailed pharmacognostic and ecological profiles of medicinal plant species possibly occurring on an agroecological farm of the Central European province.** Pharmacognostical source: EMA; European Pharmacopoeia (Ph Eur 10.0 2020); Ożarowski and Jaroniewski 1989; van Wyk and Wink 2008. Ecological source: www.floraweb.de; www.pladias.cz; www.infoflora.ch; Ellenberg et al. 1992; Landolt et al. 2010; Oberdorfer and Schwabe 2001. Abbreviations of the table headings: Ls = life span, F = plant family, Or = origin, L = light conditions, T = temperature, M = moisture, R = soil-pH, N = nutrients,

| 6 Species of riparian buffer strips and areas next to waterbodies | | | | | | | | |
|---|---|---|---|---|---|---|---|---|
| Plant name | Ls | F | Or | L | T | M | R | N |
| *Acorus calamus* L. | Perennial | Araceae | Eu, As, Am | full light | warm | wet | acid/alk | rich |
| *Epilobium angustifolium* L./ *E. parviflorum* Schreb. | Perennial | Onagraceae | Worldwide | full light | indifferent | fresh | sl acidic | saturated |
| *Filipendula ulmaria* (L.) Maxim. | Perennial | Rosaceae | Eu | light | moderate | wet | indifferent | poor |
| *Lythrum salicaria* L. | Perennial | Lythraceae | Eu, As, E-Af | light | moderate | wet | sl acidic | indifferent |
| *Mentha arvensis* L./ *M.* x *piperita* L. | Perennial | Lamiaceae | Worldwide | light | indifferent | wet | indifferent | indifferent |
| *Symphytum officinale* L. | Perennial | Boraginaceae | Eu, W-As | light | warm | wet | indifferent | saturated |
| *Valeriana officinalis* L. | Perennial | Valerianaceae | Eu, As | light | warm | wet | acid/alk | moderate |
| 7 Species of hedgerows and windbreaks | | | | | | | | |
| Plant name | Ls | F | Or | L | T | M | R | N |
| *Berberis vulgaris* L. | Perennial | Berberidaceae | Eu | light | indifferent | mod dry | moist | poor |
| *Crataegus monogyna* Jacq./*C. laevigata* (Poir.) DC. | Perennial | Rosaceae | Eu, W-As | light | moderate | mod dry | mod alk | poor |
| *Hippophae rhamnoides* L. | Perennial | Elaeagnaceae | Eu, As | full light | warm | mod dry | mod alk | poor |
| *Prunus spinosa* L. | Perennial | Rosaceae | Eu | light | moderate | mod dry | acid/alk | indifferent |
| *Rosa canina* L./ *R. centifolia*/ *R. gallica* L. | Perennial | Rosaceae | Eu, W-As | full light | moderate | mod dry | indifferent | indifferent |
| *Salix purpurea* L./ *S. fragilis* L. | Perennial | Salicaceae | Eu | full light | moderate | wet | acid/alk | moderate |
| *Usnea spec.* | - | Parmeliaceae | - | light | cool | no data | sl acidic | no data |

OU = other use. Abbreviations of the origin: Europe = Eu, Asia = As, America = Am, Africa = Af, Australia/New Zealand = Aus-New, North = N, East = E, South = S, West = W and Middle = M. Abbreviations of the ecological data: acid = acidic, alk = alkaline, acid/alk = slightly acidic/slightly alkaline, sl = slightly, mod = moderate. Abbreviations of the other uses: basketry = bas, biomass = bio, bird food = birf, culinary = cul, dye = dye, fertilizer = fer, fibre = fib, fodder = fod, melliferous = mel, oil = oil, ornamental = orn, pollinator = pol, repellent = rep, sap = sap, stabilizer = sta.

| On farm | OU | Part | Properties | Disease | How to use |
|---|---|---|---|---|---|
| 5 6 | Cul, bas, rep | Roots | Carminative, bitter | Flatulence, dyspepsia, | Tea of the dried roots WARNING: drug interactions, toxic variations |
| 5 6 | Fib, pol | Herb | Anti-inflammatory | Prostatic inflammation | Tea made from dried herb |
| 5 6 | Mel, pol | Flowers, herb | Antipyretic, anti-inflammatory | Cold, fever, articular pain | Tea made from dried flowers and herb WARNING: Allergies against salicylates are known, not during pregnancy |
| 6 8 | pol | Herb | Anti-haemorrhagic, astringent | Heavy menstrual bleeding, diarrhoea | Tea made from dried herb |
| 2 5 69 | Cul, pol | Leaves | Digestive, carminative | Dyspepsia, gastrointestinal spasms | Tea made from dried leaves |
| 5 6 8 | Fod, mel, pol | Roots | Vulnerary | Injuries | Ointment (externally) WARNING: Pyrrolizidine alkaloids in the plant |
| 5 6 9 | Pol | Roots | Antianxiety, antispasmodic | Nervous tension, sleep disorder | Tea made from dried roots WARNING: > 12 years old |

| On farm | OU | Part | Properties | Disease | How to use |
|---|---|---|---|---|---|
| 4 7 | Birf, cul, dye, fod, pol | Roots, bark, fruits | Cholagogue, antifungal | Gallbladder complaints, digestive inflammations | Decoction out of dried roots, bark, fruits |
| 4 6 7 | Birf, cul, fod, pol | Leaves, flowers | Hypotensive, cardiotonic | Heart complaints, palpitations | Tea made from the dried flowers |
| 3 4 7 | Birf, cul, dye, fer, sta | Fruits | Rich in vitamin C, vulnerary | Cold, stomach ulcer | Fresh, juiced, oil for consumption |
| 4 6 7 | Birf, cul, dye, pol | Fruits, flowers | Diuretic, astringent | Mouth inflammation, diarrhoea | Tea made from dried flowers and fruits |
| 4 7 | Birf, cul, orn, pol | Flowers, fruits | Anti-inflammatory, rich in vitamin C | Sore throat, cold | Tea made from dried flowers, fruits |
| 4 6 7 | Bas, bio, mel, pol, sta | Bark | Antipyretic, anti-inflammatory | Fever, rheumatism | Tea made from dried bark WARNING: Allergies against salicylates are known, not during pregnancy |
| 4 7 | Indicator for air quality | Whole lichen | Anti-inflammatory | Sore throat, dry cough | Tea made from the whole dried lichen |

*(continued)*

## TABLE 2 (Continued)
**Detailed pharmacognostic and ecological profiles of medicinal plant species possibly occurring on an agroecological farm of the Central European province.** Pharmacognostical source: EMA; European Pharmacopoeia (Ph Eur 10.0 2020); Ożarowski and Jaroniewski 1989; van Wyk and Wink 2008. Ecological source: www.floraweb.de; www.pladias.cz; www.infoflora.ch; Ellenberg et al. 1992; Landolt et al. 2010; Oberdorfer and Schwabe 2001. Abbreviations of the table headings: Ls = life span, F = plant family, Or = origin, L = light conditions, T = temperature, M = moisture, R = soil-pH, N = nutrients,

| 8 | | | Flowering species of flower strips | | | | | |
|---|---|---|---|---|---|---|---|---|
| Plant name | Ls | F | Or | L | T | M | R | N |
| *Borago officinalis* L. | Annual | Boraginaceae | Eu | full light | warm | mod dry | acid/alk | rich |
| *Cichorium intybus* L. | Perennial | Asteraceae | Eu, W-As | full light | warm | mod dry | mod alk | moderate |
| *Helianthus annuus* L. | Annual | Asteraceae | N-Am | full light | hot | fresh | sl acidic | rich |
| *Hyssopus officinalis* L. | Perennial | Lamiaceae | Eu, W-As | full light | warm | dry | acid/alk | poor |
| *Inula helenium* L. | Perennial | Asteraceae | W-As | light | warm | fresh | acid/alk | moderate |
| *Lavandula angustifolia* Mill. | Perennial | Lamiaceae | Eu | light | moderate | dry | sl acidic | poor |
| *Verbascum thapsus* L./*V. densiflorum* Bertol. | Biennial | Scrophula-riaceae | Eu, W-As | full light | indifferent | mod dry | acid/alk | rich |

| 9 | | | Weed species of fallow sites, crop sites and areas next to pathways | | | | | |
|---|---|---|---|---|---|---|---|---|
| Plant name | Ls | F | Or | L | T | M | R | N |
| *Capsella bursa-pastoris* (L.) Medik. | Annual, biennial | Brassicaceae | Worldwide | light | indifferent | fresh | indifferent | rich |
| *Centaurea cyanus* L. | Annual | Asteraceae | Eu | light | warm | indifferent | indifferent | indifferent |
| *Equisetum arvense* L. | Perennial | Equisetaceae | Worldwide | light | indifferent | moist | indifferent | poor |
| *Leonurus cardiaca* L. | Perennial | Lamiaceae | Eu, W-As | full light | warm | fresh | mod alk | saturated |
| *Plantago lanceolata* L. | Perennial | Plantaginaceae | Eu, W-As | light | indifferent | indifferent | indifferent | indifferent |
| *Taraxacum* sect. *Ruderalia* Wiggers | Perennial | Asteraceae | Eu | light | indifferent | fresh | indifferent | rich |
| *Urtica dioica* L. | Perennial | Urticaceae | Worldwide | indifferent | indifferent | moist | acid/alk | saturated |

OU = other use. Abbreviations of the origin: Europe = Eu, Asia = As, America = Am, Africa = Af, Australia/New Zealand = Aus-New, North = N, East = E, South = S, West = W and Middle = M. Abbreviations of the ecological data: acid = acidic, alk = alkaline, acid/alk = slightly acidic/slightly alkaline, sl = slightly, mod = moderate. Abbreviations of the other uses: basketry = bas, biomass = bio, bird food = birf, culinary = cul, dye = dye, fertilizer = fer, fibre = fib, fodder = fod, melliferous = mel, oil = oil, ornamental = orn, pollinator = pol, repellent = rep, sap = sap, stabilizer = sta.

| On farm | | | OU | Part | Properties | Disease | How to use |
|---|---|---|---|---|---|---|---|
| 1 | 2 | 8 | Cul, dye, mel, oil, orn, rep | Flowers, herb, seeds | Expectorant, anti-inflammatory | Respiratory infections, skin inflammation (seeds) | Tea made from dried flowers and herb, seed oil for consumption WARNING: Pyrrolizidine alkaloids in the plant |
| 8 | 9 | | Cul, fod, mel, pol | Roots | Hepato-protective, bitter | Liver dysfunction, dyspepsia | Tea made from dried roots |
| 1 | 3 | 8 | Birf, cul, fer, fod, mel, oil, orn | Flowers | Antipyretic, vulnerary | Fever, externally for bruises | Tea made from dried flowers, fresh flowers externally |
| 1 | 2 | 8 | Cul, pol, rep | Herb | Expectorant, carminative | Respiratory infections, stomach complaints | Tea made from dried herb WARNING: not during pregnancy |
| 6 | 8 | 9 | Cul, pol | Roots | Expectorant, antiparasitic | Respiratory infections, parasites | Tea made from dried roots |
| 2 | 8 | | Mel, orn, pol, rep | Flowers | Antianxiety, antibacterial | Mental stress, dyspepsia | Tea made from dried flowers |
| 5 | 8 | 9 | Dye, orn, pol | Flowers | Expectorant, vulnerary | Sore throat, dry cough | Tea made from dried flowers |

| On farm | | OU | Part | Properties | Disease | How to use |
|---|---|---|---|---|---|---|
| 9 | | Birf, cul, fod | Herb | Anti-haemorrhagic, astringent | Any haemorrhages, skin injuries | Tea made from dried herb WARNING: not during pregnancy |
| 8 | 9 | Dye, orn, pol | Flowers | Antifungal, diuretic | Eye inflammation, water retention | Tea made from dried flowers |
| 9 | | Cul, dye, rep | Herb | Diuretic, rich in silica | Urinary complaints, lack of silica | Tea made from dried herb |
| 9 | | Pol | Herb | Antianxiety, antiarrhythmic | Mental stress, circulatory disorders | Tea made from dried herb |
| 5 | 9 | Cul, fod, pol | Leaves | Demulcent, vulnerary | Dry cough, skin injuries | Tea made from dried leaves |
| 8 | 9 | Cul, fod, mel, pol | Roots, leaves, herb | Cholagogue, digestive, diuretic | Digestive and urinary disorders | Tea made from dried plant |
| 6 | 9 | Cul, fer, fib, fod | Leaves, herb | Antirheumatic, antianemic, diuretic | Rheumatism (herb), hypertrophy of the prostate (roots) | Tea made from dried leaves and herb |

**TABLE 3**
**List of medicinal plant species that are likely to occur on an agroecological farm in the Central European floristic province (without plant species from the plant profiles).**
Source: EMA; European Pharmacopoeia (Ph Eur 10.0 2020); Ożarowski and Jaroniewski 1989; van Wyk and Wink 2008. Numbers show where on the farm (in which vegetation site) the plant species are likely to occur, based on the identified plant groups as presented in Fig. 2.

| Medicinal plant species | Plant family | Used plant part | Medicinal properties | On the farm | | |
|---|---|---|---|---|---|---|
| *Aesculus hippocastanum* L. | Hippocastanaceae | Seeds | Antiswelling, stimulate vein tone | 2 | | |
| *Agrimonia eupatoria* L. | Rosaceae | Herb | Antidiarrhoeal, hepatic | 5 | 9 | |
| *Alchemilla vulgaris* L. | Rosaceae | Herb | Astringent, gynaecological | 5 | | |
| *Allium sativum* L. | Alliaceae | Bulb | Antiseptic, anticholesterolemic | 1 | | |
| *Althaea officinalis* L. | Malvaceae | Roots | Antitussive, emollient | 5 | 9 | |
| *Anethum graveolens* L. | Apiaceae | Fruits | Carminative, digestive | 1 | | |
| *Angelica archangelica* L. | Apiaceae | Roots | Antispasmodic, carminative | 5 | 6 | |
| *Arctium lappa* L. | Asteraceae | Roots | Antibacterial, antifungal (skin) | 6 | 9 | |
| *Arctostaphylos uva-ursi* (L.) Spreng. | Ericaceae | Leaves | Diuretic, antiseptic (urinary system) | 4 | | |
| *Arnica montana* L. | Asteraceae | Flowers | Anti-inflammatory, vulnerary | 5 | | |
| *Artemisia absinthium* L. | Asteraceae | Herb | Anthelmintic, appetizer | 9 | | |
| *Artemisia vulgaris* L. | Asteraceae | Herb | Digestive, antiemetic | 5 | 9 | |
| *Asparagus officinalis* L. | Asparagaceae | Rhizome | Diuretic | 1 | 8 | |
| *Ballota nigra* L. | Lamiaceae | Herb | Antispasmodic, sedative | 9 | | |
| *Bellis perennis* L. | Asteraceae | Flowers | Antitussive, digestive | 8 | 9 | |
| *Betonica officinalis* L. | Lamiaceae | Herb | Anthelmintic, carminative | 4 | 5 | 9 |
| *Bidens tripartita* L. | Asteraceae | Herb | Antiseptic, depurative (skin) | 5 | 6 | 9 |
| *Brassica nigra* (L.) W.D.J. Koch | Brassicaceae | Seeds | Antibacterial, digestive | 1 | 6 | 9 |
| *Calluna vulgaris* (L.) Hull | Ericaceae | Flowers | Diuretic, antirheumatic | 4 | 5 | 8 |
| *Carum carvi* L. | Apiaceae | Essential oil, fruits | Carminative, digestive | 5 | | |
| *Castanea sativa* Mill. | Fagaceae | Leaves | Astringent, antitussive | 1 | 2 | |
| *Centaurium erythraea* Rafn. | Gentianaceae | Herb | Appetizer, bitter | 5 | | |
| *Cetraria islandica* | Parmeliaceae | Whole lichen | Antitussive, antiemetic | 4 | 5 | 9 |
| *Chamaemelum nobile* (L.) All. | Asteraceae | Flowers | Anti-inflammatory | 2 | 9 | |
| *Chelidonium majus* L. | Papaveraceae | Herb | Externally for warts | 9 | | |
| *Cirsium oleraceum* (L.) Scop. | Asteraceae | Herb, roots | Anti-inflammatory, diuretic | 5 | 6 | |
| *Cnicus benedictus* L. | Asteraceae | Herb | Appetizer, bitter | 2 | | |
| *Coriandrum sativum* L. | Apiaceae | Fruits | Carminative, digestive | 1 | | |
| *Daucus carota* L. | Apiaceae | Herb, roots, fruits | Diuretic, carminative | 1 | | |
| *Elymus repens* (L.) Gould | Poaceae | Rhizome | Diuretic | 9 | | |
| *Foeniculum vulgare* Mill. | Apiaceae | Fruits | Carminative, antispasmodic | 1 | 2 | |
| *Fragaria vesca* L. | Rosaceae | Leaves | Antidiarrhoeal, astringent | 1 | 4 | |
| *Fraxinus excelsior* L. | Oleaceae | Leaves | Anti-inflammatory, antirheumatic | 4 | | |

**TABLE 3 (Continued)**
**List of medicinal plant species that are likely to occur on an agroecological farm in the Central European floristic province (without plant species from the plant profiles).**
Source: EMA; European Pharmacopoeia (Ph Eur 10.0 2020); Ożarowski and Jaroniewski 1989; van Wyk and Wink 2008. Numbers show where on the farm (in which vegetation site) the plant species are likely to occur, based on the identified plant groups as presented in Fig. 2.

| Medicinal plant species | Plant family | Used plant part | Medicinal properties | On the farm | | |
|---|---|---|---|---|---|---|
| *Fumaria officinalis* L. | Fumariaceae | Herb | Cholagogue, antispasmodic | 9 | | |
| *Galeopsis segetum* Neck. | Lamiaceae | Herb | Anti-inflammatory (lungs) | 9 | | |
| *Galeopsis tetrahit* L. | Lamiaceae | Herb | Expectorant, for pneumonia | 4 | 9 | |
| *Galium odoratum* (L.) Scop. | Rubiaceae | Herb | Antispasmodic, sedative | 4 | | |
| *Galium verum* L. | Rubiaceae | Herb | Diuretic | 5 | 8 | 9 |
| *Gentiana lutea* L. | Gentianaceae | Roots | Cholagogue, bitter | 5 | | |
| *Geranium robertianum* L. | Geraniaceae | Herb | Antidiarrhoeal | 4 | 9 | |
| *Geum urbanum* L. | Rosaceae | Roots | Antidiarrhoeal, astringent | 4 | 9 | |
| *Glechoma hederacea* L. | Lamiaceae | Herb | Digestive, diuretic | 4 | 9 | |
| *Glycine max* (L.) Merr. | Fabaceae | Seeds | Anticholesterolemic | 1 | | |
| *Hedera helix* L. | Araliaceae | Leaves | Antispasmodic, expectorant | 4 | | |
| *Helichrysum arenarium* (L.) Moench | Asteraceae | Flowers | Cholagogue | 5 | 9 | |
| *Herniaria glabra* L. | Caryophyllaceae | Herb | Diuretic | 9 | | |
| *Hieracium pilosella* L. | Asteraceae | Herb with roots | Diuretic | 5 | | |
| *Juglans regia* L. | Juglandaceae | Leaves | Astringent, anthelmintic | 1 | | |
| *Juniperus communis* L. | Cupressaceae | Pseudo-fruits | Diuretic, antiseptic | 4 | 5 | |
| *Lamium album* L. | Lamiaceae | Flowers | Gynaecological, vulnerary | 9 | | |
| *Levisticum officinale* Koch | Apiaceae | Roots | Carminative, digestive | 1 | | |
| *Linaria vulgaris* Mill. | Plantaginaceae | Herb | Sedative, laxative | 5 | 9 | |
| *Lycopus europaeus* L. | Lamiaceae | Herb | Regulating the thyroid | 6 | | |
| *Marrubium vulgare* L. | Lamiaceae | Herb | Digestive, cholagogue | 9 | | |
| *Menyanthes trifoliata* L. | Menyanthaceae | Leaves | Carminative, amara | 5 | 6 | |
| *Nasturtium officinale* R. Br. | Brassicaceae | Herb | Antiscorbutic, depurative (urinary system) | 6 | | |
| *Nepeta cataria* L. | Lamiaceae | Herb | Sedative, digestive | 2 | 8 | 9 |
| *Nigella sativa* L. | Ranunculaceae | Seeds | Immunostimulant, antispasmodic | 1 | 8 | |
| *Oenothera biennis* L./ *O. glazioviana* Micheli. | Onagraceae | Seed oil | Anticholesterolemic, anti-inflammatory (eczema) | 2 | 8 | 9 |
| *Ononis spinosa* L. | Fabaceae | Roots | Diuretic | 5 | | |
| *Origanum majorana* L. | Lamiaceae | Herb | Expectorant, antibacterial | 1 | | |
| *Origanum vulgare* L. | Lamiaceae | Herb | Expectorant | 4 | 5 | |
| *Papaver rhoeas* L. | Papaveraceae | Flowers | Sedative, expectorant | 9 | | |
| *Papaver somniferum* L. | Papaveraceae | Seeds | Anticholesterolemic, anti-inflammatory | 1 | | |
| *Parietaria officinalis* L. | Urticaceae | Herb | Diuretic, anti-inflammatory | 9 | | |
| *Petroselinum crispum* (Mill.) A. W. Hill | Apiaceae | Herb with roots | Diuretic | 1 | | |
| *Picea abies* ( L.) H. Karst. | Pinaceae | Essential oil | Antibacterial, antirheumatic (externally) | 4 | | |

(*continued*)

**TABLE 3 (Continued)**
**List of medicinal plant species that are likely to occur on an agroecological farm in the Central European floristic province (without plant species from the plant profiles).**
Source: EMA; European Pharmacopoeia (Ph Eur 10.0 2020); Ożarowski and Jaroniewski 1989; van Wyk and Wink 2008. Numbers show where on the farm (in which vegetation site) the plant species are likely to occur, based on the identified plant groups as presented in Fig. 2.

| Medicinal plant species | Plant family | Used plant part | Medicinal properties | On the farm | | |
|---|---|---|---|---|---|---|
| *Pimpinella saxifraga* L. | Apiaceae | Roots | Expectorant, antiseptic | 5 | 9 | |
| *Polygonum aviculare* L. | Polygonaceae | Herb | Diuretic, purgative | 9 | | |
| *Polygonum bistorta* L. | Polygonaceae | Herb | Astringent | 5 | 6 | |
| *Polygonum hydropiper* L. | Polygonaceae | Herb | Antihaemorrhagic | 5 | 6 | |
| *Polypodium vulgare* L. | Polypodiaceae | Rhizome | Expectorant | 4 | 6 | |
| *Populus nigra* L. | Salicaceae | Buds | Antiseptic, antirheumatic | 4 | 7 | |
| *Potentilla anserina* L. | Rosaceae | Herb | Astringent | 5 | 6 | 9 |
| *Prunella vulgaris* L. | Lamiaceae | Flower spikes | Astringent, vulnerary | 4 | 5 | 9 |
| *Pulmonaria officinalis* L. | Boraginaceae | Leaves | Expectorant | 4 | | |
| *Quercus robur* L./*Q. petraea* (Matt.) Liebl./*Q. pubescens* Willd. | Fagaceae | Bark | Antidiarrhoeal, astringent | 4 | 7 | |
| *Raphanus sativus* L. | Brassicaceae | Roots | Antibacterial, antiscorbutic | 1 | | |
| *Rhamnus cathartica* L. | Rhamnaceae | Fruits | Laxative | 4 | 7 | |
| *Rheum palmatum* L. | Polygonaceae | Roots | Astringent (small doses), laxative (high doses) | 2 | | |
| *Ribes nigrum* L. | Grossulariaceae | Leaves | Diuretic, astringent | 1 | 4 | |
| *Rosmarinus officinalis* L. | Lamiaceae | Essential oil, leaves | Digestive, antibacterial | 1 | 8 | |
| *Rubus idaeus* L. | Rosaceae | Leaves | Antipyretic, diaphoretic | 1 | 4 | |
| *Rumex crispus* L. | Polygonaceae | Roots | Laxative, astringent | 5 | 9 | |
| *Ruscus aculeatus* L. | Ruscaceae | Rhizome | Stimulate vein tone, anti-inflammatory | 2 | | |
| *Salvia sclarea* L. | Lamiaceae | Essential oil | Digestive, antispasmodic | 2 | 8 | 9 |
| *Sanguisorba officinalis* L. | Rosaceae | Herb | Antihaemorrhagic, astringent | 5 | 6 | |
| *Sanicula europaea* L. | Apiaceae | Herb | Astringent, expectorant | 4 | | |
| *Solanum dulcamara* L. | Solanaceae | Stem | Anti-inflammatory (eczema, externally) | 4 | 6 | |
| *Solidago virgaurea* L. | Asteraceae | Herb | Diuretic | 4 | 5 | 9 |
| *Sorbus aucuparia* L. | Rosaceae | Fruits | Laxative, diuretic | 4 | | |
| *Stellaria media* (L.) Vill. | Caryophyllaceae | Herb | Antipruritic (externally) | 9 | | |
| *Thymus vulgaris* L. | Lamiaceae | Herb | Expectorant, antibacterial | 1 | | |
| *Tilia cordata* Mill./ *T. platyphyllos* Scop. | Tiliaceae | Flowers | Diaphoretic, sedative | 4 | | |
| *Trigonella foenum-graecum* L. | Fabaceae | Seeds | Anti-inflammatory (ulcers), appetizer | 1 | 2 | 9 |
| *Tropaeolum majus* L. | Tropaeolaceae | Herb | Antibiotic, diuretic | 8 | | |
| *Tussilago farfara* L. | Asteraceae | Leaves | Antitussive | 6 | 9 | |
| *Vaccinium myrtillus* L. | Ericaceae | Fruits | Antidiarrhoeal, astringent | 4 | | |
| *Vaccinium oxycoccos* L. | Ericaceae | Fruits | Antibacterial, anti-inflammatory (urinary system) | 4 | 5 | |

**TABLE 3 (Continued)**
**List of medicinal plant species that are likely to occur on an agroecological farm in the Central European floristic province (without plant species from the plant profiles).** Source: EMA; European Pharmacopoeia (Ph Eur 10.0 2020); Ożarowski and Jaroniewski 1989; van Wyk and Wink 2008. Numbers show where on the farm (in which vegetation site) the plant species are likely to occur, based on the identified plant groups as presented in Fig. 2.

| Medicinal plant species | Plant family | Used plant part | Medicinal properties | On the farm | |
| --- | --- | --- | --- | --- | --- |
| *Vaccinium vitis-idaea* L. | Ericaceae | Leaves, fruits | Diuretic, antirheumatic | 4 | |
| *Verbena officinalis* L. | Verbenaceae | Herb | Antispasmodic, antitussive | 1 | 9 |
| *Veronica officinalis* L. | Plantaginaceae | Herb | Expectorant | 4 | 5 |
| *Viola tricolor* L. | Violaceae | Herb | Diuretic, purgative | 5 | 9 |
| *Viscum album* L. | Viscaceae | Herb | Hypotensive, used in cancer treatment | 4 | |
| *Zea mays* L. | Poaceae | Corn style | Hypoglycaemic, diuretic | 1 | |

management and use of MPs. Examples from literature prove that MPs are an inherent element of agroecological farming systems. Diversity of MPs on the farm and their location in the structure of the farm depend on many factors rooted in the local environmental and socio-cultural context.

Overall, the results of the exploration of agroecological systems in traditional societies corroborate the outcome of the theoretical part. First, the deliberate maintenance, cultivation and use of medicinal plants has potential to improve health care in communities for ailments that do not strictly require professional medical attention. Furthermore, diversity of practices within the agroecological paradigm can be a central driving factor towards increased empowerment in community health.

We stress the importance of knowledge exchange (Networks) as one of the core elements conducive to inclusion of MPs by design in agroecological systems. We intend that all provided information about an exemplary MP-inclusive agroecological farm within the Central European floristic province (in detail, the MPs we identified on such a farm and their medicinal and ecological profiles) can be useful as a manual for practitioners. We believe that the examples of traditional agroecological farming, combined with the most relevant vegetation sites and MPs we identified, can facilitate design and management of new farms in our focus region. Importantly, design and management need to be adapted to the geographical and ecological conditions of the farm whilst prioritising native and perennial plants and avoiding invasive species.

Overall, we strongly emphasise exploring, continuing or restoring local medicinal traditions of harvesting, maintaining, planting and using certain species that benefit local culture and can strengthen the identity of the community. Our aim here is to encourage the reader to take conscious steps towards health sovereignty in their community.

## 6. REFERENCES

Agelet, Antoni, Maria Àngels Bonet, and Joan Vallés. 2000. "Homegardens and Their Role as a Main Source of Medicinal Plants in Mountain Regions of Catalonia (Iberian Peninsula)." *Economic Botany* 54 (3): 295–309. https://doi.org/10.1007/bf02864783.

Albrecht, Matthias, David Kleijn, Neal M. Williams, Matthias Tschumi, Brett R. Blaauw, Riccardo Bommarco, Alistair J. Campbell, Matteo Dainese, Francis A. Drummond, and Martin H. Entling. "The Effectiveness of Flower Strips and Hedgerows on Pest Control, Pollination Services and Crop Yield: A Quantitative Synthesis." *Ecology Letters* 23, no. 10 (2020): 1488–98. https://doi.org/10.1111/ele.13576

Altieri, Miguel A. 2004. "Linking Ecologists and Traditional Farmers in the Search for Sustainable Agriculture." *Frontiers in Ecology and the Environment* 2 (1): 35–42. https://doi.org/10.1890/1540-9295(2004)002[0035:leatfi]2.0.co;2.

Altieri, Miguel A., M. Kat Anderson, and Laura C. Merrick 1987. "Peasant Agriculture and the Conservation of Crop and Wild Plant Resources." *Conservation Biology* 1 (1): 49–58. https://doi.org/10.1111/j.1523-1739.1987.tb00008.x.

Anderson, Colin Ray, Janneke Bruil, M. Jahi Chappell, Csilla Kiss, and Michel Patrick Pimbert. 2021. *Agroecology Now!: Transformations towards More Just and Sustainable Food Systems. Library.oapen. org*. Springer Nature. https://library.oapen.org/handle/20.500.12657/43292

Anderson, Colin Ray, Janneke Bruil, Michael Jahi Chappell, Csilla Kiss, and Michel Patrick Pimbert. 2019. "From Transition to Domains of Transformation: Getting to Sustainable and Just Food Systems through Agroecology." *Sustainability* 11 (19): 5272. https://doi.org/10.3390/su11195272.

Anderson, E. N. 2011. "Ethnobiology and Agroecology." In *Ethnobiology*. Vol. 54. Edited by E. N. Anderson et al., 305–18. Hoboken, NJ, USA: John Wiley & Sons, Inc.

Applequist, W. L., J. A. Brinckmann, A. B. Cunningham, R. E. Hart, M. Heinrich, D. R. Katerere, and T. van Andel. 2020. "Scientists' Warning on Climate Change and Medicinal Plants." *Planta Medica* 86 (1): 10–18. https://researchrepository.murdoch.edu.au/id/eprint/54348/.

Astier, Marta, Jorge Quetzal Argueta, Quetzalcóatl Orozco-Ramírez, María V. González, Jaime Morales, Peter R. W. Gerritsen, Miguel A. Escalona, et al. 2017. "Back to the Roots: Understanding Current Agroecological Movement, Science, and Practice in Mexico." *Agroecology and Sustainable Food Systems* 41 (3-4): 329–48. https://doi.org/10.1080/21683565.2017.1287809.

Azpiazu, Celeste, Pilar Medina, Ángeles Adán, Ismael Sánchez-Ramos, Pedro del Estal, Alberto Fereres, and Elisa Viñuela. 2020. "The Role of Annual Flowering Plant Strips on a Melon Crop in Central Spain. Influence on Pollinators and Crop." *Insects* 11 (1): 66. https://doi.org/10.3390/insects11010066.

Bane, Peter. 2012. *The Permaculture Handbook: Garden Farming for Town and Country. Google Books*. New Society Publishers. https://books.google.pl/books?hl=en&lr=&id=4p3zAgAAQBAJ&oi=fnd&pg=PR11&dq=Bane.

Baumflek, Michelle, Stephen DeGloria, and Karim-Aly Kassam. 2015. "Habitat Modeling for Health Sovereignty: Increasing Indigenous Access to Medicinal Plants in Northern Maine, USA." *Applied Geography* 56 (January): 83–94. https://doi.org/10.1016/j.apgeog.2014.10.012.

Bezner Kerr, Rachel, Catherine Hickey, Esther Lupafya, and Laifolo Dakishoni. 2019. "Repairing Rifts or Reproducing Inequalities? Agroecology, Food Sovereignty, and Gender Justice in Malawi." *The Journal of Peasant Studies* 46 (7): 1499–1518. https://doi.org/10.1080/03066150.2018.1547897.

Burlingame, Barbara, and Sandro Dernini. 2012. *Sustainable Diets and Biodiversity: Directions and Solutions for Policy, Research and Action; Proceedings of the International Scientific Symposium Biodiversity and Sustainable Diets: United Against Hunger, 3-5 November 2010, FAO Headquarters, Rome*. Rome: FAO.

Caballero-Serrano, Veronica, Brian McLaren, Juan Carlos Carrasco, Josu G. Alday, Luis Fiallos, Javier Amigo, and Miren Onaindia. 2019. "Traditional Ecological Knowledge and Medicinal Plant Diversity in Ecuadorian Amazon Home Gardens." *Global Ecology and Conservation* 17 (January): e00524. https://doi.org/10.1016/j.gecco.2019.e00524

Ceccolini, Lorenzo. 2002. "The Homegardens of Soqotra Island, Yemen: An Example of Agroforestry Approach to Multiple Land-Use in an Isolated Location." *Agroforestry Systems* 56 (2): 107–15. https://doi.org/10.1023/a:1021365308193.

Chappell, M. J., and Frances M. Lappé. 2018. *Beginning to End Hunger: Food and the Environment in Belo Horizonte, Brazil, and Beyond*. Baltimore, Md., Oakland, California: Project MUSE, University of California Press.

de Alwis, Lyn. 1997. "A Biocultural Medicinal Plants Conservation Project in Sri Lanka." In *Medicinal Plants for Forest Conservation and Health Care*. Vol. 11. Edited by G. Bodeker et al. 100–108, Rome: FAO.

Díaz-Reviriego, Isabel, Álvaro Fernández-Llamazares, Matthieu Salpeteur, Patricia L. Howard, and Victoria Reyes-García. 2016. "Gendered Medicinal Plant Knowledge Contributions to Adaptive Capacity and Health Sovereignty in Amazonia." *Ambio* 45 (S3): 263–75. https://doi.org/10.1007/s13280-016-0826-1.

Ellenberg, Heinz, Heinrich E. Weber, Rubrecht Düll, Volkmar Wirth, Willy Werner, and Dirk Paulißen. 1992. *Zeigerwerte von Pflanzen in Mitteleuropa. Scripta Geobotanica 18*. Göttingen, Goltze.

Fabricant, D S, and N R Farnsworth. 2001. "The Value of Plants Used in Traditional Medicine for Drug Discovery." *Environmental Health Perspectives* 109 (suppl 1): 69–75. https://doi.org/10.1289/ehp.01109s169.

Fernandes, E. C. M., A. Oktingati, and J. Maghembe. 1985. "The Chagga Home Gardens: A Multi-Storeyed Agro-Forestry Cropping System on Mt. Kilimanjaro, Northern Tanzania." *Food and Nutrition Bulletin* 7 (3): 1–8. https://doi.org/10.1177/156482658500700311.

Finerman, Ruthbeth, and Ross Sackett. 2003. "Using Home Gardens to Decipher Health and Healing in the Andes." *Medical Anthropology Quarterly* 17, no. 4: 459–82. doi:10.1525/maq.2003.17.4.459.

Geels, Frank W. 2011. "The Multi-Level Perspective on Sustainability Transitions: Responses to Seven Criticisms." *Environmental Innovation and Societal Transitions* 1 (1): 24–40. https://doi.org/10.1016/j.eist.2011.02.002.

Geels, Frank W., and Johan Schot. 2007. "Typology of Sociotechnical Transition Pathways." *Research Policy* 36 (3): 399–417. https://doi.org/10.1016/j.respol.2007.01.003

Gliessman, S. R. 2015. *Agroecology: The Ecology of Sustainable Food Systems.* 3rd ed. Boca Raton London New York: CRC Press Taylor & Francis Group.

Gliessman, Steve. 2013. "Agroecology: Growing the Roots of Resistance." *Agroecology and Sustainable Food Systems* 37, no. 1: 19–31.

Grabherr, Georg. 2009. "Biodiversity in the High Ranges of the Alps: Ethnobotanical and Climate Change Perspectives." *Global Environmental Change* 19 (2): 167–72. https://doi.org/10.1016/j.gloenvcha.2009.01.007.

Hemenway, Toby. 2009. *Gaia's Garden: A Guide to Home-Scale Permaculture, 2nd Edition. Google Books.* Chelsea Green Publishing. https://books.google.pl/books?hl=en&lr=&id=gxW0MGXha6cC&oi=fnd&pg=PR8&dq=Hemenway.

HLPE. 2019. "Agroecological and Other Innovative Approaches for Sustainable Agriculture and Food Systems That Enhance Food Security and Nutrition: Summary and Recommendations from the HLPE Report." Rome.

Holzer, Sepp. 2011. *Sepp Holzer's Permaculture: A Practical Guide to Small-Scale, Integrative Farming and Gardening. Google Books.* Chelsea Green Publishing. https://books.google.pl/books?hl=en&lr=&id=jNXK9kE-lOoC&oi=fnd&pg=PR9&dq=HLPE.+%E2%80%9CHolzer.

Hoogerbrugge, Inge, and Louise O. Fresco. 1993. *Homegarden Systems: Agricultural Characteristics and Challenges.* Gatekeeper Series sustainable agriculture and rural livelihoods programme of the IIED 39. London: International Institute for Environment and Development (IIED), Sustainable Agriculture Programme.

Ingold, Tim. 2000. *The Perception of the Environment: Essays on Livelihood, Dwelling and Skill. Google Books.* Psychology Press. https://books.google.pl/books?hl=en&lr=&id=5LpTBInNGkEC&oi=fnd&pg=PP1&dq=Ingold.

IPCC. 2019. *Climate Change and Land: An IPCC Special Report on Climate Change, Desertification, Land Degradation, Sustainable Land Management.* https://www.ipcc.ch/srccl/.

IPES-Food. 2016. "From Uniformity to Diversity: A Paradigm Shift from Industrial Agriculture to Diversified Agroecological Systems. International Panel of Experts on Sustainable Food Systems."

Kassam, Karim-Aly, Munira Karamkhudoeva, Morgan Ruelle, and Michelle Baumflek. 2010. "Medicinal Plant Use and Health Sovereignty: Findings from the Tajik and Afghan Pamirs." *Human Ecology* 38 (6): 817–29. https://doi.org/10.1007/s10745-010-9356-9.

Kickbusch, Ilona. 2000. "The Development of International Health Policies — Accountability Intact?" *Social Science & Medicine* 51 (6): 979–89. https://doi.org/10.1016/s0277-9536(00)00076-9.

Kletter, Christa, and Monika Kriechbaum. 2001. *Tibetan medicinal plants.* CRC Press.

Kumar, B.M., and P.K.R. Nair. 2004. "The Enigma of Tropical Homegardens." *Agroforestry Systems* 61-62 (1-3): 135–52. https://doi.org/10.1023/b:agfo.0000028995.13227.ca.

La Torre, Lucía de, Hugo Navarrete, Priscilla Muriel, Manuel J. Macía, and Henrik Balslev. 2008. *Enciclopedia De Las Plantas Útiles Del Ecuador (Con Extracto De Datos).* Herbario QCA de la Escuela de Ciencias Biológicas de la Pontificia Católica del Ecuador.

Landolt, E., B. Bäumler, A. Ehrhardt, O. Hegg, F. Klötzli, W. Lämmler, M. Nobis, et al. 2010. *Flora Indicativa: Okologische Zeigerwerte und biologische Kennzeichen zur Flora der Schweiz und der Alpen. Www.zora.uzh.ch.* Bern: Haupt. https://www.zora.uzh.ch/id/eprint/34105/.

Lemonnier, Nathanaël, Guang-Biao Zhou, Bhavana Prasher, Mitali Mukerji, Zhu Chen, Samir K. Brahmachari, Denis Noble, Charles Auffray, and Michael Sagner. 2017. "Traditional Knowledge-Based Medicine." *Progress in Preventive Medicine* 2 (7): e0011. https://doi.org/10.1097/pp9.0000000000000011.

Lichtfouse, Eric. 2011. *Agroecology and Strategies for Climate Change. Google Books.* Springer Science & Business Media. https://books.google.pl/books?hl=en&lr=&id=040krQwBTuAC&oi=fnd&pg=PR3&dq=Lichtfouse.

Łuczaj, Łukasz. 2013. *Dzika kuchnia.* Nasza Księgarnia.

Maclean, I. M. D., and R. J. Wilson. 2011. "Recent Ecological Responses to Climate Change Support Predictions of High Extinction Risk." *Proceedings of the National Academy of Sciences* 108 (30): 12337–42. https://doi.org/10.1073/pnas.1017352108.

Méndez, V., Christopher Bacon, and Roseann Cohen. 2015. "Introduction: Agroecology as a Transdisciplinary, Participatory, and Action-Oriented Approach." In *Agroecology.* Vol. 2. Edited by V. Méndez et al., 1–22. Advances in Agroecology. CRC Press.

Oberdorfer, Erich, and Angelika Schwabe. 2001. "Pflanzensoziologische Exkursionsflora Für Deutschland Und Angrenzende Gebiete. 8., Stark Überarb. Und Erg." Aufl. Stuttgart (Hohenheim): Ulmer.

O'kting'ati, A., J. A. Maghembe, E. C. M. Fernandes, and G. H. Weaver. 1984. "Plant Species in the Kilimanjaro Agroforestry System." *Agroforestry Systems* 2 (3): 177–86. https://doi.org/10.1007/bf00147032.

Ożarowski, A., and W. Jaroniewski. 1989. *Rośliny Lecznicze i Ich Praktyczne Zastosowanie.* Warszawa: Instytut Wydawniczy Związków Zawodowych.

Padoch, Christine, and Wil de Jong. 1991. "The House Gardens of Santa Rosa: Diversity and Variability in an Amazonian Agricultural System." *Economic Botany* 45 (2): 166–75. https://doi.org/10.1007/bf02862045.

Parmentier, Stéphane. 2014. *Scaling-up Agroecological Approaches: What, Why and How?* Belgium: Oxfam-Solidarity.

Pfadenhauer, Jörg S., and Frank A. Klötzli. 2020. *Global Vegetation: Fundamentals, Ecology and Distribution.* Springer Nature.

Ph Eur 10.0. 2020. *Europäisches Arzneibuch. Amtliche Deutsche Ausgabe.* 10. Ausgabe. Stuttgart: Deutscher Apotheker Verlag.

Pimbert, M. P. 2018. "Global Status of Agroecology: A Perspective on Current Practices, Potential and Challenges." *Economic & Political Weekly* 53, no. 41.

Qicheng, Fang. 1980. "Some Current Study and Research Approaches Relating to the Use of Plants in the Traditional Chinese Medicine." *Journal of Ethnopharmacology* 2 (1): 57–63. https://doi.org/10.1016/0378-8741(80)90031-8

Raskin, Ben and S. Osborn, eds. 2019. *The Agroforestry Handbook: Agroforestry for the UK.* 1st Edition. Bristol, UK: Soil Association Limited.

Rico-Gray, Victor, Jose G. Garcia-Franco, Alexandra Chemas, Armando Puch, and Paulino Sima. 1990. "Species Composition, Similarity, and Structure of Mayan Homegardens in Tixpeual and Tixcacaltuyub, Yucatan, Mexico." *Economic Botany* 44 (4): 470–87. https://doi.org/10.1007/bf02859784.

Roig y Mesa, J. T. 1945. *Plantas Medicinales, Aromáticas O Venenosas De Cuba.* Tomo1+2. La Habana, Cuba: Servicio de Publicidad y Debulgación. Ministerio de Agricultura.

Rothschild, Karen, ed. 2017. *The Nyéléni Peasant Agroecology Manifesto.*

Rundlöf, Maj, Ola Lundin, and Riccardo Bommarco. 2018. "Annual Flower Strips Support Pollinators and Potentially Enhance Red Clover Seed Yield." *Ecology and Evolution* 8 (16): 7974–85. https://doi.org/10.1002/ece3.4330.

Scharmer, Otto. 2018. *The Essentials of Theory U: Core Principles and Applications.* Berrett-Koehler Publishers.

Scheper, Jeroen, Riccardo Bommarco, Andrea Holzschuh, Simon G. Potts, Verena Riedinger, Stuart P. M. Roberts, Maj Rundlöf, et al. 2015. "Local and Landscape-Level Floral Resources Explain Effects of Wildflower Strips on Wild Bees across Four European Countries." Edited by Sarah Diamond. *Journal of Applied Ecology* 52 (5): 1165–75. https://doi.org/10.1111/1365-2664.12479.

Schultes, Richard E. 1979. "The Amazonia as a Source of New Economic Plants." *Economic Botany* 33, no. 3: 259–66.

Solecki, R. S. 1975. "Shanidar IV, a Neanderthal Flower Burial in Northern Iraq." *Science* 190 (4217): 880–81. https://doi.org/10.1126/science.190.4217.880.

Sunwar, Sharmila, Carl-Gustaf Thornström, Anil Subedi, and Marie Bystrom. 2006. "Home Gardens in Western Nepal: Opportunities and Challenges for On-Farm Management of Agrobiodiversity." *Biodiversity and Conservation* 15 (13): 4211–38. https://doi.org/10.1007/s10531-005-3576-0.

Universal Declaration of Human Rights. 1948. UN.

Uprety, Yadav, Hugo Asselin, Archana Dhakal, and Nancy Julien. 2012. "Traditional Use of Medicinal Plants in the Boreal Forest of Canada: Review and Perspectives." *Journal of Ethnobiology and Ethnomedicine* 8: 7. doi:10.1186/1746-4269-8-7.

van Wyk, Ben-Erik, and Michael Wink. 2008. *Rośliny Lecznicze Świata: Ilustrowany Przewodnik Naukowy Po Najważniejszych Roślinach Leczniczych Świata I Ich Wykorzystaniu.* MedPharm Polska.

Wezel, Alexander, Margriet Goris, Janneke Bruil, Georges F. Félix, Alain Peeters, Paolo Bàrberi, Stéphane Bellon, and Paola Migliorini. 2018. "Challenges and Action Points to Amplify Agroecology in Europe." *Sustainability* 10 (5): 1598. https://doi.org/10.3390/su10051598.

Wezel, Alexander, Marion Casagrande, Florian Celette, Jean-François Vian, Aurélie Ferrer, and Joséphine Peigné. 2014. "Agroecological Practices for Sustainable Agriculture. A Review." *Agronomy for Sustainable Development* 34 (1): 1–20. https://doi.org/10.1007/s13593-013-0180-7.

Wezel, Alexander, and Svane Bender. 2003. "Plant Species Diversity of Homegardens of Cuba and Its Significance for Household Food Supply." *Agroforestry Systems* 57, no. 1: 39–49.

WHO. 2013. *WHO Traditional Medicine Strategy: 2014-2023.* World Health Organization.

Wiersum, K. F. 2006. "Diversity and Change in Homegarden Cultivation in Indonesia." In *Tropical Homegardens.* Vol. 3. Edited by P. K. R. Nair and B. M. Kumar, 13–24. Advances in Agroforestry. Dordrecht: Springer Netherlands. https://doi.org/10.1007/978-1-4020-4948-4_2.

# Section II

## INSIGHTS and OVERVIEWS

# 3 Mucuna pruriens vs. Morbus Parkinson

## Making the case for medicinal supplements within Medicinal Agroecology

Immo Norman Fiebrig, Monique van de Vijver and Corrie J. van Kan

*Science means learning to say "I don't know."* – Ashok D. B. Vaidya (2011)

## 1 INTRODUCTION

Morbus Parkinson, also known as Parkinson's disease (PD), is a progressive ailment of the central nervous system (CNS) in humans (Armstrong and Okun 2020). While there is no cure for this impairing disease to date, the pharmaceutical industry provides an array of synthetic drugs to ameliorate symptoms with efforts being made towards the development of disease-modifying treatments (DMT), thus obtaining a lasting clinical benefit (Morant et al. 2019; Lang and Espay 2018). More recently, surgical procedures to the brain have become part of the treatments offered by modern medicine, such as deep brain stimulation, amongst others (Kogan et al. 2019). However, therapy is expensive and thus is not accessible to many PD patients in the world (Fothergill-Misbah et al. 2020). Even those who can afford medical specialists, plus all the prescribed medication to manage disease symptoms, may – nonetheless – still not enjoy an acceptable quality of life because of the narrow scope of conventional antiparkinson medication (Bloem et al. 2021; Behari et al. 2005).

This current chapter reports an initially singular case of a PD patient in Germany who could not be treated with conventional PD medication and whose quality of life had deteriorated dramatically after diagnosis, including long episodes of psychoses. Following therapy with an Ayurvedic (botanical) preparation from the beans of *Mucuna pruriens* (MP) the patient´s fate was improved substantially at the time (Anonymised 1997, 2010). This compelling and – rather by fortune – well-documented case was pivotal to spreading the experience with MP within the European network of PD patients, mostly by word of mouth. The PD patient community and neurologists were supported by available online information around the medicinal plant, MP, and the bean´s active components, as well as an initial clinical study (HP-200 Study Group 1995) followed by a second one (Katzenschlager et al.) in 2004. Based on this first evidence, some neurologists were willing to support their patients with MP-based complementary therapy. As far as the authors are aware, from approximately 1997 until to date such cases were treated predominantly in Germany, Switzerland, Austria (GSA[1] countries), and The Netherlands. Clinical experience is mainly based on the feedback

---

[1] GSA: acronym referring to countries whose main spoken language is German – in German they are also referred to by the acronym 'D-A-CH' or 'DACH'.

DOI: 10.1201/9781003146902-5

**45**

from patients and volunteers pushing to understand the added value of MP and prompting the availability of a safe and effective natural medicinal product that can be prescribed to PD patients. Their findings constitute the basis of this chapter.

## 2   PARKINSON`S DISEASE

Worldwide, PD is the second most common neurodegenerative disorder after Alzheimer´s disease, affecting over six million people. Prevalence is expected to double to over 12 million by 2040 and has been described as an emerging pandemic (Bloem et al. 2021; Simon et al. 2020, Neurology Collaborators 2019, Dorsey et al. 2018, GBD 2016).

PD belongs to a group of neurological disorders called parkinsonism. Therein, PD is the most common form of so-called Parkinson's syndrome, characterised by shaking (rest tremor), slowness of movement (bradykinesia), stiffness (rigidity), and postural imbalance. These are referred to as the four main motor symptoms (i.e. relevant to skeletal muscles). The main *non*-motor symptoms include constipation due to impaired bowel movements, insomnia, low mood related to depression, or cognitive impairments partly related to dementia (Armstrong and Okun 2020; Nouws 2015).

### 2.1   CAUSES AND TYPOLOGY OF PD

The causes of PD are likely multi-factorial and linked to both non-genetic (environmental) and genetic factors (Chade et al. 2006; Lau and Breteler 2006). An increased incidence of PD is partly thought to be related to the ageing of the population, although young-onset and juvenile-onset PD are expected to be likewise on the rise (Dorsey et al. 2018, Reeve et al. 2014). More recently, it has been suspected that in the majority of PD cases the disease is caused actually much earlier – at least 20 years before motoric symptoms show – after toxic substances entered the body via the gut (food?) or the nose (air?). The characteristic deterioration of specific cells in the brain are affected only in a relatively late stage of the disease (Bloem et al. 2021: 64).

There are disease variants such as the 'diffuse malignant type' with fast progression and little response to medication. Approximately 50% of PD patients, however, belong to the 'motor-predominant type', characterised by milder symptoms and better response to medication such as levodopa (LD) or dopamine (DP) agonists (see Table 1). Typically, disease progression in this group is slower. Intermediate PD types exist as well (Armstrong and Okun 2020).

The pathophysiology of PD is complex. Simplistically put, it is related, amongst others, to the degeneration or death of the brain´s basal ganglia, a group of clustered neurons of the subcortex. More specifically, they belong to dopamine-secreting neurons of the mid-brain region called *substantia nigra* and within this region they are located in the portion called *pars compacta*. The progressive 'death' or malfunction of these cells, also in other regions of the brain, is a result of mechanisms not detailed here. They lead to a deficit of dopamine (DP), a crucial transmitter in the central nervous system CNS (Davie 2008).

### 2.2   PD: DRUG TREATMENTS

Existing drugs treating PD are largely symptomatic and address a variety of disease signs and symptoms. Rascol et al. (2002) provide a practical overview of the spectrum of medication. It comprises treatments to increase so-called dopaminergic stimulation, directly or indirectly (see Table 1). A second group of drugs not listed here reduces mainly cholinergic/glutamatergic stimulation in relation to other neurotransmitter systems (Cersosimo and Micheli 2007).

Our focus on dopaminergic stimulation and related drug therapy here does not mean that other PD medication is less important. We focus on DP (and its molecular precursor, levodopa or l-dopa),

**TABLE 1**

**Treatments that increase stimulation via dopamine or in a dopamine-agonistic manner, modified after Rascol et al. (2002) and updated (Luo et al. 2020). Pro drug**: a medicinal ingredient that is converted to a pharmacologically active drug once entering the body; **MAO-B inhibitor**: monoamine oxidase inhibitor type B, a drug that inhibits the enzymatic breaking down of dopamine; inhibition of the enzyme in the CNS keeps dopamine levels high. **COMT inhibitor**: inhibits the catechol-O-methyltransferase, an enzyme which usually methylates levodopa during catabolism peripherally (i.e. not in the CNS); also here, inhibition of the enzyme keeps levodopa levels and thus dopamine levels high. **DDCI**: dopamine decarboxilase inhibitor. *Combining LD, carbidopa with entacapone as a peripherally active inhibitor in one tablet (e.g. Stalevo®) aims at improving stable plasma levels of LD.

| **Pro-drug to dopamine** substitutes the lack of endogenous dopamine in the brain | **Dopamine-agonists** act on specific receptors in a similar way to DP | **MAO-B inhibitors** prolong the availability of dopamine by inhibiting its catabolism centrally | **COMT inhibitors** prolong L-dopa bioavailability peripherally |
|---|---|---|---|
| L-dopa (+ DDCI: benserazide or carbidopa) | Apomorphine Bromocriptine Cabergoline Dihydroergocriptine Lisuride Pergolide Piribedil Pramipexole Ropinirole Rotigotine | Selegiline Rasagiline | Entacapone* Opicapone |

because a lack of DP has been the key factor to understanding and treating PD. DP substitution via LD is still considered the first choice ('gold standard') in PD therapy today, almost 50 years after the market introduction of LD-based Madopar® by Hoffmann-La Roche (LD + benserazide) and Sinemet® by Merck and Co. (today MSD; LD + carbidopa), revolutionising PD treatment as from 1973 (Paoletti et al. 2019, Hornykiewicz 2010, Amrein, 2004).

### 2.2.1   Biochemistry and pharmacology of LD and DP

At this point it seems pertinent to explain, although simplistically, the concept of a pro-drug and why combining LD either with benserazide or carbidopa was a major step in the pharmacotherapy of PD. This goes with the second question of why, if DP is lacking in the brain, is it not actually DP that is being given to the patient?

Physiologically, for a substance to get into the brain, it needs to pass what is called the 'blood-brain barrier'. This barrier protects the extra-vascular part of the brain and spinal marrow – where there is no blood – from the blood in their vessels, safeguarding the central nervous system against inappropriate messenger substances or from toxins and pathogens. This barrier is most efficient in protecting against large and hydrophilic molecules, often positively or negatively charged ones. DP, whilst being a tiny molecule, in the bodily fluids it would be positively charged due to its amino group ($-NH_3^+$) and not cross into the brain. Conversely, small, uncharged, lipophilic molecules are much better able to pass this protective barrier, as would be the case for an (amphiphilic) molecule like LD, carrying both charges in one. This means, due to LD´s amino group, plus its carboxyl group ($-NH_3^+$; $-CO_2^-$), positive and negative charges neutralise one another within the molecule, making its net charge zero at physiologic pH and the molecule more lipophilic. Once in the brain, LD gets decarboxylated (i.e. stripped of its carboxyl group ($-COOH$)), and thus converted from pro-drug LD to drug DP as an active neurotransmitter. This reaction, however, is not exclusive to the CNS beyond its blood-brain barrier. It also happens peripherally (i.e. in the rest of the body). This means that much of LD would be converted to DP by decarboxylation before reaching the CNS. The blood-brain barrier will not let DP pass whilst causing adverse reactions in the body´s periphery. To avoid such unwanted decarboxylation from happening, the DDCIs, benserazide or carbidopa, are added to LD, thus improving efficacy in the CNS, resulting in a marked improvement of akinesia and rigidity, but with less effect on tremor (Potschka 2010; Müller 2007).

For a deeper understanding of the biochemistry of dopamine in the CNS it is pertinent to look at its biosynthetic pathways. Fig. 1 shows a simplified version of this pathway described by Meiser et al. (2013), first postulated by Blaschko in 1939. It starts with L-phenylalanine, an essential amino acid. As the name indicates, it can be viewed as L-alanine – a non-essential, proteinogenic amino acid – with a phenyl group on its terminal end. Phenylalanine is a non-polar, neutral (uncharged) and hydrophobic molecule. A hydroxylation step ($-OH$) in the para-position of the phenyl group leads to 4-hydroxyphenylalanine or L-tyrosine, a non-essential amino acid that has been added a polar side group; though considered a hydrophobic amino acid, it is still more hydrophilic than its precursor L-phenylalanine and is found naturally in many foods that are high in protein – in cheese amongst others – and it is believed to promote 'deep thinking' in humans (Colzato et al. 2014). A second hydroxylation step in the meta position leads to LD (Fig. 1). LD is a normal compound in the biology of humans, some animals and plants (Hornykiewicz 2010). Decarboxylation, a common metabolic reaction (removal of $-COOH$), leads to DP as described above. In humans, DP is synthesised in the kidney and brain (Aldred and Nutt 2010) and belongs chemically to two structural families called catechol (Fig. 2A) and phenylethylamine (PEA, Fig. 2B). DP´s basic structure is thus called 'catecholamine' (Fig. 2C). Catechol and its derivatives naturally occur in small amounts in fruits and vegetables and play a central role in the browning of cuts thereof, when exposed to atmospheric oxygen (Mezquita and Queiroz 2013). Mostly synthesised by chemical industry, isolated catechol is toxic and is primarily used for the synthesis of pesticides and fine chemicals like perfumes, aromas (e.g. vanillin) and pharmaceutical compounds (Fiege et al. 2000). PEA in turn is classified as a CNS stimulant and neurotransmitter in humans, produced naturally in many plant and animal species including fungi. Food supplements sold on the market claim an improvement of mood and a benefit in weight loss, although evidence is so far rather limited (Fernstrom and Fernstrom 2007, Ueda et al. 2017).

Further steps in the biosynthetic pathway via hydroxylation of the ethyl side chain lead to noradrenaline, additional N-methylation to adrenaline (Fig. 1). They both have a function as hormones and neurotransmitters. One of the important roles in humans and other animals lies within the so-called fight-or-flight response to perceived threats (Fernstrom and Fernstrom 2007). They also exist as synthetic medications.

We have explained the related biochemistry to elicit why a DP deficit in the brain is treated with the precursor LD, but ultimately also to show that the chemical structure of important and different bioactive molecules is closely related, which may be relevant for a systems approach to primary and secondary plant metabolites and 'omic' analytical technologies outlined in the next section and towards the end of this chapter.

### 2.2.2   Disease biochemistry and DP metabolism

So-called disease modifying treatments (DMT) in PD treatment are opposed to a mere transient improvement of some of the symptoms. DMT are aimed at enduring benefit, slowing down

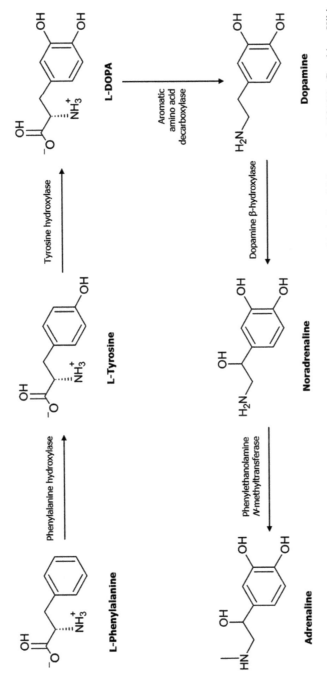

**FIGURE 1**  Biosynthetic pathways of dopamine synthesis and transformation (simplified after Meiser et al. (2013)); Graphics: Kiki Beekman/Advanced Chemistry Development Inc. (ACD/Labs).

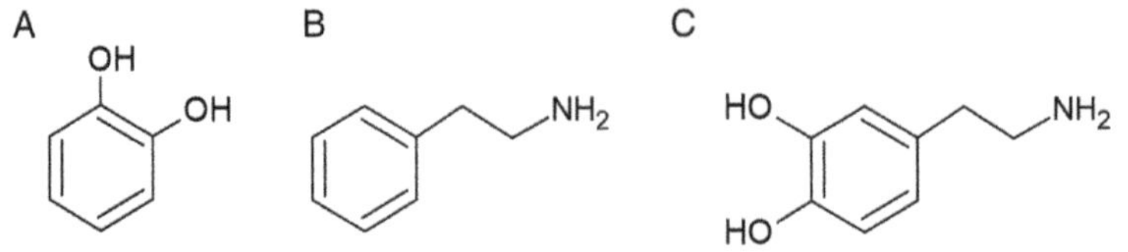

**FIGURE 2** A. Catechol, a benzenediol known as 1,2-dihydroxybenzene or benzene-1,2-diol; B. Phenylethy-lamin (PEA): precursor L-phenylalanine. C. Dopamine, together with norepinephrine (noradrenalin) and epinephrine (adrenalin), belong to the family of catecholamines. Graphics: Kiki Beekman/(ACD/Labs).

progression and ideally at stopping or even reversing the neurodegenerative process, albeit that the latter seems far out of reach. Intensification of research is to be aimed at finding root causes and pre-vention strategies to PD (Bloem et al. 2021; McFarthing et al. 2020; Morant et al. 2019).

The loss of dopaminergic neurons (neurons related to dopamine) is considered to be related largely to oxidative stress in the *substantia nigra* (Meiser et al. 2013). Within the complexity of the catecholamine metabolism, catecholaminergic neurons are believed to represent an important source of so-called 'reactive oxygen species' (ROS: e.g. hydroxyl radical, hydrogen peroxide) pro-moting membrane lipid peroxidation, which in turn is considered critical for the survival of cells (Dexter et al. 1989; Fahn and Cohen, 1992).

The second contributing neurodegenerative factor is increased mitochondrial malfunction in ageing. Mitochondria are considered the 'power plants' in the cells. With adenosine triphosphate (ATP) as the 'energy token' in an organ – the brain – (requesting around 20% of the total body oxygen and glucose whilst constituting only 2% of the total body weight) metabolic imbalances related to the mitochondria might trigger detrimental processes (Park and Larsson 2011; Purdon et al. 2002).

To better understand the complexity of dopamine metabolism, Meiser et al. (2013) propose a systems approach on a metabolic level in the cell. Here, regulation takes place on the level of genome (DNA), transcriptome (RNA/mRNA), proteome (set of proteins) and the metabolome. The latter is defined as a set of small molecules in a biological sample, for example in the cell, that can be sugars, nucleotides, amino acids, lipids, vitamins, or even exogenous substances like drugs or toxins (Wishart 2007; Weckwerth 2003). Metabolome analysis entails making a snapshot of the cell bio-chemistry. This is challenging as such because of high turnover rates. Still, the antioxidant-reducing profile of a biological sample can allegedly be measured and helps to tailor potential antioxidant therapy through drugs or food supplements (Kohen and Nyska 2002).

While in the future, metabolomic approaches might have a lot more to offer in terms of funda-mental understanding of the dynamics of live systems in general and PD pathophysiology and novel treatments in particular, we shall first revert to the classic DP-agonist treatments still used today (Table 1; Kraljevic et al. 2004; Cannon 1985).

### 2.2.3 DP agonists within R&D paradigms of pharmaceutical industry

Since Blaschko (1957) suggested dopamine to be a neurotransmitter, the principle of researching structure-activity relationships between semi-synthetic or synthetic molecules on the one hand and dopamine receptors in the human body on the other, became a major driver in pharmaceutical success stories for symptomatic PD treatments. With the advent of essential analytical methodolo-gies, chemical industry was able to 'design' new molecules that were to be more or less as effective as their natural model – with tolerable side effects and at reasonable production costs. Based on fun-damental molecular structures, such derivatives could exhibit better absorption properties following oral administration and improved distribution in the body, especially regarding the passage through the blood-brain barrier towards their site of action. Much of this type of PD-related biomimetic research was performed in the 1970s (Cannon 1985; Tolosa et al. 1998). The research principle of finding structure-activity relationships in new molecules constituted a key factor for novelty and required patentability, thus representing the very basis of the pharmaceutical industry´s business models to grant returns-on-investment (ROI) at the time and still being of high relevance today (Munos 2009).

## 2.3 ALTERNATIVE PD TREATMENT WITH *MUCUNA PRURIENS* (MP)

Moving away from more or less nature-based synthetic molecules and the pharma industries' IPR[2]-based business model, we now turn to an entirely natural treatment for PD; that is, to preparations made from *Mucuna pruriens* (MP) beans that contain LD naturally from a legume growing in most tropical and some subtropical parts of the world.

While considerable research has shown the effectiveness of MP in different ways, we first focus on an anecdotal report from a pharmacy in southeast Germany whose experience was pivotal in bringing an MP preparation from India into Europe ('Patient No. 1'). This report will be followed by a series of other PD patients' feedbacks in the context of more than 20 years of compelling and mostly unpaid, voluntary engagement to encourage, flank, and support research towards an EU-authorised medicinal product based on MP.

### 2.3.1 *Mucuna pruriens* L. (DC), its botany, diversity, and uses

MP is an annual climbing liana or shrub from the family of *Fabaceae* within the order of *Fabales*. The name of its genus *Mucuna* (Adans.) is derived from the Brazilian Tupi-Guarani[3] vernacular name *mucunã* comprising up to 150 species that can be annual or perennial and grow mostly in tropical and subtropical climates (Hutchinson 1964; Buckles 1995). Latté (2008) searched ancient literature where MP is identified and described under a variety of other genus names, such as *Stilozobium p.*, *Negretia p.*, *Dolichos p.* or *Carbopogon p.* The IITA Genebank (1987) adds *Macaranthus cochinchinensis* to this list of MP synonyms. MP can be found in Asia, Africa, and in the West Indies, Central and South America but is believed to have originated in China, Malaysia, or India (Moura 2018; Quattrocchi 2000, Eilittä et al. 2002). Pantropical (worldwide) distribution was presumably fostered by seeds adapted to oceanic dispersal (Moura et al. 2016). The plant grows well in a variety of soils and thrives vigorously, 'like a weed', mostly below 1600m (Eilittä and Carsky 2003; Eilittä et al. 2002, Buckles et al. 1998).

Two varieties or subspecies of MP are often described but not always differentiated as such. One is *Mucuna pruriens var. pruriens* whose beans are known under the trivial name cowage or cowitch bean (derived from Hindi: *Kiwach;* Manyam 1990). The second one is *var. utilis* whose seed is commonly called velvet bean. Latté (2008) in his succinct review specifies the use of *var. pruriens* mainly as a "medicinal plant with dark purple flowers like hanging racemes and red-yellow or brownish S-shaped puffy pods covered with stinging hairs". Seeds are brown and their LD content is specified with 4.0 to 4.9% based on dry weight with higher values (5.3%) in the endocarp. Itching reactions by the stinging hairs of *var. pruriens* are caused by a pruritogenic proteinase called mucunain and can be a severe problem for collectors (Shelley and Arthur 1955; Broadbent 1953).

---

2   IPR = intellectual property rights, usually referring more specifically to patents on substances, indication and production technologies but also including data protection (Ger.: *Unterlagenschutz*) (Begeroff et al. 2019).
3   Tupi-Guarani is referred to as a group of numerous indigenous languages from South America (Katzner 1996).

**FIGURE 3** A: Seed pods of MP, source: *Herbario Alfredo Marín*, Etnoflora Yucatense, Autonomous University of Yucatan UADY, Mérida, Mexico. Cultivated and collected in 1988 near Valladolid, Yucatan, cat. No. 05123. Figure 3B: Different variants of cultivated MP seeds from Yucatan, Mexico, between approx. 13 and 17 mm in length. Photos: I. N. Fiebrig.

*Var. utilis* in turn is used primarily as fodder and green manure. Flowers are smaller and either of lighter purple colour or whitish. Seed pods are covered with felt-like hairs that do not sting (Fig. 3A). Seeds are of off-white or light orange colour or black, with some intermediately mottled variants (Fig. 3B). LD content is lower and given as between 3.1 and 4.9% based on dry weight.

Patil et al. (2016) nevertheless raise general doubts over the identification and authentic classification of MP, showing the wide diversity of *Mucuna* species and MP varieties using RAPD (Random Amplified Polymorphic DNA) as a PCR-based fingerprinting method. With regard to agronomic traits, like biomass production, fertility, and yield potential, Pugalenthi and Vadivel (2005), for example, characterised agrobiodiversity of 11 accessions of '*Mucuna pruriens* (L.) DC. *var. utilis* (Wall. ex Wight) Baker ex Burck' in four districts of South India alone. Sathyanarayana et al. (2011) used AFLP fingerprinting to assess the germplasm diversity of 25 *Mucuna* accessions belonging to five species, including *Mucuna pruriens* represented by its three sub-species *var. utilis*, *pruriens* and *hirsuta*, showing overall high genetic diversity, which is good for breeding programmes. High similarities between *var. pruriens* and *var. hirsuta* however, demanded merging under one name: *var. pruriens*. Another example is that of Ezeagu et al. (2003) who characterised physio-chemical properties of seeds from 12 accessions of *Mucuna* as well as the nutrient and anti-nutrient factors such as LD, trypsin-inhibitors, or tannins in the seeds. The authors conclude that differences in the chemical compositions of the various seeds are marginal once dehulled.

## 2.3.2   MP as part of ancient, traditional medicine

The medicinal use of MP in relation to PD goes back to the ancient, traditional medicine from India called *Ayurveda,* meaning 'knowledge of life' in Sanskrit[4], and is considered to be the oldest traditional medicine in the world. Its origins date back to the Vedic period[5] on the Indian subcontinent (Lampariello et al. 2012). Gourie-Devi et al. (1991) define *Ayurveda* as "…the quintessence of ancient systems of health care…" estimated to have already been practised in India much earlier, between 5,000 to 3,000 BCE. However, a completed treatise was not available until before 1,000 BCE (Dutt and King 2018). The conceptualisation of *Ayurveda* comprises three categories (humours): (1) *Dosha,* (2) *Dhatu,* and (3) *Mala.*

*Dosha*: In International Alphabet Sanskrit Transliteration (IAST), *doṣa* can be translated as 'fault', 'defect' or 'that which causes problems'. *Doshas* might offer the approach towards healing. The concept is divided into three types of substances: 'wind, bile and phlegm' (IAST: *vāta doṣa, pitta doṣa, kapha doṣa*), the so-called *Tridosha* system. It is believed to belong to three bio-entities controlling basic physiological functions. Each is divided into five sub-*Doshas*. The *Dosha* system is eventually embedded in the five classical elements or enduring qualities: (1) *Vayu* (air), (2) *Jala* (water), (3) *Aakash* (space or ether), (4) *Prithvi* (earth), and (5) *Teja* (fire) of which the entire universe is believed to be composed, thus making us part of it (Jaiswal and Williams 2017; Gouri-Devi et al. 1991).

The second abovementioned main system, *Dhatu,* in turn (Sanskrit: *Sapta Dhatus* or 'seven tissues') relates to human physiology and anatomy and is categorised as (1) *Rasa Dhatu* (lymph, tissue fluids, plasma), (2) *Rakta Dhatu* (blood), (3) *Mamsa Dhatu* (muscles) (4) *Meda Dhatu* (adipose fat and connective tissue), (5) *Ashti Dhatu* (bones), (6) *Majja Dhatu* (bone marrow and joint fluids), and (7) *Shukra Dhatu* (semen, reproductive system).

*Ayurveda* is finally rounded off by the third category, the *Mala* concept referring to three waste products of the body, also summarised as *Trimala,* to include (1) *Purisha* (faeces), (2) *Mutra* (urine), and (3) *Sweda* (sweat), including any excreta of eyes, ears, nose, nails, or hair (Jaiswal and Williams 2017; Jonas 2005).

This *Ayurvedic* structure was probably essential to build up a knowledge base and to transmit understandings about health and healthcare across the millennia from one generation to the next. While the underlying medicinal world view contributes to a holistic perspective of health, *Ayurveda* can give new momentum towards novel therapies in today´s sense of evidence-based medicine (EBM) where "the conscientious, explicit and judicious use of current best evidence in making decisions about the care of individual patients" is essential (Sackett et al. 1996).

Still today, *Ayurveda* can be seen as a system that aids improving the balance between mind, body, and spirit, thus promoting health and preventing disease but not fighting it (Miller 2016). Not all that is traditionally offered by *Ayurveda* can be practised the same way in other healthcare systems outside of India, such as those of the EU, for regulatory issues. Advice on appropriate diets, massage therapies, yoga, meditation, breathing and relaxation techniques, or bowel cleansing measures are accepted as complementary medicine here (Niemi and Ståhle 2016; Lad, 2012). The cost of such treatments may not be covered by the usual public, non-private insurance systems. This goes also for specific herbal medicinal preparations or dietary supplements, which may have additional import restrictions related to applicable medicinal product acts.

In *Ayurvedic* literature, a condition called *Kampavata* can be put on a par with 'shaking palsy'. First described by James Parkinson in 1817 (Parkinson 2016), it was coined with its current term Morbus Parkinson or PD by French neurologist Jean-Martin Charcot around 60 years later (Manyam, 1990; Critchley 1955). The '*Vata'* in *Kampavata* relates to the first aforementioned *Dosha,* whilst

---

4   Sanskrit: sacred language of Hinduism and language of Hindu philosophy amongst others, belonging to the old Indo-Aryan branch of Indo-European languages (Woodard 2008).

5   Vedic period: approx. 1,500 to 500 BCE, i.e., late Bronze Age to early Iron Age (McClish and Olivelle 2012).

'*Kampa*' means tremor. *Vata* is responsible for mental and physical movements in and of the body. Ancient *Ayurvedic* texts give details of *Kampavata* as an illness that can be related to PD, with typical symptoms such as rigidity and stiffness, slowness of movement, tremor as well as associated depression, somnolence, stammering speech, or salivating and drooling (Manyam 1990). This indicates that in these ancient times, PD was already known to be a condition requiring treatment, although decoding *Ayurvedic* concepts and translating them into the concepts of modern medicine is hardly possible (Ovalath and Deepa 2013). On the assumption that environmental toxins are also responsible for the manifestation of PD, this may not only comprise toxins from modern synthetic chemistry, such as pesticides or organic solvents, but might also include toxic substances from less polluted, more pristine environments according to Manyam (1990), or support theories of endogenous causes of neurotoxicity by isoquinoline derivatives (Storch et al. 2002).

Apart from the evidence that ancient *Ayurveda* not only had knowledge of *Kampavata* as a neurological condition, equivalent to today´s PD, it also knew of preparations containing MP – *Atmagupta* in Sanskrit – that would have been used to treat PD patients (Blonder 2018; Ovalath and Deepa 2013). Examples of documented ayurvedic preparations are *Chyavanprasha avalcha* or *Abhayamalaki avaleha* (Manyam 1990).

The isolation of LD in plants had first been reported by Torquati, having extracted an azotic (nitrogenous) substance from *Vicia faba* seeds in 1913 (Torquati 1913). Shortly afterwards, this led to the establishment of the chemical structure by Guggenheim, a young chemist working at the pharmaceutical company F. Hoffmann-La Roche (1913), who followed-up on Torquati´s work. Research culminated years later in the abovementioned launch of Madopar (synthetic LD; Hornykiewicz 2010; Amrein 2004). Researchers from India in turn had isolated LD from *Atmagupta* seeds in 1937. Its effectiveness in the treatment of PD was not appreciated by modern medicine at the time.

### 2.3.3   Clinical trial preparation made from MP bean powder

In 1995, a customer entered a village pharmacy[6] in South Eastern Germany and handed a scientific paper – and not a prescription as would be usual – over to the pharmacist. The copy of the paper with the title '*An Alternative Medicine Treatment for Parkinson´s Disease: Results of a Multicenter Clinical Trial – HP-200 in Parkinson´s Disease Study Group*' described a clinical study administering a treatment from beans of the medicinal plant MP (HP-200 Study Group 1995). The customer asked to procure what seemed to be clinical trial medication under the code name HP-200. The natural preparation was needed for her husband who suffered from severe Parkinson´s disease and whose standard medication had failed to alleviate symptoms.

Following an enquiry to the pharmacy´s distributing wholesaler, the only available product on the German market was a homeopathic preparation from the hairy, itchy pods of *Mucuna pruriens* to treat skin disorders (itching, Herpes Zoster) and liver ailments. Neither the herbal drug (whole MP beans) nor any allopathic herbal preparations were available.

It is the very structure of Germany´s mostly owner-operated pharmacies and non-limited shop concessions that encourages the procurement of non-standard medicinal products, and with it, the development of advisory expertise. Within this favourable framework, the pharmacist in charge (i.e. the lead author of this chapter) used his personal network of professionals to find the enterprise that had produced the clinical study medication, at the time already marketed in India as Ayurvedic medicine by Zandu Pharmaceutical Works Ltd (Zandu; see bean powder in Fig. 4). Overall, it took roughly six months to find the supplier of HP-200 and to solve all the related logistical and regulatory challenges that are unfamiliar to the average pharmacy. Amongst others, this entailed getting a pertinent medical prescription that had to be issued by the treating neurologist. HP-200 was liable to be defined as 'medicinal product without marketing authorisation in the EU' by German customs

---

6    Names of places and persons remain undisclosed or are altered in order to protect the privacy of individuals as well as business interests whilst complying with statutory requirements.

**FIGURE 4**    MP bean powder Zandopa by Zandu. Photo: I. N. Fiebrig.

authorities and would then have to comply with the German drug law and its general ban on the intro-duction of unauthorised medicinal products (Deutscher Bundestag 1976, p. 123: *Verbringungsverbot* §73(3)1.). The first pot of 600g of HP-200 – later to be renamed 'Zandopa' by Zandu – sufficient for one month´s treatment, was handed over to the patient´s wife together with dosage and application instructions. At the time the patient had been unable to leave his home.

### 2.3.4    Patient No. 1: The personal diary

The following details are taken from a case documented by the patient´s wife, who was a nurse by profession as well as the main caregiver at home. The documentation was provided to the pharmacist spontaneously with the explicit consent from the wife and the patient to use it for research purposes. In the following, we reproduce a summarised version of this case documentation so as to prevent redundancies, to ease readability, and to grant anonymity, whilst taking care to be unbiased as to all relevant information related to this case.

Patient demographics are summarised as follows: born 1936, male, Caucasian, German national. Profession as stated: "worked for many years in a window factory with direct exposure to nitrocellulose paint". First Parkinson symptoms showed aged 57 (1993) as slight tremor in right hand and pain as well as rigidity in left elbow. Belief system of the patient: a distinctive dislike for consulting medics and taking medication. For 57 years, he would only take medi-cinal herbs when feeling ill whilst believing strongly in his self-healing abilities. The wife urged her husband to have a series of medical examinations performed. At their conclusion, the diagnosis of Morbus Parkinson had been established by a neurologist. The doctor prescribed various standard drugs to treat PD whilst continuously attempting to adjust medication appropriately. Four months later the patient had become a "wreckage" [sic!]. The wife describes his condition as follows: […] *He seemed collapsed, aged, constantly tired, weak, feeble. He developed depression and within a short period of time his condition worsened considerably. PD symptoms became worse and more frequent, such as difficulties swallowing, slurred speech, monotone voice, massive saliva production, plaques on the scalp, and the tremor had become even stronger. There was also the usual rigour, but only in the right arm.*

*The doctor tried this drug and that drug, but everything only got worse – until he* [patient] *suffered a psychosis. I stopped all medication and took care of him myself, without consulting the neurologist again.*

*It took many months for my husband to become viable again, although I tried my best setting up a treatment programme for him to cope with the mental health problems without taking medication and without any external help.*

*Normally, in our society, it takes the help of neurological clinics and the use of psychotropic drugs to get out of such a serious condition. I had to look after him around the clock because he was a danger to himself and to us as family members. Above all, he was afraid of noises, such as the ticking of a clock, the passing of a car or other everyday noise.*

*Often, he could neither hear nor see me, he had a look on his face that seemed to go through everything. For weeks he couldn't sleep a minute, neither during the day nor at night. He was unable to take care of himself, he didn't know if he was hungry or thirsty or when he had to go to the toilet. The sense of personal hygiene seemed to be completely missing. It had become unthinkable for him to go out with people, to go shopping or the like, in other words, everything that would seem normal no longer functioned.* [...and further psychiatric symptoms described] *Like a child, I led him by the hand and helped him to literally relearn everything. The therapeutic programme I had specially created for him included music therapy, massages, stroking, gradually more and more extended walks to deserted places* […plus spa treatment, movement exercises, speech therapy described in detail]. *Through my intensive and persistent training, patience and love, he regained more and more independence and self-confidence until he was restored. Now that he was no longer taking medication, he also became physically more stable and mobile, had a normal speech pattern again and felt generally better.* […the continued spa treatment also had a positive effect].

*In August 1995 we moved to…* [another place]. *My husband was no longer able to work and received an early retirement pension. As a result, we could no longer afford the spa treatment. This had a negative impact on his illness.*

*He started to feel exactly the same pain in his right knee joint that he had once felt in his right elbow at the beginning of the disease symptomatology. After that, the whole right leg stiffened more and more. He audibly dragged it across the floor and stumbled more and more often.* [The nurse describes details on changes in posture and gait pattern typical for PD – the patient was taken to another specialised clinic in April 1996 and diagnosed with PD for a second time. The patient was prescribed levodopa/benserazide 100/25 mg (1-0-1).]

*After a short time, more negative symptoms set in again, and not one positive sign. The dose was immediately reduced. Levodopa/benserazide 50/12.5 with slowly increasing intake and up to 3 x daily one capsule. A few weeks later, I recognised renewed symptoms of psychosis. Levodopa was discontinued, but it already seemed too late. My husband had become mentally ill again. The doctor treating him told me that no one could help him* [there] *now, that instead he had to be taken to…* [name of a city] *to the mental hospital.*

[She describes that instead of taking her husband to the psychiatric hospital, she did as she had done before, treating him at home by herself and giving him lots of sleep using a tranquiliser, because he had had many weeks of sleep deficit.]

*Again, I did the training that had done him so much good before. Step by step, it's all very slow, but every hour is a small, almost imperceptible progress. This time, my husband had already made it after about three weeks.*

[She describes his mental state and improvements towards a self-determined life in detail]

*His frequent depressive phases often caused him to sink into negative feelings and think of suicide. He was driven by an inner restlessness, was dissatisfied, shouted at his family for no reason, always found a reason to get upset and then could no longer find anything positive about life.*

*About four weeks after the last psychosis, I made a final attempt to try levodopa/benserazide 50/12.5 mg again, but only at a lower dose (1-1-0). He never seemed to be in balance again, which made living together difficult in the long run and weakened the harmony* [as it would happen] *even in the strongest family.*

*His right hand was still shaking, now continuously, very badly, except when he was asleep. But the mobility was somewhat better. He was perpetually tired and powerless. But he also stumbled less and so we stayed with this setting of levodopa for the time being.*

*Dr.* [general practitioner] *told me about a bean that produces dopamine[7] and grows in India. The only side effect is possibly flatulence[8], but it is not yet on the market in Germany.*

*For me, this meant finally getting my hands on something that would prevent my husband's psyche from getting into disarray and also prevent the devastating side effects from occurring altogether. This meant a real step forward in our case!* [the last two sentences include a retrospective look after three years' use of HP-200]

*You know yourself how I got hold of HP-200, and my gratefulness knows no bounds!!!*

[At the end of January 1997] *my husband took HP-200 for the first time and as follows: 0-0-1 HP-200 in combination with levodopa/benserazide 50/12.5 mg 1-1-0.*

What follows is a patient diary. Only changes to the previous day are described. Day 1 and 2 showed no changes. Day 3: slightly increased symptoms of fatigue. Day 4: exhaustion and such fatigue that he fell over his own legs. Day 5: still extreme fatigue although less than previous day. Day 6: a little fitter with deep and restful sleep at night. Day 7: as day 6 and good mood. Day 8: medication changed to levodopa/carbidopa 50/12.5 1-1-0 and HP-2000-0-1-1. Day 10: calm and balanced, but not tired. Day 18: restless, nervous, little sleep, but probably caused by a family tragedy that had occurred. Day 19: slightly better than previous day. Day 20: slept well but very nervous. Day 22: excited, restless, nervous, hardly slept, still affected by the terrible event that directly affected us. Day 23: Strong tremor, sometimes involuntarily drops objects from his hand. He is very distressed by this observation. Day 25: slept well, noticeable dragging of the right leg, slightly less uncontrolled dropping of objects. Day 26: slept well, calmer, less trembling of the right hand. Day 27: levodopa/carbidopa 50/12.5 mg reduced to 1-0-0-0; HP-200 increased to 0-1-1-1. Day 30: deep, restful sleep, no tremor at times, positive mood. Day 31: barely noticeable tremor, no depression, good sleep. Day 33: he feels more energy, positive mood. Day 34: greater mobility of the hand, everything else positive. Day 36: the whole right side is so soft and mobile that it almost reaches a normal state. Day 38: stronger tremor, dragging of the right leg. Day 39: more or less trembling of the hand with good mobility, less tiredness and more stamina, clean speech, not slurred and no longer monotonous, no visible coatings on the scalp and since then no more depressions and no negative psychological changes.

*This sums up the account up until day 52. Since day 50 he no longer takes levodopa*[/benzerazide], *only HP-200 as follows: 1-1-1-1 at 1.5 g per taking* [1.5 g 4 times a day].

*As has been shown so far, I can only confirm that HP-200 has had a very positive effect. It is almost as if my husband had regained his physical and mental abilities. I can say with certainty that*

---

7    This is not quite correct as you will find in more detail below. Instead MP 'contains' levodopa naturally.

8    This is not according to the side effects stated by the pharmaceutical manufacturer. MP powder does have side effects similar to those of synthetic levodopa. However, the overall side effect profile is believed to be gentler and is being tolerated much better. Flatulence as a side effect is typical of a bean meal and to be expected in a bean powder.

*my husband has become more resilient. However, I think that the effect of HP-200 will only really become apparent with time. I have given you all the background information so that you can better understand the problem we were dealing with and recognise the changes brought about by the new drug.* [the wife apologises for the brevity of the report and expresses her hope that it may be of help].

This report is further corroborated by the fact that the PD patient himself came to the pharmacy to pick up a second pot of HP-200 at the end of the first month´s treatment. Neither his gait nor posture or other symptoms made it noticeable that he suffered from PD. The patient became a regular customer for HP-200 prescriptions. At the same time, it was probably word of mouth that led to prescriptions from other patients and other doctors. A survey as part of regular customer care allowed a spatial representation of anonymised prescriptions in over 12 years in DACH countries which ceased in 2010 (Fig. 5). Here, we show the spread of patients and prescribing doctors within DACH countries, although sporadically prescriptions came from other European and non-European countries as well.

At this point, it is necessary to note that HP-200 was being considered by some customs authorities at German borders to be a 'medicinal product' without marketing authorisation in the EU. This was due to product presentation making it look like a medicinal product and additionally due to levodopa as an active ingredient, regardless of its natural provenance. Thus, not only individual prescription was, and still is, mandatory for procurement activities, as already noted above, but also any kind of advertising is illegal (Medicinal Products Advertising Act, HWG n. d.). On the other hand, the geographical spread suggests that the urge for improved medication in PD patients is such that the impetus for self-initiatives from patients and their families is strong, possibly through networking within self-help groups.

Within this follow-up, the nurse of patient No. 1 was contacted and asked if she wanted to give some more information on her husband's health in 2010. At the same time, she gave her consent to publicise these writings in anonymised form.

She goes on to explain: [as from March 1997] *my husband took the HP-200 powder independently, regularly and on time. I asked him what flavour it had, as the grey-beige colour did not look exactly appetising. He said it was relatively tasteless. It did not give him flatulence as a possible side effect and even if it had done so, it could have been easily remedied and* [as a side effect] *it would not have been in any proportion to the benefits my husband had received from* [HP-200]. *The fact is that HP-200 made my husband increasingly better in all aspects of life. He no longer suffered from psychoses, became stable, had a much better and safer mobility and was able to do small daily tasks at home again* [the wife goes on to explain in more detail what practical activities of daily life he was able to perform, including socialising with neighbours]. *It was a time of breathing and relaxation for all involved.*

*After a little over a year, significant physical deterioration appeared again. He became stiffer, stumbled more easily and began to tremble increasingly, no longer only on the right side of his body, but now on both sides and his head was also shaking. So, the dose of HP-200 was slowly increased until the positive condition was achieved again.*

*He and the whole family had had the best experience with this natural product HP-200 until 21.12.1999, i.e., for three years. It was three years of regained quality of life. Peace had returned, freed from the burden of the recurring psychoses. Then, shortly before Christmas Eve, my husband fell and was diagnosed with a fractured neck of the femur in hospital.*

[The wife gives details about the hospitalisation including the husband falling into coma, suffering from pneumonia, losing his swallowing reflex after regaining consciousness and returning home with a gastric tube]. *But I wanted him to keep getting HP-200 and not all the medicines he had already tried and had had such negative experiences with. Doctors from all over Germany were consulted. They said my behaviour was irresponsible because at this stage HP-200 would no longer be able to help, and he was considered untreatable. On the other hand, the doctors did not believe me about how he would react to any of the other medication, i.e., going from one psychosis to*

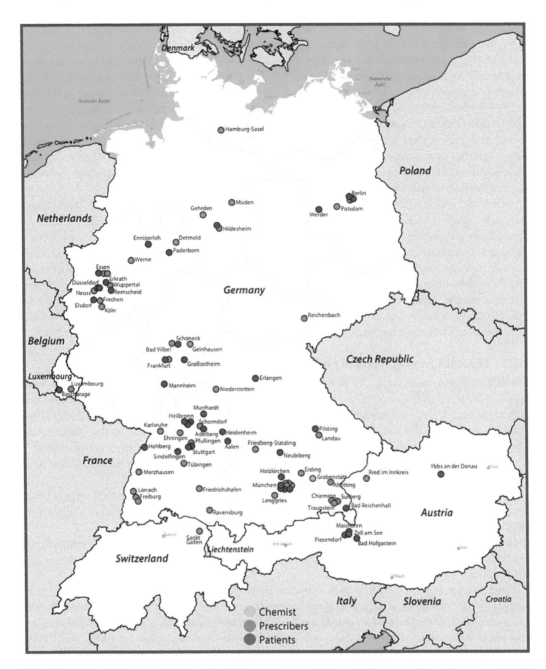

**FIGURE 5** Spatial representation of HP200/Zandopa prescriptions during more than 12 years in DACH countries. Mostly prescribers and their corresponding PD patients were not very distant from one another. In some cases, prescribers may have used the product for themselves or for treating close relatives, especially where no patients can be identified nearby. Graphics: Zoran Laufer/Lila Publishing e.K.

*another. My husband was unable to voice his wishes and from today´s perspective I believe I was too tired to continue fighting on my own and to carry all the responsibility by myself. I let them medicate him with conventional dopamine drugs. His fracture of the thigh was then operated on with spinal anaesthesia. He could never walk again. Neither could he eat nor drink by himself. He received his medication and food through a stomach tube.* [Details about the husband having to be transferred to a nursing home as a result.] *My husband got psychoses all the time since he could no longer take HP-200. He no longer recognised the children, was in a wheelchair and degraded more and more. At times, for brief moments, he recognised me, but only rarely. He spent the last two years of his life in bed, full of contractures, unable to speak, without me being able to find any meaning in this way of life. He often opened his eyes in terror and with clenched fists pressed towards the ceiling as far as he could. (I thought: now he is seeing the horror images again that had tormented him countless times!)* [the patient died several months later and his wife writes in her last sentence]: *Everyone who is affected by this disease and reads my report should do what he/she thinks is right and even if the whole world may be against it, I wish them the courage and strength to go different, new ways* [signed by the wife of a PD patient].

Within the roughly 12 years that the pharmacy imported HP-200, later to become Zandopa and then to be licensed to Emami Ltd, patients were referred to doctors, and doctors in turn received all available and pertinent information related to the application of MP powder from the pharmacy. Feedback is summarised in the next section. With increased e-commerce in the pharmaceutical sector from abroad, customs authorities became progressively zealous, retaining Zandopa for unacceptable lengths of time before releasing it to the pharmacy and thus to the patient in need, making chronic therapy unviable, regardless of ethical aspects. At the same time, other e-commerce routes within Europe may have become more practical for PD patients – yet possibly without helpful advice from experienced doctors and pharmacists.

### 2.3.5   Feedback from 10 years of pharmaceutical advice to PD patients

The selected patient experiences presented herein are related to >300 PD patients who were supplied with HP-200/Zandopa® by one pharmacy in Germany, and who received at least one prescription up to ongoing prescriptions over a period of three years or less until the end of 2011 (Table 2, patients 2 to 6 and 8). Feedback from patient No. 7 (Table 2) is from the Netherlands, where Zandopa had been defined as food supplement at the time and was available from a Dutch Ayurvedic preparations supplier and with medical prescription only. Feedback had been received between 2007 and 2012.

Our argumentation towards improved management for at least some PD patients is largely based on our own notes, observations, and spontaneous as well as prompted patient feedback in our professional roles as pharmacist or herbalist and privately as caregiver to PD patients.

Such spontaneous feedback widens the knowledge of the pharmacist regarding the practical use of a specific medication which can direct further pharmaceutical advice to medical professionals and patients.

Table 2 summarises such feedback in a structured manner and shows a series of patient feedbacks that give clear indication as to a perceived benefit in quality of life (QOL) from taking HP-200/Zandopa mostly in the form of a supplement to conventional, guideline-compliant PD medication. This led to a reduction of perceived side-effects concomitant with the reduction of PD medication other than MP preparations. One case also indicates a benefit in QOL from MP in RLS, which is often treated with synthetic LD. Patient No. 7, from the Netherlands, claimed in his self-report that after taking his first medication of the day, synthetic medication only worked after a latency period of half an hour, whereas with MP powder the effect became apparent almost immediately, with the limbs and joints also appearing to work much more smoothly. Interestingly, the patient did a model calculation of the cost of his former synthetic PD medication (€ 3,700.00 p. a.) versus the cost of HP-200 as substitute medication (€ 890.00 p. a.), claiming exemplary annual net savings of € 2,810.00 to the insurance. Regardless of these effective savings, the patient´s insurer did not agree to make an exception to their rules and refused to reimburse the cost of the prescribed MP powder at the time.

**TABLE 2**
**/d: per day; CG: caregiver giving feedback; N. d.: no date but within the given timeframe 1998–2011. ['text xyz'] = feedback summarised by editor; [...] = redundancies removed from original statements by editor.**

| No. Date and patient demography | Medication and treatment | Feedback |
|---|---|---|
| 2  Male, PD | Levodopa/benserazide 100/25 tablets 3-4/d HP-200: 17–20 g powder first doses: 8.00 to 9.00 a.m.; last doses at 8.00 p.m.; plus selegiline in the morning. | 2004, CG [The addition of HP-200 to levodopa/benserazide meant a reduction of side effects, tremor was minimised, no more spasms and no more feelings of pressure in the head. Taking entacapone could be stopped.] |

CG: „[...] *This combination* [levodopa/benserazide/HP-200] *has improved his condition a lot and he has also become much, much more agile and he can walk again very nicely. And his skin has now become completely normal, it no longer has the greasy sheen it had before. Swallowing has also got a lot better. Now he can again drink water from the glass in such a way that he holds the glass in his hand and drinks that way. In general, his whole condition has got much, much better and other people have already noticed that* [...]

| 3  Female, PD at age 46 as a result of severe surgery, five months in wheelchair | Zandopa | 2008, Pat: [Many different drugs to treat PD were tried. Through personal contacts at a self-help group she got to know about Zandopa. Four weeks after taking it, patient was much better and could leave the wheelchair, learnt to walk again.] |

"[...] *My tremors and cramps were visibly better. I just became a different person. My zest for life has grown again and I learned to be a happy person again! I am overjoyed that this Zandopa exists and that I have such great success with it. Unfortunately, my husband died [year], but my children share my happiness with me! Thank you for me becoming human again and thank you for your great help!*]"

| 4  Male, PD symptoms since 1990 | Withdrawal of lorazepam in spring 2009 MP capsules (3-5/d.) Metaldetox (ayurvedic) Ashwagandha capsules, 3/d. vegetarian diet, therapeutic massage, steam sauna | 2010, CG: [mood improved during withdrawal, the last two weeks the patient had been in high spirits before his death in late 2009. Eating a (non-vegetarian) sausage in August 2009 as an exception had caused three days of spasms and extreme Parkinson symptoms.] |
| 5  Male, age 71, PD 1999 | Pramipexole, amantidine, rasagiline, HP-200 | 2011, Pat.: [Improvement by addition of MP preparation achieved: reduced tremor, facial expression more open, better gait, reduction of all PD symptoms, patient states to be living a relatively normal life since.] |
| 6  Male, PD since 1998 | 4 scoops of Zandopa powder at night; Levodopa/carbidopa and cabergoline tablets during the day | N. d.; Pat.: see below. |

"*Taking Zandopa means a rest from the side effects of my PD tablets. [...] I wake up after a restful sleep with no muscle or back pain. [...] After taking the tablets, back pain, muscle twitching etc. start again.*"

| 7  Male, age 65, PD since age 50 | 25 to 30 g HP-200 per day at maximum combined with LD + carbidopa and pramipexole | 2008; after 14 years plus 1 year of HP-200. Gradual reduction of LD + carbidopa possible until half the original dosage with radically reduced side effects, improved sleep, faster and longer ON time; body more flexible in the morning and calmer mood. |

*(continued)*

**TABLE 2 (Continued)**
**/d: per day; CG: caregiver giving feedback; N. d.: no date but within the given timeframe 1998–2011. ['text xyz'] = feedback summarised by editor; [...] = redundancies removed from original statements by editor.**

| No. Date and patient demography | Medication and treatment | Feedback |
|---|---|---|
| | "[...] I got Parkinson's at the age of 50 and after about 15 years [taking] the number of about 25 pills per day consisting of the well-known [LD + carbidopa, LD + benserazide, LD + carbidopa + entacapone and entacapone]... and the [LD] agonists etc. [the pills] were no longer able to guarantee a whole day ON. The OFF periods continued to increase and became highly annoying. This phenomenon is well known and feared within our Parkinson's world. [...] Now after a year of using HP-200, I am very satisfied with it and I hope that the long term will also be achieved. Every day you marvel at the fact that with HP-200 you can get out of bed after 10 minutes. [...]" | |
| 8   Female, restless legs syndrome (RLS) for 40 years | HP-200/Zandopa | N. d.; Pat.: [Previously, during eight years only such drugs had been prescribed that the insurance would pay for. These medications had very strong side effects and the patient had to stop taking them. MP as alternative medication had produced significant relief "without side effects".] |

### 2.3.6   MP, PD, and clinical experience

In Table 3 of this section, we give a summary of clinical experience with MP preparations based on a report commissioned by the Dutch Parkinson Association (Parkinson Vereniging; Stegeman et al. 2017), based on previous work of their 'Project Group *Mucuna pruriens*'. The report has been produced by Cochrane Netherlands, having reviewed the central question of the efficacy and safety of MP in PD patients based on pertinent publications. To this end, evidence of the following four claims about MP was researched: (1) inhibition of PD progression, (2) absorption enhancement of synthetic LD, (3) faster onset and stronger, as well as longer, effect with fewer side effects than synthetic LD/CD, and (4) inclusion of additional substances responsible for this improved effect. Our summary in Table 3 is completed with an added publication (Nagashayana et al. 2000) and updated with more recent data (Cilia et al. 2017, 2018, 2018b).

Collected evidence comprising three randomised controlled studies (RCT) and two uncontrolled cohort studies were evaluated using the GRADE[9] system (cf Schünemann et al. 2013). The first documented clinical experience in the Cochrane review was the study by Vaidya et al. (1978) showing comparable efficacy to synthetic LD with fewer adverse effects in 33 PD patients. Major weaknesses of the study design are lack of a control group and a lack of description for the varying measurement moments of the end point (effect). However, this study may have inspired and supported the design and execution of the trial conducted by the HP-200 Study Group (1995) carried out in three centres in Mumbai (formerly Bombay) and one centre in Chennai (formerly Madras), India. Out of the 60 idiopathic[10] PD patients, 34 were LD naïve, meaning they had not taken LD previously in any form, such as synthetic LD. In both groups, the treatment was effective, with fewer side effects in the LD naïve group. Due to the lack of a placebo group, technically, how the disease would have manifested without treatment could not be measured. However, it was this very MP preparation that, once

---

9    GRADE: Grading of Recommendations Assessment, Development and Evaluation.
10   idopathic PD refers to the unknown cause of PD in these patients

**TABLE 3**

**Summary of clinical experience with MP preparations based on a report commissioned by the Dutch Parkinson Association (Parkinson Vereniging; Stegeman et al. 2017) and updated/amended accordingly.**

| Study type/end point | Patients | Preparation | Outcome | Ref. |
|---|---|---|---|---|
| Open label due to difficulties in matching the placebo to the powder, no control group, bioavailability of LD studied on 9 patients from plasma; NUDS and handwriting. | 33 PD patients included, 10 patients dropped out after 3 weeks, 18 men, 5 women, powder taken 4 weeks up to 1 year | Powder made from whole bean of MP purchased from local market, up to 40-50g of powder per day containing 4.5 to 5.5% of LD. | 23 patients treated on average for 20 weeks. Decreased incidence of adverse effects compared to synthetic LD that patients had received before entering trial. Powder was well tolerated; reduced bulk and improved taste and flavour were desired by patients. Significant absorption of LD, peak of blood plasma levels after 1 hour of admin. Significant therapeutic response, side effects infrequent and mild. Conventional side effects of LD not observed. | Vaidya et al. 1978 |
| Open label study, no control group; UPDRS scores, Hoehn and Yahr stages measured at weeks 2, 4, 8 and 12. | 60 idiopathic PD patients, 46 male, 14 female; 34 LD naïve, 12 weeks | HP-200*, powder formulation of MP endocarp in sachets, adjusted for LD content 3%; [taste correction with flavouring agent and sweetener for palaatability but not defined] | Degree of adverse reactions defined as 'mild'; type of side effects similar to those when taking synthetic LD. "...significant improvement in all the major components of parkinsonism..." | HP-200 Study Group 1995; Manyam et al. 2004 |
| Open label study, UPDRS rating | 18 PD patients, on average approx. 60 years of age, Male/female ratio 5:4. All medication stopped 15 days prior to study initiation. | Mixture of cow´s milk, MP, *Hyoscyamus reticulatus* seeds, *Withania somnifera roots* and *Sida cordifolia* roots containing hyoscyamine, somniferin, ephedrine, amongst others. | Effect of LD contained in MP and to a lower extent in *H. reticulatus* and *W. somnifera* is confirmed. However, the study's main conclusion lies on the benefit of Ayurvedic cleansing therapy ('*panchacarma*') before palliative medication. | Nagashayana et al. 2000 |

*(continued)*

**TABLE 3 (Continued)**

**Summary of clinical experience with MP preparations based on a report commissioned by the Dutch Parkinson Association (Parkinson Vereniging; Stegeman et al. 2017) and updated/amended accordingly.**

| Study type/end point | Patients | Preparation | Outcome | Ref. |
|---|---|---|---|---|
| Randomised, controlled, double-blind crossover; 3 single dose challenges on different days; UPDRS and 'brain test' with AIMS and Goetz rating scale | 9 idiopathic PD patients enrolled (QSBB criteria), 1 dropout (vomiting), 8 completed: 5 women, 4 men. | 200 mg LD/50 mg CD vs. PTX-200**: sachets with powder standardised, dosed at 500 mg and 1,000 mg LD. | MP preparation led to considerably faster onset of effect, and approx. 22% longer effect. No significant differences in dyskinesias or tolerability were found. | Katzenschlager et al. 2004 |
| Randomised, controlled, double-blind crossover; non-inferiority phase 2b study, monocentric in Santa Cruz, Bolivia by neurologist Janeth Laguna with experience in MP treatment for indigent patients; UPDRS III and AIMS. | 18 advanced idiopathic PD patients on stable anti-parkinsonian therapy for at least 30 days | Single dose, 6 treatments: (1) 100/25 mg LD/DDCI 3,5 mg/kg; (2) high-dose MP powder at 17,51 mg/kg, (3) low-dose MP powder at 12.5 mg/kg, (4) LD without DDCI 17,5 mg/kg, (5) MP + DDCI 3,5 mg/kg, (6) placebo. MP powder from peeled and roasted seeds. Ground nuts/soluble coffee were used as MP placebo to mimic texture and flavour of MP powder. | In terms of efficacy, the treatment with MP low and high dose was not inferior ('non-inferiority') compared to LD/DDCI 90 and 180 min after intake in all outcome measures. No patients dropped out of the study and adverse events were reduced in MP high and low dose. MP seeds from 'Bolivian black ecotype' with 5.7% of LD were used in a form of 'no pharmacologic processing'. Overall efficacy of MP is in line with previous findings (Katzenschlager et al. 2004): "Shorter latency to ON, longer ON duration; reduced dyskinesias as compared to LD+DDCI". | Cilia et al. 2017 ClinicalTrials.gov identifier: NCT02680977. |
| Second part of the study by Cilia et al. 2017; open-label, non-inferiority, randomised, crossover, phase 2b pilot trial study; Quality of Life questionnaire, 39 items (PDQ-39) | 14 randomised PD patients from the above group, Cilia et al. 2017. Study extension for patients having discontinued MP continued taking supernatant of aqueous suspension of MP powder instead, during between 2 and 48 weeks. | MP or commercial LD/CD: 2 dose adjustment periods before each treatment phase due to different ON times and dosing frequencies. 3.5-fold to 5-fold dose conversion factor LD/CD to LD in MP. Study extension with MP aqueous supernatant. | Proving non-inferiority of MP powder compared to LD/CD in terms of efficacy and safety over a period of 16 weeks. Efficacy of MP was similar to LD/CD. Tolerability issues of MP (gastrointestinal side effects and progressively worse motor performance) | Cilia et al. 2018; ClinicalTrials.gov identifier: NCT02680977. |

| Study | Patients | MP preparation | Results | Reference |
|---|---|---|---|---|
| Multicentre 52-week phase 2 prospective study in 3 Ghanaian hospitals; non-inferior change of Quality of Life as per PDQ-39, UPDRS I-IV, Hoehn and Yahr staging including postural instability and dysphagia. | 26 LD naïve PD patients after three months, aim of study: 90 idiopathic PD patients. | MP powder from peeled and roasted seeds, 43 ± 6 g per day, suspended in water versus 620 ± 205 mg LD + DDCI. | leading to dropouts, largely attributed to too short a switching process in the crossover design and advanced PD. MP was overall well tolerated; progressive shortening of ON time was seen after a few weeks in a few patients but no drop-out due to adverse events; some patients with long-standing untreated PD showed much improved postural stability due to MP powder. MP was successfully cultivated in a hospital's garden and in the private gardens of 2 patients. | Cilia et al. 2018b Pan African Clinical Trial Registry, ID: PACTR201611 001882367. Study evaluation in progress (Cilia 2021) |

Acronyms: AIMS, Abnormal Involuntary Movement Scale (Colosimo et al. 2010); CD: Carbidopa; NUDS: Northwestern University Disability Scale (Canter et al. 1961); PDQ: Parkinson´s Disease Questionnaire (Jenkinson et al. 1997); UPDRS: Unified Parkinson´s Disease Rating Scale I – III (UPDRS 2003; Goetz et al. 2007); QSBB: Queen Square Brain Bank criteria for PD diagnosis (Hughes et al. 1992); *HP-200: short term (48 h, female/male mice and albino rats) and long-term toxicology studies (52 w) with rats preceded the clinical studies to assure safety; **PTX-200: improved formulation compared to HP-200: stability, solubility, taste.

commercially available, was pivotal in the improvements of Patient No. 1 in Germany and many other patients who followed, some having provided the aforementioned patient feedbacks.

According to the late Dr. Krishnakant M. Parikh, who at the time was the owner of Zandu Pharmaceutical Works Ltd. (as per personal communication with the first author in 1997), HP-200 had been developed and was being marketed by Zandu to provide the Indian population with an affordable *Ayurvedic* alternative to comparatively expensive synthetic LD/CD preparations, which were unaffordable to many. It had also been Zandu´s aim to introduce the preparation as a nutraceutical or dietary supplement within Western medicine and to meet the local specifics of licensing regulations (in particular, the United States). This endeavour, however, did not materialise. The German start-up enterprise CMI AG (Centres for Medical Innovation AG), later to become Phytrix AG, had picked up the idea of developing an MP preparation, in order for it to attain regular marketing authorisation as a medicinal product in Europe and the United States at the beginning of the 21st century (Van der Giessen et al. 2004).

Meanwhile, Nagashayanas et al.´s (2000) study on 18 PD patients using an MP containing an *Ayurvedic* preparation added probably nothing of much relevance to the development of a modern herbal preparation outside of *Ayurveda*, whilst emphasising the complementary benefit of *Ayurvedic* cleansing or 'eliminative' therapy that included "[…] oleation, sudation, purgation, enema and errhines by administering prescribed *Ayurveda* drugs. […]". It is mentioned here for completeness.

The clinical trial by Katzenschlager et al. (2004) in turn is to be considered the first randomised, double-blind, crossover study on MP, according to the standards of EBM (cf Sackett et al. 1996) with trial medication PTX-200, similar to HP-200 (ground MP seeds), but produced according to GMP (Good Manufacturing Practice) medicinal product standards and with an improved pharmaceutical formulation to increase patient acceptance (taste, solvability, etc.). At the end of the trial with eight included patients and sponsored by CMI/Phytrix, the authors concluded: "The rapid onset of action and longer ON-time without concomitant increase in dyskinesias on *mucuna* seed powder formulation [PTX-200] suggest that this natural source of L-dopa might possess advantages over conventional L-dopa preparations in the long-term management of PD". Stegeman et al. (2007) critically concluded that the study showed no significant differences with regard to UPDRS scores between the three groups. Onset of effect in the MP treatment group, however, was roughly half an hour faster than in the LD/CD group (34.6 versus 68.5 min; p = 0.021) and lasted approximately 22% longer. The limited difference in UPDRS scores between groups may not come as a surprise due to the small cohort and the briefness of the interventions. The budgetary constraints of a start-up company may have put limits on a sizeable study in terms of the number of patients and duration of the study. Although speculative, the main intention of the clinical trial may rather have been to show sufficient effect of the herbal treatment with an improved, patent-protected formulation as part of the business model. This in turn may have been considered to be a strategic prerequisite to attract large investors able to finance the extensive clinical trials required until marketing authorisation was gained. Unfortunately, PTX-200 never reached the European market and Phytrix changed its research and business activities before the study was published.

By contrast, a philanthropic approach seemed at the heart of research endeavours of the research group around Prof. Gianni Pezzoli, supported by the Italian non-profit foundation *Fondazione Grigioni per il Morbo di Parkinson*. They argued that while LD is the cheapest treatment in PD, in many low-income countries it is nevertheless neither available to, nor affordable for, many patients (Fothergill-Misbah et al. 2020). Thus, to start with, a low-cost preparation method based on MP beans was developed jointly in Italy, Bolivia, and Ghana (Cilia et al. 2011; Cassani et al. 2016). Beans of known LD content were roasted for ca. 15 minutes until the tegument popped. This would ensure microbial safety and ease the peeling. Subsequently, the beans were ground (e.g. with coffee grinders), sieved, and dosed as a powder in soup or water for intake. A subsequent two-part study was performed at *Clínica Niño Jesús*, a neurology clinic in Santa Cruz, Bolivia. The neurologist Janeth Laguna here had already had long-standing experience with indigent patients using MP.

The first part of the study included 18 PD patients on stable anti-parkinsonian therapy who were to receive six different single-dose treatments in varying sequence, depending on randomisation. The six groups comprised (1) LD+DDCI (3.5mg LD/kg) as reference, (2) MP high dose (17.5mg LD/kg), (3) MP low dose (12.5mg LD/kg), (4) MP reference dose (3.5mg LD/kg + DDCI), (5) LD without DDCI (17.5mg/kg), and (6) placebo. For the first time, this study explored the established treatment with synthetic LD and DDCI versus an adequate corresponding dose of LD in MP. This had been expected to be significantly higher due to the lack of DDCI in study preparations like HP-200 (Zandopa) or PTX-200 (by Phytrix) but were never systematically matched with synthetic LD plus DDCI, nor referenced additionally to the well-studied effect of synthetic LD without DDCI. It is proposed that MP preparations contain one or more natural substances with an activity similar to synthetic DDCIs, such as genistein and its precursor genistin (Cassani et al. 2016; Kasture et al. 2013; Hussain and Manyam 1997). MP was shown to be non-inferior to synthetic LD+DDCI and to be a potentially effective and safe alternative to marketed medication in indigent populations who cannot afford a regular medicinal product. The second part of the study on 14 of the 18 initial patients, however, points towards issues with gastrointestinal side effects (these are well known) and a reduction of ON time (tachyphylaxis) during the eight-week treatment period (Cilia et al. 2018). The authors conclude that a longer titration period (conversion phase) between one treatment and the next one and longer follow-up times (3–9 months) are needed to improve tolerability and plan a study on a larger population of PD patients. This planned 52-week follow-up trial with MP from home grown seeds vs. synthetic LD/DDCI in three Ghanaian hospitals has so far been presented as a conference poster with no peer-reviewed paper published, yet with data analysis being performed at the time of writing of this chapter (Cilia 2021; Cilia et al. 2018b).

The Cochrane team´s conclusion regarding the efficacy and safety of MP qualified the conducted research as 'limited' with low levels of evidence and very low levels of confidence in the effects found – based on the GRADE system (Stegeman et al. 2017). The Cochrane review did include the Bolivian trials, at the time unpublished, but the small number of additional patients may add little to their verdict of raising their emphasis on the need for curative or progression-inhibiting PD treatments. However, the patient feedbacks cited above in this chapter have shown MP preparations to be beneficial at least to a few patients who would either not accept some of the conventional medication or who would suffer too much from their side effects. It is possible that precisely these critical patients (e.g. suffering from cognitive impairment, psychotic symptoms, or other advanced symptoms and impairments) would have been typically ineligible for any of the clinical studies conducted so far, but conversely might have particularly benefitted from them, had they taken part (e.g. Cilia et al. 2017).

In addition, the hope of at least a reduction of disease progression is an important one, jointly with curative PD therapy. Thus, the neuroprotective activity of MP and a resulting reduction of oxidative stress in the brain, including oxidative stress caused by long-term therapy with synthetic LD, has been suggested by various authors, whilst remaining controversial and requiring further long-term clinical research (Tharakan et al. 2007, Radad et al. 2005, Berg et al. 2004).

## 2.4   OTHER MEDICINAL USES AND PHARMACOLOGIC EFFECTS

MP has not only been a medicinal plant of particular interest for PD, including its neuroprotection through a presumed anti-oxidative effect. Latté (2008) lists numerous other indications from folk medicine, also comprising parts of the plant other than the beans, such as leaves or roots, with some research conducted on seed extracts regarding anti-diabetic, anti-venom, anti-microbial, or anti-tumor effects. A more recent review by Rai et al. (2020) evaluates research conducted into anti-ischemic, anti-inflammatory, anti-epileptic, or anti-hypertensive effects, pointing out the potential of MP in other neurodegenerative diseases such as Alzheimer´s and Huntington´s disease. Much of the research on indications other than PD has so far been conducted in *in vitro* or in animal models, including probably the most popular indication, that of an aphrodisiac to enhance male virility, in rats (e.g. Suresh and Prakash 2012). It is likely that folk medicine claims around aphrodisiac properties of MP seeds have created an astonishing wealth of marketed dietary supplements (Google 2021). Whilst recent reviews on folk aphrodisiacs do not include MP, its aphrodisiac activity has been shown with MP seed powder in male rats and, although speculative, may find recommendation amongst males within electronic social networks (Ashidi et al. 2019; West and Krychman 2015; Sandroni 2001). Such dietary supplement preparations are usually made from MP extracts (as powder) and filled into capsules, as opposed to the whole ground MP bean preparations used in clinical trials (Table 3). We mention this here, because from spontaneous, as well as prompted, patient or caregiver feedback, we know that such extracts are factually being used by PD patients instead of clinically tested Zandopa for three main reasons: (1) lack of availability of Zandopa, as it is considered an unauthorised medicinal product by EU customs authorities; (2) ease of taking a capsule versus the bulk of powder suspended in a substantial amount of liquid; and (3) none of the gastrointestinal symptoms typical of a bean meal. Although such patient feedback has been too scarce to derive any level of clinical evidence, we grouped the feedback we gathered into three main categories for MP extract in capsules: (1) 'beneficial in combination with other PD medication to lower side effects of synthetic levodopa and other medication'; (2) 'no benefit' (because side effects are equal to taking synthetic LD); and (3) 'no effect'.

### 2.5  MP AS FOOD SUPPLEMENT

In the EU and other jurisdictions, food supplements fall under food-related laws and regulations. Legislation pertinent to medicinal products and their marketing authorisation does not apply (Anadón et al. 2021; Silano et al. 2011). Thus, quality standards in terms of active ingredients are not as strict as they would be under drug laws. In the case of MP, this concerns LD or other secondary plant metabolites thought to be responsible for the putative benefits of MP bean preparations as well as their extracts (Hussain and Manyam 1997; Latté 2008). Such extracts may not contain enough LD, lack other unidentified anti-parkinsonian compounds, or lack natural efficacy-enhancing adjuvants or side effect-reducing components because they were not extracted adequately (i.e. with the appropriate extraction process or such compounds were not determined by quantitative analysis during production or, even though extracted, these active compounds may not be stable enough over the shelf life of the product). For example, aqueous extraction of LD may lead to polymerisation and thus inactivation of the LD, whilst extraction of the putative neuroprotective, anti-oxidative components may rely on extraction with fewer polar solvents (Misra and Wagner 2007). In summary, food supplements from MP may have been more easily available to consumers and patients than Zandopa, but quality – including partial or total lack of natural active ingredients, the presence of toxic substances as well as illicit health claims, and 'prohibited endorsements' in advertising – present additional risks to PD patients (Muela-Molina et al. 2021, Jairoun et al. 2020, Low et al. 2017; Latté 2008). Furthermore, food supplements are not subject to the quality regulations laid down in pharmacopoeias or required by pharmaceutical production standards, such as GMP (Good Manufacturing Practice; Daue 2017) or GACP (Good Agricultural and Collection Practice; Nieber and Dohm 2013), while specific GMP standards of their own may apply (FSE n. d.).

Food supplements, however, do not require previous proof of efficacy and safety, and their declared ingredients including LD content are allowed to vary up to 50% in the actual product. Medicinal (pharmaceutical) products, in contrast, have set this margin of deviation to 5% (BVL n.d.)

### 2.6  MP AS FOOD

With the closeness of food supplements to food and food regulations, it is pertinent to mention that MP beans with their high protein content serve as a valuable legume to some ethnic groups in tropical countries in South America (e.g. Brazil), Africa (e.g. Ghana, Nigeria, Malawi), or Asia (e.g. India, Philippines) (Pugalenthi 2005; Onweluzo and Eilittä 2003; Eilittä et al. 2002). However, LD in the seeds and other alkaloids, including hallucinogenic substances in various parts of the plant, are considered anti-nutritional (Pathania et al. 2020; Pugalenthi et al. 2005; Szabo 2003). MP beans consumed as food require the inactivation of LD and possibly other anti-nutritionals through cooking (wet heating) or fermentation (Eilittä et al. 2003).

In fact, it is LD, indolealkylamines, and the hallucinogenic potential of some components that have led to food supplements with 'MP extract' being considered 'novel food' in the category of "Part A – forbidden substances" according to an assessment commissioned by the German Ministry for Climate Protection, Environment, Agriculture, Nature and Consumer Protection of North Rhine-Westphalia (*Ministerium für Klimaschutz, Umwelt, Landwirtschaft, Natur- und Verbraucherschutz des Landes Nordrhein-Westfalen*; Clausen et al. 2011, p. 58). While MP beans as raw material are not considered a drug substance according to this report, MP extracts as a food supplement are viewed as a non-authorised novel food that cannot be legally traded in the EU. In principle, this would also apply to the whole unextracted bean if traded as foodstuff (NFL 2015). The European Food and Safety Authority (EFSA 2012) lists MP along the same lines (i.e. as a plant of possible concern regarding all parts of the plant and substances like LD, as well as indole alkaloids, therein). A further risk assessment report by the Spanish agency AECOSAN (2016) concludes: "due to the presence of L-Dopa, and other biologically active substances which may in turn have a synergic action, the voluntary intake of seeds of *Mucuna pruriens* in uncontrolled and unassessed conditions is a 'risk factor' to be considered for the health of consumers".

It seems prudent to position 'MP as medicinal plant material to treat PD patients', whether as raw material (beans), as semi-finished bulk (e.g. powder), or as part of ready-to-take preparation (e.g. capsules) within the realm of pharmaceutical business activity. This could be, for example, retail pharmacies that would warrant originality, quality, safety, and availability alongside qualified advice regarding the product. MP as food or food supplement on the shelves of supermarkets or drugstores in the EU would certainly not be the right distribution channel.

## 2.7 MP AND AGROECOLOGY

From our personal view, if a nomination existed for 'Plant of the Year' in Medicinal Agroecology, MP would qualify as one of the first nominees. Lampariello et al (2012) wrote their review on 'The Magic Velvet Bean' concentrating on medicinal properties and concluded: "*Mucuna pruriens* is an exceptional plant" both as a food – regarding the beans – and as medicine – regarding all parts of the plant. Eilittä and Carsky (2003), on the other hand, introduce MP from the agricultural perspective as being "an intriguing crop" and point out that MP has been viewed "as a potential miracle crop" to "alleviate decreasing soil fertility in tropical regions".

Beside its medicinal and nutritional properties for humans, MP also plays a functional role in agroecosystems such as through providing a livestock fodder rich in protein, carbohydrates, and fat, preventing soil erosion in arable systems and fertilising the soil through nitrogen fixation. There is extensive literature around the use of MP mostly in traditional and small-scale farming for subsistence (Eilittä et al. 2004; Eilittä and Carsky 2003). Here, we shall refer to a few selected examples for illustration only.

In view of climate change and the fragility of supply chains such as experienced as a result of the CoViD-19 pandemic crisis, it is worth noting that MP is fairly resistant to low soil fertility, drought, and soil acidity whilst being less tolerant to frost, water logging, and cold climate (Galanakis 2020; Pugalenthi et al. 2005). MP's use as fodder has been explored in monogastric animals such as fish, poultry, and swine. It is grown and used to feed ruminants like cattle with protein and amino acids on islands such as Mauritius or Madagascar, thus promoting self-sufficiency and decoupling from soybean-based corporate feed supplies (Pugalenthi et al. 2005, Muinga et al. 2003; Buckles 1995). It has also been studied as feed supplement in small ruminants like goats and sheep (Eilittä et al. 2003).

From the perspective of cultivation and the promotion of mixed cultures for improving agri-cultural sustainability, MP helps to suppress weeds. MP will also protect the soil from erosion by rainfall. In inter-row cropping systems, for example, it has been shown to benefit main crop yield, while in practice, MP should be sown well after the first crop, such as corn, to avoid overgrowth of the companion species (Buckles et al. 1998; Buckles 1995).

In her plea, van de Vijver emphasizes the potential of MP and medicinal plants in general from a socio-economic and ecological perspective (see Foreword One of this book). With a predicted growing need for accessible and affordable anti PD medication, the controlled cultivation of MP becomes urgent for PD patients in low- and middle-income countries. At the same time, producing medicinal plants for high value markets can provide an important source of income to smallholder farmer communities as part of an agro ecological farming system and a driver for farmers to transi-tion from industrial to sustainable or regenerative agriculture (van de Vijver 2022).

## 3   DISCUSSION

We have traced a perspective of MP as a medicinal plant able to contribute to various aspects of health in humans and sketched to a lesser extent the role of MP in agroecosystems including animals, plants, and soil. Our emphasis lay on clinical experiences with PD patients. These comprised – some more, some less – controlled trials with relatively small numbers of patients, and short trial periods together with spontaneous or prompted patient feedback; the latter outside of the formality of clin-ical trial settings. As to its role in Traditional Medicine (TM), such as *Ayurveda*, MP falls under the scope of the latest World Health Organization Traditional Medicine Strategy (WHO 2014). The strategy paper aims at "…supporting Member States in harnessing the potential contribution of T&CM [Traditional and Complementary Medicine] to health, wellness and people-centred health care…".

We argue that while this strategy paper is indeed an important starting point, it may promote less in terms of manifesting the desired "safe and effective use of TM [Traditional Medicine] by regulating, researching and integrating TM products, practitioners and practice into health systems, where appropriate" (WHO 2014). We further argue that if national governments are comfortable leaving such changes mainly to market forces, too little shall happen in terms of 'appropriateness' and developments in the best interest of the patient. For MP, this may mean that research shall remain scattered over long periods of time, with clinical trials not being part of a coherent and sustained clinical development programme leading to marketing authorisation of a safe, effective, and affordable medicinal product. We propose various reasons why this may be so and how to over-come the challenges through the suppositional perspective of the industry and so-called 'market forces'.

### 3.1   TOP GLOBAL PHARMA COMPANIES

The top global pharmaceutical companies are generally active in the market of prescription ('Rx') drugs, with an R&D of their own typically feeding medicinal product innovation into a neurologic product portfolio. Such portfolio may comprise medicines to treat PD, some of them including LD (cf. Pharmaceutical Technology 2020).

*Pros*: Such multi-nationals are assumed to be in a financial position that would permit the long-term investments needed for clinical development towards marketing authorisation. For MP, this would ideally be an extract. Such an extract would need to present all the benefits of the admix-ture of active components in MP seeds and conform to the high-quality standards of the pharma-ceutical industry in terms of pharmaceutical stability as well as clinical safety and efficacy. The lesser volume of extract would be more acceptable to patients, as opposed to bulky and unpalatable volumes of ground MP seed.

*Cons*: (1) A natural LD preparation within an existing portfolio of other businesswise successful synthetic LD-containing PD medications may represent undesired competition. It might position the MP product in a niche with too small an ROI. (2) Intellectual Property (IP) protection may be limited to patents on specifics of the extraction process or plant variety rights, if any. This stands in opposition to IP rights on a New Chemical Entity (NCE, new synthetic drug substance), more clearly defined and an essential part of well-established and lucrative pharma business models (cf. Saha and Bhattacharya 2011; Kartal 2007). (3) Aspects of equitable sharing of benefits arising from the use of genetic resources (cf. Nagoya Protocol 2011; Mishra 2005) may add to (4) the complexity of batch-to-batch quality assurance of the MP product, if not one active component, but rather a combination, is shown to be responsible for improved efficacy and neuroprotection as well as curbed side effects. The stability of all active components needs to be warranted on the one hand, whilst on the other, the quality of MP seeds will vary from harvest to harvest including climate and soil conditions, and depend on the accessions cultivated (Kroes 2014). (5) If these are not already enough imponderabilia, a new MP-based PD medication would need to show its benefits over pre-existing synthetic LD preparations, so as to be acceptable for reimbursement by health insurers (Gerber-Grote and Windeler 2014).

## 3.2  HERBAL PHARMA COMPANIES

In Germany, pharmaceutical companies specialising in the production and marketing of herbal medicinal products have historically grown from small, family-owned enterprises such as small-scale manufacture in retail chemist shops, going back to a time before the rise of synthetic drugs in the last century (Kraft 2001). In 1976, there were around 78,000 herbal medicinal products available on the German market (Beer et al. 2013) while the second half of the last century has seen a rising public demand for herbal preparations as alternatives to synthetic medication in the Global North (Sewell and Rafieian-Kopaei 2014; Harrison, 1998), with Germany and France leading in Europe regarding total over-the-counter (OTC) sales (De Smet 2005; Steinhoff 1993). Nonetheless, amongst herbal medicines in Europe, there has probably never been any blockbuster, as there have been with synthetic molecules that bring significant profits.

Since 2004, new social legislation in Germany has severely restricted the reimbursement of OTC medication by public health insurance in order to cut public spending in health care. Continuous harmonisation of drug legislation and marketing authorisation requirements, which demand proof of safety and efficacy within the EU on the other hand, have made clinical investments towards the registration or authorisation of herbal medicines too risky and expensive for some products, with innovations becoming rare (Armbrüster 2016). By the end of 2015, only 1,613 single herbal preparations and 201 combination products were left on the national pharmaceutical market (Beer et al. 2013). Armbrüster and Roth-Ehrang (2021) analysed the legal situation in Germany and advocated for various improvements to make herbal medicines more attractive to investors, thus safeguarding therapeutic diversity in future.

Turning back to PD, LD treatment, whether herbal or not, requires medical supervision, making Rx status mandatory. Prescription drugs, however, is a market that herbal pharma companies do not customarily target. In 2015, more than 98% of all herbal medicines sold in German chemist shops were prescription free (Armbrüster 2016). An authorised MP preparation for treating PD would thus require additional marketing investments addressing prescribers instead, and not patients. One such successful example of a herbal medicinal product is Laif® 900, a highly dosed dry extract of St. John´s wort (*Hypericum perforatum*) authorised for the treatment of moderate depression. It received Rx status once it had proven its efficacy for the stated indication and is being reimbursed by health insurers (Uebelhack et al. 2004). Laif 900 is currently marketed by Bayer Vital GmbH, part of global player, Bayer AG, which acquired a typical German family-owned herbal pharma company, Steigerwald Arzneimittelwerk GmbH, in 2013, to help widen Bayer´s OTC portfolio

(Communications Bayer AG 2013). Acquisitions of this kind can make one preparation a herbal 'cash cow', whilst the loss of other, less profitable herbal preparations may in turn be 'accepted with approval'. The resulting (strategic) 'death' of a pre-existing diverse herbal product portfolio usually happens 'silently' and to the detriment of individualised patient care.

## 3.3  CURRENT AND FUTURE TRENDS

Where does all of this take us? In present times, on the one hand, the WHO publishes an updated list of essential medicines defined as "those [medicines] that satisfy the priority health care needs of a population..." and includes evidence regarding efficacy, safety and comparative cost-effectiveness (WHO 2021). The model list contains medicines to treat conditions posing the greatest threat to public health; it contains more than 2,000 unique medicines for 137 member countries (Persaud et al. 2019).

The list contains some plant-derived molecules, such as atropine (e.g. *Atropa belladonna*), caffeine (e.g. *Coffea arabica, Thea sinensis, Ilex paraguariensis*), codeine and morphine (e.g. *Papaver somniferum*), paclitaxel (e.g. *Taxus brevifolia* or via a precursor from *Taxus baccata* or biotechnologically from *Taxus* cell cultures), pilocarpine (e.g. *Pilocarpus jaborandi*), and quinine (*Cinchona* species). Herbal preparations are the exception, such as podophyllum resin (from e.g. *Podophyllum peltatum*) or senna (from *Senna* species) as sennosides or 'traditional dosage forms'. Most of the essential medicines are mono-substances of semi-synthetic, synthetic, or biotechnological origin, sometimes in combination with up to three other active substances, not forgetting levodopa (+ carbidopa or benserazide). They can be considered mainstream (WHO 2021a). On the other hand, the 62nd World Health Assembly in its agenda item 12.4, 'Traditional Medicines', included TM to strengthen health systems whilst defining TM as covering a wide variety of therapies and practices, to include herbal medicines, thus being implied as complementary to WHO´s essential medicines list. The importance of plants and plant extracts and their medicinal value as part of TM is undisputed, with TM still being a mainstay of healthcare in some regions of the world (WHO 2009, 2014, 2021a). What changes are needed to get safe, efficacious, and affordable herbal medicinal innovations to the market swiftly and make them, if not deemed essential, at least to be considered highly desirable? How could such herbal medicine from MP become a fully integrated and affordable complementary therapy for the treatment of PD in both affluent and low-income countries? An important step forward on policy level is the initiative of the WHO in setting up an expert committee for an international herbal pharmacopoeia. Its aim is to create a harmonised compilation of entries from national (herbal) pharmacopoeias that ensure the quality and safety of herbs and herbal (medicinal) products in a global market (WHO 2020). In our view, this pharmacopoeia must also comprise GACP, including best agroecological practices to ensure freedom from pesticides within land regeneration processes that re-establish or maintain environmental integrity.

### 3.3.1  Pipeline of herbal treatments against PD

In their recent review, McFarthing et al. (2020) analysed trial data from the ClinicalTrials.gov database on 145 registered and, at the time, ongoing clinical trials (21 Jan 2020) targeting PD, either in a symptomatic (ST) or disease-modifying (DMT) manner. They assigned the trials to one of 14 groups and included one group called 'botanicals', referring to "agents derived from herbal extracts where the mechanism of action was unknown or unclear" and found in both the ST and the DMT category. Three out of the total of four studies were in clinical trial phase II (WIN-1001X, DA-9805, and SQJZ herbal mix), one in phase III (Lingzhi), with 'SQJZ herbal mix' belonging to the ST category, the others to the DMT category. The authors refer to the botanicals as follows: "[...] they are mixtures with a number of potential active agents and have attracted a lot of attention in the Parkinson´s community as a source of current and future medicines". A 2021 update on PD-related clinical trials by Prasad and Hung (2021), with a selection of 293 registered clinical trials as from

**TABLE 4**
**Current herbal anti-Parkinson clinical trials registered on ClinicalTrials.gov, a registry run by the US National Library of Medicine at the National Institutes of Health, considered the world´s largest clinical database. Summaries have been drawn from McFarthing et al. (2020), Prasad and Hung (2021) or directly from ClinicalTrials.gov (accessed 03/11/2021) and the 'herbal' dimension includes fungi.**

| Botanicals'ID | Trial Phase | Treatment Category | Composition | ClinicalTrials. gov ID | Sponsor |
|---|---|---|---|---|---|
| WIN-1001X | II | DMT: anti-oxidant, improved autophagy and reduced neuro-inflammation | Herbal extract from *Angelica tenuissima Nakai*, *Dimocarpus longan (L.)*, and *Polygala tenuifolia* | NCT04220762 | Medihelpline Co., Ltd. (CRO), Seoul, Republic of Korea |
| DA-9805 | II | DMT: anti-oxidant, anti-inflammatory | Three main herbal materials, not specified | NCT03189563 | Dong-A ST Co., Ltd., pharmaceutical manufacturer, Seoul, Republic of Korea |
| SQJZ herbal mix | II | ST: anti-oxidant | Herbal mixture of 'Chinese herbs' containing extracts of *Rehmannia glutinosa Libosch, Astragalus membranaceus (Fisch.) Bunge* etc. | | Dongzhimen Hospital, Beijing, China |
| Linghzhi (fungus!) | III | DMT: neuro-protective | *Ganoderma lucidum* extract | NCT03594656 | Xuanwu Hospital, Beijing, China |
| Hypoestoxide | I/II | DMT: anti-oxidant, anti-inflammatory | Diterpene from *Hypoestes rosea*, dry powder | NCT04858074 | Prof. Adesola Ogunniyi, University of Ibadan, Nigeria |

2008 till 16 June 2021 from the same database, shows two active studies in relation to antioxidant botanical-based medication hypoestoxide (phase I/II) and the abovementioned WIN-1001X. Studies on botanicals from both publications are summarised in Table 4. It is striking that none of these studies seems to stem from Europe, but rather from countries with a long tradition of ethnomedicine (TM), such as China, Korea, or Nigeria.

The most recent addition to the ClinicalTrials.gov database regarding PD and herbal treatments is a clinical study sponsored by Hong Kong Baptist University (Identifyer: NCT05001217). Its aim is to compare conventional medication regimes, including LD, with conventional treatment *plus* Chinese herbal treatments – in four patient subgroups according to a Chinese medicinal pattern: (1) *Huanglian Wendan* decoction ('phlegm-heat stirring wind subgroup'); (2) *Jin Gui Shen Qi* pill ('spleen- and kidney-*Yang* subgroup'); (3) *Qi Ju Dihuang* pill plus *Zhen Gan Xi Feng* decoction ('deficiency of liver- and kidney-*Yin* subgroup'); and (4) *Bu Yang Huan Wu* decoction combined with *Chang Yuan Wendan* decoction ('*Qi* deficiency and stasis of blood subgroup'). Herbal drugs follow the instructions of the China pharmacopoeia. The trial start date was 1/12/2021 and the trial end date is estimated for 31/01/2023 for this single-blind (outcomes assessor), randomised study that exemplifies a combination of conventional medicine with a patient-centred, individualised, herbal add-on treatment.

The scope for treating neurologic diseases seems to offer a lot of potential for research (a potential that currently does not seem to be recognised by 'Big Pharma') as expounded, for example, by Balkrishna and Misra (2017). Calling it 'The Herbal Hope', they had compiled 31 *Ayurvedic* medicinal herbs with traditional therapeutic answers to eight common brain disorders, including PD. The authors argue that, as opposed to the 'allopathic system', *Ayurveda* is in principle more focused on treating the cause of illness rather than the symptoms, whilst admitting that comparative studies are still needed to show this. In the wake of the CoViD-19 pandemic crisis an editorial in the *Journal of Ayurveda and Integrative Medicine*, Vaidya et al. (2020) emphasised the importance of integrated healthcare (IHC) within a synergic vision. It must combine modern allopathy with the medicine from traditional healers. The author lists 14 of the most important *Ayurvedic* drugs to be implemented urgently – with MP in PD being No. 1 on this list.

We conclude that future research should bring MP further, in whatever form, as a (co)treatment – a 'medicinal supplement' and not a food supplement – for PD patients around the globe, with the potential to help slow down progression and treat PD (cf. van de Vijver, Foreword One to this book).

## 3.4   WHAT COULD BE THE FUTURE ROLE OF METABOLOMICS?

Being focused on the European context, as is in this chapter, it seems worthwhile looking at what the European Medicines Agency (EMA) and its Committee on Herbal Medicinal Products (HMPC) have to say regarding the challenges of quantitative and qualitative analysis of (traditional) herbal medicinal products. Such challenges must be resolved in order to gain marketing authorisation. In HMPC´s reflection paper on markers used for quantitative and qualitative analysis, active substances in *herbal* medicinal products are defined as consisting of:

> "…complex mixtures of phytochemical constituents. […] further complicated when two or more herbal substances and/or herbal preparations are combined in a herbal medicinal product. A limited number of herbal substances and herbal preparations possess constituents which are generally accepted to contribute substantially to their therapeutic activity. These are defined as '*constituents with known therapeutic activity*'. However, for the majority of herbal substances and herbal preparations, the constituents or groups of constituents responsible for the therapeutic activity are not known. In some cases, certain constituents or groups of constituents may be generally accepted to contribute to the therapeutic activity but are not responsible for the full therapeutic effect. Such constituents or groups of constituents are useful for control purposes and are defined as '*active markers*'. *Analytical markers* are constituents or groups of constituents that serve solely for analytical purposes […]" (HMPC 2008).

For constituents with known therapeutic activity, this means that they need to be determined quantitatively. For MP bean preparations – as described above – their favourable efficacy and tolerance profile is strongly believed to be related to additional substances other than LD. Exactly which ones they are and what quantities are required for such beneficial profile remains largely speculative.

In this context, new analytical methods are hoped to "create a type of 'holistic' fingerprint", but HMPC's related concept paper raises doubts over their usefulness in the quality control of herbal medicinal products and regards them as 'optional'. One of the examples given is hyphenated techniques (HPLC-MS or LC-NMR)[11], which couple two different analytical procedures, a separating technique and a detection technique (HMPC 2018). Hyphenated techniques, however, are said to have improved dimensional changes in a remarkable manner in natural product analysis with a 'systems approach' to biochemical profiling using metabolomics within the so-called 'omic techniques' (Wilson et al. 2021, Weckwerth 2003). Metabolomics allow a non-targeted

---

11   HPLC-MS: High Performance Liquid Chromatography coupled with Mass Spectrometry; LC-NMR: Liquid Chromatography coupled with Nuclear Molecular Resonance.

identification and quantitation of small-molecular-weight metabolites and the determination of relationships among components of plant systems, thus supporting researchers by providing a fuller understanding of functional relationships of multi-component systems or 'biochemical networks'. Such 'metabolic fingerprinting' may lead to a better understanding of how exactly MP may have advantages over mono-substances like synthetic LD.

Mono-substances, be they for example pharmaceuticals – or pesticides – from synthetic organic chemistry and regardless of whether or not they produce any harmful effect, are *per se* xenobiotic to living systems, plants, and animals alike, including humans. Our evolutionary history of life on earth, from unicellular to multicellular life forms, is a historically complex mixture of components where no such thing as 'highly purified chemical entities' existed; they are alien to our evolutionary metabolisms. Animals have surely co-evolved with the plant kingdom, having had to deal with plant defence mechanisms all along. Today, all plants together are estimated to host more than 200,000 different metabolites (Mawalagedera et al. 2019; Weckwerth 2003). Although speculative, MP seeds may lend themselves particularly well to a starting point of a metabolomic approach of *Mucuna* species, not only because of the wondrously numerous medicinal aspects of the plant as such, but also because of seeds representing a metabolically 'frozen' or dormant state that is to kick-start plant metabolism once the seed starts germinating. With increased analytical reproducibility, we could gain insights regarding accession-to-accession and harvest-to-harvest variations as well as developing a better understanding of a metabolite-efficacy profile. Additionally, metabolomics may elucidate possible benefits of new accessions of MP or newly identified *Mucuna* species and their seeds; for example, the seeds of '*Mucuna sanjappae* Aitawade et Yadav' with a remarkable LD content allegedly in excess of 7% (Patil et al. 2015). The usefulness of metabolomic approaches is further supported by a recent study on various extracts from neuroactive medicinal herbs with sedative and anxiolytic effect. A correlation between comprehensive chemical fingerprints and their bioactivity was investigated and found to have commonalities for *Hypericum perforatum*, *Melissa officinalis*, *Passiflora incarnata*, and *Valeriana officinalis* (Gonulalan et al. 2020). Certainly, and more than ever, the future of herbal medicine requires a systems biology approach that looks at the complex interactions of components and their synergistic or antagonistic effects, as argued in a review including relevant experimental approaches (Williamson, 2001).

## 4   CONCLUSIONS

The CoViD-19 pandemic has shown the world to what astonishing degree pharmaceutical development can be speeded up, in terms of time to market, by financial interests and obviously by pressing medical need. Where profits are expected to soar, while multiple patents secure the investments sufficiently long term, the underlying technology of one of the first vaccines had been developed almost entirely through public and philanthropic funding (The Lancet 2020). Herbal and traditional (ethno)medicine, so far, can only dream of such copious financial support (Laird 2013). Large investments are much more likely to go into gaining IPR for so-called biologicals – such as vaccines or antibodies – or 'advanced therapies', such as medicines for gene therapy, somatic cell therapy, or tissue-engineered medicines. They fit the paradigms of high-tech IPR, so-called evidence-based medicine (EBM), and the imperative of statistical significance much better. For successful business models this is primordial, not least because mostly, health insurers and national healthcare systems, for reasons of their own, have a tendency to reimburse nothing less than EBM treatments – with evidence meaning 'statistical significance' regarding end points within a defined clinical trial population when the aim should *also and not least* be "improved quality of life for the world´s (over 6 million) different individual Parkinson cases" (Bloem et al. 2021).

The current R&D mechanisms, however, tend to lend themselves to grant first and foremost corporate ROI and speculation within global financial markets (Pandharinath 2011). In their opinion paper, Jureidini and McHenry (2022) argue that EBM has created an illusion from "[evidence] *corrupted by corporate interests, failed regulation, and commercialisation of academia*" (cf. Jureidini and McHenry 2020). We believe, as such, that the underlying mindset will hardly support UN Sustainable Development Goals (SDG), especially SDG 3 'Good Health and Wellbeing' and SDG 10 'Reduced Inequalities' "*[ensuring] healthy lives and [promoting] well-being for all at all ages*" (UN 2021; WHO 2015). Current inequalities in health and wellbeing between the Global North and the Global South are largely related to colonial mindsets from the past, maintained and replicated in our current R&D systems and the protection of knowledge, technologies, and markets. This mindset is very much blocking R&D and business development for a sound herbal medicine sector in line with and supportive to SDGs 3 and 10. For this to happen, local and international R&D and market players in the herbal medicine sector are to collaborate on eye level. The negative impact of the current system shows the dramatic decrease of available herbal medicinal preparations in a country like Germany, once referred to as the 'Pharmacy of the World', in the limited freedom of choice for patients as well as in the current inadequacy of public health systems in low- and middle-income countries.

What is overlooked greatly by those promoting herbal medicine as a crucial healthcare strategy for low- and middle-income countries is the availability of the right quality of herbal material, collected from the wild in a sustainable manner or produced with appropriate inputs and training for domestication and cultivation. Only by creating short and transparent supply chains can the quality and safety of the herbal material be guaranteed (see also Chapter 9 of this book). In addition, the provision of the right quality planting material, appropriate and low-cost methods to analyse the spectrum and amount of important secondary metabolites as well as identifying potential toxic substances or contaminants has to be secured. Ecosystem degradation and biodiversity depletion is already undermining wild sources of medicinal plants. At the same time, the majority of governments are not supporting the domestication nor promoting the sustainable cultivation of the same. In addition, the capacity to process and produce herbal medicinal products according to internationally approved standards is generally lacking in the countries of origin of the herbal raw material. This increases the loss of quality and favours adulteration.

Therefore, we advocate for a more patient- and nature-centred approach in herbal medicine R&D that should include supportive policies backed by public funding and investments, whilst collaborating with pharmaceutical industries where fruitful. The idea is to integrate herbal medicine within allopathic treatment approaches as a *medicinal* supplement and not as a food supplement.

Delivery bottlenecks of allopathic, synthetic medicines in Germany, due to repeated 'healthcare reforms', with spending cuts by the insurance system have been described by Weidenauer (2020). The author compares the detrimental effects on the medicines market with the López effect[12] in the automotive industries. Weidenauer suggests a new model of healthcare based on a not-for-profit foundation that acts in the interest of the public in general and patients in particular. This could make use of the local small-scale production infrastructure of chemist shops that is commonplace not only in Germany (*formula magistralis, formula officinalis*) and would benefit from strengthening the pharmacist's role in individual patient care to start with. Such an R&D model should run under the umbrella of a trans-national organisation like WHO, seeking to implement R&D models in different countries, suitable to their specific cultural, structural, and economic realities. It should include open-source production protocols and be linked to generic marketing authorisation protocols, including proof measures for environmental protection to avoid the ruthless exploitation of resources.

## 4.1 MUCUNA PRURIENS AND MEDICINAL SOVEREIGNTY

We believe that this chapter is emblematic of an emergent field of research in a quest for what we call 'medicinal sovereignty', where peoples and populations of the world are granted the human right of

---

12   López effect: synonymous with cheap and often faulty components related to López de Arriortúa, a former executive in the automotive industries and his negotiation methods with suppliers, considered to have been ruthless (Bergmann 1998).

easy access to quality healthcare that incorporates traditional, regional, ethnobotanical medicine, and herbal medicinal products whose ingredients are sourced regeneratively, whilst taking into account all pertinent dimensions of sustainability. The idea of medicinal sovereignty has been inspired by the concept of 'food sovereignty' (www.foodfirst.org) and could translate to the right of peoples to (1) define their traditional herbal medicines by (2) promoting regulatory frameworks supporting marketing authorisation of such medicines in line with (3) a quality that assures therapeutic benefit and safety whilst (4) being sufficiently pragmatic and flexible to accommodate the idiosyncrasies of a specific health tradition (*Ayurveda*, TCM, etc.) and finally taking into account (5) aspects of a holistic sustainability and regeneration in agriculture and forestry comprising ecological, economic, and social integrity. This is in fact already being piloted by *Solidaridad* (solidaridadnetwork.org) in India through a collaborative effort of government, science, civil society, businesses, and farmer organisations (cf. van de Vijver, 2022).

For that matter, MP has not only been an *Ayurvedic* medicinal plant, but it has also been food for humans, fodder for animals, and a functional crop in agroecological systems. Ironically, amongst other environmental and genetic factors, the cause of PD has been linked to the ever-increasing exposure to and the chronic uptake of pesticides used by workers in conventional agriculture and passed on mainly through fruits, vegetables, and processed food or through private use of pesticides at home (cf. Epilogue One of this book, Leu and Shiva 2014; Keikotlhaile 2010; Dick 2006; Stephenson 2000). The link between occupational pesticide use and the cause of PD has been studied in various countries (Narayan et al. 2017). Furthermore, PD is now being recognised as an occupational disease in France, a country claimed to use 80,000 tonnes of pesticides per year and thus being the third largest pesticide user in the world (Giorgio 2018).

All of the above contributes to our plea for holistic research approaches as well as inclusive and regenerative business models. We would expect them to take us beyond concepts of reductionist singular components and towards increased knowledge of effective component systems, intrinsically more in tune with human physiology. With this in mind, we find it pertinent to cite Schwabl and van der Valk (2019) who challenge the notion of 'active substances' in biomedicines from Tibetan medical formulas altogether. They describe a flexibility in the herbal ingredients used as long as functional qualities are maintained. In Tibetan Traditional Medicine (TTM), medicinal plants from different plant species may be used interchangeably for the same formula, depending on their availability on local markets and as long as the Tibetan formula exhibits a comparable "[...] signature of action, especially when combined into multi-target 'network medicines' which mirror the complexity of chronic diseases". We believe it is important to get a better in-depth understanding of this functional complexity and of how it supports health and healing processes. Eventually, this may not only support TM and traditional therapies but also inform synthetic chemical ('allopathic'/conventional) pharmacology and overall clinical research. From patient experiences and research on MP it may be important for scientists to say "I don´t know…why MP seems to provide additional benefits to yet unmet medical needs", calling for increased and alternative, more holistic research approaches. For a final outlook, Epilogue Two in this book gives the reader a glimpse of the iconography and the idiosyncrasies of Tibetan (Buddhist) medicine.

## 5  ACKNOWLEDGEMENTS

We are most grateful to the patients and their caregivers who came forward and allowed publication of testimonials on their experience with *Mucuna pruriens* (herbal) preparations against PD. Without their trustful communication, this chapter would not have been written and indeed, this entire book project would not have received the impetus required to give birth to "Medicinal Agroecology". Our gratitude goes to community pharmacist Jochen Bischoff († 11.11.2021) who initiated and supported the import of HP200/Zandopa within his business activity over many years. Thanks to his open-hearted humaneness, MP found the grounds to 'grow' for patients the way it did outside of tropical climates. We are truly sorry that he did not live to see this chapter nor the finished book. We are also very grateful to researchers who willingly gave support with pertinent literature and editorial improvements: Carlos Pastor, agronomist, University of Valencia, Valencia, Spain; Doreen Ramogola-Masire, Professor O&G, Faculty of Medicine, University of Botswana, Botswana.

# 6  REFERENCES

AECOSAN. 2016. Review of *Report of the Scientific Committee of the Spanish Agency for Consumer Affairs, Food Safety and Nutrition (AECOSAN) on the Risk of the Use of Seeds of Mucuna Pruriens in Craft Products*. Agencia Española de Consumo, Seguridad Alimentaria y Nutrición. https://www.mscbs.gob.es/consumo/vigilanciaMercado/organosAsesores/docs/rcc24_07mucuna.pdf.

Aldred, J., and J.G. Nutt. 2010. "Levodopa." *Encyclopedia of Movement Disorders*, 132–37. https://doi.org/10.1016/b978-0-12-374105-9.00340-3.

Amrein, Roman. 2004. "The History of Madopar." Essay. In *Focus on Parkinson´s Disease: L-Dopa 30 Years on and Still the Gold Standard* 16, Supplement A, 16: A7–A12. Proceedings of the Madopar 30th Anniversary Symposium.

Anadón, Arturo, Irma Ares, María-Rosa Martínez-Larrañaga, and María-Aránzazu Martínez. 2021. "Evaluation and regulation of food supplements: European perspective." *Nutraceuticals*, 1241–71. https://doi.org/10.1016/b978-0-12-821038-3.00073-2.

Anonymised. 1997. "Bericht und Krankheitsverlauf." *Manuscript*: 1983 words, 10,989 characters. Language: German. On file.

Anonymised. 2010. "Fortsetzungsbericht über den Verlauf der Behandlung mit HP-200 seit März 1997." *Manuscript*: 889 words, 4,810 characters. Language: German. On file.

Armbrüster, Nicole. 2016. "Phytopharmaka im gesundheitspolitischen und regulatorischen Umfeld." *Journal of Medicinal and Spice Plants* 21 (2): 80–82.

Armbrüster, Nicole, and René Roth-Ehrang. 2021. "Fit für die Zunkunft! Phytopharmaka müssen attraktiver werden." *Zeitschrift für Phytotherapie* 42: 95–99.

Armstrong, Melissa J., and Michael S. Okun. 2020. "Diagnosis and Treatment of Parkinson Disease." *JAMA* 323 (6): 548. https://doi.org/10.1001/jama.2019.22360.

Ashidi, Joseph Senu, Folarin Ojo Owagboriaye, Funmilola Balikis Yaya, Deborah Eyinjuoluwa Payne, Olubukola Ireti Lawal, and Stephen Olugbemiga Owa. 2019. "Assessment of Reproductive Function in Male Albino Rat Fed Dietary Meal Supplemented with Mucuna Pruriens Seed Powder." *Heliyon* 5 (10): e02716. https://doi.org/10.1016/j.heliyon.2019.e02716.

Balkrishna A., and LN Misra. 2017. "Ayurvedic Plants in Brain Disorders: The Herbal Hope." *Journal of Traditional Medicine & Clinical Naturopathy* 06 (02). https://doi.org/10.4172/2573-4555.1000221.

Beer, André-Michael, Heinz Schilcher, and Dieter Loew. 2013. "Phytotherapie in Not? Die derzeitige Lage in Deutschland." *Deutsche Apotheker Zeitung* 153 (34): 54–59.

Beggerow, Elisa, Maximilian Kuhn, and Antje Haas. 2019. "AMNOG – und dann? Arzneimittel zwischen Erstattungs- und Festbetrag." *GKV 90 Prozent* 13: 1–14. https://www.gkv-90prozent.de/bilder/ausgabe_13/tiefer-geblickt_arzneimittel-zwischen-erstattungs-und-festbetrag.pdf.

Behari, M., Achal K. Srivastava, and R.M. Pandey. 2005. "Quality of Life in Patients with Parkinson's Disease." *Parkinsonism & Related Disorders* 11 (4): 221–26. https://doi.org/10.1016/j.parkreldis.2004.12.005.

Berg, Daniela, Moussa B. H. Youdim, and Peter Riederer. 2004. "Redox Imbalance." *Cell and Tissue Research* 318 (1): 201–13. https://doi.org/10.1007/s00441-004-0976-5.

Bergmann, Rainer. 1998. "Der Fall Lopez." In *Organisation: Ressourcenorientierte Unternehmensgestaltung*, edited by Heike Nolte, 228–56. München: Oldenbourg. https://www.researchgate.net/publication/303676616_Der_Fall_Lopez.

Blascko, H. 1939. "The specific action of L-dopa decarboxylase." *J. Physiol.*, 96 (50): 50–51.

Bloem, Bastiaan R, Jorrit Hoff, and Ray Dorsey. 2021. *De Parkinsonpandemie: Een Recept Voor Actie*. Koog Aan De Zaan: Poiesz Uitgevers.

Blonder, Lee Xenakis. 2018. "Historical and Cross-Cultural Perspectives on Parkinson's Disease." *Journal of Complementary and Integrative Medicine* 15 (3). https://doi.org/10.1515/jcim-2016-0065.

Broadbent, J. L. 1953. "Observations on itching produced by cowhage, and on the part played by histamine as a mediator of the itch sensation." *British Journal of Pharmacology and Chemotherapy* 8 (3): 263–70. https://doi.org/10.1111/j.1476-5381.1953.tb00792.x.

Buckles, Daniel. 1995. "Velvetbean: A 'New' Plant with a History." *Economic Botany* 49 (1): 13–25. https://doi.org/10.1007/bf02862271.

Buckles, Daniel, B. Triomphe, and G. Sain. 1998. *Cover Crops in Hillside Agriculture Farmer Innovation with "Mucuna."* México, DF. International Maize and Wheat Improvement Center – CIMMYT and Ottawa International Development Research Centre – IDRC.

BVL. n.d. www.bvl.bund.de. https://www.bvl.bund.de/DE/Arbeitsbereiche/01_Lebensmittel/03_Verbraucher/04_NEM/01_NEM_Arzneimittel/NEM_Arzneimittel_node.html. Accessed October 10, 2021.

Cannon, Joseph G. 1985. "Dopamine Agonists: Structure-Activity Relationships." In *Progress in Drug Research*, edited by E. Jucker. Basel: Birkhäuser Verlag.

Canter, G. J., R. De La Torre, and M. Mier. 1961. "A Method for Evaluating Disability in Patients with Parkinson's Disease." *The Journal of Nervous and Mental Disease* 133: 143–47. https://doi.org/10.1097/00005053-196108000-00010.

Cassani, Erica, Roberto Cilia, Janeth Laguna, Michela Barichella, Manuela Contin, Emanuele Cereda, Ioannis U. Isaias, et al. 2016. "Mucuna Pruriens for Parkinson's Disease: Low-Cost Preparation Method, Laboratory Measures and Pharmacokinetics Profile." *Journal of the Neurological Sciences* 365 (June): 175–80. https://doi.org/10.1016/j.jns.2016.04.001.

Cersosimo, M, and F Micheli. 2007. "Antiglutamatergic Drugs in the Treatment of Parkinson's Disease." *Parkinson's Disease and Related Disorders, Part II*, 127–36. https://doi.org/10.1016/s0072-9752(07)84036-x.

Chade, A. R., M. Kasten, and C. M. Tanner. 2006. "Nongenetic Causes of Parkinson's Disease." *Parkinson's Disease and Related Disorders*, 147–51. https://doi.org/10.1007/978-3-211-45295-0_23.

Cilia, Roberto, Albert Akpalu, Momodou Cham, Alba Bonetti, Marianna Amboni, Elisa Faceli, and Gianni Pezzoli. 2011. "Parkinson's Disease in Sub-Saharan Africa: Step-By-Step into the Challenge." *Neurodegenerative Disease Management* 1 (3): 193–202. https://doi.org/10.2217/nmt.11.28.

Cilia, Roberto, Janeth Laguna, Erica Cassani, Emanuele Cereda, Nicolò G. Pozzi, Ioannis U. Isaias, Manuela Contin, Michela Barichella, and Gianni Pezzoli. 2017. "Mucuna Pruriens in Parkinson Disease." *Neurology* 89 (5): 432–38. https://doi.org/10.1212/wnl.0000000000004175.

Cilia, Roberto, Janeth Laguna, Erica Cassani, Emanuele Cereda, Benedetta Raspini, Michela Barichella, and Gianni Pezzoli. 2018. "Daily Intake of Mucuna Pruriens in Advanced Parkinson's Disease: A 16-Week, Noninferiority, Randomized, Crossover, Pilot Study." *Parkinsonism & Related Disorders* 49 (April): 60–66. https://doi.org/10.1016/j.parkreldis.2018.01.014.

Cilia, Roberto, Francesca Del Sorbo, Fred S. Sarfo, Momodou Cham, Albert Akpalu, Serena Caronni, Erica Cassani, et al. 2018b. *Long-Term Intake of Mucuna Pruriens in Drug-Naïve Parkinson's Disease in Sub-Saharan Africa: A Multicentre, Non-Inferiority, Randomised, Controlled Clinical Trial.* Accessed September 27, 2021. https://www.bm-association.it/sites/default/files/nutrizione-parologie/mds_2018_completo.pdf.

Cilia, Roberto. Letter to Immo Fiebrig. 2021. Review of *Mucuna Pruriens, Clinical Study Progress*. E-mail, October 5, 2021.

Clausen, Angela, Katrin von Nida, and Carolin Semmler. 2011. *Marktcheck: Internethandel mit Nahrungsergänzungsmitteln*. www.vzbv.de. Verbraucherzentrale Nordrhein-Westfalen e.V. https://www.vzbv.de/sites/default/files/downloads/Nahrungsergaenzungsmittel-marktcheck-vz-nrw-2011.pdf.

Colosimo, Carlo, Pablo Martínez-Martín, Giovanni Fabbrini, Robert A. Hauser, Marcelo Merello, Janis Miyasaki, Werner Poewe, et al. 2010. "Task Force Report on Scales to Assess Dyskinesia in Parkinson's Disease: Critique and Recommendations." *Movement Disorders* 25 (9): 1131–42. https://doi.org/10.1002/mds.23072.

Colzato, Lorenza S., Annelies M. de Haan, and Bernhard Hommel. 2014. "Food for Creativity: Tyrosine Promotes Deep Thinking." *Psychological Research* 79 (5): 709–14. https://doi.org/10.1007/s00426-014-0610-4.

Communications, Bayer AG. 2013. "Bayer will Steigerwald Arzneimittelwerk GmbH übernehmen." Accessed 25/10/2021. https://www.media.bayer.de/baynews/baynews.nsf/id/7A66EB690DDFFB5EC1257B6D00311CEF?open&ref=irrefndcd.

Connolly, Barbara S., and Anthony E. Lang. 2014. "Pharmacological treatment of Parkinson disease." *JAMA* 311 (16): 1670. https://doi.org/10.1001/jama.2014.3654.

Critchley M. 1955. *James Parkinson (1755-1824): A Bicentenary Volume of Papers Dealing with Parkinsons Disease, Incorporating the Original Essay on Shaking Palsy.* Macmillan.

Daue, Raphael. 2017. "EudraLex – Volume 4 – Good Manufacturing Practice (GMP) Guidelines." Gesundheitswesen – European Commission. December 15, 2017. https://ec.europa.eu/health/docume nts/eudralex/vol-4_de. Accessed 10 October 2021.

Davie, C. A. 2008. "A Review of Parkinson's Disease." *British Medical Bulletin* 86 (1): 109–27https://doi.org/ 10.1093/bmb/ldn013.

De Smet, Peter A.G.M. 2005. "Herbal Medicine in Europe — Relaxing Regulatory Standards." *New England Journal of Medicine* 352 (12): 1176–78. https://doi.org/10.1056/nejmp048083.

Deutscher Bundestag. 1976. "Gesetz über den Verkehr mit Arzneimitteln (Arzneimittelgesetz – AMG)." *Jahrbuch Für Wissenschaft Und Ethik* 10 (1). https://doi.org/10.1515/9783110182521.433. https://www. gesetze-im-internet.de/amg_1976/AMG.pdf. Accessed 27 July 2021.

Dexter, D T, C J Carter, F R Wells, F Javoy-Agid, Y Agid, A Lees, P Jenner, and C D Marsden. 1989. "Basal lipid peroxidation in *substantia nigra* is increased in Parkinson's disease." *Journal of Neurochemistry* 52 (2): 381–89. https://doi.org/10.1111/j.1471-4159.1989.tb09133.x.

Dick, F. D. 2006. "Parkinson's Disease and Pesticide Exposures."*British Medical Bulletin* 79–80 (1): 219–31. https://doi.org/10.1093/bmb/ldl018.

Dorsey, E. Ray, Todd Sherer, Michael S. Okun, and Bastiaan R. Bloem. 2018. "The Emerging Evidence of the Parkinson Pandemic." Edited by Patrik Brundin, J. William Langston, and Bastiaan R. Bloem. *Journal of Parkinson's Disease* 8 (s1): S3–8. https://doi.org/10.3233/jpd-181474.

Dutt, Udoy Chand and George King. 2018. *The Materia Medica of the Hindus.* Calcutta Thacker, Spink and Co.

EFSA. 2012. "Compendium of botanicals reported to contain naturally occurring substances of possible con- cern for human health when used in food and food supplements." *EFSA Journal* 10 (5): 2663. https:// doi.org/10.2903/j.efsa.2012.2663.

Eilittä, M, R Bressani, LB Carew LB, RJ Carsky, M Flores, R Gilbert, L Huyck, L St-Laurent, NJ Szabo. 2002. "Mucuna as a food and feed crop: an overview." In Flores, B, M Eilittä, R Myhrman, LB Carew, RJ Carsky (Eds). *Mucuna as a Food and Feed: Current Uses and the Way Forward.* Workshop held April 26-29, 2000 in Tegucigalpa, Honduras. CIDICCO, CIEPCA, and World Hunger Research Center. Tegucigalpa, Honduras. 18–46.

Eilittä, M., and R.J. Carsky. 2003. Review of *Efforts to Improve the Potential of Mucuna as a Food and Feed Crop: Background to the Workshop.* In *Increasing Mucuna´s Potential as a Food and Feed Crop. Proceedings of a Workshop Organized by KAKI and CIEPCA-IITA Mombasa, Kenya, September 23- 26, 2002,* edited by M. Eilittä, R. Muinga, J. Mureithi, C. Sandoval-Castro, and N. Szabo, 47–55. Mexico: Facultad de Medicina Veterinaria y Zootecnia Universidad Autónoma de Yucatan UADY. https://www.redalyc.org/articulo.oa?id=93911288002.

Eilittä, M., R. Miunga, J. Mureithi, C. Sandoval-Castro, and N. Szabo, eds. 2003. *Increasing Mucuna´s Potential as a Food and Feed Crop. Tropical and Subtropical Agroecosystems* 1 (2-3): 1–343. https://www.vete rinaria.uady.mx/publicaciones/journal/2002-2-3/indice-1(2-3).html. Special Issue (proceedings of a workshop).

Ezeagu, I.E., B. Maziya-Dixon and G. Tarawali. 2003. "Seed characteristics and nutrient and antinutrient composition of 12 Mucuna accessions from Nigeria." *Tropical and Subtropical Agroecosystems* 1, no. 2-3:129–139. https://www.redalyc.org/articulo.oa?id=93911288014.

Fahn, Stanley, and Gerald Cohen. 1992. "The oxidant stress hypothesis in Parkinson's Disease: evidence supporting it." *Annals of Neurology* 32 (6): 804–12. https://doi.org/10.1002/ana.410320616.

Fernstrom, John D., and Madelyn H. Fernstrom. 2007. "Tyrosine, Phenylalanine, and Catecholamine Synthesis and Function in the Brain." *The Journal of Nutrition* 137 (6): 1539S1547S. https://doi.org/10.1093/jn/ 137.6.1539s.

Fiege, Helmut, Heinz-Werner Voges, Toshikazu Hamamoto, Sumio Umemura, Tadao Iwata, Hisaya Miki, Yasuhiro Fujita, Hans-Josef Buysch, Dorothea Garbe, and Wilfried Paulus. 2000. "Phenol Derivatives." *Ullmann's Encyclopedia of Industrial Chemistry.* 2002 Wiley-VCH Verlag. https://doi.org/10.1002/ 14356007.a19_313.

Fothergill-Misbah, Natasha, Harshvadan Maroo, Momodou Cham, Gianni Pezzoli, Richard Walker, and Roberto Cilia. 2020. "Could Mucuna Pruriens Be the Answer to Parkinson's Disease Management in Sub-Saharan Africa and Other Low-Income Countries Worldwide?" *Parkinsonism & Related Disorders* 73 (April): 3–7. https://doi.org/10.1016/j.parkreldis.2020.03.002.

FSE. n. d. "Food Supplements Europe Guide to Good Manufacturing Practice for Manufacturers of Food Supplements." https://foodsupplementseurope.org/wp-content/themes/fse-theme/documents/publi cations-and-guidelines/good-manufacturing-practice-for-manufacturers-of-food-supplements.pdf. Accessed 10 October 2021.

GACP. 2018. "Good Agricultural and Collection Practice for Starting Materials of Herbal Origin – European Medicines Agency." European Medicines Agency. September 17, 2018. https://www.ema.europa.eu/en/ good-agricultural-collection-practice-starting-materials-herbal-origin.

Galanakis, Charis M. 2020. "The Food Systems in the Era of the Coronavirus (COVID-19) Pandemic Crisis." *Foods* 9 (4): 523. https://doi.org/10.3390/foods9040523.

GBD 2016 Neurology Collaborators. 2019. "Global, Regional, and National Burden of Neurological Disorders, 1990–2016: A Systematic Analysis for the Global Burden of Disease Study 2016." *The Lancet Neurology* 18 (5): 459–80. https://doi.org/10.1016/s1474-4422(18)30499-x.

Gerber-Grote, Andreas, and Jürgen Windeler. 2014. "What is the contribution of health economic evaluations to decision-making in health care? Experiences from 7 selected countries." *Zeitschrift für Evidenz, Fortbildung und Qualität im Gesundheitswesen* 108 (7): 358–59. https://doi.org/10.1016/ j.zefq.2014.08.018.

Giorgio, Marie-Thérèse. 2018. "Régime agricole: La maladie de Parkinson provoquée par les pesticides est désormais reconnue en maladie professionnelle" Atousante.com. 2018. https://www.atousante.com/act ualites/regime-agricole-maladie-parkinson-pesticides-maladie-professionnelle/. Accessed 06/01/2022.

Goetz, Christopher G., Stanley Fahn, Pablo Martinez-Martin, Werner Poewe, Cristina Sampaio, Glenn T. Stebbins, Matthew B. Stern, et al. 2007. "Movement Disorder Society-Sponsored Revision of the Unified Parkinson's Disease Rating Scale (MDS-UPDRS): Process, Format, and Clinimetric Testing Plan." *Movement Disorders* 22 (1): 41–47. https://doi.org/10.1002/mds.21198.

Gonulalan, Ekrem M., Emirhan Nemutlu, Omer Bayazeid, Engin Koçak, Funda N. Yalçın, and L. Omur Demirezer. 2020. "Metabolomics and proteomics profiles of some medicinal plants and correlation with BDNF activity." *Phytomedicine*, April, 152920. https://doi.org/10.1016/j.phymed.2019.152920.

Google. 2021. Review of *Mucuna Pruriens: Google Pictures.* October 6, 2021. https://www.google. com/search?q=Mucuna+pruriens&client=firefox-b-d&source=lnms&tbm=isch&sa=X&ved= 2ahUKEwjV0tzDtrXzAhWw_7sIHZ6WA1oQ_AUoA3oECAEQBQ&biw=1536&bih= 739&dpr=1.25.

Gourie-Devi, M, M G Ramu, and B S Venkataram. 1991. "Treatment of Parkinson's Disease in 'Ayurveda' (Ancient Indian System of Medicine): Discussion Paper." *Journal of the Royal Society of Medicine* 84 (8): 491–92. https://doi.org/10.1177/014107689108400814.

Guggenheim., M. 1913. "Dioxyphenylalanin, eine neue Aminosäure aus Vicia faba."*Hoppe-Seyler´s Zeitschrift für physiologische Chemie* 88 (4): 276–84. https://doi.org/10.1515/bchm2.1913.88.4.276.

Harrison, Pam. 1998. "Herbal medicine takes root in Germany." *Canadian Medical Association Journal* 158 (5): 637–39. https://www.cmaj.ca/content/cmaj/158/5/637.full.pdf.

HMPC. 2008. "Reflection paper on markers used for quantitative and qualitative analysis of herbal medicinal products and traditional herbal medicinal products." https://www.ema.europa.eu/en/documents/scienti fic-guideline/reflection-paper-markers-used-quantitative-qualitative-analysis-herbal-medicinal-products _en.pdf. Doc. Ref. EMEA/HMPC/253629/2007. Accessed 03/11/2021.

HMPC. 2018. "Concept paper on the development of a reflection paper on new analytical methods/technolo gies in the quality control of herbal medicinal products." https://www.ema.europa.eu/en/documents/sci entific-guideline/concept-paper-development-reflection-paper-new-analytical-methods/technologies- quality-control-herbal-medicinal-products_en.pdf. Doc. Ref. EMA/HMPC/541422/2017. Accessed 03/ 11/2021.

Hornykiewicz, Oleh. 2010. "A Brief History of Levodopa." *Journal of Neurology* 257 (Suppl 2): S249–52. https://doi.org/10.1007/s00415-010-5741-y.

HP-200 Study Group. 1995. "An Alternative Medicine Treatment for Parkinson's Disease: Results of a Multicenter Clinical Trial." *The Journal of Alternative and Complementary Medicine* 1 (3): 249–55. https://doi.org/10.1089/acm.1995.1.249.

Hughes, A J, S E Daniel, L Kilford, and A J Lees. 1992. "Accuracy of Clinical Diagnosis of Idiopathic Parkinson's Disease: A Clinico-Pathological Study of 100 Cases." *Journal of Neurology, Neurosurgery & Psychiatry* 55 (3): 181–84. https://doi.org/10.1136/jnnp.55.3.181.

Hussain, Ghazala, and Bala V. Manyam. 1997. "Mucuna Pruriens Proves More Effective than L-DOPA in Parkinson's Disease Animal Model." *Phytotherapy Research* 11 (6): 419–23. https://doi.org/3.0.co;2-q">10.1002/(sici)1099-1573(199709)11:6<419::aid-ptr120>3.0.co;2-q.

Hutchinson, John. 1964. *The Genera of Flowering Plants: Angiospermae. Vol. 1, Dicotyledones.* Oxford: Clarendon Press.

"HWG – Gesetz über die Werbung auf dem Gebiete des Heilwesens." n.d. Www.gesetze-Im-Internet.de. Accessed July 30, 2021. https://www.gesetze-im-internet.de/heilmwerbg/BJNR006049965.html.

ITTA Genebank. 1987. "Taxonomy – GRIN-Global Web v 1.10.6.1." Gringlobal.iita.org. Accessed August 17, 2021. https://gringlobal.iita.org/gringlobal/taxonomydetail.aspx?id=311952.

Jairoun, Ammar Abdulrahman, Moyad Shahwan, and Sa'ed H. Zyoud. 2020. "Heavy Metal Contamination of Dietary Supplements Products Available in the UAE Markets and the Associated Risk." *Scientific Reports* 10 (1). https://doi.org/10.1038/s41598-020-76000-w.

Jaiswal, Yogini S., and Leonard L. Williams. 2017. "A Glimpse of Ayurveda – the Forgotten History and Principles of Indian Traditional Medicine." *Journal of Traditional and Complementary Medicine* 7 (1): 50–53. https://doi.org/10.1016/j.jtcme.2016.02.002.

Jenkinson, Crispin, Ray Fitzpatrick, Viv Peto, Richard Greenhall, and Nigel Hyman. 1997. "The Parkinson's Disease Questionnaire (PDQ-39): Development and Validation of a Parkinson's Disease Summary Index Score." *Age and Ageing* 26 (5): 353–57. https://doi.org/10.1093/ageing/26.5.353.

Jonas, Wayne B. 2005. *Mosby's Dictionary of Complementary & Alternative Medicine.* St. Louis, MO. Elsevier Mosby.

Jureidini, Jon, and Leemon B. McHenry. 2020. *The Illusion of Evidence-Based Medicine: Exposing the Crisis of Credibility in Clinical Research.* 1st ed. Mile End, Australia: Wakefield Press.

Jureidini, Jon, and Leemon B. McHenry. 2022. "The illusion of evidence based medicine." *BMJ* 376 (March): o702. https://doi.org/10.1136/bmj.o702.

Kartal, Murat. 2007. "Intellectual Property Protection in the Natural Product Drug Discovery, Traditional Herbal Medicine and Herbal Medicinal Products." *Phytotherapy Research* 21 (2): 113–19. https://doi.org/10.1002/ptr.2036.

Kasture, Sanjay, Mahalaxmi Mohan, and Veena Kasture. 2013. "Mucuna Pruriens Seeds in Treatment of Parkinson's Disease: Pharmacological Review." *Oriental Pharmacy and Experimental Medicine* 13 (3): 165–74. https://doi.org/10.1007/s13596-013-0126-2.

Katzenschlager, R, A. Evans, A. Manson, P. Patsalos, N. Ratnaraj, H. Watt, L. Timmermann, R. van der Giessen, and A. Lees. 2004. "Mucuna Pruriens in Parkinson's Disease: A Double Blind Clinical and Pharmacological Study." *Journal of Neurology, Neurosurgery & Psychiatry* 75 (12): 1672–77. https://doi.org/10.1136/jnnp.2003.028761.

Katzner, Kenneth. 1996. *The Languages of the World.* London: Routledge.

Keikotlhaile, B.M., P. Spanoghe, and W. Steurbaut. 2010. "Effects of Food Processing on Pesticide Residues in Fruits and Vegetables: A Meta-Analysis Approach." *Food and Chemical Toxicology* 48 (1): 1–6. https://doi.org/10.1016/j.fct.2009.10.031.

Kogan, Michael, Matthew McGuire, and Jonathan Riley. 2019. "Deep Brain Stimulation for Parkinson Disease." *Neurosurgery Clinics of North America* 30 (2): 137–46. https://doi.org/10.1016/j.nec.2019.01.001.

Kohen, Ron, and Abraham Nyska. 2002. "Invited Review: Oxidation of Biological Systems: Oxidative Stress Phenomena, Antioxidants, Redox Reactions, and Methods for Their Quantification." *Toxicologic Pathology* 30 (6): 620–50. https://doi.org/10.1080/01926230290166724.

Kraft, Karin. 2001. "History of Herbal Medicinal Use in Germany with a Treatise on Present Day Practice." *Journal of Herbal Pharmacotherapy* 1 (2): 43–49. https://doi.org/10.1080/j157v01n02_05.

Kraljevic, Sandra, Peter J. Stambrook, and Kresimir Pavelic. 2004. "Accelerating Drug Discovery." *EMBO Reports* 5 (9): 837–42. https://doi.org/10.1038/sj.embor.7400236.

Kroes, Burt. 2014. "The Legal Framework Governing the Quality of (Traditional) Herbal Medicinal Products in the European Union." *European Journal of Integrative Medicine* 6 (6): 701. https://doi.org/10.1016/j.eujim.2014.09.039.

Lad, Vasant D. 2012. *Textbook of Ayurveda. Volume Three, General Principles of Management and Treatment.* Albuquerque, N.M.: Ayurvedic Press.

Laird, Sarah A. 2013. "Bioscience at a Crossroads: Access and Benefit Sharing in a Time of Scientific, Technological and Industry Change: The Pharmaceutical Industry." *Secretariat of the Convention on Biological Diversity*, 2013. www.cbd.int.

Lampariello, Lucia Raffaella, Alessio Cortelazzo, Roberto Guerranti, Claudia Sticozzi, and Giuseppe Valacchi. 2012. "The Magic Velvet Bean of Mucuna Pruriens." *Journal of Traditional and Complementary Medicine* 2 (4): 331–39. https://doi.org/10.1016/s2225-4110(16)30119-5.

Lang, Anthony E., and Alberto J. Espay. 2018. "Disease Modification in Parkinson's Disease: Current Approaches, Challenges, and Future Considerations." *Movement Disorders* 33 (5): 660–77. https://doi.org/10.1002/mds.27360.

Latté, Klaus Peter. 2008. "Mucuna Pruriens (L.) DC. Die Juckbohne." *Zeitschrift Für Phytotherapie* 29: 199–206.

Lau, Lonneke ML de, and Monique MB Breteler. 2006. "Epidemiology of Parkinson's Disease." *The Lancet Neurology* 5 (6): 525–35. https://doi.org/10.1016/s1474-4422(06)70471-9.

Leu, André and Vandana Shiva. 2014. *The Myths of Safe Pesticides*. Austin, T.X.: Acres U.S.A.

Low, Teng Yong, Kwok Onn Wong, Adelene L. L. Yap, Laura H. J. De Haan, and Ivonne M. C. M. Rietjens. 2017. "The Regulatory Framework across International Jurisdictions for Risks Associated with Consumption of Botanical Food Supplements." *Comprehensive Reviews in Food Science and Food Safety* 16 (5): 821–34. https://doi.org/10.1111/1541-4337.12289.

Luo, Dan, Maarten Reith, and Aloke K. Dutta. 2020. "Dopamine Agonists in Treatment of Parkinson's Disease: An Overview." *Diagnosis and Management in Parkinson's Disease*, 445–60. https://doi.org/10.1016/b978-0-12-815946-0.00026-0.

Manyam, Bala V., Muralikrishnan Dhanasekaran, and Theodore A. Hare. 2004. "Effect of Antiparkinson Drug HP-200 (Mucuna Pruriens) on the Central Monoaminergic Neurotransmitters." *Phytotherapy Research* 18 (2): 97–101. https://doi.org/10.1002/ptr.1407.

Mawalagedera, Sundara M. U. P., Damien L. Callahan, Anne C. Gaskett, Nina Rønsted, and Matthew R. E. Symonds. 2019. "Combining Evolutionary Inference and Metabolomics to Identify Plants with Medicinal Potential." *Frontiers in Ecology and Evolution* 7 (July). https://doi.org/10.3389/fevo.2019.00267.

McClish, Mark and Patrick Olivelle. 2012. *The Arthaśāstra: Selections from the Classic Indian Work on Statecraft*. Indianapolis, Ind.: Hackett Pub. Co.

McFarthing, Kevin, Sue Buff, Gary Rafaloff, Thea Dominey, Richard K. Wyse, and Simon R. W. Stott. 2020. "Parkinson's Disease Drug Therapies in the Clinical Trial Pipeline: 2020." *Journal of Parkinson's Disease* 10 (3): 757–74. https://doi.org/10.3233/JPD-202128.

Meiser, Johannes, Daniel Weindl, and Karsten Hiller. 2013. "Complexity of Dopamine Metabolism." *Cell Communication and Signaling* 11 (1): 34. https://doi.org/10.1186/1478-811x-11-34.

Mesquita, V. L. V., and C. Queiroz. 2013. "Enzymatic Browning." In *Biochemistry of Foods*, edited by N. A. M. Eskin and F. Shahidi, 387–418. London: Elsevier Inc.

Miller, Kelli. 2016. "What Is Ayurveda?" WebMD. WebMD. December 13, 2016. https://www.webmd.com/balance/guide/ayurvedic-treatments. Accessed August 18, 2021.

Mishra, Jai Prakash. 2005. "Biodiversity, Biotechnology and Intellectual Property Rights." *The Journal of World Intellectual Property* 3 (2): 211–24. https://doi.org/10.1111/j.1747-1796.2000.tb00124.x.

Misra, Laxminarain, and Hildebert Wagner. 2007. "Extraction of bioactive principles from Mucuna pruriens seeds." *Indian Journal of Biochemistry and Biophysics* 44 (February): 56–60.

Morant, Anne Vinther, Vivien Jagalski, and Henrik Tang Vestergaard. 2019. "Labeling of Disease-Modifying Therapies for Neurodegenerative Disorders." *Frontiers in Medicine* 6 (October). https://doi.org/10.3389/fmed.2019.00223.

Moura, Tania M., Mohammad Vatanparast, Ana M. G. A. Tozzi, Félix Forest, C. Melanie Wilmot-Dear, Marcelo F. Simon, Vidal F. Mansano, Tadashi Kajita, and Gwilym P. Lewis. 2016. "A Molecular Phylogeny and New Infrageneric Classification of Mucuna Adans. (Leguminosae-Papilionoideae) Including Insights from Morphology and Hypotheses about Biogeography." *International Journal of Plant Sciences* 177 (1): 76–89. https://doi.org/10.1086/684131.

Moura, Tânia M. de, Gwilym P. Lewis, Vidal F. Mansano, and Ana M. G. A. Tozzi 2018. "A Revision of the Neotropical Mucuna Species (Leguminosae—Papilionoideae)." *Phytotaxa* 337 (1): 1. https://doi.org/10.11646/phytotaxa.337.1.1.

Muela-Molina, C., S. Perelló-Oliver, and A. García-Arranz. 2021. "Health-related claims in food supplements endorsements: a content analysis from the perspective of EU regulation." *Public Health* 190 (January): 168–72. https://doi.org/10.1016/j.puhe.2020.10.020.

Müller, Thomas. 2007. Review of *The Role of Levodopa in the Chronic Neurodegenerative Disorder – Parkinson´s Disease*. In *Oxidative Stress and Neurodegenerative Disorders*, edited by G. Ali Qureshi and S. Hassan Parvez, 237–46. Elsevier Science.

Muinga, R.W., H.M. Saha and J.G. Mureithi. 2003. *The Effect of Mucuna* (Mucuna pruriens) *forage on the performance of lactating cows. Proceedings of a Workshop Organized by KAKI and CIEPCA-IITA Mombasa, Kenya, September 23-26, 2002*, edited by M. Eilittä, R. Muinga, J. Mureithi, C. Sandoval-Castro, and N. Szabo, 87–91. Mexico: Facultad de Medicina Veterinaria y Zootecnia Universidad Autónoma de Yucatan UADY. https://www.redalyc.org/articulo.oa?id=93911288007.

Munos, Bernard. 2009. "Lessons from 60 Years of Pharmaceutical Innovation." *Nature Reviews Drug Discovery* 8 (12): 959–68. https://doi.org/10.1038/nrd2961.

Nagashayana, N, P Sankarankutty, M.R.V Nampoothiri, P.K Mohan, and K.P Mohanakumar. 2000. "Association of L-DOPA with Recovery Following Ayurveda Medication in Parkinson's Disease." *Journal of the Neurological Sciences* 176 (2): 124–27. https://doi.org/10.1016/s0022-510x(00)00329-4.

Nagoya Protocol. 2011. "Nagoya Protocol on Access to Genetic Resources and the Fair and Equitable Sharing of Benefits Arising from their Utilization to the Convention on Biological Diversity – Text and Annex." https://www.cbd.int/abs/doc/protocol/nagoya-protocol-en.pdf. Accessed 21/10/2021.

Narayan, Shilpa, Zeyan Liew, Jeff M. Bronstein, and Beate Ritz. 2017. "Occupational pesticide use and Parkinson's disease in the Parkinson Environment Gene (PEG) Study." *Environment International* 107 (October): 266–73. https://doi.org/10.1016/j.envint.2017.04.010.

NFL. 2015. *REGULATION (EU) 2015/2283 of the EUROPEAN PARLIAMENT and of the COUNCIL of 25 November 2015 on Novel Foods.* https://eur-lex.europa.eu/legal-content/EN/TXT/PDF/?uri=CELEX:32015R2283&from=DE The European Parliament and the Council of the European Union.

Nieber, M, and C Dohm. 2013. "Von GACP Zu GMP - Der Weg Vom Kontrollierten Anbau Zur Validierten Extraktion Der Wirkstoffe." *Zeitschrift Für Phytotherapie* 34 (S 01). https://doi.org/10.1055/s-0033-1338238.

Niemi, Maria, and Göran Ståhle. 2016. "The use of Ayurvedic medicine in the context of health promotion – a mixed methods case study of an Ayurvedic centre in Sweden." *BMC Complementary and Alternative Medicine* 16 (1). https://doi.org/10.1186/s12906-016-1042-z.

Nouws, A. 2015. *Mijn denken stottert vaak meer dan mijn benen: Hoe mensen Parkinson beleven.* Koog Aan De Zaan: Poiesz Uitgevers, Oktober.

Onweluzo, J., and M. Eilittä. 2003. *Surveying Mucuna´s utilization as a food in Enugu and Kogi states of Nigeria. Tropical and Subtropical Agroecosystems* 1: 213–25. https://www.redalyc.org/pdf/939/9391 1288022.pdf.

Ovallath, Sujith, and P. Deepa. 2013. "The History of Parkinsonism: Descriptions in Ancient Indian Medical Literature." *Movement Disorders* 28 (5): 566–68. https://doi.org/10.1002/mds.25420.

Pandharinath, RavindraR. 2011. "'Science Means Learning to Say 'I Don't Know': An Interview with Dr. Ashok D.B. Vaidya." *Journal of Ayurveda and Integrative Medicine* 2 (4): 211. https://doi.org/10.4103/0975-9476.90771.

Panova, A. S., D. S. Dergachev, M. A. Subotyalov, and V. D. Dergachev. 2020. "Review of Mucuna Pruriens L. Therapeutic Potential for Parkinson's Disease." *Meditsinskiy Sovet = Medical Council*, no. 8 (July): 82–87. https://doi.org/10.21518/2079-701x-2020-8-82-87.

Paoletti, Federico Paolini, Nicola Tambasco, and Lucilla Parnetti. 2019. "Levodopa Treatment in Parkinson's Disease: Earlier or Later?" *Annals of Translational Medicine* 7 (S6): S189–89. https://doi.org/10.21037/atm.2019.07.36.

Park, Chan Bae, and Nils-Göran Larsson. 2011. "Mitochondrial DNA Mutations in Disease and Aging." *The Journal of Cell Biology* 193 (5): 809–18. https://doi.org/10.1083/jcb.201010024.

Parkinson, James. 2016. *An Essay on the Shaking Palsy.* London: Wentworth Press.

Pathania, Ruhi, Prince Chawla, Huma Khan, Ravinder Kaushik, and Mohammed Azhar Khan. 2020. "An Assessment of Potential Nutritive and Medicinal Properties of Mucuna Pruriens: A Natural Food Legume." *3 Biotech* 10 (6). https://doi.org/10.1007/s13205-020-02253-x.

Patil, R. R., A. R. Gholave, J. P. Jadhav, S. R. Yadav, and V. A. Bapat. 2014. "*Mucuna sanjappae* Aitawade et Yadav: A new species of Mucuna with promising yield of anti-Parkinson's drug L-DOPA." *Genetic Resources and Crop Evolution* 62 (1): 155–62. https://doi.org/10.1007/s10722-014-0164-8.

Patil, Ravishankar R., Kiran D. Pawar, Manali R. Rane, Shrirang R. Yadav, Vishwas A. Bapat, and Jyoti P. Jadhav. 2016. "Assessment of Genetic Diversity in Mucuna Species of India Using Randomly Amplified Polymorphic DNA and Inter Simple Sequence Repeat Markers." *Physiology and Molecular Biology of Plants* 22 (2): 207–17. https://doi.org/10.1007/s12298-016-0361-3.

Persaud, Nav, Maggie Jiang, Roha Shaikh, Anjli Bali, Efosa Oronsaye, Hannah Woods, Gregory Drozdzal, et al. 2019. "Comparison of Essential Medicines Lists in 137 Countries." *Bulletin of the World Health Organization* 97 (6): 394–404C. https://doi.org/10.2471/blt.18.222448.

Pharmaceutical Technology. 2020. "The World's Biggest Pharmaceutical Companies: Top Ten by Revenue." www.pharmaceutical-Technology.com. October 1, 2020. https://www.pharmaceutical-technology.com/features/top-ten-pharma-companies-in-2020/. Accessed 20/10/2021.

Potschka, Heidrun. 2010. "Targeting the brain – surmounting or bypassing the blood-brain barrier." *Handbook of Experimental Pharmacology*, 197: 411–31. https://doi.org/10.1007/978-3-642-00477-3_1.

Prasad, E. Maruthi, and Shih-Ya Hung. 2021. "Current Therapies in Clinical Trials of Parkinson's Disease: A 2021 Update." *Pharmaceuticals* 14 (8): 717. https://doi.org/10.3390/ph14080717.

Pugalenthi, M., V. Vadivel, and P. Siddhuraju. 2005. "Alternative Food/Feed Perspectives of an Underutilized Legume Mucuna Pruriens Var. Utilis—a Review." *Plant Foods for Human Nutrition* 60 (4): 201–18. https://doi.org/10.1007/s11130-005-8620-4.

Pugalenthi, M., and V. Vadivel. 2006. "Agrobiodiversity of Eleven Accessions of Mucuna Pruriens (L.) DC. Var. Utilis (Wall. Ex Wight) Baker Ex Burck (Velvet Bean) Collected from Four Districts of South India." *Genetic Resources and Crop Evolution* 54 (5): 1117–24. https://doi.org/10.1007/s10722-006-9003-x.

Purdon, A. D., T. A. Rosenberger, H. U. Shetty, and S. I. Rapoport. 2002. "Energy consumption by phospholipid metabolism in mammalian brain." *Neurochemical Research* 27 (12): 1641–47. https://doi.org/10.1023/a:1021635027211.

Quattrocchi, Umberto. 2000. *CRC World Dictionary of Plant Names: Common Names, Scientific Names, Eponyms, Synonyms, and Etymology/Vol. II*D-L.* Boca Raton: CRC Press.

Radad, Khaled, Gabriele Gille, and Wolf-Dieter Rausch. 2005. Review of *Short Review on Dopamine Agonists: Insight into Clinical and Research Studies Relevant to Parkinson's Disease. Pharmacological Reports* 57: 701–12. http://if-pan.krakow.pl/pjp/pdf/2005/6_701.pdf.

Radder, Danique L.M., Andreas T. Tiel Groenestege, Inge Boers, Eline W. Muilwijk, and Bastiaan R. Bloem. 2019. "Mucuna Pruriens Combined with Carbidopa in Parkinson's Disease: A Case Report." *Journal of Parkinson's Disease* 9 (2): 437–39. https://doi.org/10.3233/jpd-181500.

Rai, Sachchida Nand, Vivek K. Chaturvedi, Payal Singh, Brijesh Kumar Singh, and M. P. Singh. 2020. "Mucuna Pruriens in Parkinson's and in Some Other Diseases: Recent Advancement and Future Prospective." *3 Biotech* 10 (12). https://doi.org/10.1007/s13205-020-02532-7.

Rascol, Olivier, Christopher Goetz, William Koller, Werner Poewe, and Cristina Sampaio. 2002. "Treatment Interventions for Parkinson's Disease: An Evidence Based Assessment." *The Lancet* 359 (9317): 1589–98. https://doi.org/10.1016/s0140-6736(02)08520-3.

Reeve, Amy, Eve Simcox, and Doug Turnbull. 2014. "Ageing and Parkinson's Disease: Why Is Advancing Age the Biggest Risk Factor?" *Ageing Research Reviews* 14 (March): 19–30. https://doi.org/10.1016/j.arr.2014.01.004.

Sackett, D. L, W. M C Rosenberg, J A M. Gray, R B. Haynes, and W S. Richardson. 1996. "Evidence Based Medicine: What It Is and What It Isn't." *BMJ* 312 (7023): 71–72. https://doi.org/10.1136/bmj.312.7023.71.

Saha, Chandra Nath and Sanjib Bhattacharya. 2011. "Intellectual Property Rights: An Overview and Implications in Pharmaceutical Industry." *Journal of Advanced Pharmaceutical Technology & Research* 2 (2): 88. https://doi.org/10.4103/2231-4040.82952.

Sandroni, Paola. 2001. "Aphrodisiacs Past and Present: A Historical Review." *Clinical Autonomic Research* 11 (5): 303–7. https://doi.org/10.1007/bf02332975.

Sathyanarayana, N., M. Leelambika, S. Mahesh, and M. Jaheer. 2011. "AFLP Assessment of Genetic Diversity among Indian Mucuna Accessions." *Physiology and Molecular Biology of Plants* 17 (2): 171–80. https://doi.org/10.1007/s12298-011-0058-6.

Schünemann, Holger, Jan Brożek, Gordon Guyatt, and Andrew Oxman, eds. 2013. *GRADE Handbook*. https://gdt.gradepro.org/app/handbook/handbook.html.

Schwabl, Herbert and Jan M. A. van der Valk. 2019. *"Challenging the Biomedical Notion of "Active Substance": The Botanical Plasticity of Tibetan Medical Formulas." HIMALAYA, the Journal of the Association for Nepal and Himalayan Studies* 39 (1): 208–18.

Sewell, Robert D. E., and Mahmoud Rafieian-Kopaei. 2014. "The history and ups and downs of herbal medicines usage." *Journal of HerbMed Pharmacology* 3 (1): 1–3.

Shelley, Walter B. and Robert P. Arthur. 1955. "Studies on Cowhage (Mucuna Pruriens) and its Pruritogenic Proteinase, Mucunain." *Archives of Dermatology* 72 (5): 399. https://doi.org/10.1001/archderm.1955.037 30350001001.

Shop-Apotheke.com. n.d. "DHU Dolichos Pruriens D6 10 G – Shop-Apotheke.com." Www.shop-Apotheke. com. Accessed August 5, 2021. https://www.shop-apotheke.com/homoeopathie/4215329/dhu-dolichos-pruriens-d6.htm.

Silano, Vittorio, Patrick Coppens, Ainhoa Larrañaga-Guetaria, Paola Minghetti, and René Roth-Ehrang. 2011. "Regulations applicable to plant food supplements and related products in the European Union." *Food & Function* 2 (12): 710. https://doi.org/10.1039/c1fo10105f.

Simon, David K., Caroline M. Tanner, and Patrik Brundin. 2020. "Parkinson Disease Epidemiology, Pathology, Genetics, and Pathophysiology." *Clinics in Geriatric Medicine* 36 (1): 1–12. https://doi.org/10.1016/j.cger.2019.08.002.

Stegeman, Inge, René Spijker, Rob Scholten, and Lotty Hooft. 2017. *Mucuna pruriens bij mensen met de ziekte van Parkinson. Een literatuurstudie naar de werkzaamheid en veiligheid.* Edited by Parkinson Vereniging. www.parkinson-Vereniging.nl. Utrecht: Cochrane Netherlands. www.parkinson-vereniging. nl/l/library/download/urn:uuid:2f20b4cf-9838-478c-b543-59e0b73185d2/mucuna_pruriens_final_rapp ort.pdf?ext=.pdf. Accessed 13 September 2021.

Steinhoff, Barbara. 1993. "Europa ist bei der Phyto-Bewertung noch nicht vereint." *Ärztliche Allgemeine der Ärzte Zeitung* 4: 35–37.

Stephenson, Joan. 2000. "Exposure to Home Pesticides Linked to Parkinson Disease." *JAMA* 283 (23): 3055. https://doi.org/10.1001/jama.283.23.3055.

Storch, Alexander, Stefanie Ott, Yu-I Hwang, Rainer Ortmann, Andreas Hein, Stefan Frenzel, Kazuo Matsubara, Shigeru Ohta, Hans-Uwe Wolf, and Johannes Schwarz. 2002. "Selective Dopaminergic Neurotoxicity of Isoquinoline Derivatives Related to Parkinson's Disease: Studies Using Heterologous Expression Systems of the Dopamine Transporter." *Biochemical Pharmacology* 63 (5): 909–20. https://doi.org/10.1016/s0006-2952(01)00922-4.

Suresh, Sekar, and Seppan Prakash. 2012. "Effect of Mucuna Pruriens (Linn.) on Sexual Behavior and Sperm Parameters in Streptozotocin-Induced Diabetic Male Rat." *The Journal of Sexual Medicine* 9 (12): 3066–78. https://doi.org/10.1111/j.1743-6109.2010.01831.x.

Tharakan, Binu, Muralikrishnan Dhanasekaran, Janna Mize-Berge, and Bala V. Manyam. 2007. "Anti-Parkinson Botanical Mucuna Pruriens Prevents Levodopa Induced Plasmid and Genomic DNA Damage." *Phytotherapy Research* 21 (12): 1124–26. https://doi.org/10.1002/ptr.2219.

The Lancet. 2020. "Global Governance for COVID-19 Vaccines." *The Lancet* 395 (10241): 1883. https://doi.org/10.1016/s0140-6736(20)31405-7.

Tolosa, E, M J Martí, F Valldeoriola, and J L Molinuevo. 1998. "History of Levodopa and Dopamine Agonists in Parkinson's Disease Treatment." *Neurology* 50 (6 Suppl 6): S2-10; discussion S44-8. https://doi.org/10.1212/wnl.50.6_suppl_6.s2.

Torquati, T. 1913. "*Sulla presenza di una sostanza azotata nei germogli del semi di 'Vicia Faba'*". *Archivio di Farmacologia Sperimentale e Scienze Affini* 15: 213–23.

Uebelhack, Ralf, Joerg Gruenwald, Hans-Joachim Graubaum, and Regina Busch. 2004. "Efficacy and Tolerability of Hypericum Extract STW 3-vi in Patients with Moderate Depression: A Double-Blind, Randomized, Placebo-Controlled Clinical Trial." *Advances in Therapy* 21 (4): 265–75. https://doi.org/10.1007/bf02850158.

Ueda, Keisuke, Chiaki Sanbongi, Makoto Yamaguchi, Shuji Ikegami, Takafumi Hamaoka, and Satoshi Fujita. 2017. "The Effects of Phenylalanine on Exercise-Induced Fat Oxidation: A Preliminary, Double-Blind, Placebo-Controlled, Crossover Trial." *Journal of the International Society of Sports Nutrition* 14 (1). https://doi.org/10.1186/s12970-017-0191-x.

UN. "The 17 Goals." United Nations. 2021. https://sdgs.un.org/goals.

UPDRS. 2003. "The Unified Parkinson's Disease Rating Scale: Status and Recommendations." *Movement Disorders* 18 (7): 738–50. https://doi.org/10.1002/mds.10473.

Vaidya, A. B., T. G. Rajgopalan, N. A. Mankodi, D. S. Antarkar, P. S. Tathed, A. V. Purohit, and N. H. Wadia. 1978. "Treatment of Parkinson´s disease with the cowhage plant – *Mucuna pruriens* Bak." *Neurology India* 26 (4): 171–76.

Vaidya, Ashok DB., Rama Vaidya, and Ashwinikumar Raut. 2020. "Post-COVID-19 rethinking for a synergic vision of health-care." *Journal of Ayurveda and Integrative Medicine* 11 (3): A4–5. https://doi.org/10.1016/j.jaim.2020.09.006.

Van der Giessen, Rob, Waren C. Olanov, Andrew Lees, and Hildebert Wagner. 2004. "Mucuna pruriens and extracts thereof for the treatment of neurological diseases." WIPO, Geneva, Patent IPN: WO 2004/039385 A3, issued May 13.

Vijver, Monique van de. 2022. *Exploring the Potential of Medicinal Plants in Sustainable Development.* ResearchGate. https://www.researchgate.net/profile/Monique-Vijver/publication/361173853_EXPLORING_THE_POTENTIAL_OF_MEDICINAL_PLANTS_FOR_SUSTAINABLE_DEVELOPMENT/links/62a0c87055273755ebdd5b97/EXPLORING-THE-POTENTIAL-OF-MEDICINAL-PLANTS-FOR-SUSTAINABLE-DEVELOPMENT.pdf?origin=publication_detail. Accessed 9/6/2022.

Vijiaratnam, Nirosen, Tanya Simuni, Oliver Bandmann, Huw R Morris, and Thomas Foltynie. 2021. "Progress towards Therapies for Disease Modification in Parkinson's Disease." *The Lancet Neurology* 20 (7): 559–72. https://doi.org/10.1016/s1474-4422(21)00061-2.

Weckwerth, Wolfram. 2003. "Metabolomics in Systems Biology." *Annual Review of Plant Biology* 54 (1): 669–89. https://doi.org/10.1146/annurev.arplant.54.031902.135014.

Weidenauer, Uwe. *Nicht Lieferbar! Ausverkauf Des Deutschen Arzneimittelmarkts.* Stuttgart: Deutscher Apotheker Verlag, 2020.

West, Elizabeth, and Michael Krychman. 2015. "Natural Aphrodisiacs — a Review of Selected Sexual Enhancers." *Sexual Medicine Reviews* 3 (4): 279–88. https://doi.org/10.1002/smrj.62.

WHO. 2009. "SIXTY-SECOND WORLD HEALTH ASSEMBLY WHA62.13 Agenda Item 12.4 Traditional Medicine." https://apps.who.int/gb/ebwha/pdfwe_files/A62/A62_R13-en.pdf.

WHO. 2014. *World Health Organization Traditional Medicine Strategy: 2014-2023.* https://www.who.int/publications/i/item/9789241506096. Accessed 27/10/2021.

WHO. 2015. *Health in 2015: From MDGS, Millennium Development Goals, to SDGS, Sustainable Development Goals.* Geneva: World Health Organization. https://apps.who.int/iris/bitstream/handle/10665/200009/?sequence=1. Accessed 30/05/2022.

WHO. 2020. *Expert Committee for the International Herbal Pharmacopoeia.* https://www.who.int/news-room/articles-detail/expert-committee-for-the-international-herbal-pharmacopoeia. Accessed 28/10/2021.

WHO. 2021. *WHO Model List of Essential Medicines – 22nd List* https://www.who.int/publications/i/item/WHO-MHP-HPS-EML-2021.02. Accessed 27/10/2021.

Williamson, E. M. 2001. "Synergy and other interactions in phytomedicines." *Phytomedicine* 8 (5): 401–9. https://doi.org/10.1078/0944-7113-00060.

Wilson, Jero Victor, L. V. Karthikeyan, Shubham Kumar Parida, Bikash Chandra Nath, and M. R. Jeyaprakash. 2021. "Recent Advances and Developments in Hyphenated Techniques and Their Applications." *Journal of Pharmaceutical Research International*, January, 58–68. https://doi.org/10.9734/jpri/2020/v32i4331070.

Wishart, D. S. 2007. "Current Progress in Computational Metabolomics." *Briefings in Bioinformatics* 8 (5): 279–93. https://doi.org/10.1093/bib/bbm030.

Woodard, Roger D. 2008. *The Ancient Languages of Asia and the Americas.* Cambridge: Cambridge University Press.

Wujastyk, Dagmar, and Frederick M Smith. 2008. *Modern and Global Ayurveda: Pluralism and Paradigms.* Albany: State University Of New York Press.

# 4 Valuing Hedgerows

## Their Political Ecology and 'Medicinal' Role in Transitions to Agroecological Farming Systems in Lowland Britain

*Mark Tilzey*

## 1 INTRODUCTION

Hedgerows are an iconic feature of the lowland British, and, particularly, lowland English land-scape. Today, with the almost universal loss of 'infield' biodiversity in lowland farmed environments, they often represent the only element of natural/cultural heritage remaining on most conventional farms. Yet, these surviving elements of biodiversity and cultural heritage seem to be woefully under-appreciated, and are indeed being slowly degraded on most conventional farms through inappropriate 'management' or through neglect, rather than through outright destruction. Most are subjected to annual flailing 'from above' and, in livestock areas, browsing 'from below'. Where not grazed out, hedgerow bottom flora is commonly subjected to annual mowing and species loss through herbicide drift, eutrophication from artificial fertilizers, and consequent invasion by nutrient respon-sive species. This annual 'management' leads to loss of hedge structure (gappiness), destruction of fruits, nuts, and berries, reduction of food and habitat for birds, mammals, and invertebrates, uses considerable quantities of $CO_2$ emitting fossil fuel, and prevents the maturation of trees in the hedgerow (Tilzey 2000; Dover 2019). 'Appropriately' managed hedgerows, that is, hedgerows that are 'laid' on an approximately ten yearcycle, or are flailed no more frequently than once every three years (together with trees at frequent intervals in the hedge), are an increasing rarity in the lowland British landscape (Carey et al. 2009). Similarly, hedgerows that are afforded appropriate protection from livestock browsing are an increasingly rare sight.

Yet hedgerows have massive potential, with changed and/or relaxed management, to contribute to biodiversity conservation, soil conservation and enhancement, carbon sequestration (e.g. by allowing far more trees to grow to maturity), water retention and flood alleviation, climate change mitigation, shelter for crops and livestock, and cost savings (and reduced $CO_2$ emissions) for the farmer and land manager (Montgomery et al. 2020). In short, hedgerows can make an important potential contribution to agroecological transitions, and to medicinal agroecology through herbal/food/medicinal products (see Chapter 5 of this book), and an overall contribution to multifunctional agro-ecosystems with multiple beneficial contributions to biodiversity, climate change mitigation, soil health, human health and wellbeing.

This chapter will explore these elements of multifunctionality and assess the contribution of hedgerows to Ecosystem Services (ES), and whether this contribution may be enhanced through changes towards more permissive management as envisaged above. The chapter will then explore the relationship between ES provision and the requirements of agroecology, again looking at

DOI: 10.1201/9781003146902-6

whether the adoption of more permissive hedgerow management can contribute to agroecological production practices and transitions. We will suggest that the 'standard' management of hedgerows as described above is constraining the optimum delivery of ES by hedgerows and is, moreover, an integral part of a productivist management of the countryside that is contributing to the delivery of ecosystem *disservices* and is, therefore, antithetical to the adoption of agroecological production practices. Finally, we will outline what appears to be required, in policy and political terms, for the adoption of an agroecological framework enabling the sustainable management of hedgerows and maximising their potential for ES delivery. Overall, we will be conceptualising hedgerows as comprising an 'agroecological medicine for the landscape', and will address, through political ecology, their various and wider historical, political, and economic dimensions. This we see as a 'meta-perspective', complementing the stance adopted in Chapter 5 of this book, for example, where 'hedgerows are seen more as elements where medicinal shrubs and trees should be grown'[1].

## 2  HEDGEROWS: DEFINITION, POLITICAL ECOLOGY, AND RELATIONSHIP TO AGRARIAN CAPITALISM/PRODUCTIVISM

### 2.1  *DEFINITION*

Strictly speaking, 'hedge' and 'hedgerow' have different definitions, a 'hedge' being the woody component of a field boundary, whilst a 'hedgerow' comprises the herbaceous (ground floristic) component together with the usual bank and ditch that are constructed in tandem with the establishment of the hedge (Dover 2019). In common parlance, however, the terms 'hedge' and 'hedgerow' are used interchangeably (Forman and Baudry 1984) and Pollard et al. (1974: 24) state that 'we do not feel that there is any useful distinction to be made between hedgerow and hedge'. In this chapter, 'hedge' is employed to refer to linear strips of managed or unmanaged woody vegetation – that is, shrubs and/or lines of trees, otherwise termed woody linear features, for example, by Maskell et al. (2008).

The primary purpose behind the establishment of hedgerows was the confining/exclusion of livestock in order to protect livestock and define their ownership, facilitate the control and rotation of grazing, and to prevent livestock from causing damage to grazing/trampling-sensitive crops, notably arable and grass crops (the latter traditionally hay meadows needed for the provision of winter fodder). Hedgerows were also established to define boundaries, whether between individual owners of land or between administrative entities, such as parishes/manors. Functionally, hedgerows have now been almost entirely superseded by barbed-wire fencing (or temporary electric fencing), and this loss of contemporary functionality due to the substitution by fencing is an important factor behind the degradation and loss of hedgerows, since there is no longer an imperative for them to be stock-proof or, indeed, to exist at all.

Forman and Baudry (1984) defined three main origins of hedgerows:

- **Planted** – deliberately created, typically using a single species, and usually planted on a bank with an associated ditch, most characteristic of Rackham's (1986) 'planned countryside' (commonly associated in England with enclosure of formerly common land from the late 15th century and, especially, with the Parliamentary Enclosure of the 18th and 19th centuries [see below]);
- **Remnant** – typically the result of woodland clearance where a strip of trees/shrubs is retained along ownership boundaries. These are usually older hedges, especially characteristic of Rackham's (1986) 'ancient countryside', and are commonly species-rich due to the fact that they have not been planted and their age has facilitated colonisation by additional species according to 'Hooper's rule' (whereby he maintained that hedges could be dated by counting

---

1    Thanks to Immo Fiebrig for suggesting this way of viewing the present chapter in relation to Chapter 5 in this book.

the number of shrubs in 30-yard lengths on the assumption that one species was added for each hundred years of the hedgerow's existence [Hooper 1970]);

* **Spontaneous** – trees and shrubs colonise naturally pre-existing structures such as field margins, banks, etc., through dispersal of seeds by animals (including birds) or wind. These can be of any age, but the older they are the more difficult it becomes to differentiate this category from remnant hedgerows.

## 2.2 POLITICAL ECOLOGY AND RELATIONSHIP TO AGRARIAN CAPITALISM/PRODUCTIVISM

Older field boundaries tend to be curvilinear or irregularly shaped, probably mainly due to piecemeal assarting[2], or the clearance of woodland, leaving strips of the original woodland cover – these tend to be differentially 'remnant' hedgerows and are the most species-rich hedges, as noted. With the emergence of capitalist agriculture and the privatisation of land in England from the 16th century onwards, and especially from the mid-1700s, new field boundaries defining new private land units tended to be much more rectilinear in form. From around 1750 these resulted especially from Private or General Parliamentary Enclosure Acts which enabled the division of former commonly held land (both the arable 'open fields', meadows, and permanent pasture of the manorial 'waste') into discrete fields (Dover 2019). During this period of the 'Great Enclosures', from approximately 1750–1850, hedges were planted around newly privatised (enclosed) land, and within these new holdings to demarcate fields, using generally a single species, the hawthorn or quickthorn (*Crataegus monogyna*). English elm (*Ulmus procera*) was also widely planted, however, the timber of which was needed for the Navy and for industry, especially for water wheels and lock gates due to its durability under water. So profitable was elm at this time that all landowners with an eye to profit planted it, so that the hedgerow elm became, with the hawthorn hedge, the principal defining feature of the new enclosures (Pollard et al. 1974). Rackham (1986) estimates that these new plantings amounted to some 322,000 km, thereby at least doubling the entire length of hedgerows planted over the course of the previous 500 years. 'Enclosure', as implied, was not simply the establishment of new field boundaries; it was the essential counterpart of the absolute or exclusive right to property, or privatisation (farming 'in severalty'), asserted by the new capitalist farmers (yeoman farmers[3]) and their landlords (Yerby 2016; Tilzey 2018). This implied the extirpation of common rights previously enjoyed by smallholders, and signalled the latter's death-knell since the small plots allocated to them (if they were lucky to receive land at all) were no longer viable as discrete units, especially since their livestock no longer had access to the broad pastures of the 'waste' and fallow open fields. All the other resources that had been part of common right, such as fuel, wild plants, herbs, and medicines, were now also closed off to the smallholders (Neeson 1993). Many had to supplement their income by selling their labour to the new capitalist farmers, and many more were obliged to move away altogether to the burgeoning industrial cities. The latter process Marx (1972) termed 'primitive accumulation', a principal foundation stone of the rise of industrial capitalism in England (Tilzey 2021).

With the emergence of industrial capitalism in England and Scotland, wool gave way to cotton as the basic material for clothing. For yeoman farmers and landlords, this implied a decline in the importance of sheep in terms of wool production in the lowlands (although production for mutton gained in importance and production for wool shifted increasingly from the lowlands to the uplands, leading to the infamous 'highland clearances' from the late-18th century until the mid-19th century) (Tilzey 2018). The lowland 'sheep rancher', whose production of wool for the export market had been the principal cause of enclosure during the Tudor and Stuart eras, now conceded to the grower

---

2   'Assarting' was the term used in medieval times to denote piecemeal woodland clearance, and the strips of woodland left as boundary markers were/are termed 'assart hedges'.

3   'Yeoman farmer' was the term given to former members of the 'upper peasantry' who became capitalist family farmers, either tenants of landlords or independent freeholders. Their agricultural produce was sold on the market for profit.

of cereals and potatoes for the urban industrial worker (Pollard et al. 1974). Enclosed fields of the 'wool era' of 60, 80, or even 100 acres (24, 32, or 40 hectares), were now divided up to smaller units typically of around 20 acres (8 hectares) suitable for horse-dependent arable husbandry, and these were planted with new hedges (Pollard et al. 1974).

The 19th and early 20th centuries could be described as the 'heyday' of the hedgerow in lowland Britain, both in terms of length (abundance) and in terms of management (pre-mechanical in nature), features conducive to the encouragement of biodiversity and other multifunctional benefits (see below under Ecosystem Services). As suggested, this coincided with a shift in the balance between arable and permanent pasture after 1750, with now more lucrative arable rotations increasing in the south and east of England, especially, at the expense of purely pastoral systems (Overton 1996). Fodder supplies did not fall, however, since the loss of permanent pasture was compensated by new fodder crops, particularly turnip and clover, in the new arable rotations, classically based on the Norfolk Four Course Rotation, developed in the mid-18th century by Lord ('Turnip') Townshend. This was a rotation of a winter-sown corn (wheat), followed by roots (usually turnip), and then by barley undersown with grasses and clover, the latter producing a ley in the fourth year. Thus, four 'large' fields were needed, each being about 20 acres (8 hectares), together with a few smaller fields to facilitate stock management (Pollard et al. 1974). This was subject to regional variations, with more rotations often being employed, this in turn implying smaller fields and more hedgerows.

Not only did these new crops result in increased fodder yields, which meant more livestock manure for the land, they were also instrumental in the 'improvement' of much former manorial 'waste' in the lowlands, and its conversion from rough pasture to 'productive' mixed farms. This was achieved by the same means, that is, as a fodder crop for livestock, the manure from which then raised the fertility of the 'improved' land (Tilzey 2018). The 'enclosure' and 'improvement' of formerly common 'waste' implied the appearance of new hedgerows in these areas, in addition to those of the previously 'open fields'. Until the introduction of artificial nitrogen, this 'improved' system of farming, together with its dense network of hedgerows, was, ecologically, relatively sustainable, and could be said to conform to many of the key principles of agroecology (see below) in its biophysical dimension. Output of food was thus increased dramatically and without central reliance on fossil fuels, synthetic fertilisers, or inputs imported from overseas. However, the increased demand for nitrogen was a function of the huge export of nutrients from the land to the cities (the so-called 'metabolic rift'), contingent upon the termination of locally circulated nutrients characteristic of self-subsistence economies, and the need to remove all straw from the arable fields for livestock. Moreover, this system was set on a trajectory of ever-expanding output desired by capitalism. Accordingly, from 1830, even this ostensibly ecologically sustainable 'first' agricultural revolution began to unravel, as it was replaced by a farming system dependent increasingly on energy-intensive inputs derived from fossil fuels, and upon the importation of both fertiliser and food, particularly cereals, from abroad (Tilzey 2018). Socially this system, additionally, failed to meet agroecological criteria in many respects. Access to food, especially on the part of the new urban working classes, was dependent on the ability to purchase it, and, therefore, on the ability to find waged employment. Food security for the majority was not assured, therefore, as long as food access was determined by the market. Moreover, access to fresh and nutritious food (fresh vegetables, milk, eggs, meat, etc.) was extremely limited for the urban working classes.

British agriculture became increasingly unsustainable from a biophysical perspective (let alone a social one) as the 19th century progressed, therefore. With the abolition of the Corn Laws in 1846 (which had protected domestic production from overseas competition), British grain producers attempted for the next quarter century or so to remain competitive with overseas producers by means of the increased importation of nitrates and phosphates (particularly in the form of guano) to sustain grain production in the face of falling soil nutrient levels. There was also some degree of mechanisation with the invention and adoption of the reaping machine from the mid-century, for example, and therefore a small degree of pressure to increase field size (and remove hedgerows) again from this time (Pollard et al. 1974). However, this machinery was not widely adopted, and manual harvesting remained prevalent at least until the 1870s. The hedgerow system remained basically intact, therefore, throughout the 19th century. Nonetheless, agroecologically, British farming was on an increasingly unsustainable trajectory. The 1830s and 1840s were characterised by an increasing soil fertility crisis due to lack of fertilisers to replace nutrients, as the 'metabolic rift' took the form of the burgeoning movement of food (and nutrients) from the countryside to the city. These developments implied that grain production, under pressure from domestic economic growth and exacerbated by competition from abroad, had reached the limits of the 'organic' four-course rotation system that had before formed the bedrock of wheat and barley output for the national market. While this system might have been sustainable in the context of a steady state economy, it was incompatible with the continuous demand to increase profit, particularly in the face of competition. The only way for capitalist producers to respond to these pressures was to augment the rotational system with artificial fertilisers and imported manures. This constituted an 'unsustainable' overlay of intensive energy inputs on the four-course rotation, representing a shift to the 'second' agricultural revolution, or the period of so-called 'high farming' (Overton 1996). This involved a shift to what has been termed the 'high feeding' of livestock, particularly cattle, to produce more meat and milk, but also to produce more dung as a vital input into the arable rotation system (Overton 1996). There was a significant move away from grain and towards pastoralism during the third quarter of the 19th century (Tilzey 2018). This shift, however, remained compatible with the retention of the dense network of hedgerows that still characterised lowland Britain.

Following the abolition of the Corn Laws in 1846, however, Britain progressively lost food self-sufficiency as the century wore on (self-sufficiency being a key criterion of agroecology) and this was implicit in the increased emphasis on meat and dairy at the expense of grain production and the progressive decline of more sustainable mixed farming systems typified by the four-course rotation system (Foster 2016). The advent of refrigeration in the late-19th century meant, however, that even domestic meat production was now subject to competition from cheap overseas imports given Britain's commitment to free trade. The result was that Britain sank 'into lasting food dependence' and agricultural depression (Mazoyer and Roudart 2006: 370), a situation that was to persist, excepting the brief interlude of the First World War, until at least the beginnings of state support for agriculture in the 1930s. During this period, there was little incentive to change field sizes and, indeed, a great deal of land went temporarily out of cultivation, with hedges spreading unmanaged into the fields (Pollard et al. 1974). Such land abandonment at home was the counterpart, however, of a 'frontier' of largely unsustainable 'extractivist' export agriculture overseas, providing Britain with 'cheap' food staples premised on the externalisation of ecological and social costs, the antithesis of agroecology (Tilzey 2018).

The year 1950 may be taken as the time when the densely hedged landscape of lowland Britain, largely intact since the parliamentary enclosures and, in the 'ancient countryside' (Rackham 1986) since the late medieval period, was poised on the brink of profound change that would adversely affect not only hedgerows but also the biodiversity status of all agriculturally managed ecosystems (Pollard et al. 1974). The post-Second World War period was one in which Britain, along with other countries in western Europe, attempted, following the severe food insecurity of the war and the disruptions to food imports, to become self-sufficient in the production of principal food staples. This brought home to Britain the ecological contradictions of productivist agriculture that had previously been externalised onto the spaces of export agriculture abroad. Far from sustaining the biodiversity and landscape resource as before (through organic rotational systems or as an inadvertent result of economic depression), agriculture now became the central factor in its loss and decline (Tilzey 2000). A massive acceleration in the rate of biodiversity loss and decline followed, attributable structurally to the impacts of a particular model of capitalist development termed 'national developmentalism' (Tilzey 2020a). As applied to the agriculture sector, we may refer to this model as 'political productivism', a state-managed policy framework to which an acceleration of the processes of 'appropriationism' and 'substitutionism'[4] are central. 'Political productivism', embodied in UK post-war policy and subsequently in the Common Agricultural Policy (CAP), was implemented by employing the instruments of guaranteed prices, investment grants, input subsidies, state regulation of major commodity markets, and their insulation from overseas competition (Tilzey 2000). The result was to 'hothouse' agrarian capitalism through a policy framework in which higher net farm income could be secured only by means of productivity (increased output per unit of labour) and production (increased output per unit area) increases. This acted as a massive incentive to cut costs through the substitution of machinery for labour, enlarging holdings, and borrowing money for land purchase and capital projects. This, in turn, created indebtedness, further reinforcing the imperative to cut costs and increase output. The environmental impacts of such productivist policies can be enumerated as series of generic issues, affecting the whole of the agricultural landscape but, especially, the lowlands, the site of most intensive production and where machinery could be deployed without constraint (see Tilzey 2000: 280–281). The generic issues of relevance to hedgerows are as follows:

- Loss or mismanagement of 'interstitial' habitats (hedgerows, field margins, ditches, etc.) due to field enlargement and mechanical management, especially the mechanical, and usually annual, flailing of hedgerows and hedge-bottom vegetation, severely reducing their biodiversity value;
- Loss of crop rotations and arable-pasture mosaics, with arable specialisation in the south and east and pastoral specialisation in the north and west, leading to the loss of functionality of hedgerows, especially in the former;
- Universal application of artificial fertiliser leading to the loss or degradation of characteristic hedgerow bottom vegetation;
- Increased grazing pressure within the pastoral zone, leading to increased browsing of hedgerows 'from below', combined with the virtual cessation of non-mechanical management of hedgerows, resulting in severe 'gappiness' in the hedgerow bottom and grazing out of herbaceous hedge-bottom flora;
- Strong trend toward contractualisation of hedgerow management, leading to the 'simplification' of management, typically entailing annual mechanical flailing of the whole hedgerow irrespective of the presence of potential trees;
- Massive increase in the use of fossil fuels to manage hedgerows (together with all farm management, exacerbated by application of synthetic fertilisers, a major source of greenhouse gas), contributing to the climate crisis.

## 3   CURRENT STATUS OF HEDGEROWS IN LOWLAND BRITAIN

Pollard et al. (1974) estimated the stock of hedges in Britain (i.e. England, Scotland and Wales) to be about 804,672 km at the end of the 1950s. Dowdeswell (1987) calculated that 230,000 km of hedgerow had been removed between 1946 and 1974, a calculation most likely based on the estimated annual loss of 8,047 km cited in Pollard et al. (1974). Estimates of hedgerow loss and length have increased considerably in accuracy with the advent of the UK Countryside Survey, based on random stratified sampling of land-use in Britain and Northern Ireland. Employing the rough estimate of the length of hedgerows from Pollard et al. (1974) and the stock of hedgerows from the latest Countryside Survey (Carey et al. 2009), and accepting that the different methodologies applied generate uncertainties, it is nonetheless possible to suggest that, since the late 1950s,

---

4   Appropriationism and substitutionism refer to the undermining of discrete elements of the agricultural production process, their transformation into industrial activities, and their re-incorporation into agriculture as inputs (e.g. human labour by machinery, animal traction by the tractor, manure by synthetic fertilizers) (Goodman et al. 1987).

Britain has lost about 41 percent of its net stock of managed hedgerows (i.e. after the balance between losses of existing and planting of new hedges has been taken into account) (Dover 2019). Between 1984 and 2007, the length of lines of trees has skyrocketed from 32,000 to 114,000 km (+ 256 percent), reflecting the cessation of management of many hedgerows, such that they are now considered to be lines of trees or relict hedges (Carey et al. 2009; Dover 2019). This also reflects the predominant dichotomy emerging between mechanically managed hedgerows, on the one hand (often 'over-managed', that is non-selective annual flailing along entire lengths), and complete neg-lect on the other, with only a miniscule percentage subject to manual management by traditional 'laying'. The herbaceous vegetation of hedgerows has also undergone drastic change under agricul-tural productivism, with significant decreases in species richness between 1978 and 1998, although no subsequent change was detected by the Countryside Survey in 2007, suggesting perhaps that the damage had effectively been done between those earlier dates (species characteristic of 'unim-proved' habitats, that is, low nutrient substrates, are highly vulnerable to increases in nutrients and will disappear very soon after first exposure to synthetic fertilisers – once nutrient levels are elevated, these species will not return [Grime 1979]). This is indeed reflected in changes in species compos-ition, with the proportion of more competitive species (of which there are very few compared to the large numbers of low-nutrient adapted species) and those characteristic of fertile, shaded, or less acidic soil conditions increasing during the same period. The increase in shade-tolerant species, a trend that continued to increase through to 2007, is considered to reflect the continuing increase in unmanaged hedgerows as these mature (Carey et al. 2009).

## 4  VALUING HEDGEROWS

Since the 1980s, the regime of 'political productivism' has conceded gradually to a more neoliberal regime of accumulation within the CAP, with commodity support giving way to direct payments, supplemented by discretionary budgets for agri-environmental measures (Tilzey and Potter 2007). This has led to a modest 'greening' of the CAP, although such 'greening' falls far short of measures required to stop continued decline in biodiversity and deterioration of the biophysical resource base (water, soil, atmosphere), let alone to undo to the damage wrought by productivism over a period of some 70 years. The demise of 'political productivism' has coincided with the rise of more holistic visions of land management, integrating low input agriculture with resource and biodiversity con-servation, thus moving attention away from the previous preoccupation with the 'reserve seques-tration' model of nature conservation towards a problematisation of policy frameworks underlying ecological decline in the 'wider countryside' (Tilzey 2000, 2011). Within this new 'sustainability phase' (see Tilzey 2011), the advocacy of an integrated 'whole countryside' approach (Tilzey 2000) has emerged that explicitly challenges the view that nature can be conserved effectively on an isolated or fragmented basis, whether spatially or in terms of individual species (Tilzey 2000; Perfecto et al. 2009). This view has problematised the sustainability of productivist agriculture itself, whether overtly state-supported or neoliberal ('market productivist'). A change is required, it is argued, towards ecological (and social) sustainability in the character of mainstream agriculture itself, seeking to generate both food security and biodiversity/resource conservation through a shift to agroecologically based production, in which food and biodiversity/resource conservation are *joint products*, that is, are *co-produced*, rather than being seen as antithetical objectives. Key elements of this new holistic sustainability paradigm are agroecology itself and the ES approach.

ES theory (UN 2005) recognises that human wellbeing is ultimately wholly dependent on the ser-vices afforded by ecosystems. These services are often described as provisioning services: providing food, fresh water, wood, fibre, and fuel; regulating services: climate, flood, disease, water purifica-tion; cultural: aesthetic educational, recreational; and supporting nutrient cycling, soil formation, primary production. To these may be added biodiversity services. Combined, all these beneficial services may be considered to be preventative and curative medicines for ourselves and the land-scape. The ES provided by hedgerows are presented in Table 1.

Sadly, the full potential of ES is being actively subverted through co-optation into hegemonic neoliberal theory (particularly neoclassical economic theory) and policy, with the rationale princi-pally being to reduce costs for capital and the state by substituting natural processes ('natural cap-ital') for substituted ('non-natural') processes (e.g. water purification and regulation by peatlands rather than by water purification and artificial flood prevention engineering) where feasible, but without reducing the overall impulse towards ecological degradation in the search for economic growth. Within this neoclassical/neoliberal framework, there is, then, an active struggle (cost-benefit analysis) around which 'bits' of nature can be usefully retained to reduce 'costs' for capital/state, and which 'bits' are expendable in the process of transforming nature into commodities (Tilzey 2011). In attempting to derive spuriously 'objective' valuations of ES arising from 'cost-benefit'

## TABLE 1
## Ecosystem Services Delivered by Hedgerows (adapted from Dover [2019]).

| Type of Service | Specific Service | Function/example of Service |
|---|---|---|
| **Supporting Services** | Composting | Soil formation |
| | Photosynthesis | Production of oxygen |
| | Primary production | Chemical energy as organic matter |
| | Nutrient cycling | Essential element recycling such as carbon, phosphorus, nitrogen |
| | Water cycling | Recycling of water |
| **Regulating services** | Air quality regulation | Air pollution control (removal of particulates and nitrogen dioxide); reducing agro-chemical drift |
| | Climate regulation | Carbon sequestration (above and below ground), renewable energy, temperature and humidity moderation, shelter for livestock and crops |
| | Water regulation | Run-off and flood control, increasing infiltration |
| | Erosion regulation | Soil retention, reduction of wind erosion |
| | Water quality/purification | Removal of sediments and pollutants, prevention of agro-chemical drift into watercourses |
| | Pest control | Source of predators and parasites of crop and livestock pests |
| | Pollination | Nectar, pollen, and nesting sites for pollinators |
| | Agricultural management | Containment of livestock |
| | Sense of well-being | Benefits for human mental and physical health |
| **Provisioning services** | Food | Human: fruit, nuts, foraged salad and vegetables, 'wildflower' and berry wine, flavourings (sloe gin). Stock: fodder (most species, ash, elm especially palatable) |
| | Fibre | Timber, fenceposts, wood for turnery, tools |
| | Fuel | Wood (from trees, coppicing, pollarding) |
| | Biochemicals and pharmaceuticals | Dyes, medicinal plants |
| | Genetic resources | Seeds for production of plants of local provenance |
| **Biodiversity services** | Corridors, stepping stones | Facilitating movement through landscape |
| | Habitat and refuges | Nesting, feeding, hibernation, aestivation, shelter from predators |
| **Cultural services** | Recreation and ecotourism | Walking, enjoying nature |
| | Cultural heritage values | Cultural and historical artefacts |
| | Education | Outdoor classroom (nature, history, etc.) |
| | Aesthetics, sense of place | Appreciation of landscape, regional identity, inspiration |

analysis, neoclassical economists generate often meaningless quantitative valuations (e.g. 'willing-ness to pay'), thereby transmuting the irreducible qualitative dimension of ecological services (use values) into a reductive calculus of monetary 'value' (exchange value).

Agroecology (Altieri 1995, 1998), for its part, is the science of applying ecological concepts and principles to the design, development, and management of sustainable agricultural systems. It is a whole systems approach to agriculture that embraces environmental health and social equity, using their synergy, rather than economic growth and capital accumulation, as the basis for defining eco-nomic wellbeing. The goal is long-term sustainability for all living organisms, not merely humans. The concept of agroecology has been incorporated into the work of the UN Food and Agriculture Organisation (FAO 2018) and comprises an important element of the International Assessment of Agricultural Knowledge, Science and Technology for Development (IAASTD 2009). Agroecology's emergence from the 1960s was largely in response to productivism in the form of agro-chemical inputs, mechanisation, and intensification/specialisation, since these were found to deplete soils, reduce agro- and wider biodiversity, impair water quality and quantity, and generate wider adverse environmental impacts (Pimentel 2006; Gomiero et al. 2011). A more holistic food system lens in agroecology emerged in the early 2000s, incorporating the social dimension of food and agriculture, including dietary diversity and nutrition for consumers, equity in food distribution, control over resources, and other key means or production required to establish agroecological food systems (Wach 2021), bringing agroecology into close alignment with food sovereignty (Tilzey 2018). Like ES, however, agroecology has become a contested concept as it has gained in importance, reflecting different understandings of the root causes of ecological unsustainability and social inequity in food systems (Wach 2021).

First, the concept has suffered co-optation by certain state agriculture departments, notably in France, where, as official policy, agroecology is divested of its holism and social dimension, and deployed in an entirely instrumental fashion (see Levidow et al. 2014), much as ES above, to reduce input costs through 'ecological modernisation' to boost the competitiveness of otherwise conven-tional farms. Second, agroecology has been invoked by those advocating economic diversification, local markets, 'ecologisation' of inputs to reduce costs and dependence on upstream suppliers, and a shift to 'post-productivist' 'economies of scope' rather than 'economies of scale' (Tilzey and Potter 2016). Such producers generally supply the high-end market demand created by 'reflexive' middle-class consumers, and such 'niche' 'post-productivist' trends have been supported by CAP Pillar 2 funding. While theorists such as van der Ploeg (2008) have described such producers as the 'new peasantry', opposed to capitalism, the reality is that these are (on the downstream side) market-dependent petty commodity producers embroiled in 'relentless micro-capitalism' (Bernstein 2014). This trend comprises a form of market segmentation, involving the co-optation of selected agroecological principles into capitalism, failing to generate the transformations required to ensure that food systems are *both* ecologically sustainable and fulfil human dietary, social, and cultural needs on an equitable basis (Wach 2021).

Increasingly, proponents of a 'political' approach to agroecology, aligning with 'radical' food sov-ereignty (Tilzey 2018), recognise the potential of, and need for, this theory and practice to address the ecological precarity and social inequity of capitalist food systems (Wach 2021). Agroecological pro-duction can meet humanity's food needs, they argue, on condition that production is determined by societal food needs, not capitalist imperatives. An understanding of the concept of 'market depend-ency' is key here (Wood 2002; Tilzey 2017, 2018; Wach 2021). This concept not only considers the commodification of agricultural *inputs* to be defining of capitalist or market-dependent agriculture, but also the compulsion to sell *outputs* into markets in order to secure economic reproduction of the producer. This approach has a strong focus on *what* is produced by farmers and, therefore, a stronger food system lens, arguing that when producers are dependent on selling their outputs into markets, even where local and small-scale, market imperatives affect not only how foods are produced but also *which* foods are produced and how they are distributed (Wach 2021). In other words, market dependency means that the focus is upon generating exchange value (maximising monetary return) and not upon meeting social need and ecological sustainability.

As we have indicated, however, the strong sustainability (ecological and social) embodied in ES and agroecology has been co-opted and subverted by the much weaker interpretations of sustainability by the state/capital nexus. The 'greening' of the CAP is largely an exercise in 'greenwashing' and the discretionary budgets for agri-environmental Pillar 2 measures continue to be dwarfed, and essentially cancelled out, by the largesse that continues to support productivism in Pillar 1 (Tilzey and Potter 2016). The counterpart of this is a continuing emphasis on the sequestration of special habitats and on the preparation of action plans to address the decline in individual species (Tilzey 2011). Necessary as such initiatives might be as short-term 'fire-fighting' measures given the drastic decline in many formerly characteristic farmland species and habitats, they continue to fail to address the basic causes of loss and decline as a result of capitalist, productivist land use throughout much of the countryside. For these reasons, productivism, and especially 'political productivism' in the form of the CAP, has long been the environmental *bete noire* of the mainstream conservation movement. For these same reasons, however, the latter has tended to be beguiled by neoliberal arguments for the freer play of 'market forces' as putatively the best means to assure greater environmental sustainability, a position long-held by the UK government. The CAP has thus been a relatively easy target for the mainstream conservation movement, precisely because such critiques sit comfortably with the new neoliberal economic agenda and its calls for the dismantling of market/direct support structures (Tilzey 2011). Such is, indeed, now coming to pass with Brexit and the proposed phasing out of inherited CAP supports through Pillar 1 over a period of seven years.

## 5    CURRENT AND ANTICIPATED AGRI-ENVIRONMENTAL POLICY FRAMEWORK FOR HEDGEROW CONSERVATION

### 5.1    CURRENT POLICY

In the face of prevailing productivist farm management, hedgerows currently are afforded a modicum of protection through the following means. First, through statutory protection under the Hedgerow Regulations 1997: these regulations prohibit the removal of any hedgerow over a stipulated length that is over thirty years old and meets an additional listed criterion (e.g. contains certain woody species) (Natural England 2014). While these regulations in theory protect most hedgerows from outright removal, they do very little to protect hedgerows from inappropriate management as described earlier in this chapter, management that remains the norm rather than the exception. Second, through cross-compliance requirements placed on the receipt of Basic Payment Scheme monies (direct payments), received by all eligible farmers (the great majority), and requiring farmers inter alia to desist from hedge trimming between the end of March and the end of August (Rural Payments Agency 2020). These requirements are minimalistic and do little to discourage inappropriate management outside these dates. Moreover, the cross-compliance requirements will discontinue with the cessation of direct payments, scheduled for 2027. Third, there are hedgerow conservation/renovation and establishment options available within the agri-environment schemes (the various tiers of the Environmental Stewardship scheme) (Rural Payments Agency 2015). These schemes, however, are voluntary and competitive due to constrained agri-environment budgets, so that they are not, in practice, even available to all farmers who might wish to take them up. Moreover, being non-mandatory, there is no obligation for farmers outside these schemes to follow their management guidelines, for example, for rotational rather than annual trimming, hedge-laying, etc. Thus, within the current policy context, appropriate hedgerow management, one that maximises the ES delineated above and minimises ecosystem disservices (perhaps notably the emission of huge quantities of greenhouse gas in the course of hedgerow 'over'-management), remains the exception rather than the rule.

## 5.2 ANTICIPATED POLICY

The UK government wishes to replace, through phased withdrawal, the inherited support structures of the CAP with a system that affords no direct economic support to farmers, and confines public subvention to 'public goods' payments – that is, to 'goods and services' (many of the ES listed above) that productivism cannot effectively commoditise, and which are therefore destroyed, degraded, or neglected. It should be noted, however, that the neoclassical economic theory on which 'public goods' arguments are made is deeply flawed, since it is assumed that it is the non-commodification of 'public goods' that causes their loss and degradation, when it is evident that commoditisation itself is equally subversive of ecological and social use values in the farmed environment through the pressures of market dependency, leading to appropriationism and substitutionism. In this way, it is proposed to effectively confine public subvention to Environmental Land Management Schemes (ELMs) of which there will be three main components: The Sustainable Farming Initiative (SFI) (to support 'environmentally sustainable farming' across the landscape), Local Nature and Recovery (to support local environmental priorities and recovery), and Landscape Recovery (to support longer-term land use change projects, including 'rewilding') (DEFRA 2021). These are to be 'supported' by a set of regulatory standards, the configuration of which remains as yet unclear. ELMs, however, will remain voluntary and discretionary (competitive), so whether they will differ in any significant way from the current Environmental Stewardship scheme, and its inadequacies, remains to be seen. The fundamental problem is that they will be competing against adverse pressures flowing, not now from 'political productivism', but rather from the 'market productivism' embodied in the 'free trade agreements' (FTAs) that the UK government is committed to concluding with countries that often have significantly lower environmental and social standards than the UK, and which will therefore exert further downward pressure on prices, forcing farmers to further externalise ecological and social costs.

Under the UK government's post-Brexit 'global Britain' scenario, therefore, enhanced competitive pressures will oblige farmers to accelerate 'market productivism' in an attempt to supply the most lucrative markets, including the externalisation of fossil-fuel dependent transportation costs entailed in meeting distant demand. We can, therefore, anticipate a perpetuation of productivism, driven now by market imperatives rather than by the overtly political objectives of the CAP (Tilzey 2000). In the resulting competitive 'race to the bottom' the high opportunity costs of diverting land, investment, and management to conservation use or agroecology mean that agri-environmental 'policy reach' will be limited. The new ELM incentive scheme, unless endowed with very generous budgets, will be able to meet such opportunity costs only on marginal land and will, therefore, 'cherry pick' those sites considered to be of the highest conservation value, leaving the bulk of the countryside to the tender mercies of cost-externalising productivism. Throughout much of the wider countryside, therefore, the only means of securing compliance with environmental objectives will be by means of tighter regulation – as noted, however, these regulations are currently the subject of intense contestation for precisely this reason. These same pressures will also lead to further farm amalgamation (loss of small and medium farms) and the further substitution of machinery for human labour. Absent stronger regulation and enhanced agri-environmental incentives, the future of hedgerows under this scenario does not look bright.

Alternatively, marginal land, threatened with farm abandonment, may well become differential recipients of agri-environment support, including for 're-wilding' projects. Farms here will become primarily producers of 'public goods' rather than food.

The result, overall, will be a dichotomous countryside, the lowlands dominated by 'de-natured' market productivism, and the uplands reverting to 'de-socialised' 'wilderness'. This represents a 'land-sparing', not a 'land-sharing', approach, and represents the antithesis of agroecology. Questions of the co-production of food and biodiversity, without recourse to fossil fuels, and of the supply of, and access to, locally grown and nutritious food for all are wholly neglected in this neoliberal scenario espoused by the UK government. Rather, the realisation of ES, the elimination of ecosystem disservices, and the adoption of 'political' agroecology, all imperatives for real sustainability, require the elimination of capitalist market dependency and the adoption of an entirely new post-capitalist mode of production. This mode would comprise a farming and food policy, integrating environmental policy, premised on the concept of Sustainability through Agroecology.

## 6   THE NEED FOR A HOLISTIC PERSPECTIVE: SUSTAINABILITY THROUGH 'POLITICAL' AGROECOLOGY

An agroecological policy framework would be designed in such a way as to achieve conservation of, and sustainable food production in, the broader fabric of the countryside whilst, simultaneously, delivering 'additionality' on special sites (such as Sites of Special Scientific Interest). This might take a tiered form, with basic tiers for wider countryside management and agroecological food production with higher tiers to deliver more demanding wildlife, resource, and landscape objectives. Farm management options would address, then, three basic situations, from higher to lower tiers: first, sensitive sites (e.g. maintenance and enhancement of semi-natural habitats); second, diversion/ reversion (semi-natural habitat expansion and creation); third, agroecological production focused on most fertile land. All farms delivering these benefits would receive an area payment, graduated according to tier, and subject to degressivity in the lowest tier for farms over a certain hectarage. A strong regulatory baseline would prescribe statutory standards of land management and farming, including proscription of agri-chemicals and artificial fertilisers, strong protection for hedgerows including a proscription of annual flailing except for health and safety purposes. Generally, such regulations would enforce an internalisation of costs currently externalised in productivist farming, including fossil fuel usage, which would no longer be subsidised. This would provide the needed stimulus to farmers to move from conventional to organic and thence to agroecological production. In addition to agri-environmental area payments, farmers would receive, at least initially as a production stimulus, guaranteed prices for food produced agroecologically. Such food would be purchased by local/regional public authorities, effectively severing market dependency and competition, being distributed equitably through stipulated price mechanisms, with free food available to those on lowest incomes. Given the increased labour intensity of agroecological production and conservation management, there would be a policy of rural re-population and diminution in the size of landholdings to encourage new entrants to farming.

What might agroecological production look like and what would be the role of hedgerows in this radically transformed production? The UK government currently pursues an agricultural and food policy that is entirely market-dependent (export and import-dependency to maximise exchange value), making it extremely vulnerable to politico-economic and environmental disruption, whilst itself contributing to those very ecological and food supply insecurities through its agri-chemical and market productivist orientation. UK agriculture achieves very high productivity, but only with massive quantities of fossil fuel, synthetic fertiliser and agro-chemicals, together with extraordinarily expensive equipment and infrastructure, also dependent on fossil fuel. Agri-chemicals derive, of course, from oil, while immense amounts of fuel and electricity are required to synthesise artificial fertilisers from natural gas. Their production and use release huge quantities of carbon dioxide and (much more potent) nitrous oxide into the atmosphere for every kilogram of food commodity that is produced, making industrially-produced bread baked with conventional flour, for example, one of the most climate-destroying foods available. Even this pails into insignificance, however, by comparison to the ecological inefficiencies of raising livestock fed with these grains (beef, pork, and chicken) – overall, 60 percent of the grain grown in the UK is fed to animals (AHDB 2019).

We urgently need, therefore, both for food security and for ecological sustainability, to, firstly, eliminate all grain-based livestock rearing, and to confine livestock farming to pastureland free of artificial fertilisers and agro-chemicals. Secondly, neither productivist nor 'rotational' organic production systems can generate the quantity of grain needed to supply UK consumption in a secure and ecologically sustainable way. Grain can be grown in an agroecologically-based way, however, that increases output whilst minimising fossil fuel usage, enhancing biodiversity, and sequestering greenhouse gases. The solution here is to grow genetically-diverse populations of 'heritage' grains in the same fields, continuously, without animal manure or tillage, following a low-input approach known as Continuous Grain Cropping (CGC) (see Letts 2020). These cereals can be grown in this way so long as the crops are genetically diverse, have tall stems (like traditional wheat) to help suppress weeds, and all the straw is left in the field post-harvest. The nitrogen removed with the grain each year is replaced by nitrogen fallout from the atmosphere, by the mineralisation of plant tissues above and below ground, and by the fixation of nitrogen by an under-sown layer of clover. CGC production yields about 2.5–3.0 t/ha even on fairly poor soils (Letts 2020), which means that current national demand could be met from approximately 2 million hectares of land (current field crop hectarage in the UK [mostly cereals] is over 6 million, but a large percentage goes to feed animals and the crop land also needs to be rotated). If diets were to become increasingly vegetarian/vegan, this area would need to expand further, but would still be less than the current field crop hectarage. The remaining area of non-cultivated land could then be devoted to grass-based agroecological livestock/dairy and other multifunctional uses such as carbon sequestration, 'rewilding', public recreation, etc.

The above determines the basic agricultural land use configuration under agroecology in which hedgerows could thrive and realise their full potential in the provision of ES. Grain production would need to be decentralised to all areas, and likewise, livestock production. This would lead to much more mixed production landscapes in which hedgerows would regain their original functionality, separating livestock from arable crops. New hedgerows would also be planted to create, once more, smaller fields to facilitate rotational grazing of livestock. More extensive production would enable hedgerows to be managed more permissively, with a re-peopled rural landscape and incentives for traditional management enabling hedges to be tended again by skilled human labour. Permissive management would enable hedgerows to realise their huge carbon sequestration potential, conservatively estimated at 40 million news trees, with no planting required (Tilzey 2020b), whilst providing shelter for crops and livestock, biodiversity benefits such as the recovery of many now threatened farmland bird species such as the turtle dove (*Streptopelia turtur*), and a wealth of wild produce for human consumption and for medicinal use (see Chapter 5 of this book). The moral of this tale is that, if only we can throw off the yoke of productivism and market-dependency, we can reclaim, through agroecology and with the help of hedgerows, a landscape that affords us with all that we *really* need – healthy and nutritious food, uncontaminated water, air, and soil, wildlife-rich countryside, fulfilling livelihoods, healing and inspiring environments – all the while helping to save our planet, our only home.

## 7 REFERENCES

AHDB 2019. *Brexit prospects for UK cereals and oilseeds trade*. Agriculture and Horticulture Development Board, 21–44. Kenilworth: Warwickshire.
Altieri, Miguel. 1995. *Agroecology: The Science of Sustainable Agriculture*. Boulder: Westview.
Altieri, Miguel. 1998. "Ecological Impacts of Industrial Agriculture and the Possibilities for Truly Sustainable Farming." *Monthly Review* 50 (3): 60. https://doi.org/10.14452/mr-050-03-1998-07_5.
Bernstein, Henry. 2014. "Food Sovereignty via the 'Peasant Way': A Sceptical View." *The Journal of Peasant Studies* 41 (6): 1031–1063. https://doi.org/10.1080/03066150.2013.852082.
Carey, Peter, et al. 2009. *Countryside Survey: UK Results from 2007*. Wallingford: Centre for Ecology and Hydrology.
DEFRA 2021. Environmental Land Management Scheme: Overview. https://www.gov.uk/government/publications/environmental-land-management-schemes-overview/environmental-land-management-scheme-overview
Dover, John W. 2019. *The Ecology of Hedgerows and Field Margins*. Abingdon: Routledge.

Dowdeswell, HW. 1987. *Hedgerows and Verges*. London: Allen and Unwin.

FAO. 2018. *FAO's Work on Agroecology: A Pathway to Achieving the SDGs*. Food and Agriculture Organisation of the United Nations. Rome: FAO.

Forman, Richard, and Jacques Baudry. 1984. "Hedgerows and Hedgerow Networks in Landscape Ecology." *Environmental Management* 8 (6): 495–510. https://doi.org/10.1007/bf01871575.

Foster, John Bellamy. 2016. "Marx as a Food Theorist," *Monthly Review* 68 (7). https://doi.org/10.14452/mr-068-07-2016-11_1

Gomiero, Tiziano, David Pimentel, and Maurizio Paoletti. 2011. "Is There a Need for a More Sustainable Agriculture?" *Critical Reviews in Plant Sciences* 30 (1-2): 6–23. https://doi.org/10.1080/07352 689.2011.553515.

Goodman, David, Bernardo Sorj, and John Wilkinson. 1987. *From Farming to Biotechnology: A Theory of Agro-Industrial Development*. Oxford: Basil Blackwell.

Grime, John Philip. 1979. *Plant Strategies and Vegetation Processes*. Wiley: Chichester.

Hooper, M D. 1970. "Hedges and History." *New Scientist* 48: 598–600.

IAASTD. 2009. *Synthesis Report with Executive Summary: A Synthesis of the Global and Sub-Global IAASTD Reports*. International Assessment of Agricultural Knowledge, Science and Technology for Development. IAASTD.

Letts, John. 2020. "Continuous Grain Cropping." *Land* 27, 28-34.

Levidow, Les, Michel Pimbert, and Gaetan Vanloqueren. 2014. "Agroecological Research: Conforming or Transforming the Dominant Agro-Food Regime?" *Agroecology and Sustainable Food Systems* 38 (10): 1127–1155. https://doi.org/10.1080/21683565.2014.951459.

Marx, Karl. 1972. *Capital Volumes 1 and 3*. London: Lawrence and Wishart.

Maskell, L.C. et al. 2008. *CS Technical Report No. 1/07 Field Mapping Handbook*. Centre for Ecology and Hydrology: Wallingford.

Mazoyer, Marcel, Laurence Roudart, and James Membrez. 2006. *A History of World Agriculture: From the Neolithic Age to the Current Crisis*. London: Earthscan.

Montgomery, Ian, Tancredi Caruso, and Neil Reid. 2020. "Hedgerows as Ecosystems: Service Delivery, Management, and Restoration." *Annual Review of Ecology, Evolution, and Systematics* 51: 81–102. https://doi.org/10.1146/annurev-ecolsys-012120-100346.

Natural England. 2014. Countryside Hedgerows: Protection and Management. https://www.gov.uk/guidance/countryside-hedgerows-regulation-and-management.

Neeson, JM. 1993. *Commoners: Common Right, Enclosure and Social Change in England, 1700-1820*. Cambridge: Cambridge University Press.

Overton, Mark. 1996. *The Agricultural Revolution in England, 1550-1850*. Cambridge: Cambridge University Press.

Perfecto, Yvette, John Vandermeer, and Angus Wright. 2009. *Nature's Matrix: Linking Agriculture, Biodiversity Conservation and Food Sovereignty*. Abingdon: Routledge.

Pimentel, David. 2006. "Soil Erosion: A Food and Environmental Threat." *Environment, Development, and Sustainability* 8 (1): 119–137. https://doi.org/10.1007/s10668-005-1262-8.

Pollard, E., MD Hooper, and NW Moore. 1974. *Hedges*. London: Collins.

Rackham, Oliver. 1986. *The History of the Countryside*. London: JM Dent and Sons Ltd.

Rural Payments Agency 2015. Environmental Stewardship: Guidance and Forms for Agreement Holders. https://www.gov.uk/government/collections/environmental-stewardship-guidance-and-forms-for-exist ing-agreement-holders

Rural Payments Agency 2020. Guide to Cross-Compliance in England. https://www.gov.uk/guidance/guide-to-cross-compliance-in-england-2021 .

Tilzey, Mark. 2000. "Natural Areas, the Whole Countryside Approach and Sustainable Agriculture." *Land Use Policy* 17 (4): 279–294. https://doi.org/10.1016/s0264-8377(00)00032-6.

Tilzey, Mark. 2011. *Conservation and Sustainability*. In *Sustainability of Rural Systems: Geographical Perspectives*, edited by Ian Bowler, Raymond Bryant, and Chris Cocklin, 247–268. Dordrecht: Springer.

Tilzey, Mark. 2017. "Reintegrating Economy, Society, and Environment for Cooperative Futures: Polanyi, Marx, and Food Sovereignty." *Journal of Rural Studies* 53 (July): 317–334. https://doi.org/10.1016/j.jrurstud.2016.12.004.

Tilzey, Mark. 2018. *Political Ecology, Food Regimes, and Food Sovereignty: Crisis, Resistance, and Resilience.* Cham: Palgrave Macmillan.

Tilzey, Mark. 2020a. "From Neoliberalism to National Developmentalism? Contested Agrarian Imaginaries of a Post-Neoliberal Future for Food and Farming." *Journal of Agrarian Change* 21: 180–201. https://doi.org/10.1111/joac.12379.

Tilzey, Mark. 2020b. "How the Humble Hedgerow Can Help Us Breathe | Letters." The Guardian. September 30, 2020. https://www.theguardian.com/environment/2020/sep/30/how-the-humble-hedgerow-can-help-us-breathe.

Tilzey, Mark. 2021. *Peasant Counter-hegemony Towards Post-Capitalist Food Sovereignty: Facing Rural and Urban Precarity.* In *Resourcing an Agroecological Urbanism: Political, Transformational and Territorial Dimensions,* edited by Chiara Tornaghi and Michiel Dehaene, 202–219. London: Routledge.

Tilzey, Mark and Clive Potter. 2007. *Neoliberalism, Neo-Mercantilism and Multifunctionality: Contested Political Discourses in European Post-Fordist Rural Governance.* In *International Perspectives on Rural Governance: New Power Relations in Rural Economies and Societies,* edited by Linda Cheshire, Vaughn Higgins, and Geoffrey Lawrence, 115–129. Abingdon: Routledge.

Tilzey, Mark, and Clive Potter. 2016. *Productivism versus Post-Productivism? Modes of Agri-Environmental Governance in Post-Fordist Agricultural Transitions.* In *Sustainable Rural Systems: Sustainable Agriculture and Rural Communities,* edited by Guy Robinson, 41–63. London: Routledge.

United Nations. 2005. *Millennium Ecosystem Assessment: Ecosystem and Human Wellbeing, a Framework for Assessment.* Washington DC: Island Press.

Van der Ploeg, Jan Douwe. 2008. *The New Peasantries: Struggles for Autonomy and Sustainability in an Era of Empire and Globalization.* Abingdon: Routledge.

Wach, Elise. 2021. "Market Dependency as Prohibitive of Agroecology and Food Sovereignty—A Case Study of the Agrarian Transition in the Scottish Highlands." *Sustainability,* *13*(4), 1927. https://doi.org/10.3390/su13041927.

Wood, Ellen. 2002. "The Question of Market Dependence." *Journal of Agrarian Change,* 2 (1): 50–87. https://doi.org/10.1111/1471-0366.00024.

Yerby, George. 2016. *The English Revolution and the Roots of Environmental Change: The Changing Concept of the Land in Early Modern England.* London: Routledge.

# 5 Medicinal tree products offer therapeutic benefits and potential diversification for small scale agroforestry and farm woodlands

## *An overview with indications for further studies*

*Anne Stobart, Jose Prieto Garcia, Anja Vieweger, Sally Westaway and Lindsay Whistance*

## 1. INTRODUCTION

There are many uses for non-wood forest products (NWFPs), including the harvest of woody perennials for phytotherapeutic products ranging from whole spectrum plant materials to medicinal extracts with standardised constituents. Such harvests from trees and shrubs are of therapeutic relevance in varied commercial products from herbal tinctures and teas to sports supplements, botanical drinks, cosmetic and veterinary preparations. As demand continues to grow worldwide, there are ongoing concerns about the quality and sustainability of wild-harvested sources. Hence, market opportunities are emerging for domestic cultivation. However, there is a need for research to assist farmers in their decision-making about diversification with medicinal trees, particularly to inform about choice of plant species, appropriate cultivation designs, harvesting and post-processing methods and means to ensure relevant quality. For agroecological farmers, there is further interest in how medicinal trees may provide benefits at the environmental level and for the local community. The net effect of a lack of research data availability is that diversification into medicinal agroforestry and related projects is poorly supported. In this overview we consider these issues through selected research studies and examples, covering key areas of demand and choice of species, supply and cultivation approaches, quality control and additional benefits alongside approaches to research.

### 1.1 DEMAND AND CHOICE OF SPECIES

Today, there is recognition of both the traditional and potential medicinal value of plants worldwide, and an active research culture of pharmaceutical discovery (Bottoni et al. 2020; Chen and Sun 2018; Howes et al. 2020). Meanwhile there are increasing incentives for the inclusion of trees in

DOI: 10.1201/9781003146902-7

agriculture, related to international strategic development goals (United Nations 2015), particularly developing agroecological approaches. The use of agroforestry, integrating trees with animals or crops, has been proposed as transformative for agriculture (Rosati et al. 2020). In addition to benefits for human health, trees provide further ecosystem services of regulation and support such as soil formation and erosion prevention, pollinator habitat, water management, pollution and wind reduction (Coutts and Hahn 2015; Torralba et al. 2016). But how do farmers access information about market demand and choice of medicinal trees suitable for their specific setting?

## 1.2    *Supply and cultivation approaches*

Supply problems affecting wild-harvested medicinal supplies include the loss of habitats due to agricultural development. Of 25,791 plant species worldwide documented as having medicinal use, many have not yet been assessed for threats to sustainability; of those assessed so far, at least 13% are considered threatened species (Antonelli et al. 2020). Increasing pressure is felt by manufacturers to ensure provenance of supplies from sustainable sources (Armbrecht 2021). As demand increases, cultivation of medicinal trees provides a way forward in both temperate and tropical contexts. Use of trees in agroforestry designs can provide additional benefits such as shade and shelter for crops and livestock, soil conservation and water catchment management. Integration of many tree species with other food plants is characteristic of permaculture design approaches which often model young and biodiverse woodland edges. But how can farmers find out which cultivation approaches will work best for them and likely harvest returns from medicinal trees?

## 1.3    *Quality assessment*

Most medicinal plant supplies in trade are purchased dried and may be whole or fragmented. Quality assessment methods are focused on authentication and analysis based on key markers. The approach to quality control is multilayered in that observations are made on the whole or plant parts using a number of techniques. Most reputable herb companies carry out their own testing on batches of herbs, from organoleptic (appearance, smell, etc.) assessment to microbiological testing and chemical analysis techniques such as high-performance thin-layer chromatography (HPTLC). Plant material is also checked for contamination (heavy metals and pesticides). The organizations carrying out these tests base their results on official and/or professional monographs (British Pharmacopoeia Commission 2021; American Herbal Products Association 2014). Without relevant research studies and analytical tools, small-scale farmers are unable to identify or confirm good practices in cultivation and harvest specific to medicinal plants. Issues include a lack of knowledge of the effects on key plant constituents of combining different species, how growth and spacing impact harvest returns. How can farmers better identify whether they are using good practices in cultivating quality plant materials?

## 1.4    *Research approaches*

Our overview focuses on cultivation in a temperate context of the Northern hemisphere, with examples of research relevant to biodiversification with medicinal trees. Cultivation designs of interest include alley cropping with trees, forest gardens and farm woodlands, alongside plant management techniques such as coppicing and pollarding to provide sustainable harvesting. Here, we consider (a) studies related to identifying demand and supply of medicinal trees; (b) relevant research techniques in the literature from examples of quality analysis and further benefits; (c) illustrative case study examples of small-scale growing of medicinal trees. We aim to highlight some key research areas worthy of further encouragement and support.

## 2. OVERVIEW OF RESEARCH LITERATURE AND METHODS

The above questions relating to demand, supply and quality suggest a lack of research answers in connection with agroecological approaches to medicinal tree cultivation and harvest. Pettenella et al. (2019) surveyed the situation of NWFPs in Europe and noted the lack of detailed data and benefits and barriers to further commercialization for small-scale and seasonal activity. In agroforestry a need for more research on medicinal plants has previously been identified in relation to propagation, improvement and quality control (Rao et al. 2004). In permaculture, designs are biodiverse projects imbued with ethical considerations of people and planet care and often, though not necessarily, based on a smaller scale. However, there is little available research on permaculture (Fiebrig et al. 2020). It has been argued by Ferguson and Lovell (2014) that a separation has emerged between agroecological scientific thinking and the movement associated with permaculture, although both have much in common, from support for diversity to interaction between elements of a design, and adaptive management. Both design approaches seek to promote species biodiversity which, in food security terms, is a key step towards increasing resilience (Ulian et al. 2020).

A search for citations including 'medicinal' and 'plant' in the PubMed database readily reveals a rise in reports of studies involving plants with therapeutic uses in the last decade. Between 2009 and 2020, these words in the title or abstract were identified in a total 37,459 published English language results in journals. An additional search for agroforestry-related contexts gives 398 results for 'medicinal'+'agroforestry' mentions in the text between 1997 and 2020. The majority of these studies are related to tropical and subtropical, rather than temperate, climates (see also Chapter 7 in this book). The studies reflect varied research paradigms and emphasise the interdisciplinary nature of our focus, drawing in researchers based in agricultural science, economic botany, environmental studies, forestry, herbal medicine, phytochemistry and pharmaceutical industries. Given the wide-ranging focus of studies involving medicinal plants, it is not surprising that there are varied examples of research techniques, from laboratory analysis to agricultural trials. For medical purposes, botanical identification of plant material is vital. However, it should be noted that the provenance of plant materials is not always clearly reported in the research literature. Frequently, there is no mention of vouchers or other identification methods (see Nesbitt et al. 2010). This kind of omission can considerably reduce the value of these studies.

Further research studies supporting the benefits of trees to economic and environmental health range from ecological surveys to clinical studies. Considerable advantages have been identified in agroforestry systems that provide greater resilience, from both economic and environmental perspectives. Integration of crops is associated with evidence of greater profitability of agroforestry in the United States (Grado and Husak 2004). With regard to environmental benefits, studies suggest that a range of wildlife and pollinators are better supported by biodiversity including trees (Alexander et al. 2020; Donkersley 2019; Varah et al. 2020). There are additional therapeutic benefits of planting trees, including improvement of mental and physical wellbeing related to being in a natural environment. However, the many ways in which forest management can promote wellbeing are still to be fully investigated (Doimo et al. 2020), and hence the specific features of agroforestry initiatives and their contribution are not well understood. European examples of high levels of self-help herbal medicine use (Welz et al. 2019) suggest that there would be interest in developing more links between farmers and users. These closer links between producers and consumers could provide opportunities for raising knowledge levels of the positive effects and risks of herbal medicines.

Increasing awareness of the importance of trees to benefit human health has promoted studies for clinical evidence (Wen et al. 2019). Additionally, studies have shown that the presence of trees in our landscape plays a crucial role in improving people's mental and physical health, and that they are essential for healthy and sustainable communities (Turner-Skoff and Cavender 2019). Donovan (2017) goes further, stating that the positive effects of trees on public health might be significantly larger than their biophysical benefits. In urban settings, a literature study (Wolf et al. 2020) showed that the presence of trees stimulates our restoring capacities, such as attention restoration, mental health and stress reduction. Thus, we might expect similar positive outcomes for farmers, land workers and local communities in addition to the direct benefits of harvests from diversification with medicinal trees.

Due to their environmental and socio-economic benefits, agroforestry systems represent an important value for society in general, providing recreational activities that can diversify income for the famer whilst improving health and enjoyment more widely (McAdam et al. 2009). However, the available evidence for the societal benefits of agroforestry is fragmented. Kay et al. (2019) assessed marketable and non-marketable ecosystem services in 11 European Agroforestry landscapes, their results show that there is a critical gap in economic assessments that fails to account for ecological and social benefits.

## 2.1 CHOOSING MEDICINAL TREE SPECIES

A key need for farmers interested in choosing suitable medicinal tree species is to clarify the likely market alongside other benefits. Potential uses of medicinal tree products could vary from self-sufficiency to commercial harvest alongside other benefits such as biodiversity support and eco-system services. Trees can be an expensive crop to establish and require years of commitment; thus, suitability to the site, ease of ongoing management, integration with existing agricultural activities and likely harvest returns are important issues. Many introduced species, including woody ornamentals, have medicinal uses and can grow well in a variety of temperate contexts (see examples in Foster 2017; Stobart 2020). However, in some situations the choice of trees may be restricted by the requirements of available funding, such as grants based on planting only native species.

Identification of market demand is hampered by a lack of trade data specific to each plant species. Within the European Union, trade records are grouped together under headings such as 'medicinal and aromatic plants', so that individual species are not easily disaggregated (Vasisht et al. 2016). Only where there are species-specific records can the data be readily analysed for particular plants and markets clearly identified. In the North American context, Kruger et al. (2020) were able to draw on legally required records of ginseng (*Panax quinquefolius*) sales in order to investigate other woodland plants in demand by licensed harvesters, finding that at least 47 other NWFPs were traded for medicinal use. Commercial data on natural product sales may be available but only accessed on a subscription basis. In the nutraceutical trade in the United States, there is information on herbal supplements sold in different contexts, from supermarkets to wholefood shops and online (Smith et al. 2018), enabling the top 40 species in terms of sales value to be identified. For example, elder (*Sambucus* species) berries are highly placed in the top five products, having total sales of US$75 m in 2018.

Markets appear to be more readily developed for certain NWFP foods with noted therapeutic benefits, including tree fruits such as those of elder and hawthorn (*Crataegus* species). The elder has been confirmed as beneficial in treating influenza (Torabian et al. 2019), the berries having multiple therapeutic effects of reducing inflammation, reducing viral transmission and modulating cytokine release. The positive benefits of elder have also been reported as having relevance as an adjuvant treatment for COVID-19 (Silveira et al. 2020). Research is ongoing into suitable elder cultivars for berry production (Thomas et al. 2015). In the UK elder flowers have traditionally been wild-harvested (Jones 2020). However, wild collection can be unpredictable, and the supply is not sufficient for the growing global market for both elder flowers and fruits, so that there is increasing interest in methods of cultivation. A study by Romero-Franco et al. (2018) reviewed the potential European elder market, highlighting increasing potential for organically certified crops, but also a lack of studies to evaluate differences between wild and cultivated populations.

Not only is it beneficial to see that certain plant products are rated highly in the present market, but also for many farmers there is interest in anticipating future market growth. Many woody plants offer potential benefit in major health concerns such as dealing with old age, degenerative conditions, cancer and infectious diseases (Porras et al. 2021; Seca and Pinto 2018; Vikrant and Arya 2011). Predictions of future demand may be based on indicative studies of the potential for antibiotic alternatives, anti-inflammatories and antioxidants, all relevant in such problematic complaints. Although studies of the effects of tree products such as bark can be found, which are laboratory or in-vitro based (Tanase et al. 2019; Vikrant and Arya 2011), relatively few proposals for medicinal tree products move forward to clinical studies. However, there are still many markets in which tree products can have 'phytotherapeutic' value, offering general improvements in health and wellbeing based on traditional experience, in addition to cosmetic and skincare uses (Hoffmann et al. 2020).

Whilst the above examples have focused on human health, much can be gained from use of medicinal trees in animal care; for example, the use of willow bark with chickens for parasite control (Panaite et al. 2020). Trees, and other plants rich in condensed tannins, are also effective treatments for intestinal parasite burdens and bloat in ruminants (Mueller-Harvey 2006). For cattle, browsing on osier willow (*Salix viminalis*) has been shown to be more effective than drenching (Anderson et al. 2012). The secondary metabolite salicylic acid is predominantly recognised as a pain suppressant, but it also has anti-inflammatory, antipyretic, antimicrobial and fungicidal properties and is abundant in trees, particularly willow (*Salix* species) and poplar (*Populus* species). Although salicylic acid is related to the active component in aspirin, a widely used human pain killer, its potential is yet to be fully realised in animal medicine and care.

## 2.2 *Cultivation and harvest returns*

Increasing interest in agroforestry is bringing more trees into cultivation within agricultural holdings (Raskin and Osborn 2019). Techniques of tree management are developing well beyond plantation style monocropping for timber. These include design systems such as alley cropping, forest gardens and silvopasture, and may include associated harvesting regimes for woody crops such as coppicing and pollarding. Here we briefly outline some relevant tree management approaches and a number of projects to illustrate agroecological design possibilities with trees, ranging from alley cropping to forest gardening and farm woodlands.

General guidelines on good agricultural and collection practices (GACP) in cultivation and harvest of medicinal field crops are available from bodies such as the European Medicines Agency (2006) and others. Organic cultivation is generally desirable, being characterised by an emphasis on renewable resources and the conservation of soil and water, using methods that promote biodiversity, enrich the soil, minimise pollution and optimise soil, animal and human health. There are organic certification possibilities for farmers based on recognised criteria for different contexts such as farm crops and wild-harvesting. There is an associated cost that may be prohibitive for smaller-scale producers and/or those offering a diverse catalogue of products.

With regard to sources of information on cultivation and harvest of medicinal trees, there have been publications in recent years on cultivation of useful plants including speciality trees and shrubs suited to integrated plantings (Crawford 2015; Fern 2000). In some temperate regions there are ongoing efforts to share knowledge about NWFPs with medicinal uses; for example, the Incredible Innovation Networks for Cork, Resins and Edibles (2021) project includes a focus on southern European aromatic and medicinal plants. In the United States, an innovative programme of research projects highlights forest floor medicinal plants that have high value and can offer good returns, and Davis and Scott (2016) provide information about ginseng cultivation and other possibilities in woodland contexts.

Coppicing offers much potential in promoting vigorous regrowth of woody plants by cutting near to the ground, applicable to many broadleaved species and a few coniferous species. Based on short cycles, coppicing can produce considerable leafy mass and offers a means to harvest bark from branches rather than stripping the main stem, which is very damaging, even killing the tree.

Coppicing is used in the production of ginkgo (*Ginkgo biloba*) leaves, cultivated in plantations in China, France and the United States, supplying a well-established market for a widely used herbal supplement (Isah 2015). This management technique can be used with a wide range of species including elder for flower and fruit production (see Figure 1). Coppice management of woodlands is associated with biodiversity including beneficial effects on flora and wildlife (Broome et al. 2011; Mason and MacDonald 2002).

Pollarding, like coppicing, involves the cutting of the main stem but at a higher level, from 1–2 m or more. This technique was once widespread in traditional farming systems in Europe and still continues in some parts (Haeggström 1998; Williamson et al. 2017). It is now being used to produce fuel, animal litter and forage, with production of up to 90 kg per year of fresh biomass from established trees such as ash and willow (Colin et al. 2017). Pollards can be harvested regularly over a prolonged period resulting in a range of economic products. Pollard production from 332 ash and oak trees was measured in a French study by Gabory (2018) over a range of different ages, finding that the average annual production rates were similar at around 29 kg dry weight per tree per year. Production varied significantly depending on the spacing between trees, age of the trunk and height of the tree.

Read (2003) conducted an extensive survey of pollarding in northern Europe, and noted that shredding (where the side branches are cut repeatedly whilst the top branches are left intact) was often practised alongside pollarding as a method of increasing leaf production. Some figures on pollard productivity were reported: in Norway elm (*Ulmus glabra*) and ash (*Fraxinus excelsior*) trees pollarded over a 5-year cycle approximately 9 kg of foliage was produced per tree per year and in Italy it was estimated that each ash tree took 3–4 hours to pollard and yielded around 0.5 m³

**FIGURE 1**    Elder (*Sambucus* species) in coppice production for fruits (photo credit: A. Stobart).

of fresh leaves. These examples demonstrate the possibility of coppicing and pollarding to optimise production, although the availability of reliable evidence on which to base decisions of medicinal tree management is lacking.

Following harvest, further processing is required in most growing systems since fresh herbal produce deteriorates very rapidly (Müller and Heindl 2006). The efficiency and effectiveness of different drying regimes is an area of interest (Thamkaew et al. 2020). For small-scale farmers, research to support the development of portable means of drying produce on site would be highly beneficial.

### 2.2.1 Alley cropping

Alley cropping provides a design approach introducing trees into field systems. Typically, rows of trees are interspersed with a number of lower and short-lived crops grown in the alleys between the trees. Alleys of a minimum of width are designed so that the area between trees can be tilled, allowing mechanisation. The expectation is that the use of multidimensional plants enhances the combined productivity of the tree and crop to a level equal or superior to usual agricultural yields. The establishment of new alley cropping agroforestry systems can add a high level of diversity to agricultural lands, which is important for system resilience against environmental variation (Wolfe and Smith 2013). This diversity can also contribute to improving crop health and significantly higher abundance of beneficial insects (pest predators and pollinators) has been found to be associated with temperate silvoarable agroforestry systems (Staton et al. 2019).

Well-designed agroforestry systems can be more productive than monocultures. Different components of an agroforestry system can be complementary in their use of solar radiation and water (Smith et al. 2012) and together with improved shelter and microclimatic conditions this may result in higher overall system productivity. Within the European AGFORWARD programme, some benefits have been indicated in productivity of herb crops based on alley cropping designs. For example, lemon balm (*Melissa officinalis*) grown between rows of cherry trees proved more productive than usual crops in terms of essential oil (Mosquera-Losada et al. 2017). The increased productivity was likely to be a consequence of the delayed flowering of the plants growing in light shade. However, competitive interactions between the trees and crops for resources can also result in overall yield reductions, especially at the interface between the trees and the crop. Field experiments and yield modelling in three European countries indicate that agroforestry can increase overall yields in arable systems by up to 40% relative to monoculture arable and woodland systems (Graves et al. 2007). Yield varies widely depending on species, site and growth conditions, and overall yields (tree plus crop) ranged from –2% lower to +140% higher in the agroforestry system when compared to the trees and crops grown in monocultures.

Further studies of ongoing projects may help to understand the benefits and disadvantages of different alley cropping designs. In south-west UK, the Dartington Hall Trust has an innovative development providing long-term access to land for speciality tree crops. This design proposed that tree crops would be maintained in alleys and harvested for elderflowers, apples and Sichuan peppers to supply a number of manufacturers. Planting commenced in 2017 and much practical experience was gained of the issues faced (Bell 2016). The possibility of evaluating further agroforestry combination plantings of this kind for specialist medicinal tree and ground crops is enticing.

### 2.2.2 Forest gardening

Forest gardening draws on permaculture design, maximising space and resources for identified purposes ranging through food and fuel production to other needs and generating many different projects (Remiarz 2017). Based on observations of natural growth processes, these projects often resemble young woodland edges with multiple layers of different species, known as 'food forests'. In south-west UK, the Holt Wood Project (2005-2020) was focused on providing sustainable supplies for a UK-based practitioner of herbal medicine. This project also demonstrated redesign and replanting of a 1 ha redundant sitka spruce plantation. The site was replanted in sections interspersed with native and introduced woody plants for medicinal use (see examples in Table 1). Quarter-acre sections facilitated a management approach based on coppicing or pollarding at regular intervals. Supplies from this medicinal forest garden proved abundant for a herbal clinical practitioner (now retired) and additional harvests were used in body care products which were sold online (Stobart 2020). Scaling up such a forest garden design could provide for a range of health-promoting and artisanal products to be marketed locally or further afield.

### 2.2.3   Farm woodlands and hedgerows

There is increasing interest in productive use of more traditional methods of integrating trees into agricultural systems such as wood pasture and mixed field boundary hedgerows (Westaway and Smith 2019). These can be managed to be multipurpose with potential relevance for production of medicinal tree harvests. For example, the most common hedgerow species in the UK is haw-thorn, which is in demand for its anti-oxidant and other properties (Benabderrahmane et al. 2019) The berries could be harvested for medicinal purposes (Figure 2, see also Chapter 4 on the value of hedgerows). Further advice is needed on suitable species and standards of harvest to encourage farmers to consider management systems maximising production of fruits. A collaborative approach between growers to compare styles of management would provide useful data.

### 2.3   ASSESSING QUALITY – CERTIFICATION AND ANALYSIS

Quality management systems are essential for assessing plant materials for therapeutic purposes, and the highest level of quality is desirable. Due to the twin drivers of customer demand and manufac-turer need for provenance, there is a widespread preference for organic certification as a minimum requirement. Demand for organic products is increasing in Europe, and a Certificate of Inspection is required to underpin organic status. Additional sets of organic criteria and standards are available for wild-harvesting and for natural cosmetic product manufacturers and processors (COSMOS 2020).

Depending on the buyer specifications, plant harvests may be purchased from farmers in whole dried form or after further processing as cut or powdered herb. With regard to the analysis of medicinal plants, whole plant materials are sometimes referred to as full-spectrum products, thus containing a wide range of constituents. On receiving plant supplies many manufacturers will fur-ther check and authenticate species and key constituents, as well as screen for pesticides, heavy metals and other contaminants. If a batch of plant material does not conform to required standards, then it is rejected, and in the United States there are steps towards ensuring that the batch cannot re-enter the supply chain (American Botanical Council 2021). Despite the analytical controls avail-able, there are still herbal products reaching the market that can be found to be adulterated (Booker et al. 2016).

Further extracts may be made according to specific criteria based on key active constituents. The transformation of raw agroforestry materials into modern medicinal products implies the pro-duction of pure, chemically defined and reproducible extracts. Assuming that the identity of the plant materials is not a concern (i.e. they have been certified to be the right botanical species and part of the plant), preliminary quality control of the raw materials is necessary to decide if they are suitable for medicinal purposes. If this test is satisfactory the material is processed (cleaned, dried, cut and or ground) and extracted (solid-liquid extraction) to yield a concentrated substance containing the bioactive primary and secondary metabolites of interest. To certify its pharmaceutical quality, the material must then be analysed for both the presence of active principles (bioactive plant metabolites) and absence of contaminants by means of suitable quantitative analytical techniques (quality control). In the spirit of this chapter, we must also ensure that the whole operation is as environmentally friendly and sustainable as possible.

**TABLE 5.1**

**Medicinal tree and shrub harvesting programme at Holt Wood: management and potential pharmaceutical added value.**

| Name | Species | Tree management cycle (years)[a] | | | Parts used | Medicinal constituents | Quality control test/assay[b] |
|---|---|---|---|---|---|---|---|
| | | 1–2 | 3–6 | 7–10 | | | |
| Barberry | *Berberis aquifolium* Pursh[c] | | Pollard | | Bark | Alkaloids (berberine) | – |
| Birch | *Betula pendula* Roth[d] | Coppice | | | Bark, leaves | Flavonoids (>10.5%) – | TLC/UV – |
| Hawthorn | *Crataegus* spp.[e] | | | Pruning | Flowers, fruit | Vitexin-2-Rha der. (>0.2%) Procyanidins (>0.06%) | TLC/UV |
| Alder buckthorn | *Frangula alnus* Mill. | | Pollard | | Bark | Anthraquinones (>0.002%) | TLC/UV |
| Ash | *Fraxinus excelsior* L.[f] | | Coppice | | Bark, leaves | Hydroxycinnamic acids | TLC/UV |
| Ginkgo | *Ginkgo biloba* L. | | Coppice/pruning | | Leaves | Flavonoids (>0.5%) | TLC/HPLC |
| Witch hazel | *Hamamelis virginiana* L. | Pruning | | | Leaves and twigs | Tannins (>5%) | TLC/gravimetric |
| Black walnut | *Juglans nigra* L. | | | Pollard | Bark | Tannin | – |
| Juniper | *Juniperus communis* L. | Pruning | | | Leaves | Essential oil (>10 ml/kg) | TLC/steam distillation |
| Bird cherry | *Prunus padus* L. | | Pollard | | Bark | – | – |
| White willow | *Salix alba* L.[g] | Pollard | | | Bark, leaves | Salicylates (>1.5%) | TLC/HPLC |
| Violet willow | *Salix daphnoides* Vill. | Pollard | | | Bark, leaves | Salicylates (>1.5%) | TLC/HPLC |
| Elder | *Sambucus nigra* L. | | Coppice | | Flowers | Flavonoids (>0.80%) | TLC/UV |
| Lime | *Tilia cordata* Mill. | | | Pollard | Flowers | Phenolics | TLC/– |
| Cramp bark | *Viburnum opulus* L. | Coppice | | | Bark | Coumarins, tannins | – |
| Prickly ash | *Zanthoxylum americanum* Mill. | | Pruning | | Berries and twigs | Alkaloids, lignans, amides and coumarins | – |

–: Not established; HPLC: high-performance liquid chromatography; TLC: thin-layer chromatography; UV: ultraviolet.

[a] Unpublished data at Holt Wood Herbs regarding frequency of management for harvesting supplies.

[b] According to the British Pharmacopoeia Commission (BP 2023).

[c] The pharmacopeial species *Berberis aristata* D.C. should not contain less than 1.4% of berberine.

[d] Other accepted pharmacopeial species such as *Betula pubescens* Ehrh. as well as hybrids of both species.

[e] Accepted pharmacopeial species are *Crataegus monogyna* Jacq. or *C. laevigata* (Poir.) DC. (syn. *C. oxyacantha* L.) or their hybrids or a mixture of these false fruits.

[f] Accepted pharmacopeial species are *Fraxinus excelsior* L. or *Fraxinus angustifolia* Vahl (syn. *Fraxinus oxyphylla* M. Bieb) or hybrids of these two species or a mixture.

[g] Accepted pharmacopeial species include other *Salix* species such as *S. purpurea* L. and *S. fragilis* L.

**FIGURE 2**    Hawthorn (*Crataegus* species) berries in hedgerows (photo credit: A. Stobart).

What is the best strategy to obtain the extract/s from relatively small quantities of raw materials produced by a number of scattered fields? There are two main approaches to this scenario: the first implies the collection of all raw materials on site and subsequent transport to a central manufacturing plant fully equipped with industrial-scale extraction and laboratories with complex analytical instrumentation (i.e. HPTLC, high-performance liquid chromatography-ultraviolet (HPLC-UV) and gas chromatography-mass spectrometry (GC-MS)). The second approach would imply on-site extraction of each small batch after preliminary quality control using portable instrumentation.

The extraction of agroforestry materials will always generate waste. If we are to devise a 'green and sustainable solution', valorisation of the waste products (exhausted plant materials) is a key factor (see also Chapter 6). When extracted in a central facility, the waste is produced in large quantities usually forcing incineration, landfill or further transport to a recycling centre. When extraction is performed on the farm, the waste is produced in manageable quantities for the farmer to produce animal bedding, compost or energy. The 'on-site' choice seems the way to go, despite the associated challenges imposed by 'moving' the industrial processes of extraction and analysis to the farm. We suggest applying two well-established yet still evolving multidisciplinary approaches in the area of applied chemistry, biology and technology: 'green-extraction' (Chemat et al. 2012) and 'pharmacognosy in a suitcase' (Cordell 2014).

Chemat and co-workers (2012) suggest that 'green extraction' must be based on six principles to innovate all aspects of solid-liquid extraction: innovation by selection of varieties and use of renewable plant resources; use of alternative solvents and principally water or agro-solvents; reduce energy consumption by energy recovery and using innovative technologies; production of co-products instead of waste to include the bio- and agro-refining industry; reduce unit operations and favour safe, robust and controlled processes; and aim for a non-denatured and biodegradable extract without contaminants.

The use of solvents with minimal ecological impact is the firstmost important aspect in this approach. The list includes (in order of decreasing polarity) water, (bio)ethanol, terpenes, glycerol and esters of fatty acids of vegetable oil (Chemat et al. 2012). However, this list may be simplified in our case as most of the traditional medicinal extracts can be obtained with the solely use of hot water (infusion, decoction, percolation) and/or water-ethanol mixtures (tinctures). The use of pure ethanol allows for the extraction of less polar compounds. However, it can be effectively substituted by 'subcritical water' (water at 250°C and a pressure just over 4 MPa). Under these conditions water has a polarity close to ethanol and is suitable for extraction of low-polarity compounds (Teo et al. 2010). Obtaining the dry residue after water extraction calls for either 'freeze drying' or 'spray drying' techniques that currently have to be undertaken in an industrial setup. The choice of extraction-drying combo will depend on the thermal lability of the active molecules of interest. The advantage of spray drying is that nanoencapsulation of the bioactive molecules can be achieved in the same step, thus generating highly bioavailable medicinal extracts (Assadpour and Jafari 2019).

Regardless of the place of extraction, in-field analysis of medicinal plants aims to reduce waste of resources in bringing low quality or 'non-conforming' plant materials from the field to the laboratory for chemical and biological analysis. One way to overcome some of these issues is the use of portable analytical instruments, 'lab-on-a-chip' or remote sensing devices that can be economically used on small plant samples in the field. This approach has been previously called 'pharmacognosy in a suitcase' (Cordell 2014). Two of the best techniques to suit the mobility, sensitivity and precision requirements include near-infrared spectroscopy (Beć et al. 2014) and hyperspectral imaging (Adão et al. 2017; Kiani et al. 2019). They have the advantage of being either non-destructive or accessing a minimum amount of plant material with maximum impact.

The ultimate step to endorse the medicinal value of the extracts has to be carried out at a well-equipped, certified instrumental laboratory that can only be located in a central facility. A series of tests for foreign matter, ashes, dry matter, pesticides and others will ensure the plant material is pure and has been well processed. To analyse if the plant extract contains the right chemistry, an integrated approach for using high-performance liquid chromatography (HPLC) coupled with ultraviolet photodiode array detector (UV-DAD), spectrophotometric, near infrared spectroscopy and chemometric techniques is favoured. Due to its affordability and versatility, thin layer chromatography (TLC) and HPLC-UV-DAD have been the techniques of choice in the field of medicinal natural products, and most of the pharmaceutical monographs call for qualitative TLC and quantitative HPLC-UV analyses of the extracts (Heinrich et al. 2018). However there has been a 'renaissance' of TLC with the development of high-performance instruments (HPTLC; see also Chapters 12-14). This technique automates the classic TLC and adds scanning and hyphenation (mass spectrometry) capabilities to enable complete documentation of the chromatographic information from complex mixtures of pharmaceuticals, natural products and food stuffs (Attimarad et al. 2011).

## 3.  DISCUSSION

### 3.1  *CHOICE OF SPECIES*

Overall, choice of medicinal tree species appears to be supported by limited information on market demand. Much depends on purpose for the harvest, which can vary from herbal medicine preparations to over-the-counter wholefood supplements and from artisan herbal teas to veterinary supplements. Efforts to provide market data in more accessible form for farmers should be considered, as could networking possibilities where collaborative planning of harvests might match increasing market demand. Although farmers may be encouraged with more information to make decisions about individual plant species, there may be further issues regarding provenance of plants, and whether efforts should be made to source plants from climates more akin to future expected climates, such as more southerly locations. Of particular interest would be examples of combinations of medicinal trees and other crops that have been found by experience to be profitable. Collaborative networks of organic farmers could share such information as demand rises.

## 3.2 CULTIVATION

There is a particular absence in the research literature of studies to underpin new or experimental methods of cultivation of medicinal trees and specific NWFPs. Regarding the supply of medicinal plant materials there are advantages in medicinal agroecological approaches using trees, from productiveness of young woodland to sustainability of the harvest. Leaf and other harvests can be substantial from coppiced and pollarded trees. However, productivity of agroforestry and permaculture designs with biodiverse planting is largely based on self-report, especially for herbal crops. In terms of productivity, the wider level of information on harvest returns for medicinal trees remains sparse. This is of critical importance on matters related to scaling up cultivation where economies of scale may be required to increase viability. For organic farmers wishing to integrate trees and shrubs, inclusive approaches to gathering data such as audit systems may be possible to set up. Participative self-audit in which farmers can record an agreed upon set of data is one way that considerable information can be shared and further analysed. In the European context there are a number of initiatives promoting collaborative efforts of farmers to identify best practices, and the AGFORWARD project has promoted much sharing of experience (e.g. see Colin et al. 2017).

## 3.3 QUALITY ISSUES

High quality of produce underpins the successful expansion of demand, whether plant materials are intended for self-help purposes or the wider market for medicinal NWFPs. Organic certification offers reassurance on provenance and quality, but further standards may be required by buyers of plant supplies and checked with quality assessments. The availability of analytical techniques is largely dependent on specialist facilities, though increasingly widespread and available at lower cost. Greater opportunities for on-site analysis could be further developed. Links between farmers and researchers could be highly productive and benefit quality of medicinal plant supplies by providing a means to identify promising cultivars, agroecological designs and harvest techniques, as well as clarifying the effects of environmental stresses and post-harvest practices.

Due to all the variables involved in the raw material production from many small farmers in geographically scattered places, site-to-site and batch-to-batch consistency is one of the biggest challenges in the development of an active herbal medicinal ingredient from sustainable agroforestry practices. The pharmacological value of a herbal formulation depends on its exact, reproducible chemical composition in both qualitative and quantitative terms. The main active ingredients (or phytomarkers) must be within pharmacopoeial acceptable levels to ensure efficacy and safety.

Development and validation of robust production, traceability, processing, extraction and analytical methods, comprising qualitative and quantitative evaluation of phytochemical profile and markers, remain the most significant challenges. However, without proper quality assurance, to expect a consistent therapeutic effect or the assurance of safety is not possible.

## 3.4 FURTHER BENEFITS

As noted above there are wider benefits of agroecological approaches that integrate trees. It is important to recognise interdisciplinary networking, since relevant research studies can be found in many contexts, from ethnobotany to forestry and environmental science and from clinical medicine to social sciences. It is apparent that greater shared understanding of concepts and key terminology would be helpful for all involved, improving accessibility of relevant studies. The small scale of growing sites can have implications for the comparability and reliability of data, unless there is some common understanding on the nature of records to be kept (including voucher samples). Studies designed to capture beneficial and other outcomes could also be improved with guidance on suitable monitoring records and support for collaborative networking. Some training in botanical identification and data collection would assist the development of data compatibility. Such projects can not only enable the compilation of self-audited data, but also promote development of relevant research skills amongst all involved.

## 4. CONCLUSION

There is potential for cultivating medicinal trees in temperate contexts as demand increases for high quality herbal supplies. Farmers who seek to diversify by planting medicinal trees need information, particularly in regard to species choice and potential market demand or other benefits. Agroecological approaches to integrating medicinal trees include agroforestry and permaculture designs. Farm woodlands and hedgerows may also have a role. Management approaches involving coppicing and pollarding offer productive and sustainable harvests. For quality control and improvement there is a need to enable closer links and access to analytical facilities. On-site analytical techniques could provide a way forward for small-scale farmers to meet required standards. Participative approaches to research such as collaborative audits could also provide data to inform species choices and cultivation approaches. Ideally, funding schemes need to enable collaboration in an interdisciplinary manner, linking experts from the range of sectors along the value chain to jointly elaborate future solutions for more diverse growing systems for medicine production.

## REFERENCES

Adão, T., J. Hruška, L. Pádua, J. Bessa, E. Peres, R. Morais, and J. J. Sousa. 2017. "Hyperspectral imaging: A review on UAV-based sensors, data processing and applications for agriculture and forestry." *Remote Sensing* 9:1110.

Alexander, K., J. Butler, and T. Green. 2020. "The value of tree and shrub species to wildlife." *ARB Magazine* 190:57–64.

American Botanical Council. 2021. *ABC AHP NCNPR Botanical Adulterants Program.* https://www.herbalgram.org/resources/botanical-adulterants-prevention-program/(accessed 25 March 2021)

American Herbal Products Association (AHPA). 2014. *Welcome and about the AHPA Botanical Identity References Compendium.* http://www.botanicalauthentication.org/index.php/Main_Page (accessed 13 January 2021)

Anderson, C. W. N., B. H. Robinson, D. M. West, L. Clucas, and D. Portmann. 2012. "Zinc-enriched and zinc-biofortified feed as a possible animal remedy in pastoral agriculture: Animal health and environmental benefits." *Journal of Geochemical Exploration* 121:30–35.

Antonelli, A., C. Fry, R. J. Smith, M. S. J. Simmonds, P. J. Kersey, and H. W. Pritchard. 2020. *State of the world's plants and fungi 2020.* Kew, UK: Royal Botanic Gardens.

Armbrecht, A. 2021. *The business of botanicals: Exploring the healing promise of plant medicines in a global industry.* White River Junction, VT: Chelsea Green.

Assadpour, E., and S. M. Jafari. 2019. "Advances in spray-drying encapsulation of food bioactive ingredients: From microcapsules to nanocapsules." *Annual Review of Food Science and Technology* 10:103–31.

Attimarad, M., K. K. Ahmed, B. E. Aldhubaib, and S. Harsha. 2011. "High-performance thin layer chromatography: A powerful analytical technique in pharmaceutical drug discovery." *Pharmaceutical Methods* 2:71–5.

Beć, K. B., J. Grabska, and C. W. Huck. 2021. "NIR spectroscopy of natural medicines supported by novel instrumentation and methods for data analysis and interpretation." *Journal of Pharmaceutical and Biomedical Analysis* 193:113686.

Bell, H. 2016. *Food and farming blog: Will our unique new approach bear fruit?* https://www.dartington.org/will-our-unique-new-approach-bear-fruit/(accessed 15 January 2021).

Benabderrahmane, W., M. Lores, O. Benaissa, J. P. Lamas, T. de Miguel, and A. Amrani. 2019. "Polyphenolic content and bioactivities of *Crataegus oxyacantha* L. (Rosaceae)." *Natural Product Research* 35: 627–632.

Booker, A. A. A., D. A. Frommenwiler, E. Reich, S. Horsfield, and M. Heinrich. 2016. "Adulteration and poor quality of Ginkgo biloba supplements." *Journal of Herbal Medicine* 6:79–87.

Bottoni, M., F. Milani, L. Colombo, K. Nallio, P.S. Colombo, C. Giuliani, P. Bruschi, and G. Fico. 2020. "Using medicinal plants in Valmalenco (Italian Alps): From tradition to scientific approaches." *Molecules* 25:4144.

British Pharmacopoeia Commission. 2023. *British Pharmacopoeia.* London: TSO. Available at https://www.pharmacopoeia.com (accessed 30 January 2023).

Broome, A., S. Clarke, A. Peace, and M. Parsons. 2011. "The effect of coppice management on moth assemblages in an English woodland." *Biodiversity and Conservation* 20:729–49.

Chemat, F., M. A. Vian, and G. Cravotto. 2012. "Green extraction of natural products: Concept and principles." *International Journal of Molecular Sciences* 13:8615–27.

Chen, G., and W. Sun. 2018. "The role of botanical gardens in scientific research, conservation, and citizen science." *Plant Diversity* 40:181–88.

Colin, J., P. Van Lerberghe, and F. Balaguer. November 2017. *Farming with pollards: A productive way of pruning.* Innovation Leaflets: vol. 26, AGFORWARD. https://www.agforward.eu/index.php/en/Innovation-leaflets.html (accessed 16 January 2021).

Cordell, G. A. 2014. "Ecopharmacognosy: Exploring the chemical and biological potential of nature for human health." *Biology, Medicine, and Natural Product Chemistry* 3:1–14.

COSMOS. 2020. *Organic and natural products certification.* https://www.cosmos-standard.org/cosmos-certification (accessed 14 January 2021).

Coutts, C., and M. Hahn. 2015. "Green infrastructure, ecosystem services, and human health." *International Journal of Environmental Research and Public Health* 12:9768–98.

Crawford, M. 2015. *Trees for gardens, orchards and permaculture.* East Meon, UK: Permanent Publications.

Davis, J., and W. Scott Persons. 2016. *Growing and marketing ginseng, goldenseal and other woodland medicinals.* Gabriola Island, Canada: New Society Publishers.

Doimo, I., M. Masiero, and P. Gatto. 2020. "Forest and wellbeing: Bridging medical and forest research for effective forest-based initiatives." *Forests* 11:791.

Donkersley, P. 2019. Trees for bees. *Agriculture, Ecosystems and Environment* 270-271:79-83.

Donovan, G. H. 2017. "Including public-health benefits of trees in urban-forestry decision making." *Urban Forestry and Urban Greening*, 22:120–123.

European Medicines Agency. 2006. *Guidelines on good agricultural and collection practice for starting materials of herbal origin.* https://www.ema.europa.eu/en/good-agricultural-collection-practice-starting-materials-herbal-origin (accessed 25 March 2021).

Ferguson, R. S., and S. T. Lovell. 2014. "Permaculture for agroecology: Design, movement, practice, and worldview. A review." *Agronomy for Sustainable Development* 34:251–74.

Fern, K. 2000. *Plants for a future: Edible and useful plants for a healthier world.* East Meon, Hampshire: Permanent Publications.

Fiebrig, I., S. Zikeli, S. Bach, and S. Gruber. 2020. "Perspectives on permaculture for commercial farming: Aspirations and realities." *Organic Agriculture* 10:379–94.

Foster, S. 2017. "Forest gems: Exploring medicinal trees in American forests." *HerbalGram* 116:50–71.

Gabory, Y. 2018. *Productivity of pollards in agroforestry systems (ash and oak).* Proceedings of the 2nd European Symposium on Pollarding, Sare, France. Association Française d'Agroforesterie. https://www.agroforesterie.fr/colloque_europeen_trognes_2018/documents/ACTES-COLLOQUE-TROGNES-intervention-Yves-Gabory-Mission-Bocage-Maine-et-Loire-France.pdf (accessed 25 March 2021).

Grado, S., and A. L. Husak. 2004. "Economic analysis of a sustainable forestry system in the southeastern United States." In: *Valuing agroforestry systems*, edited by J. R. R. Alavalapati and D. E. Mercer. New York: Kluwer, 39–57.

Graves, A. R., P. J. Burgess, J. H. N. Palma, F. Herzog, G. Moreno, M. Bertomeu, C. Dupraz, F. Liagre, K.J. Keesman, W. van der Werf, A. Koeffeman de Nooy, and J.P. Briel. 2007. "Development and application of bio-economic modelling to compare silvoarable, arable, and forestry systems in three European countries'." *Ecological Engineering* 29:434–49.

Haeggström, C.-A. 1998. "Pollard meadows: Multiple use of human-made nature." In *The ecological history of European forests*, edited by K. J. Kirby and C. Watkins. Wallingford: CABI, 33–41.

Heinrich, M., J. Barnes, J. Prieto-Garcia, S. Gibbons, and E.M. Williamson. 2018. *Fundamentals of pharmacognosy and phytotherapy.* 3rd ed. Edinburgh: Elsevier.

Hoffmann, J., F. Gendrisch, C. M. Schempp, and U. Wölfle. 2020. "New herbal biomedicines for the topical treatment of dermatological disorders." *Biomedicines* 8:27.

Howes, M. R., C. L. Quave, J. Collemare, E.C. Tatsis, D. Twilley, E. Lulekal, A. Farlow, L. Li, M.-E. Cazar, D.J. Leaman, T.A.K. Prescott, W. Milliken, C. Martin, M. Nuno De Canha, N. Lall, H. Qin, B.E. Walker, C. Vásquez-Londoño, B. Bob Allkin, M. Rivers, M.S.J. Simmonds, E. Bell, A. Battison, J. Felix, F. Forest, C. Leon, C. Williams, and E.N. Lughadha. 2020. "Molecules from nature: Reconciling biodiversity conservation and global healthcare imperatives for sustainable use of medicinal plants and fungi." *Plants, People, Planet* 2:463–81.

Incredible Innovation Networks for Cork, Resins and Edibles. 2020. *INCREDIBLE offers Networks for Mediterranean Non-Wood Forest Products that foster innovation in science and practice exchange.* https://www.incredibleforest.net (accessed 16 January 2021).

Isah, T. 2015. "Rethinking *Ginkgo biloba* L.: Medicinal uses and conservation." *Pharmacognosy Reviews* 9:140–48.

Jones, A. 2020. "Cultivating elders for the UK processing industries. Nuffield International Farming Scholars." https://www.nuffieldscholar.org/reports/gb/2019/cultivating-elders-uk-processing-industries (accessed 21 March 2021).

Kay, S., A. Graves, J. H. N. Palma, G. Moreno, J.V. Roces-Díaza, S. Aviron, D. Chouvardas, J. Crous-Duran, N. Ferreiro-Domínguez, S García de Jalón, V. Măcicăşan, M.R. Mosquera-Losada, A. Pantera, J.J. Santiago-Freijanes, E. Szerencsits, M. Torralba, P.J. Burgess, and F. Herzog. 2019. "Agroforestry is paying off – economic evaluation of ecosystem services in European landscapes with and without agroforestry systems." *Ecosystem Services* 36:100896.

Kiani, S., S. M. van Ruth, L. W. D. van Raamsdonk, and S. Minaei. 2019. "Hyperspectral imaging as a novel system for the authentication of spices: A nutmeg case study." *LWT* 104: 61–9.

Kruger, S. D., J. F. Munsell, J. L. Chamberlain, J.M. Davis, and R.D. Huish. 2020. "Describing medicinal non-timber forest product trade in eastern deciduous forests of the United States." *Forests* 11:435.

Mason, C. F., and S. M. MacDonald. 2002. "Responses of ground flora to coppice management in an English woodland – a study using permanent quadrats." *Biodiversity and Conservation* 11:1773–89.

McAdam, J., P. J. Burgess, A. R. Graves, A. Rigueiro-Rodríguez, and M. R. Mosquera-Losada. 2009. "Classifications and functions of agroforestry systems in Europe." In: Rigueiro-Rodríguez, A. J. McAdam, and M. R. Mosquera-Losada (eds.) *Agroforestry in Europe: Current status and future prospects.* Dordrecht: Springer Netherlands, 21–41.

Mosquera Losada, M. R., N. Feirreiro-Dominguez, R. Romero-Franco, and A. Rigueiro-Rodriguez. 2017. *Intercropping medicinal plants under cherry timber trees: Understory planting to improve productivity of plantations.* Innovation Leaflets Vol. 29. AGFORWARD. https://www.agforward.eu/index.php/en/Innovation-leaflets.html (accessed 14 January 2021).

Mueller-Harvey, I. 2006. "Unravelling the conundrum of tannins in animal nutrition and health." *Journal of the Science of Food and Agriculture.* 86:2010–37.

Müller, J., and A. Heindl. 2006. "Drying of medicinal plants." In *Medicinal and aromatic plants*, edited by R. J. Bogers, L. E. Craker and D. Lange. Dordrecht: Springer, 237–52.

Nesbitt, M., R. P. H. McBurney, M. Broin, and H.J. Beentje. 2010." Linking biodiversity, food and nutrition: The importance of plant identification and nomenclature." *Journal of Food Composition and Analysis* 23:486–98.

Panaite, T. D., M. Saracila, C. P. Papuc, C.N. Predescu, and C. Soica. 2020. "Influence of dietary supplementation of *Salix alba* bark on performance, oxidative stress parameters in liver and gut microflora of broilers." *Animals (Basel)* 10:958.

Pettenella, D., G. Corradini, R. Da Re, M. Lovric, and E. Vidale. 2019. "NWFPs in Europe – consumption, markets and marketing tools." In: *Non-wood forest products in Europe: Seeing the forest around the trees*, edited by B. Wolfslehner, I. Prokofieva and R. Mavsar. Joensuu, Finland: European Forest Institute, 31–53.

Porras, G., F. Chassagne, J. T. Lyles, L. Marquez, M. Dettweiler, A.M. Salam, T. Samarakoon, S. Shabih, D.R. Farrokhi, and C.L. Quave. 2021. "Ethnobotany and the role of plant natural products in antibiotic drug discovery." *Chemical Reviews* 121:3495–560.

Rao, M. R., M. C. Palada, and B. N. Becker. 2004. "Medicinal and aromatic plants in agroforestry systems." *Agroforestry Systems* 61–62:107–22.

Raskin, B., and S. Osborn, eds. 2019. *The agroforestry handbook: Agroforestry for the UK.* Bristol: Soil Association.

Read, H. 2003. *A study of practical pollarding techniques in northern Europe.* https://www.ancienttreefo rum.co.uk/wp-content/uploads/2017/05/A-study-of-practical-pollarding-techniques-in-northern.pdf (accessed 25 March 2021).

Remiarz, T. 2017. *Forest gardening in practice: An illustrated practical guide for homes, communities and enterprises.* East Meon, Hampshire: Permanent Publications.

Romero-Franco, R., A. Rigueiro-Rodríguez, N. Ferreiro-Domìnguez, M. González-Hernández, F. Rodriguez-Rigueiro, M. Rosa Mosquera. 2018. *Production of medicinal and culinary plants in agroforestry systems:* Sambucus nigra *L. AFINET Factsheet 23.* https://euraf.isa.utl.pt/files/pub/20190615_factshee t_23_en_web.pdf (accessed 25 March 2021).

Rosati, A., R. Borek, and S. Canali. 2020. "Agroforestry and organic agriculture." *Agroforestry Systems.* doi: 10.1007/s10457-020-00559-6. Epub 20 October 2020.

Seca, A. M. L., and D. C. G. A. Pinto. 2018. "Plant secondary metabolites as anticancer agents: Successes in clinical trials and therapeutic application." *International Journal of Molecular Sciences* 19:E263.

Silveira, D., J. -M. Prieto-Garcia, F. Boylan, O. Estrada, Y.M. Fonseca-Bazzo, C.M. Jamal, P.O. Magalhães, E.O. Pereira, M. Tomczyk, and M. Heinrich. 2020. "COVID-19: Is there evidence for the use of herbal medicines as adjuvant symptomatic therapy?" *Frontiers in Pharmacology* 11:581840.

Smith J., B. D. Pearce, and M. S. Wolfe. 2012. "A European perspective for developing modern multifunctional agroforestry systems for sustainable intensification." *Renewable Agriculture and Food Systems* 27:323-32.

Smith, T., K. Kawa, V. Eckl, C. Morton, and R. Stredney. 2018. "Herbal supplement sales in US increased 8.5% in 2017, topping $8 billion." *HerbalGram* 119:62–71.

Staton T., Walters R. J., Smith J., and Girling R. D. 2019. "Evaluating the effects of integrating trees into temperate arable systems on pest control and pollination." *Agricultural Systems* 176:102676.

Stobart, A. 2020. *The medicinal forest garden handbook: Growing, harvesting and using healing trees and shrubs in a temperate climate.* East Meon, Hampshire, UK: Permanent Publications.

Tanase, C., S. Cos, and D. -L. Muntean. 2019. "A critical review of phenolic compounds extracted from the bark of woody vascular plants and their potential biological activity." *Molecules* 24:1182.

Teo, C. C., S. N. Tan, J. W. H. Yong, C. S. Hew, and E. S. Ong. 2010. "Pressurized hot water extraction (PHWE)." *Journal of Chromatography A* 1217:2484–94.

Thamkaew, G., I. Sjöholm, and F. G. Galindo. 2020. "A review of drying methods for improving the quality of dried herbs." *Critical Reviews in Food Science and Nutrition.* 61:1763–86.

Thomas, A. L., P. L. Byers, Jr. J. D. Avery, M. Kaps, and S. Gu. 2015. "Horticultural performance of eight American elderberry genotypes at three Missouri locations." *Acta Horticulture* 1061:237–44.

Torabian, G., P. Valtchev, Q. Adil, and F. Dehghan. 2019. "Anti-influenza activity of elderberry (*Sambucus nigra*)." *Journal of Functional Foods* 54:353–60.

Torralba, M., N. Fagerholm, P. J. Burgess, G. Moreno, and T. Plieninger. 2016. "Do European agroforestry systems enhance biodiversity and ecosystem services? A meta-analysis." *Agriculture, Ecosystems and Environment* 230:150–61.

Turner-Skoff, J. B., and N. Cavender. 2019. "The benefits of trees for livable and sustainable communities." *Plants, People, Planet* 1:323–35.

Ulian, T., M. Diazgranados, S. Pironon, S. Padulosi, U. Liu, L. Davies, M.-J.R. Howes, J.S. Borrell, I. Ondo, O.A. Pérez-Escobar, S. Sharrock, P. Ryan, D. Hunter, M.A. Lee, C. Barstow, L. Łuczaj, A. Pieroni, R. Cámara-Leret, A. Noorani, C. Mba, R.N. Womdim, H. Muminjanov, A. Antonelli, H.W. Pritchard, and E. Mattana. 2020. "Unlocking plant resources to support food security and promote sustainable agriculture." *Plants, People, Planet* 2:421–45.

United Nations. 2015. *Sustainable development goals: Knowledge platform: Forests.* https://sustainabledeve lopment.un.org/topics/forests (accessed 13 January 2021).

Varah, A., H. Jones, J. Smith, and S.G. Potts. 2020. "Temperate agroforestry systems provide greater pollination service than monoculture." *Agriculture, Ecosystems & Environment* 301:107031.

Vasisht, K., N. Sharma, and M. Karan. 2016. "Current perspective in the international trade of medicinal plants material: An update." *Current Pharmaceutical Design* 22:4288–336.

Vikrant, A., and M. L. Arya. 2011. "A review on anti-inflammatory plant barks." *International Journal of PharmTech Research* 3:899–908.

Welz, A. N., A. Emberger-Klein, and K. Menrad. 2019. "The importance of herbal medicine use in the German health-care system: Prevalence, usage pattern, and influencing factors." *BMC Health Services Research* 19:952.

Wen, Y., Q. Yan, Y. Pan, X. Gu, and Y. Liu. 2019. "Medical empirical research on forest bathing (Shinrin-yoku): A systematic review." *Environmental Health and Preventive Medicine* 24:70.

Westaway, S., and J. Smith. 2019. *Productive hedges: Guidance on bringing Britain's hedges back into the farm business. Technical note.* doi:10.5281/zenodo.2641807 (accessed 21 March 2021).

Williamson, T., G. Barnes, and T. Pillatt. 2017. *Trees in England: Management and disease since 1600.* Hatfield, UK: University of Hertfordshire Press.

Wolf, K. L., S. T. Lam, J. K. McKeen, G. R. A. Richardson, M. van den Bosch, and A. C. Bardekjian. 2020. "Urban trees and human health: A scoping review." *International Journal of Environmental Research and Public Health* 17:4371.

Wolfe, M. S., and J. Smith. 2013. "Darwin, diversity and future land use." *Aspects of Applied Biology* 121:11–6.

# 6 The potential use of steam distillation residues from medicinal plant material as a natural agricultural agent

*Sibylle Kümmritz, Bettina Klocke and Andrea Krähmer*

## 1 INTRODUCTION

Medicinal plants harbour broad spectra of valuable bioactive substances that are not only widely applied in human and veterinary medicine but are also beneficial for plant health. Essential oils (EOs) are traditionally produced via steam distillation and mainly contain low-molecular terpenoids. The percentage of EO content in dry plant material mostly lies in the range of single to double digits. During the production of EOs via steam distillation, several streams of by-products are not fully used and typically discarded. In addition to the EOs in the condensed phase, these comprise the aqueous phase (so-called hydrolate), distilled biomass or pomace and in the case of distillation under reflux, residual water (the 'tea-extract'). The hydrolates, also known as hydrosols, contain small amounts of residual EO, which cannot be completely separated (at reasonable cost) and mainly consist of polar, water-soluble components (D'Amato et al. 2018; Rao 2012).

Consideration of these by-products might increase the economic value of the plant resources and thus their processing. The beneficial effects of selected EOs on plant health are known (Srivastava et al. 2015; Zimmermann 2012; Raveau et al. 2020). They have advantageously high volatility and therefore low persistence in comparison with synthetic pesticides. Moreover, pests have developed resistance to some synthetic pesticides, but this has not yet been described for EOs, possibly because they have multiple components. However, EO composition and quality depend on the plant variety, diverse environmental factors and processing conditions.

EO production has considerable economic importance for several countries such as orange oil from Brazil (Barbieri and Borsotto 2018) and lavender oil from Bulgaria and France (Giray 2018). Due to their antimicrobial, antioxidant or antiinflammatory activity, EOs have application in various sectors including the flavour and fragrance, cosmetics, food and pharmaceutical industries. Furthermore, there is growing interest in the use of EOs for pest and disease management in agroecosystems (Srivastava et al. 2015), due to both the lack of pests' resistance to them (associated with their complex composition) and greater ecological friendliness than synthetic pesticides.

Since the production of medicinal and aromatic plants does not currently meet global demand, the commercial use of EOs and their by-products is mainly restricted to human and veterinary medicine, resulting in limited application in crop protection (Isman 2020). In recent years, increasing socio-political pressures for a reducing the use of synthetic pesticides has coincided with a decrease in the number of licensed agents. This has raised awareness of plant-based alternatives. There are also increasing pressures for multiple and sustainable use of resources, especially agricultural raw material materials, as espoused, for example, in "A European Green

DOI: 10.1201/9781003146902-8

Deal" (European Commission 2021). Concerning distillation residues, some studies have focused on the antibacterial effects against *Staphylococcus* sp. or *Escherichia coli*, which are largely due to antioxidative activities and of particular interest for the food industries (D'Amato et al. 2018). Steam distillation residues also have known agricultural applications, but they have not been systematically studied. This chapter aims to provide a summary of applications of these so-called waste-streams for pest-management and plant fortification demonstrated *in vitro*, with some examples of *in vivo* application. These examples and explanations highlight the bioactivities of hydrolates (e.g., antifungal effects on plant pathogenic fungi and repellent effects on insects). Furthermore, important findings concerning application procedures and regulations are discussed. In recent years, the focus has shifted to include effects of EOs as biopesticides not only on target species but also on non-target organisms (Pavela and Benelli 2016). Thus, this chapter also discusses the initial findings concerning hydrolates.

The use of by- and waste-products of EO production increases the agricultural value of aromatic and medicinal plants, and thus could improve the profitability of distillation and expand the cultivation of these plants. This would result in higher plant diversity in the fields, thereby contributing to the enhancement of biodiversity and resilient agriculture. Further investigation of the protective effects of these secondary plant metabolites will support the desired transformation to a sustainable, bio-based economy.

## 2   AVAILABILITY AND CHEMICAL FACTS ABOUT HYDROLATES

Concerning the economic value of hydrolates, the global hydrolate market is forecast to grow at 5.17% per annum from 2019 to 2024 (MRFR 2021). By the end of 2024, it might reach a market value of USD 437 million. Hydrolates used in the cosmetic industry are generally made of fresh plant material and the most prominent plant sources include (among others) rose, roman chamomile, neroli and lavender. Europe dominates the supply of hydrolates for the cosmetics industry, accounting for about 40% of the global share, with Italy in the leading position, but growth in production is fastest in the Asia-Pacific region. Regarding the global hydrolate market, Germany, France, Italy, Spain and UK are important European countries, among others (MRFR 2021). For comparison, EO production was estimated as exceeding 150,000 t per annum in 2017, with a market value of USD 6 billion and is projected to reach 370,000 t by the end of the 2020s with a market value higher than USD 10 billion (Barbieri and Borsotto 2018). China and India, followed by Indonesia, Sri Lanka and Vietnam are the main global producers. According to the European Federation of Essential Oils, global EO production accounts for approximately 600,000 ha of the total 1.6 billion ha of land used for agricultural production and a few crops (such as orange, mint and lemon) account for more than two-thirds of total EO crop production (Barbieri and Borsotto 2018). Table 1 lists the plant species providing hydrolates of interest for plant fortification, pest management as well as other agricultural uses and their production characteristics.

Regarding composition, the EO content in a hydrolate typically ranges from less than 1 g/L (D'Amato et al. 2018) to 4 g/L (Petrakis et al. 2015), depending on the methods used to separate EO from the hydrolate and characterize it. However, we have obtained values of up to 1.2 g/L for industrially produced hydrolates (unpublished data). A recent publication by Aćimović et al. (2020) provided an overview of the content and composition of volatile organic compounds in hydrolates from certain plants compared to the corresponding EOs. The composition of the two condensed phases may be similar or dissimilar. With pH ranging from 4.5 to 5.5, hydrolates are acidic and their volatile components mainly comprise alcohols (including sesquiterpene alcohols), aldehydes and ketones, which are more polar and hydrophilic than pure EOs. The amounts and composition of volatile components determine the quality of the hydrolate and pleasantness of its odour. Hydrocarbon monoterpenes present in EOs that have less than 5 mg/L solubility at pH 7 are not present in hydrolates or below the limit of detection.

**TABLE 1**
**Important plant species for essential oil production including metabolites protecting plant crops and managing pest infestations in cultures and products.**

| Plant | Cultivation | Oil production [t per year] | Reference |
|---|---|---|---|
| *Thymus vulgaris* | Extensively in Spain, Germany, France, England | 20-30 globally | (Andrés et al. 2018) |
| *Lavandula* sp. | Main producers | | (Giray 2018) |
| | Bulgaria | 200 | |
| | France | 109 | |
| | Spain | 80 | |
| | China | 10-40 | |
| | Ukraine | 10-15 | |
| *Citrus sinensis* | Extensively in Brazil, United States, Israel, Italy, Spain | 20,000 globally | (Krammer 2004) |
| *Citrus aurantium subsp. bergamia* | Calabria, Brazil, Ivory Coast | 100 – 200 | (Locher and Hartmann-Schreier 2008) |
| *Eucalyptus globulus* *E. smithii* *E. polybractea and E. radiata* | Spain, Portugal, China, Chile, South Africa, Australia | 2,000 – 3,000 globally | (Hartmann-Schreier 2003) |

However, they can occur as oxygenated compounds due to oxidation processes. For example, linalool dominates *Lavandula angustifolia* and *L. intermedia* hydrolates, while the linalyl acetate component of EO derived from these plants is absent. Moreover, the sensitivity of hydrolate components to hydration processes and their intrinsic antimicrobial activity strongly affects hydrolates' stability. Lavandin hydrolates tested by Politi et al. (2020) remained stable and resistant to microbial deterioration for up to 12 months without additional stabilizers, and some hydrolates are relatively stable for up to 2 years (Aćimović et al. 2020).

The geographical origin, cultivation area, harvesting and extraction conditions, as well as the EOs from which hydrolates are derived, influence their composition (Aćimović et al. 2020). Hydrolates obtained from plants such as *Ocimum vulgare* and *Thymus vulgaris*, containing thymol and carvacrol, have high contents of oxygenated monoterpenes. Based on their well-described anti-microbial activity, they could provide added value in various applications (Aćimović et al. 2020). Table 2 describes the composition of selected hydrolates along with the method used to extract the compound. This table also illustrates the diversity of extraction conditions used for analysing hydrolate' composition. A caveat is that there is no standard method for the analysis of hydrolates, which hampers direct comparison of their composition, and hence there are no quality standards for these products.

To illustrate the scale of potential hydrolate production, even by small producers, distillation of one kg of fresh harvested flowers and stems of *Lavandula x intermedia* Emeric ex Loisel. yields approximately 830 ml of hydrolate according to Politi et al. (2020). Moreover, each of a number of traditional small family businesses and small- to medium-sized companies that produce EOs in Germany reportedly generates approximately one million litres of hydrolate per year (personal communication). Most of this remains unused and is discarded. For example, VER Reichstädt GmbH (Germany) produces 1.5 million litres of *Pinus sylvestris* hydrolate per year (Fischer and Fischer 2021).

**TABLE 2**
**Composition of selected hydrolates and methods used for hydrolate extraction.**

| Plant | Extractant/method | Most abundant components (identified) | Reference |
|---|---|---|---|
| *Calendula arvensis*, aerial parts | diethylether | zingiberenol 1 & 2, (*E,Z*)-farnesol, erem oligenol | (Belabbes et al. 2017) |
| *Carum carvi*, fruits | hexane | (*S*)-(+)-carvone | own results[b] |
| *Cinnamomum verum*, bark | none, directly analysed | *trans*-cinnamaldehyde | (Smail Aazza 2011) |
| *Artemisia absinthium*, flowering plants | activated carbon and Soxhlet extraction | (5Z)-2,6-dimethylocta-5,7-diene-2,3-diol | (Julio et al. 2017) |
| *Laurus nobilis*, leaves and stalks with or without flowers or fruits | hexane | 1,8-cineole, methyl eugenol, α-terpineol, (*R*)-linalool, eugenol | (Di Leo Lira et al. 2009) |
| *Lavandula angustifolia*, flowers | pentane | (*R*)-linalool, linalool oxide, borneol | (Prusinowska et al. 2016) |
| *Lavandula x intermedia*, flowers and stems | HS-SPME[a] PDMS fibre | (*R*)-linalool, 1,8-cineole, camphor | (Politi et al. 2020) |
| *Lavandula × intermedia* var. super, flowers and leaves of aerial parts | dichloromethane | α-terpineol, (*R*)-linalool, camphor, borneol | (Andrés et al. 2018) |
| *Lavandula luisieri*, flowers and leaves of aerial parts | dichloromethane | camphor, 2,3,4,4-tetramethyl-5-methylidenecyclopent-2-en-1-one, 5-hydroxymethyl-2,3,4,4-tetramethylcyclopent-2-en-1-one | (Andrés et al. 2018) |
| *Lavandula officinalis*, flowers | none, directly analysed | (*R*)-linalool, camphor, 1,8-cineole, borneol, α-terpineol | (Smail Aazza 2011) |
| *Matricaria chamomilla*, flowers with stems | hexane | α-bisabolol, β-farnesene, α-bisabolol oxide B, chamazulene, bisabolol oxide A | own results[b] |
| *Melissa officinalis*, dried leaves | diethylether | carvacrol, (*Z*)-citral (neral), (*E*)-citral (geranial), | (Petrakis et al. 2015) |
| *Mentha pulegium*, aerial parts | diethylether | carvacrol, piperitenone, 1,8-cineole | (Zekri et al. 2016) |
| *Mentha pulegium*, aerial parts fully flowered, dried | diethylether | piperitone | (Petrakis et al. 2015) |
| *Mentha suaveolens*, both aerial parts | diethylether | piperitenone oxide | (Zekri et al. 2016) |
| *Origanum onites*, aerial parts | hexane, chloroform; SPME[a] | carvacrol, thymol, *cis*-linalool hydrate, *cis*-p-menth-4-ene-1,2-diol, *cis*-p-menth-3-ene-1,2-diol, *cis*-p-menthan1,8-diol, (*R*)-linalool | (Aydin et al. 1996) |
| *Origanum majorana*, aerial parts fully flowered | diethylether | carvacrol, terpinen-4-ol | (Petrakis et al. 2015) |
|  | none, directly analysed | terpinen-4-ol, α-terpineol | (Smail Aazza 2011) |
| *Pinus cembra*, chopped branches with needles | hexane | α-terpineol, terpinen-4-ol, verbenone, *cis-p*-Menth-2-en-1-ol, (+)-borneol | (Chizzola et al. 2021) |
| *Pinus sylvestris*, wood cut | hexane | α-terpineol, terpinen-4-ol, verbenone, camphor, (+)-borneol | own results[b] |
| *Rosmarinus officinalis*, leaves and flowers | HS-SPME[a] PDMS/ DVB fibre | camphor, 1,8-cineole | (Hay et al. 2018) |
|  | none, directly analysed | camphor, 1,8-cineole, verbenone | (Smail Aazza 2011) |

**TABLE 2 (Continued)**
**Composition of selected hydrolates and methods used for hydrolate extraction.**

| Plant | Extractant/method | Most abundant components (identified) | Reference |
|---|---|---|---|
| *Salvia officinalis,* leaves and flowers of aerial parts | none, directly analysed | camphor, 1,8-cineole, *β*-thujone | (Smail Aazza 2011) |
| *Satureja montana* | hexane | carvacrol, thymol | (Pino-Otín u. a. 2022) |
| *Syzygium aromaticum,* leaves | none, directly analysed | eugenol, 1,8-cineole, camphor | (Smail Aazza 2011) |
| *Thymus vulgaris,* leaves and flowers of aerial parts | none, directly analysed dichloromethane | 1,8-cineole, verbenone, camphor, thymol, | (Smail Aazza 2011) (Andrés et al. 2018) |
| *Thymus vulgaris,* leaves and flowers | hexane | thymol | own results[b] |
| *Thymus zygis,* leaves and flowers of aerial parts | dichloromethane | thymol | (Andrés et al. 2018) |

a (HS-)SPME – (head space) solid phase microextraction
b Hydrolates were extracted with six repetitions, using fresh extractant each time. Combined extracts were dried using a speed vac concentrator and resolved in hexane. Extracted components were identified via GC-MS and GC-FID-analysis according to European Pharmacopoeia 8.1 (2015).

In comparison with pure EO, hydrolates have several advantages:

- Easier application and rinsing from surfaces due to their aqueous nature
- Mild odour
- A balanced concentration near saturation
- Currently, hydrolates are typically unused by-products of EO production (exept for distillation of roses).

These characteristics enable direct application of hydrolates wherever aqueous solutions with high concentrations of bioactive compounds in non-toxic solvents are required. Hence, hydrolates can be used to reduce levels of microorganisms in the food industry and for other hygienic purposes (e.g., medical or veterinary cleaning processes). Hydrolates also have great potential utility in plant fortification and protection (e.g., against bacteria, fungi and insects). Although they have lower concentrations of the active compounds than pure EOs, and hence weaker activity (depending on their composition), they also pose lower risks to the applicants and the environment (Aćimović et al. 2020).

## 2.1 HYDROLATES IN PLANT FORTIFICATION AND PROTECTION

Within tritrophic plant, insect and microorganism systems, EOs have broad spectra of ecological and physiological activities. EOs have both attractant effects (e.g., for beneficial insects) and deterrent effects along with associated inhibitory effects on oviposition or larval development for harmful arthropods (Tripathi et al. 2009; Chaudhari et al. 2021). EOs also have protective activities for plants against pathogenic microorganisms. Zimmermann (2012) and Raveau et al. (2020) presented comprehensive descriptions of EOs against harmful organisms (including insects, snakes, mites and weeds) and phytopathogenic bacteria and fungi. However, most studies have concentrated on *in vitro* experiments, and few have investigated effects of applying EOs directly in the field. This could be due to the higher amounts of EO and formulation needed for homogeneous application, but it could also be at least partly due to economic and potential eco-toxicological aspects of EO use. However, as large volumes of hydrolates are produced as by-products during the isolation of EOs via steam distillation and they carry low concentrations of metabolites, they likely have substantially weaker eco-toxicological effects than EOs themselves.

The composition, main components and functional group of EOs are well known to determine the biological activities of hydrolates. Investigations of their biological activities have mainly focused on antimicrobial, antioxidant and antiinflammatory effects. Compared to the substantial body of literature on the beneficial effects of EOs against plant pests, little is known about the efficacy of hydrolates for plant protection. Such efficacy depends on several factors including the (phyto-) pathogen or target organism, the part of the plant used to acquire EO for distillation, harvesting and processing conditions, and the formulation of the applied hydrolate solution. Some hydrolates reportedly have higher antimicrobial effectiveness than the source EOs due to their aqueous nature, and consequently higher terpene availability (e.g., *Calendula arvensis* towards fungi) (Belabbes et al. 2017). Moreover, Garzoli et al. (2020) found that nanoemulsions of *Lavandula intermedia* had a minimum inhibitory concentration of 0.7% towards *Escherichia coli* and 0.06% towards *Bacillus cereus*, while the corresponding pure hydrolate was inactive.

Some positive and supporting effects of hydrolates on plants health are listed in Table 3 and described below in detail.

## 2.2 Nematicidal effects

Andrés et al. (2018) found that hydrolates of *Thymus vulgaris, T. zygis, Lavandula* x *intermedia* and *L. luiseri* had strong nematicidal potential against *Meloidogyne javanica* both *in vitro* and *in vivo*. The dichloromethane fraction of the extract derived from *Thymus* sp., with thymol as the main organic component, had the strongest nematicidal effects *in vitro*. In contrast, the active components of *Lavandula* extracts might be in the aqueous fraction, but they remain unknown. *In vivo* experiments with tomato seedlings revealed that treatment with the dichloromethane fraction of *Thymus* sp. hydrolate had strongly suppressive effects on nematode egg production, infection frequencies and multiplication rates. In addition, Julio et al. (2017) observed strong nematicidal effects of undiluted hydrolate of *Artemisia absinthum* on *Meloidogyne javanica*.

## 2.3 Antibacterial effects

EOs can be used to disinfect seeds. For example, EOs of *Satureja spicigera* containing carvacrol and thymol have reported bioactivity towards *Clavibacter michiganensis*, *Pseudomonas syringae* and *Xanthomonas axonopodis* (Zimmermann 2012). However, due to differences in the structure of their cell membranes, Gram-negative bacteria are less sensitive to EOs than Gram-positive bacteria.

## 2.4 Antifungal effects

*Calendula arvensis* hydrolate extract reportedly has stronger effects *in vitro* against *Penicillium expansum* and *Aspergillus niger* than the source EO. Moreover, in *in vivo* experiments, the hydrolate provided 100% protective activity against *P. expansum* in pears for up to 7 days at 0.02 mg dry weight/L, and a higher concentration of the EO was required for comparable protection (Belabbes et al. 2017). The stronger antifungal effect of the hydrolate is probably due to its higher content of farnesol, which has known inhibitory effects on microorganisms and antifungal properties (Belabbes et al. 2017).

Mattos et al. (2019) investigated the fungicidal effects of hydrolates and aqueous extracts of *Corymbia citriodora*, *Cymbopogon citratus*, *Cymbopogon flexuosus* and *Curcuma longa*. In *in*

**TABLE 3**
**Examples of applications of hydrolates from indicated plant sources against phytopathogens and harmful organisms.**

| Plant | Parasite/Pathogen/Pest | *In vitro or in vivo* assay | Active Component/ mechanism of action | Reference |
|---|---|---|---|---|
| *Artemisia absinthium,* flowering plants | *Meloidogyne javanica* (root-knot nematode) | *in vitro* and *in vivo* | (5Z)-2,6-dimethylocta-5,7-diene-2,3-diol main component, activity N/A | (Julio et al. 2017) |
| *Calendula arvensis* | *Penicillium expansum* (on pears) *Aspergillus niger* (fungi) | *in vitro* | farnesol inhibits microorganisms, antifungal properties | (Belabbes et al. 2017) |
| *Curcuma longa* and *Corymbia citriodora* | *Alternaria steviae Botryosphaeria dothidea Sclerotium rolfsii* (fungi) | *in vitro* | N/A | (Mattos et al. 2019) |
| *Cymbopogon citratus* DC. Stapf | *Colletotrichum lagenarium* (Cucumber anthracnose, fungi) | *in vivo* | citral may cause rupture of the cell membrane integrity and extravasation of cellular components of microorganisms | (Mattos et al. 2019) |
| *Lavandula × intermedia* Emeric ex Loisel. var. super and *L. luisieri* (Rozeira) Rivas-Martínez | *Meloidogyne javanica* (root-knot nematode) | *in vitro* and *in vivo* | N/A | (Andrés et al. 2018) |
| *Lavandula x intermedia* Emeric ex Loisel. | *Tribolium confusum* (flour beetle, insect pest) | *in vitro* | linalool, 1,8-cineole causes insects to turn away and reject flavoured food | (Politi et al. 2020) |
| *Lavandula x intermedia* Emeric ex Loisel. | *Raphanus sativus* (weed) | *in vitro* | 1,8-cineole, camphor and linalool phytotoxic | (Politi et al. 2020) |
| *Melissa officinalis* and *M. pulegium* | *Myzus persicae* (peach aphid) | *in vitro* | N/A | (Petrakis et al. 2015) |
| *Mentha pulegium* L. and *M. suaveolens* | *Toxoptera aurantii* (*Aphididae*) (black aphids prominent on citrus) | *in vitro* | piperitenone oxide and carvacrol insecticidal activity | (Zekri et al. 2016) |
| *Origanum majorana* | *Myzus persicae* (peach aphid) | *in vitro* | carvacrol responsible for mortality, also known for anti-oviposition effects in *Thrips* sp. | (Petrakis et al. 2015) |
| *Thymus vulgaris* L. and *T. zygis* Loefl ex L. | *Meloidogyne javanica* (root-knot nematode) | *in vitro* and *in vivo* | thymol and carvacrol | (Andrés et al. 2018) |

N/A – data not available

*vitro* tests of effects of *C. citriodora* and *C. longa* hydrolates on *Alternaria steviae, Botryosphaeria dothidea, Colletotrichum gloeosporioides* and *Sclerotium rolfsii*, they had the strongest antifungal effects on *B. dothidea,* inhibiting its mycelial growth by up to 64%.

## 2.5  INSECTICIDAL EFFECTS

EOs may have substantial utility in pest management, due to effects on insect growth and development as semiochemicals, and as antifeedants, attractants and/or chemosterilants (Srivastava et al. 2015). The hydrolates may also contain effective EO components and thus possess these activities, *inter alia* against aphids (which are widespread vectors for various pathogens of numerous crops and hence among the most serious plant pests). For example, hydrolates of *Origanum majorana, Mentha pulegium* and *Melissa officinalis* have varying effects on the settling and motion behaviour of *Myzus persicae* (peach aphid) on eggplant (*Solanum melongena.* L. 'Bonica') leaves *in vitro* (Petrakis et al. 2015). In the cited study, application of *O. majorana* hydrolate resulted in 10 to 15% mortality of aphids on leaf discs, but had no inhibitory effect on their settling behaviour. *Mentha pulegium* hydrolate had a minor effect on aphid mortality (< 5%), but a strong inhibitory effect on their settling. *Melissa officinalis* hydrolate had no effect on aphid mortality, but a strong inhibitory effect on their settling behaviour, and increased the average duration of their movement on eggplant leaf discs up to four-fold (Petrakis et al. 2015). Zekri et al. (2016) evaluated the activity of *Mentha pulegium* L. and *M. suaveolens* hydrolates against *Toxoptera aurantii* (black aphid) *in vitro*, and found that concentrations that were lethal to 50% of the insects ($LC_{50}$ values) were 0.01 ml and 0.5 ml diluted in 10 ml of distilled water. However, the effects also depend on aphid age (Petrakis et al. 2015). Volatile components of EOs of members of the Lamiaceae, such as spearmint, are thought to have insecticidal effects through gustatory and olfactory responses (Hori 1999). Thus, piperitenone oxide (Tripathi et al. 2004) and carvacrol (Sedy and Koschier 2003) (Table 2) might make major contributions to these effects as they are volatile and have known toxic effects on insect pests. High carvacrol content might be responsible for mortality (Can Baser 2008), and also has antioviposition effects in *Thrips* ssp. (Sedy and Koschier 2003). Kumar et al. (2011) have reviewed action mechanisms of these and other insecticidal properties of EOs and extracts of *Mentha* species. In addition, Isman (2020) has provided an overview of the commercial development of EOs and derived products as bio-based insecticides. Various publications show (*inter alia*) that when hydrolates and source EOs have similar composition (Table 2) the hydrolates are likely to have comparable effects to the EOs.

## 2.6  WEED CONTROL

EOs from various plants have allelopathic and toxic effects on plants. Thus, they represent natural alternatives to chemical synthetic herbicides. Raveau et al. (2020) and Abd-ElGawad et al. (2021) have provided comprehensive overviews concerning the phytotoxic effects of EOs, including EOs from important sources such as *Eucalyptus* ssp., *Thymus* ssp., *Mentha* ssp. and *Carum carvi*. To date, there have been no comparable reviews concerning hydrolates. Attempts to commercialize these plant-based herbicides have failed due to high cultivation costs and the limited availability of sufficient raw material. However, the use of waste and by-products of EO production for the manufacture of weed control products increases the financial viability and availability of the plant resources. Effects of *Artemisia annua* EO on germination of *Amaranthus retroflexus* and *Setaria viridis in vitro* and spraying on seedlings of both plants *in vivo* have demonstrated the potential utility of EOs or associated products as herbicidal agents (Benvenuti et al. 2017).

In a study of Lavandin flower and stem hydrolates, Politi et al. (2020) showed that both completely inhibited *Raphanus sativus* seed germination, the flower hydrolate most strongly, suggesting it has potential utility for weed control. Monoterpene components of the hydrolate, such as

1,8-cineole and camphor, might be largely responsible for these phytotoxic effects, as various open-chain and monocyclic alcohols (e.g., linalool and 4-terpineol) are more active than hydrocarbons (Politi et al. 2020). Özkan and Tunctürk (2021) studied *in vitro* effects of EOs and hydrosols of *Salvia fruticosa, Ocimum basilicum, Dracocephalum moldavica, Mentha spicata, Salvia officinalis, Melissa officinalis, Origanum onites* and *Thymus kotschyanus* on the germination of *A. retroflexus* seeds. They found that the germination rate decreased with increasing concentrations. Of the tested pure hydrolates, *T. kotschyanus* had the strongest and *S. fructicosa* the weakest inhibitory effects on germination, respectively. To extend the spectrum of hydrolate application for weed control, further research is needed.

## 3  HYDROLATES FOR STORED PRODUCT PROTECTION

Hydrolates have apparent suitability for application not only in agricultural and horticultural production but also in protection of products post-harvest and during storage. Their antioxidant properties suggest that they can reduce spoilage due to oxidation processes (e.g., lipid peroxidation). For example, Hay et al. (2018) detected antioxidant activity of *Thymus vulgaris* hydrolate using the 2,2'-azino-bis(3-ethylbenzothiazoline-6-sulphonic acid) (ABTS) method for measuring free radical scavenging capacity, and obtained a prospective 50% inhibitory concentration ($IC_{50}$) value of about 3000 µl/L. In addition, Belabbes et al. (2017) found that hydrolate extracts of *Calendula arvensis* had similar 2,2-diphenyl-1-picrylhydrazyl (DPPH) radical-quenching activities to butylated hydroxytoluene (BHT), and promising $\beta$-carotene bleaching activity, albeit lower than that of BHT.

This activity might be related to zingiberenols and other oxygenated sesquiterpenes found in the hydrolate (Belabbes et al., 2017) (see Table 2). Among hydrolates of *Lavandula officinalis, Origanum majorana, Rosmarinus officinalis, Salvia officinalis, Thymus vulgaris, Cinnamomum verum* and *Syzygium aromaticum,* Smail Aazza (2011) found that the *S. aromaticum* extract had the highest antioxidative activity, followed by the *Thymus vulgaris* extract. These findings are based on analyses including the thiobarbituric acid reactive species (TBARS) assay as an indicator of prevention of lipid peroxidation, determination of hydroxyl radical and superoxide anion scavenging activities and ABTS and oxygen radical activity capacity (ORAC) assays. Eugenol and carvacrol, as the main components of these hydrolates (see Table 2), might be responsible for such antioxidant activity (Smail Aazza 2011). However, *Lavender angustifolia* and *S. officinalis* hydrolates, respectively, tested by Prusinowska et al. (2016) and Smail Aazza (2011) had negligible antioxidant capacities according to DPPH assays.

Some hydrolates have shown inhibitory effects on tyrosinase, an enzyme responsible for the browning reaction in fruits and vegetables. Moreover, some hydrolates have suitability as natural food sanitizers, as demonstrated by antibacterial effects on *E. coli* and *Staphylococcus* strains in experiments with fresh-cut apples and carrots (Tornuk et al. 2011), salad (Ozturk et al. 2016), tomato and cucumber (Sagdic et al. 2013), wheat, lentil and mung bean seeds (Sahan and Tornuk 2016) and parsley (Törnük and Dertli 2015). Hydrolates, derived from oregano, for example, are also used as flavouring agents in beverages, they are regarded as health-promoting and safe for human consumption (Can Baser 2008). Further possible applications of some hydrolates, of thyme, for example, lie in preventing the deterioration of fish and fish products due to inhibitory effects against *Listeria monocytogenes* (Tornuk et al. 2011), *Aeromonas hydrophila* and *Pseudomonas fluorescens* (Oral et al. 2008). In addition, their antifungal effects (previously described in relation to plant protection, see section 2.4) have proven applicability in preventing the undesired deterioration of fermented products such as sausages. Aćimović et al. (2020) extensively reviewed this field of hydrolates' application as natural food sanitizers. Hydrolates also have applicability as sanitizing solutions for tools, machines and surfaces in the food industry, with significant advantages for preventing biofilm formation over EOs (Aćimović et al. 2020) as they can be more easily rinsed away and are less persistent (D'Amato et al. 2018).

The antifungal properties of hydrolates can reduce undesirable or toxic effects on human health caused by mycotoxins produced by various moulds (e.g. *Aspergillus* spp.) found in food commodities. Amongst others, thyme EO (rich in thymol and carvacrol) can inhibit the growth of a wide spectrum of food-spoiling and mycotoxin-producing fungi (e.g. *Fusarium* and *Aspergillus* species). Therefore, it has potential use in preventing mycotoxin formation, as a food preservative and a potent antimicrobial agent (Srivastava et al. 2015).

*Lavandula* x *intermedia* Emeric ex Loisel. hydrolate extract tested by Politi et al. (2020) had repellent effects on *Tribolium confusum* (confused flour beetle), one of the most widespread and destructive insect pests of economic importance for stored grains in warm temperate regions (Chaudhari et al. 2021).

The application of hydrolates for stored product protection may contribute to food security and safety through reducing losses (material and economic) of crop products, time and costs for quality monitoring and decontamination.

## 4   ISSUES REGARDING HYDROLATE APPLICATION IN AGRICULTURAL PRACTICE

The efficacy of EOs as sources of alternative plant protection products has been extensively described. They are starting points for the development of hydrolate-based products for similar purposes, but a number of factors currently severely limit the commercial use of EOs in agriculture. Apart from efficacy, the most important factor is persistence (referring in this context to how long an EO remains biologically effective in the field and prevents re-infestation with a target organism). This depends on both the volatility of the EO and its susceptibility to degradation (Chandler et al. 2011). Volatility is often a limiting factor for field application of EOs (Raveau et al. 2020), and/or a major cause of requirements for several well-scheduled treatments for effectiveness (Isman et al. 2011). Failure to apply EOs at appropriate rates could lead to phytotoxicity. Even at low concentrations, any EO can have phytotoxic effects (Basaid et al. 2020).

However, appropriate formulation of an EO could mitigate the disadvantages of lower persistence and phytotoxicity described here (Yang et al. 2009). These include the techniques of emulsion and encapsulation to improve EO stability. An emulsion is a mixture of two immiscible and suspended phases in a liquid state, often with a surfactant added to act as a stabilizer (Raveau et al. 2020). Depending on particle diameter and thermodynamic stability, they are referred to as macroemulsions, nanoemulsions or microemulsions. In microencapsulation, a membrane encloses small particles of solid, liquid or gas. The aim is to protect the core material from adverse environmental conditions, such as undesirable levels of light, moisture and exposure to oxygen (Bertolini et al. 2001). This is also suitable for combining EO with different chemical compositions in order to increase the durability of the product and enable controlled, longer-term release in the field (Moretti et al. 2002). It should be noted that the optimal formulation is highly dependent on the intended indication and method of application, the target pathogen and potential environmental factors (Ahmed Salim 2017; Kfoury et al. 2016). With regard to hydrolates, little on the effects of these factors has been described in the literature.

The costs of registering an EO as a plant protection product also require consideration. Often the cost is too high for many producers, as virtually identical investigations are required for the registration of biochemicals and conventional chemical synthetic plant protection products. In addition, EOs are typically described as niche products with very specific applications, reducing the possibility of widespread use (Chandler et al. 2011). The limited use of EOs in agriculture is also partly due to the current low availability of practical results. Additional information concerning their impact on non-targeted organisms or the interactions between different EOs would be desirable (Pavela and Benelli 2016). Large-scale application at the field level is limited by the insufficient efficacy of EOs (Basaid et al. 2020), because yield losses in years with high disease pressure can only be avoided by using products with high efficacy.

There is also a need to optimize knowledge transfer between researchers and producers to promote the faster use of EOs in practice. Farmers can only be convinced if solutions are offered that are not only effective but also economical, even when they have clear ecological advantages. For practical purposes, alternative products based on EOs often require larger quantities, leading to considerable application costs that are increased by the need for several applications (Koul et al. 2008). Therefore, they cannot currently compete with conventional chemical synthetic plant protection products in terms of cost. To compensate for this, it is necessary to invest in innovative and more efficient extraction and formulation processes that enable higher extraction yields (Basaid et al. 2020) and improve the stability and durability of EOs in the field. Using production streams that are not currently considered like hydrolates reduces competition with EO usage in the pharmaceutical and food industries. However, research in this area is challenging and must deliver results for the use of EO-based products in the field that are at least as efficient as under controlled laboratory conditions.

In Germany, only a few plant protection products based on EOs can be found on the market. Examples include Prev-Gold® and Prev-Am® from Oro Agri, which are based on orange (*Citrus* x *aurantium* L.) EO, and have been approved as acaricides, fungicides and insecticides. The product BIOX-M® (XEDA international S. A.) is used to inhibit sprouting of potatoes and is based on spearmint oil. Although the supply of available products is currently low, this is expected to change over the next few years. As registrations for further plant protection products end (largely due to sustainability concerns), farmers will have fewer resources at their disposal in the future. Consistent resistance management will only be possible then with additional alternatives. The awareness of society as a whole of the need for sustainable crop protection has already favoured greater use of alternative methods.

The use of EOs here could considerably reduce the application of chemical synthetic pesticides, in accordance with the principles of Integrated Pest Management (an increasingly favoured approach, involving multiple strategies to control pests in an ecologically friendly manner).

## 5   EFFECTS OF HYDROLATES ON NON-TARGET ORGANISMS (IN FIELD APPLICATIONS)

In general, biopesticides are regarded as eco-friendly alternatives to synthetic pesticides and expected to be less harmful to the environment and human health. Some studies have already evaluated toxicological effects of some EOs on non-target organisms, but more data are required to inform risk assessments for the natural environment and humans (Chaudhari et al. 2021). There is also insufficient information on environmental risks associated with hydrolates and derived products. Most Application Notes do not consider the phytotoxic effects of hydrolates, which thus require further research.

For example, thymol is widely used as active ingredient of commercial formulations against the parasitic mite *Varroa destructor* in European honeybee colonies. Carvacrol also showed positive effects towards honeybee diseases (Wiese et al. 2018, Can Baser 2008). Glavan et al. (2020) determined mortality on honey bees at concentrations of 1% (w/w) for thymol and 5% (w/w) for carvacrol, but its effects on honeybee learning, behaviour and the productivity of entire colonies are unknown (Glavan et al. 2020; Colin et al. 2019). Recently, *Satureja montana* hydrolate, containing mainly carvacrol and thymol, showed strong phytotoxic effect on *Allium cepa* roots with $LC_{50}$ value of 0.05%. It also has a high toxicity against water non-target organisms such as *Daphnia magna* and *Vibrio fisheri* with $LC_{50}$ less than 1% and a $LC_{50}$ value of 4.23% towards river periphyton communities (Rosa Pino-Otín 2022). With $LC_{50}$ value of 4.25% for *Eisenia fetida*, *S. montana* hydrolate showed lowest toxicity to earthworms (Rosa Pino-Otín 2022). Rosa Pino-Otín (2019) determined that *Artemisia* hydrolate inhibited the growth of *A. cepa* roots (with an $LC_{50}$ of 3.9%, v/v) and strongly affected the survival of *E. fetida*. It also decreased the growth of a bacterial community in natural soil at concentrations above 1% (v/v), with a $LC_{50}$ of 25.7% (v/v) after 24 h exposure. However, no significant effect on the physiological diversity of the microbial community was observed. *Lavandula* hydrolate extract showed acute toxicity towards *Allium cepa* (decreasing root growth) but did not increase mortality of *Eisenia fetida* and had minor effects on the worm at doses between 1 and 100% (v/v). However, doses above 1% (v/v) significantly decreased the growth of a bacterial community in natural soil (Pino-Otín, Val, Ballestero, Navarro, Sánchez, González-Coloma, et al. 2019).

## 6 FURTHER FIELDS OF APPLICATION

In addition to the promising agricultural and sanitizing uses of hydrolates derived from plant EOs, they can be potentially used as food additives. For example, *Cymbopogon citratus* hydrolate has been added at 3.5% to create new naturally flavoured ice-cream in attempts to generate so-called new foods (Aćimović et al. 2020).

Hydrolates have also shown therapeutic and prophylactic potential in animal husbandry. Moreover, they can be used as veterinary drugs to treat infections, thus adding to obtain sustainable animal farming practices. For example, *Thymus vulgaris* and *Nepeta cataria* hydrolates have significantly reduced porcine reproductive and respiratory syndrome loads due to antiviral activity *in vitro* (Aćimović et al. 2020).

Besides hydrolates, solid wastes such as distilled biomass should be considered when discussing the value-adding possibilities for EO production by-products as they also contain bioactive compounds (Santana-Méridas et al. 2012). For example, Tiliacos (2008) found that the main components of cyclohexane extracts of distilled lavandin waste were coumarin and its methyl derivative herniarin. Coumarins have diverse biological activities (Stringlis, de Jonge, and Pieterse 2019) and play important roles in plant nutrition and health (e.g. improving iron uptake (Robe et al. 2021)). Like hydrolates, the distilled biomass also often has antioxidant and radical scavenging activities, as described for residues of *Lavandula latifolia* (Méndez-Tovar et al. 2015), *Achillea millefolium*, *Artemisia dracunculus*, *Lavandula latifolia*, *L. latifolia* x *L. angustifolia* Miller, *Melilotus officinalis*, *Foeniculum vulgare* (Parejo et al. 2002), *Hyssopus officinalis*, *Santolina chamaecyparissus*, *Lavandula* x *intermedia* var. *Super Satureja montana* (Ortiz de Elguea-Culebras et al. 2017), *Poliomintha longiflora* (Cid-Pérez et al. 2019), *Origanum sp.*, *Rosmarinus officinalis*, *Satureja thymbra* (Oreopoulou et al. 2018) and *Salvia lavandulifolia* Vahl (Sánchez-Vioque et al. 2018). In some cases, reported antioxidant and radical scavenging activities have even been higher for distilled biomass than for the untreated plant material (Parejo et al. 2002; Ortiz-de Elguea-Culebras et al. 2017). Amongst the distilled biomass of *Achillea millefolium*, *Artemisia dracunculus*, *Foeniculum vulgare*, *Melilotus officinalis* and *Lavandula latifolia* x *Lavandula angustifolia*, *M. officinalis* showed the highest free radical scavenging activity, as determined by the *β*-carotene blenching method, and highest total phenolic content. The highest superoxide radical scavenging activity (measured by the nitro-blue tetrazolium reduction method) was exhibited by extracts of distilled *A. ranunculus*. Additionally, *L. latifolia* provided the highest (and *F. vulgare* showed promising) antioxidant activity, according to assays with 2,2'-azobis(2-methylpropionamide) dihydrochloride (Parejo et al. 2002). Thus, these solid wastes should also be considered when discussing by-products of EO distillation as sustainable products for plant fortification and product protection.

In addition, distilled biomass of *Laurus nobilis* can be used as a fibrous feed, particularly for ruminants (e.g. cattle and sheep), as it has suitable characteristics in terms of nutritional pattern and both energetic and digestibility values (Di Leo Lira et al. 2009).

## 7 CONCLUSION AND FUTURE PROSPECTS

Hydrolates are produced in large volumes as by-products of the industrial steam distillation process. Besides being used as flavour ingredients and for aromatherapy, the application of hydrolates in agriculture is also both possible and prospective. This circumvents a major disadvantage of EOs: lack of cost-effectiveness for agricultural use due to sophisticated and mostly small-scale cultivation. However, differences in the composition of EOs and hydrolates, due to differences in polarity of the components, should be taken into account. Many hydrolates have similar positive effects to EOs against pests, but most evaluations to date have been based on *in vitro* assays. Thus, the applicability of promising hydrolates in fields or greenhouses requires further investigation, as their practical efficacy is a major determinant of their potential contribution to agricultural and broader sustainability. Further research should also focus on the risk assessment of hydrolates towards both the environment and people, as it is a mandatory requirement for the registration of a hydrolate as a plant protection product. There are also clear needs for the optimization and standardization of methods for compositional analyses of hydrolates, preparation of formulations and other key elements of quality assurance and comparability.

In summary, value-adding through the use of EO by-products has clear potential to enhance the importance of medicinal and aromatic plants in agriculture globally. It may also contribute to a more sustainable approach to plant protection and higher plant diversity in the field, but substantial further research is required.

## 8 ACKNOWLEDGMENT

This work was financially supported by the Federal Ministry of Food and Agriculture (BMEL) by decision of the German Bundestag and managed by the Agency for Renewable Resources (FNR) under the program "Renewable Resources", project number: 22021517.

## 9 REFERENCES

Abd-ElGawad, Ahmed M., Abd El-Nasser G. El Gendy, Abdulaziz M. Assaeed, Saud L. Al-Rowaily, Abdullah S. Alharthi, Tarik A. Mohamed, Mahmoud I. Nassar, Yaser H. Dewir, and Abdelsamed I. Elshamy. 2021. "Phytotoxic Effects of Plant Essential Oils: A Systematic Review and Structure-Activity Relationship Based on Chemometric Analyses." *Plants-Basel* 10 (1). Basel: Mdpi: 36. doi:10.3390/plants10010036.

Aćimović, Milica G., Vele V. Tešević, Katarina T. Smiljanić, Mirjana T. Cvetković, Jovana M. Stanković, Biljana M. Kiprovski, and Vladimir S. Sikora. 2020. "Hydrolates: By-Products of Essential Oil Distillation: Chemical Composition, Biological Activity and Potential Uses." *Advanced Technologies* 9 (2): 54–70. doi:10.5937/savteh2002054A.

Ahmed Salim, EL Rasheed. 2017. "Formulation of Essential Oil Pesticides Technology and Their Application." *Agricultural Research & Technology: Open Access Journal* 9 (2). doi:10.19080/ARTOAJ.2017.09.555759.

Andrés, Maria Fe, Azucena González-Coloma, Ruben Muñoz, Felipe De la Peña, Luis Fernando Julio, and Jesus Burillo. 2018. "Nematicidal Potential of Hydrolates from the Semi Industrial Vapor-Pressure Extraction of Spanish Aromatic Plants." *Environmental Science and Pollution Research* 25 (30): 29834–40. doi:10.1007/s11356-017-9429-z.

Aydin, S., K.H.C. Baser, and Y. Öztürk. 1996. "The Chemistry and Pharmacology of Origanum (Kekik) Water." In *Essential Oils: Basic and Applied Research, Proceedings of the 27th International Symposium on Essential Oils*. Vienna, Austria.

Barbieri, Cinzia, and Patrizia Borsotto. 2018. "Essential Oils: Market and Legislation." In *Potential of Essential Oils*, edited by Hany A. El-Shemy. InTech. doi:10.5772/intechopen.77725.

Basaid, Khadija, Bouchra Chebli, El Hassan Mayad, James N. Furze, Rachid Bouharroud, François Krier, Mustapha Barakate, and Timothy Paulitz. 2020. "Biological Activities of Essential Oils and Lipopeptides Applied to Control Plant Pests and Diseases: A Review." *International Journal of Pest Management*, January, 1–23. doi:10.1080/09670874.2019.1707327.

Belabbes, Rania, Mohammed El Amine Dib, Nassim Djabou, Faiza Ilias, Boufeldja Tabti, Jean Costa, and Alain Muselli. 2017. "Chemical Variability, Antioxidant and Antifungal Activities of Essential Oils and Hydrosol Extract of *Calendula arvensis* L. from Western Algeria." Edited by Paul Hatcher. *Chemistry & Biodiversity* 14 (5): e1600482. doi:10.1002/cbdv.201600482.

Benvenuti, S., P. L. Cioni, G. Flamini, and A. Pardossi. 2017. "Weeds for Weed Control: Asteraceae Essential Oils as Natural Herbicides." *Weed Research* 57 (5): 342–53. doi:10.1111/wre.12266.

Bertolini, A. C., A. C. Siani, and C. R. F. Grosso. 2001. "Stability of Monoterpenes Encapsulated in Gum Arabic by Spray-Drying." *Journal of Agricultural and Food Chemistry* 49 (2): 780–85. doi:10.1021/jf000436y.

Can Baser, K. H. 2008. "Biological and Pharmacological Activities of Carvacrol and Carvacrol Bearing Essential Oils." *Current Pharmaceutical Design* 14 (29): 3106–19. doi:10.2174/138161208786404227.

Chandler, David, Alastair S. Bailey, G. Mark Tatchell, Gill Davidson, Justin Greaves, and Wyn P. Grant. 2011. "The Development, Regulation and Use of Biopesticides for Integrated Pest Management." *Philosophical Transactions of the Royal Society B: Biological Sciences* 366 (1573): 1987–98. doi:10.1098/rstb.2010.0390.

Chaudhari, Anand Kumar, Vipin Kumar Singh, Akash Kedia, Somenath Das, and Nawal Kishore Dubey. 2021. "Essential Oils and Their Bioactive Compounds as Eco-Friendly Novel Green Pesticides for Management of Storage Insect Pests: Prospects and Retrospects." *Environmental Science and Pollution Research* 28 (15): 18918–40. doi:10.1007/s11356-021-12841-w.

Chizzola, Remigius, Felix Billiani, Stefan Singer, and Johannes Novak. 2021. 'Diversity of Essential Oils and the Respective Hydrolates Obtained from Three *Pinus Cembra* Populations in the Austrian Alps'. *Applied Sciences* 11 (12). Multidisciplinary Digital Publishing Institute: 5686. doi:10.3390/app11125686.

Cid-Pérez, Teresa Soledad, Raúl Ávila-Sosa, Carlos Enrique Ochoa-Velasco, Blanca Estela Rivera-Chavira, and Guadalupe Virginia Nevárez-Moorillón. 2019. "Antioxidant and Antimicrobial Activity of Mexican Oregano (*Poliomintha longiflora*) Essential Oil, Hydrosol and Extracts from Waste Solid Residues." *Plants* 8 (1): 22. doi:10.3390/plants8010022.

Colin, Théotime, Meng Yong Lim, Stephen R. Quarrell, Geoff R. Allen, and Andrew B. Barron. 2019. "Effects of Thymol on European Honey Bee Hygienic Behaviour." *Apidologie* 50 (2): 141–52. doi:10.1007/s13592-018-0625-8.

D'Amato, Serena, Annalisa Serio, Clemencia Chaves López, and Antonello Paparella. 2018. "Hydrosols: Biological Activity and Potential as Antimicrobials for Food Applications." *Food Control* 86 (April): 126–37. doi:10.1016/j.foodcont.2017.10.030.

Di Leo Lira, P., D. Retta, E. Tkacik, J. Ringuelet, J. D. Coussio, C. van Baren, and A. L. Bandoni. 2009. "Essential Oil and By-Products of Distillation of Bay Leaves (*Laurus nobilis* L.) from Argentina." *Industrial Crops and Products* 30 (2): 259–64. doi:10.1016/j.indcrop.2009.04.005.

European Commission. 2021. "A European Green Deal." Text. Accessed April 8. https://ec.europa.eu/info/strategy/priorities-2019-2024/european-green-deal_en.

European Pharmacopoeia (Ph. Eur.) *DVD-ROM 8. Ausgabe, 1. Nachtrag (Ph. Eur. 8.1) Amtliche deutsche Ausgabe*. 2015. Stuttgart: Deutscher Apotheker Verlag.

Fischer, Robert, and Siegfried Fischer. 2021. "Re-Inventing Pine Oil, a Well-Known Natural Product." Accessed March 3, 2021. https://www.sofw.com/en/news/interviews/1987-re-inventing-pine-oil-a-well-known-natural-product.

Garzoli, Stefania, Stefania Petralito, Elisa Ovidi, Giovanni Turchetti, Valentina Laghezza Masci, Antonio Tiezzi, Jordan Trilli, et al. 2020. "*Lavandula* x *intermedia* Essential Oil and Hydrolate: Evaluation of Chemical Composition and Antibacterial Activity before and after Formulation in Nanoemulsion." Industrial Crops and Products 145: 112068. doi:10.1016/j.indcrop.2019.112068.

Giray, Fatma Handan. 2018. "An Analysis of World Lavender Oil Markets and Lessons for Turkey." *Journal of Essential Oil Bearing Plants* 21 (6). Taylor & Francis: 1612–23. doi:10.1080/0972060X.2019.1574612.

Glavan, Gordana, Sara Novak, Janko Božič, and Anita Jemec Kokalj. 2020. "Comparison of Sublethal Effects of Natural Acaricides Carvacrol and Thymol on Honeybees." Pesticide Biochemistry and Physiology 166. Elsevier: 104567.

Hartmann-Schreier, Jenny. 2003. "Eucalyptusöle." Edited by F. Böckler, B. Dill, U. Dingerdissen, G. Eisenbrand, F. Faupel, B. Fugmann, T. Gamse, R. Matissek, G. Pohnert, and G. Sprenger. Thieme Gruppe. RÖMPP. https://roempp.thieme.de/lexicon/RD-05-02123.

Hay, Yann-Olivier, Miguel A. Abril-Sierra, Luis G. Sequeda-Castañeda, Catherine Bonnafous, and Christine Raynaud. 2018. "Evaluation of Combinations Essential Oils and with Evaluation of Combinations Essential Oils and with Hydrosols on Antimicrobial and Antioxidant Activities." *Journal of Pharmacy & Pharmacognosy Research* 6 (3): 216. doi:https://hal.inrae.fr/hal-02628131.

Hori, Masatoshi. 1999. "Antifeeding, Settling Inhibitory and Toxic Activities of Labiate Essential Oils against the Green Peach Aphid, *Myzus persicae* (Sulzer) (Homoptera: Aphididae)." *Applied Entomology and Zoology* 34 (1): 113–18. doi:10.1303/aez.34.113.

Isman, Murray B. 2020. "Commercial Development of Plant Essential Oils and Their Constituents as Active Ingredients in Bioinsecticides." *Phytochemistry Reviews* 19 (2): 235–41. doi:10.1007/s11101-019-09653-9.

Isman, Murray B., Saber Miresmailli, and Cristina Machial. 2011. "Commercial Opportunities for Pesticides Based on Plant Essential Oils in Agriculture, Industry and Consumer Products." *Phytochemistry Reviews* 10 (2): 197–204. doi:10.1007/s11101-010-9170-4.

Julio, Luis F., Azucena González-Coloma, Jesus Burillo, Carmen E. Diaz, and Maria Fe Andrés. 2017. "Nematicidal Activity of the Hydrolate Byproduct from the Semi Industrial Vapor Pressure Extraction of Domesticated *Artemisia absinthium* against *Meloidogyne javanica*." *Crop Protection* 94 (April): 33–37. doi:10.1016/j.cropro.2016.12.002.

Kfoury, Miriana, Anissa Lounès-Hadj Sahraoui, Natacha Bourdon, Frédéric Laruelle, Joël Fontaine, Lizette Auezova, Hélène Greige-Gerges, and Sophie Fourmentin. 2016. "Solubility, Photostability and Antifungal Activity of Phenylpropanoids Encapsulated in Cyclodextrins." *Food Chemistry* 196 (April): 518–25. doi:10.1016/j.foodchem.2015.09.078.

Koul, Opender, Suresh Walia, and G.S. Dhaliwal. 2008. "Essential Oils as Green Pesticides: Potential and Constraints." *Biopesticides International* 4 (1): 63–84.

Krammer, Gerhard. 2004. "Orangenschalenöle." Edited by F. Böckler, B. Dill, U. Dingerdissen, G. Eisenbrand, F. Faupel, B. Fugmann, T. Gamse, R. Matissek, G. Pohnert, and G. Sprenger. Thieme Gruppe. RÖMPP. https://roempp.thieme.de/lexicon/RD-15-00769.

Kumar, Peeyush, Sapna Mishra, Anushree Malik, and Santosh Satya. 2011. "Insecticidal Properties of *Mentha* Species: A Review." *Industrial Crops and Products* 34 (1): 802–17. doi:10.1016/j.indcrop.2011.02.019.

Locher, Sanja, and Jenny Hartmann-Schreier. 2008. "Bergamottöl." Edited by F. Böckler, B. Dill, U. Dingerdissen, G. Eisenbrand, F. Faupel, B. Fugmann, T. Gamse, R. Matissek, G. Pohnert, and G. Sprenger. Thieme Gruppe. RÖMPP. https://roempp.thieme.de/lexicon/RD-02-00963.

Mattos, Amanda P., Fabricio P. Povh, Bruna B. Rissato, Vítor V. Schwan, and Kátia R. F. Schwan-Estrada. 2019. "In Vitro Antifungal Activity of Plant Extracts, Hydrolates and Essential Oils of Some Medicinal Plants and Control of Cucumber Anthracnose." *European Journal of Medicinal Plants*, August, 1–9. doi:10.9734/ejmp/2019/v28i430139.

Méndez-Tovar, Inés, Baudilio Herrero, Silvia Pérez-Magariño, José Alberto Pereira, and M. Carmen Asensio-S.-Manzanera. 2015. "By-Product of *Lavandula latifolia* Essential Oil Distillation as Source of Antioxidants." *Journal of Food and Drug Analysis* 23 (2): 225–33. doi:10.1016/j.jfda.2014.07.003.

MRFR. 2021. "Hydrosols Market Size, Share and Global Analysis, 2024". Accessed February 19, 2021. https://www.marketresearchfuture.com/reports/hydrosols-market-4789

Moretti, Mario D. L., Giovanni Sanna-Passino, Stefania Demontis, and Emanuela Bazzoni. 2002. "Essential Oil Formulations Useful as a New Tool for Insect Pest Control." *AAPS PharmSciTech* 3 (2): 64–74. doi:10.1208/pt030213.

Oral, Nebahat, Leyla Vatansever, Abamüslüm Güven, and Murat Gülmez. 2008. "Antibacterial Activity of Some Turkish Plant Hydrosols." *Kafkas Univ Vet Fak Derg* 14 (2): 205–9.

Oreopoulou, Antigoni, Eleni Papavassilopoulou, Haido Bardouki, Manolis Vamvakias, Andreas Bimpilas, and Vassiliki Oreopoulou. 2018. "Antioxidant Recovery from Hydrodistillation Residues of Selected Lamiaceae Species by Alkaline Extraction." *Journal of Applied Research on Medicinal and Aromatic Plants* 8 (March): 83–89. doi:10.1016/j.jarmap.2017.12.004.

Ortiz de Elguea-Culebras, Gonzalo, María I. Berruga, Omar Santana-Méridas, David Herraiz-Peñalver, and Raúl Sánchez-Vioque. 2017. "Chemical Composition and Antioxidant Capacities of Four Mediterranean Industrial Essential Oils and Their Resultant Distilled Solid By-Products." European Journal of Lipid Science and Technology 119 (12): 1700242. doi:10.1002/ejlt.201700242.

Özkan, Reyyan Yergin, and Murat Tunçtürk. 2021. "Effect of Essential Oils and Hydrosols from Some Selected Lamiaceae Species on Redroot Pigweed (*Amaranthus retroflexus* L.)." Romanian Biotechnological Letters 26 (2): 2471–75. https://doi.org/10.25083/rbl/26.2/2471.2475.

Ozturk, Ismet, Fatih Tornuk, Oznur Caliskan-Aydogan, M. Zeki Durak, and Osman Sagdic. 2016. "Decontamination of Iceberg Lettuce by Some Plant Hydrosols." *LWT* 74 (December): 48–54. doi:10.1016/j.lwt.2016.06.067.

Parejo, Irene, Francesc Viladomat, Jaume Bastida, Alfredo Rosas-Romero, Nadine Flerlage, Jesús Burillo, and Carles Codina. 2002. "Comparison between the Radical Scavenging Activity and Antioxidant Activity of Six Distilled and Nondistilled Mediterranean Herbs and Aromatic Plants." *Journal of Agricultural and Food Chemistry* 50 (23): 6882–90. doi:10.1021/jf020540a.

Pavela, Roman, and Giovanni Benelli. 2016. "Essential Oils as Ecofriendly Biopesticides? Challenges and Constraints." *Trends in Plant Science* 21 (12): 1000–1007. doi:10.1016/j.tplants.2016.10.005.

Petrakis, E.A., A.C. Kimbaris, D.P. Lykouressis, M.G. Polissiou, and D.C. Perdikis. 2015. "Hydrosols Evaluation in Pest Control: Insecticidal and Settling Inhibition Potential against *Myzus persicae* (Sulzer)." *Journal of Applied Entomology* 139 (4): 260–67. doi:10.1111/jen.12176.

Pino-Otín, Mª. Rosa, Jonatan Val, Diego Ballestero, Enrique Navarro, Esther Sánchez, Azucena González-Coloma, and Ana M. Mainar. 2019. "Ecotoxicity of a New Biopesticide Produced by *Lavandula luisieri* on Non-Target Soil Organisms from Different Trophic Levels." *Science of The Total Environment* 671 (June): 83–93. doi:10.1016/j.scitotenv.2019.03.293.

Pino-Otín, Mª. Rosa, Jonatan Val, Diego Ballestero, Enrique Navarro, Esther Sánchez, and Ana M. Mainar. 2019. "Impact of *Artemisia absinthium* Hydrolate Extracts with Nematicidal Activity on Non-Target Soil Organisms of Different Trophic Levels." *Ecotoxicology and Environmental Safety* 180 (September): 565–74. doi:10.1016/j.ecoenv.2019.05.055.

Pino-Otín, María Rosa, Juliana Navarro, Jonatan Val, Francisco Roig, Ana M. Mainar, und Diego Ballestero. 2022. "Spanish *Satureja montana* L. Hydrolate: Ecotoxicological Study in Soil and Water Non-Target Organisms". *Industrial Crops and Products* 178 (April): 114553. doi:10.1016/j.indcrop.2022.114553.

Politi, Matteo, Luigi Menghini, Barbara Conti, Stefano Bedini, Priscilla Farina, Pier Luigi Cioni, Alessandra Braca, and Marinella De Leo. 2020. "Reconsidering Hydrosols as Main Products of Aromatic Plants Manufactory: The Lavandin (*Lavandula × intermedia*) Case Study in Tuscany." *Molecules* 25 (9): 2225. doi:10.3390/molecules25092225.

Prusinowska, Renata, Krzysztof Śmigielski, Agnieszka Stobiecka, and Alina Kunicka-Styczyńska. 2016. "Hydrolates from Lavender (*Lavandula angustifolia*) – Their Chemical Composition as Well as Aromatic, Antimicrobial and Antioxidant Properties." *Natural Product Research* 30 (4): 386–93. doi:10.1080/14786419.2015.1016939.

Rao, B.R. Rajeswara. 2012. "Hydrosols and Water-Soluble Essential Oils of Aromatic Plants: Future Economic Products." *Indian Perfum* 56: 29–33.

Raveau, Robin, Joël Fontaine, and Anissa Lounès-Hadj Sahraoui. 2020. "Essential Oils as Potential Alternative Biocontrol Products against Plant Pathogens and Weeds: A Review." *Foods* 9 (3): 365. doi:10.3390/foods9030365.

Robe, Kevin, Esther Izquierdo, Florence Vignols, Hatem Rouached, and Christian Dubos. 2021. "The Coumarins: Secondary Metabolites Playing a Primary Role in Plant Nutrition and Health." *Trends in Plant Science* 26 (3): 248–59. doi:10.1016/j.tplants.2020.10.008.

Sagdic, Osman, Ismet Ozturk, and Fatih Tornuk. 2013. "Inactivation of Non-Toxigenic and Toxigenic *Escherichia coli* O157:H7 Inoculated on Minimally Processed Tomatoes and Cucumbers: Utilization of Hydrosols of Lamiaceae Spices as Natural Food Sanitizers." *Food Control* 30 (1): 7–14. doi:10.1016/j.foodcont.2012.07.010.

Sahan, N., and F. Tornuk. 2016. "Application of Plant Hydrosols for Decontamination of Wheat, Lentil and Mung Bean Seeds Prior to Sprouting." *Quality Assurance and Safety of Crops & Foods* 8 (4): 575–82. doi:10.3920/QAS2016.0858.

Sánchez-Vioque, Raúl, María Elena Izquierdo-Melero, María Quílez, David Herraiz-Peñalver, Omar Santana-Méridas, and María José Jordán. 2018. "Solid Residues from the Distillation of *Salvia lavandulifolia* Vahl as a Natural Source of Antioxidant Compounds." *Journal of the American Oil Chemists' Society* 95 (10): 1277–84. doi:10.1002/aocs.12128.

Santana-Méridas, Omar, Azucena González-Coloma, and Raúl Sánchez-Vioque. 2012. "Agricultural Residues as a Source of Bioactive Natural Products." *Phytochemistry Reviews* 11 (4): 447–66. doi:10.1007/s11101-012-9266-0.

Sedy, K. A., and E. H. Koschier. 2003. "Bioactivity of Carvacrol and Thymol against *Frankliniella occidentalis* and *Thrips tabaci*." *Journal of Applied Entomology* 127 (6): 313–16. doi:10.1046/j.1439-0418.2003.00767.x.

Smail Aazza. 2011. "Antioxidant Activity of Some Morrocan Hydrosols." *Journal of Medicinal Plants Research* 5 (30). doi:10.5897/JMPR11.1176.

Srivastava, Bhawana, Anand Sagar, Nawal Kishore Dubey, and Lipika Sharma. 2015. "Essential Oils for Pest Control in Agroecology." In *Sustainable Agriculture Reviews: Volume 15*, edited by Eric Lichtfouse, 329–52. Sustainable Agriculture Reviews. Cham: Springer International Publishing. doi:10.1007/978-3-319-09132-7_8.

Stringlis, Ioannis A., Ronnie de Jonge, and Corné M. J. Pieterse. 2019. "The Age of Coumarins in Plant–Microbe Interactions." Plant and Cell Physiology 60 (7). Oxford Academic: 1405–19. doi:10.1093/pcp/pcz076.

Tiliacos, Christophe, Emile M. Gaydou, Jean-Marie Bessière, and Raymond Agnel. 2008. "Distilled Lavandin (*Lavandula intermedia* Emeric Ex Loisel) Wastes: A Rich Source of Coumarin and Herniarin." *Journal of Essential Oil Research* 20 (5): 412–13. doi:10.1080/10412905.2008.9700043.

Tornuk, Fatih, Hasan Cankurt, Ismet Ozturk, Osman Sagdic, Okan Bayram, and Hasan Yetim. 2011. "Efficacy of Various Plant Hydrosols as Natural Food Sanitizers in Reducing *Escherichia coli* O157:H7 and *Salmonella typhimurium* on Fresh Cut Carrots and Apples." *International Journal of Food Microbiology* 148 (1): 30–35. doi:10.1016/j.ijfoodmicro.2011.04.022.

Törnük, Fatih, and Enes Dertli. 2015. "Decontamination of *Escherichia coli* O157:H7 and *Staphylococcus aureus* from Fresh-Cut Parsley with Natural Plant Hydrosols." *Journal of Food Processing and Preservation* 39 (6): 1587–94. doi:10.1111/jfpp.12387.

Tripathi, Arun K., Veena Prajapati, Ateeque Ahmad, Kishan K. Aggarwal, and Suman P. S. Khanuja. 2004. "Piperitenone Oxide as Toxic, Repellent, and Reproduction Retardant Toward Malarial Vector *Anopheles stephensi* (Diptera: Anophelinae)." *Journal of Medical Entomology* 41 (4): 691–98. doi:10.1603/0022-2585-41.4.691.

Tripathi, Arun K., Shikha Upadhyay, Mantu Bhuiyan, and P. R. Bhattacharya. 2009. "A Review on Prospects of Essential Oils as Biopesticide in Insect-Pest Management." *Journal of Pharmacognosy and Phytotherapy* 1 (5): 052–063. doi:10.5897/JPP.9000003.

Wiese, Natalie, Juliane Fischer, Jenifer Heidler, Oleg Lewkowski, Jörg Degenhardt, and Silvio Erler. 2018. "The Terpenes of Leaves, Pollen, and Nectar of Thyme (*Thymus vulgaris*) Inhibit Growth of Bee Disease-Associated Microbes". *Scientific Reports* 8 (1). Nature Publishing Group: 14634. doi:10.1038/s41598-018-32849-6.

Yang, Feng-Lian, Xue-Gang Li, Fen Zhu, and Chao-Liang Lei. 2009. "Structural Characterization of Nanoparticles Loaded with Garlic Essential Oil and Their Insecticidal Activity against *Tribolium castaneum* (Herbst) (Coleoptera: Tenebrionidae)." *Journal of Agricultural and Food Chemistry* 57 (21): 10156–62. doi:10.1021/jf9023118.

Zekri, Nadia, Nadia Handaq, Abdelhamid El Caidi, Touria Zair, and Mohamed Alaoui El Belghiti. 2016. "Insecticidal Effect of *Mentha pulegium* L. and *Mentha suaveolens* Ehrh. Hydrosols against a Pest of Citrus, *Toxoptera aurantii* (Aphididae)." *Research on Chemical Intermediates* 42 (3): 1639–49. doi:10.1007/s11164-015-2108-0.

Zimmermann, Gisbert. 2012. "Übersichtsarbeit-Wirkung Und Anwendungsmöglichkeiten Ätherischer Öle Im Pflanzenschutz: Eine Übersicht." *Journal Für Kulturpflanzen-Journal of Cultivated Plants* 64 (1): 1.

# 7 "Farmacies" are integral to agroecological systems

*Marion Johnson*

## 1 DRAWING TOGETHER THE KNOWLEDGE

### 1.1 AGROECOLOGY AND HEALTH

If we are to have a future as a biodiverse planet, our agricultural practices must work with nature, not against it. We should envision our farm lands as a reflection of the diverse ecology in which they are, or once were, situated. In 1939, Hanson discussed the concept of linking agriculture and ecology. He saw how important it was that ecology was the guiding force, 'to ferret out relationships with the environment, so that man using this knowledge in conjunction with that obtained from other fields can strive intelligently to secure balance and stabilisations...we need agroecologists'.

Our farms should be complex biodiverse structures, recycling and holding the blocks of life – plants, soils, water and nutrients. Agroecological systems encourage closed production cycles, the minimisation of external inputs, and health and resilience (FAO 2018). Agroecology uses natural processes such as biological control, the enhancement of biota, polyculture – optimisation within the given parameters rather than maximisation. Agroecology values health. Healthy soils, healthy water, healthy plants and healthy people.

Agroecological practices value diversity. Diversity underpins health and resilience. Within the farm, diversity is a web of complexity to be nurtured and supported, from the soil through the pastures, through woody areas, through waterways, through all manner of flora and fauna. Which in turn supports an agricultural community of diverse species, breeds and strains. For livestock, agroecology presents a diversity of forage bestowing upon them a broad diet, allowing them to express natural behaviours, indulge in curiosity and maintain homeostasis through zoopharmacognosy.

Agroecological management optimises sunlight and water, drawing them into plants and expressing them as rich pastures, hedges and forests, wetlands and waterways. We have a diversity of species, structure, function and genetics wrapped in a landscape of diversity, of traditions, cultures and knowledges. Species richness ensures the resilience of the whole (Tomich et al. 2011) and the whole is more than the sum of the parts (Buhner 2004).

The pressure for continued intensification at all costs, paid for by species, soils, water, ecology, livestock and farmers themselves, and the concept that production should be based upon a number of highly (and questionably), selected pasture species propped up with nitrogen has sickened our land and water and animals. Gliessman (1998) suggests that agroecology, with all its diversity and social contracts, supports an ecosystem's immunity – inducing a return to health.

DOI: 10.1201/9781003146902-9

## 1.2 Livestock and health

Ecological systems cannot function effectively unless they are biodiverse, and current losses in bio-diversity are threatening Earth and the ability of our systems to adapt (Kok et al. 2020). Livestock managed sensitively have always contributed to biodiversity, not degraded it, not least by supporting the fertility of soils and pastures through deposition of unadulterated dung. Healthy soil micro-bial populations underpin the cycling of the elements through the soil to plants, where they are rearranged, and then cycled through animals, wherein they are rearranged again, before returning back to the soil, promoting the health of microbes. (Teague 2017). The elements, combined with sunlight, provide us with life.

Healthy livestock are a key component of a balanced, mutually dependent system. Their grazing and dunging contribute to the growth of pastures and crops. Grazing animals can be managed in various ways, to enhance grasslands, increase carbon stocks and improve soil health. 'With appro-priate grazing management ruminant livestock can increase the carbon sequestered to more than offset their Greenhouse Gas emissions and can support and improve local ecosystem services' (Teague et al. 2016). When trees are introduced into the mix, carbon sequestration increases con-comitantly, as does animal health and welfare (Broom 2013). Unfortunately, there is a pervasive belief that meat and milk are bad. Livestock and farmers are victims of models that fail to differ-entiate between industrialised intensive production and extensive agroecologically based farming systems (Houzer and Scoones 2021).

Steps need to be taken to halt the worldwide genetic erosion of our animal species, and to rear and conserve breeds. They may play a vital role as farming adjusts and adapts to future challenges. Local breeds have been selected to survive and thrive under local conditions. As such, they have resistance to local diseases and possess instinctive knowledge. They are adapted to local landscapes, and granted, these landscapes are changing with the changing climate, but openness of communi-cation and sharing of knowledge, one of the principles of agroecology, will support local shifts and ecological re-matches.

In addition, more than a billion people depend upon livestock for their livelihoods and security (PAEPARD 2021). In terms of survival, food security should underpin all thinking (Falvey 2015). Well-managed and integrated livestock underpin resilience, providing food, draught power, energy, manure, social security and biodiversity functions. Agroecology is the key to successful livestock management – healthy, locally adapted animals support populations of plants and people.

It is argued that ruminants utilise resources that could be used for human consumption. As discussed by Eisler et al. (2014), Wilkinson and Lee (2017) and Houzer and Scoones (2021) ruminants managed under agroecological systems with access to pastures and diverse forages including shrubs and trees are not competing, but rather complimentary. Interestingly, when livestock are part of a diverse system, even just a simple silvopasture system, local biodiversity increases (Broom 2013).

## 1.3 Botanical medicine and livestock

Livestock that are offered an array of plant choices will eat a diet to balance their 'internal milieu', achieving a state of equilibrium in their systems, acknowledging the 'wisdom of the body' (Villalba and Provenza 2007). Animals have a genetic component and a behavioural component to the choices they make; there is an innate knowing and they can learn that certain foods/nutrients/tastes/chemicals increase their fitness or will help alleviate the indications of disease (Villalba and Provenza 2007). In effect, animals can self-medicate. It is accepted that animals are more likely to ingest plants higher in secondary compounds when exposed to them with their mother's guidance (Provenza 2003, Sanga et al. 2011), but experiments by Provenza (1995), Arsenos et al. (2000) and Juhnke et al. (2012) have shown that sheep can learn to choose their food and remember foods (and flavours) that prove beneficial to them. Villalba et al. (2006) showed in a series of experiments that lambs learned to choose diets balancing a possible negative effect of one species with a species that would ameliorate the effect. They also showed that animals that did not have the opportunity to learn that a certain food balanced another, i.e. they didn't have the opportunity to explore those food choices together, and thus never learned to associate cause and effect. Animals learn to balance their intakes and to mix their foods through experience, provided they are offered a wide choice. An animal that eats a species high in secondary compounds knows to eat a different plant to negate any toxic effects and balance their intake for the day. They ingest a range of plants to achieve the desired levels of nutrition and self-care but to avoid toxicosis. Diversity is key.

Self-medication can be classified as prophylactic – choosing a diet to maintain homeostasis and optimise health – or therapeutic (Alvaro et al. 2019) – a deliberate choice of a species to cure an affliction. Therapeutic activity by an animal has been clearly demonstrated by Huffman (1997) in his studies on chimpanzees and their use of *Veronia amygdalina* in Mahale, Tanzania. Alvaro et al. (2019) list numerous studies providing evidence for self-medication in a wide range of species from geese to bears and also provide a fascinating discourse on observations of self-medication from the 4th century BC onwards.

Traditional farming systems are a rich source of knowledge as to the benefits of species and their contributions to managing animal health. For example, goatherds in Karamoja, Uganda know that parasitised goats forage on *Albizia anthelmintica*, after which the worms are expelled and the goat improves. A further survey amongst the pastoralists elucidated that they observed 50 self-medicating behaviours covering a range of diseases and plant species (Gradé et al. 2009). Animals can also self-medicate by changing their behaviour, for example, simply moving away from an area or by applying a plant to themselves, as Kodiak bears have been observed to do with chewed *Ligusticum* roots (Huffman 2002), or starlings lining their nests with aromatic herbs to reduce chick mortality (Clark and Mason 1985).

Agroecological system diversity provides further health benefits as livestock are able to scratch and rub, freeing their coats of ectoparasites, excess hair and itches. By finding shade or shelter they can maintain their temperature in the optimum thermal zone.

## 1.4 Traditional and Local Knowledge of Animal Health

Drawing on the accumulated wisdom of previous generations we can develop foragescapes that benefit livestock and contribute to a self-sustaining medicinal agroecology. By introducing bio-diversity back into our farmed lands, we not only support the health of our animals, but also the health of the ecosystem in ways ranging from shade, shelter and habitat for other species, to soil health, erosion control and water quality. We support local people through the provision of timber, firewood, food and medicines rekindling culture and a chance to reconnect with a therapeutic land-scape (Wilson 2003).

Research into traditional knowledge surrounding the rearing and management of animals, or ethnoveterinary medicine was defined by Constance McCorkle (1986) as 'systematic research and development which takes as its principle subject or major departure point folk knowledge and belief, practices, techniques and resources, social organisation and so forth pertaining to any aspects of animal health among species raised or managed by human beings'. Ethnoveterinary knowledge is a rich store of plant, animal and cultural information upon which agroecology should draw.

There are an increasing number of surveys and records of 'ethnoveterinary knowledge' in the literature, but too frequently a paper simply records 'this plant is used for that'. This methodology misses the nuances of management and culture, thus the knowledge is factually lost. Ethnoveterinary knowledge is often recorded incidentally as part of an ethnobotanical survey, but again usually in the form of "this… [plant] is used to treat that… [ailment]". Ethnoveterinary practice, or folk veterinary practice, is much richer than this and there is a deep knowledge developed by farmers in their locale of different grazings providing different benefits at different times of year, of areas for shelter and shade, of the plants their livestock select and of plants that have alleviated ailments in the past. Given the caveat that there is more to health than simply providing a plant, many species that are mentioned in the folk veterinary knowledge systems have a global distribution so can be drawn upon as a first step in creating a medicinal planting. For example, garlic (*Allium sativum*) is widely used to treat poultry and ruminants in India (Ghotge and Ramdas 2008), and is used by camel owners and healers in Rajasthan (Gupta et al. 2015), pig owners in Canada (Lans et al. 2007), livestock farmers in Europe (Mayer et al. 2014) and in rabbits by Nigerian farmers (Fajimi and Taiwo 2005). Societies have long traditions of the connections between animals, plants and the land and have developed management systems that reflect this. In some Mediterranean silvopastoral systems, chestnut (*Castanea* spp.) leaves and fruit form a proportion of the diet of goats and pigs. Other trees such as oak (*Quercus* spp.) alder (*Alnus* spp.) and *Pinus radiata* are included in the woodlands (Broom 2013). All these species have a recorded medicinal activity. In Africa graziers have long-managed silvopastoral systems, and trees provide a vital feed source at key times of the year. Commonly utilised species include Acacia, Calliandra, Commiphora, Morus and Leucaena. Farmers, when questioned acknowledge an improvement in animal health when livestock can access the broad range of fodder available (Franzel et al. 2014).

Leucaena, introduced into Australia in 1890, is now an important silvopastoral beef production system (The Leucaena Network n.d.). The tree, noted by farmers for its benefits to animal health, contains high levels of condensed tannins (Marie-Magdeleine et al. 2020).

## 1.5  Science and botanical medicine

Science has a role in assessing the possibilities of plants for inclusion, after all, many of today's drugs are derived from plants (Farnsworth et al. 1985) – but more often than not science examines a plant through a reductionist lens.

Laboratories are assessing plants to find the next drug, and publications abound, but frequently plants are evaluated out of context or chemicals are extracted in the search for an active principle, completely ignoring the activity of the whole. Often when scientists aim to elucidate the active component of a plant, they "apply extracts to pathogens directly at high concentrations", which is unrealistic in the real-world setting (Masé 2017). Short-term trials conducted indoors to elucidate the response of a particular species to a particular plant bear no resemblance to holistic agroecological management. They frequently return a 'non-significant' response and the plant being trialed is deemed ineffective and all the knowledge surrounding it disparaged. Despite the fact that the plant, animal and their interaction were being studied in an alien environment.

Plant choices can be studied in the context of animal behaviours and physiology. However, it is of the utmost importance to understand that an animal may not choose a particular plant that is on offer, because it simply does not need it at the time. Allelochemicals often taste bitter and plants observed to be avoided by healthy animals are often sought out when needed, presumably in response to a physiological change and a requirement for homeostatic regulation (Hart 2005, Villalba et al. 2007).

As Enioutina et al. (2017) explain, herbs may be recognised as 'antimicrobial', but the plant being examined may also have antifungal, antiparasitic or antiviral properties, or it may be tonic or immune boosting, improving resistance to infection. Plants act over a broad spectrum and when there is a diversity of planting the spectra overlap, promoting the health of the animals ingesting them.

There are effective 'anti' or 'immuno' compounds in many plants, but plants contain a plethora of chemicals and chemistries, and perhaps we should view them rather as bolstering the internal ecosystem of our animals, strengthening their resilience through a broad mixed diet, nutrition and medicine.

## 2   CREATING THE 'FARMACY'

### 2.1   CONVERSATIONS

On any farm or range local knowledge and tradition should inform the planting of grasses, forbs, shrubs and trees. It is important to take the time to talk with people who have plant knowledge and livestock knowledge, if they are available. For example, Meuret and Provenza (2014), through conversations with shepherds explore the art of shepherding in the mountains of France, and the connections of livestock to the landscape, all of which form a part of the rich tapestry of medicinal agroecology. Without the knowledge of the shepherds the sheep would not be so well fed, and the area would neither be used as effectively nor conserved as well.

If the only advice available is from the written word, it is still vital to include farmers and land managers in all discussions and plans. Farmers and practioners should be understood as experts in their field, scientific information meshing with local knowledge. 'Agroecological processes require participation, and enhancement of farmers ecological literacy' (Altieri 2009). On farms where we have worked in New Zealand, the stock manager's input is vital; their knowledge of how the animals move across the land, where they like to cross creeks and where they like to camp. All plantings should complement, augment and enrich the daily life of stock. We held Kaputī korero (cup of tea casual conversations) with a number of farmers who either farmed organically or had farmed pre-1950. All agreed, that in their view, for an animal to be healthy it had to be able to select a mixed diet and that the healthiest animals were often those that had access to the bush and thus broader nutrition and opportunity to self-medicate. One farmer stated that 'contrary to popular conservationist opinion the animals rarely damaged the bush'; they simply selected what they needed, as his stock were reared on 'weedy' pastures and thus had a broad diet. Onions, garlic, carrots, kale, comfrey, chou moellier, turnips, swedes and mixed pastures or 'weeds' all featured in the discussions. All of the farmers agreed that small herds and flocks on smaller farms made a difference, as 'you knew your animals and they knew you' (Johnson 2015a).

Developing a 'farmacy' in an agroecological context encourages a process of co-creation: what would the farmer like, what would the community like, are there species people like to see return to the area? For example, when working with a farm on Banks Peninsula they expressed a desire to encourage the return of tui (*Prosthemadera novaeseelandiae*). A report by Cunningham (2012) identified that tui thrived upon both native and introduced species eating fruit and nectar as well as small invertebrates. Flax, harakeke (*Phormium tenax*) is important in bringing tui to breeding condition and species such as kowhai (*Sophora microphylla*), mahoe (*Melicytus ramiflorus*), titoki (*Alectryon excelsus*) and karamu (*Coprosma robusta*) are important contributors to diet throughout the year. Additionally, they have properties contributing to livestock health.

### 2.2   LAYERS

With the medicinal aspect of agroecology foremost, we are aiming for optimal health by growing and providing access to a wide variety of species. We should endeavour to provide the broadest range of choice – but working within the constraints imposed by locality such as climate, soils and the highest level of welfare. We should study the landscape – the foragescape – and understand where spatially divergent species will grow and how they fit together embracing the layers that make up the food and medicinal choices available to an animal. The pasture grasses, herbs and forbs layer include leaves, seeds, flowers, stems and bulbs. The shrub and tree layers offer leaves, twigs, flowers and fruit. Climbing plants, traversing the foragescape, offer leaves, stems, flowers, fruits or berries.

Different livestock species also have different physiology. For example, the saliva of deer differs from that of sheep and cattle allowing them to choose a diet higher in tannins (Rochfort et al. 2008). Different animals have different preferences and movements. Different species utilise both the foragescape and the plants within it differently (Duncan and Gordon 1999) thus creating further layers of diversity.

## 2.3 Visualisation

When contemplating how to diversify a foragescape we always begin by mapping the farm. Thus, we understand the topography, the constraints (waterways, tracks, buildings) and how the land is currently used by both the farmer and the stock. We then identify areas that require particular treatment, such as highly eroded areas or areas of unstable land that should be taken out of routine grazing and planted for recovery.

In New Zealand, species such as tree lucerne (*Chamaecytisus palmensis*), willow (*Salix* spp), poplar (*Populus* spp), wineberry (*Aristotelia serrata*), mahoe (*Melicytus ramiflorus*) and titoki (*Alectryon excelsus*), will stabilise the soil and provide food for birds and insects and fodder for livestock. Matagouri (*Discaria toumatou*) will cover the ground. Although it is thorny (once used by Māori for tattooing) it provides habitat for native species and fixes nitrogen. Ogle (2018) quotes a farmer from the Ahuriri valley, saying that he encouraged matagouri 'because sheep sheltered under it, especially during heavy snow. During summer drought, the lightly shaded pasture under matagouri continues growing, benefiting from the added nitrogen'.

As the land stabilises the plantings can be browsed gently and eventually the area can be underplanted and sensitively grazed.

Traditional healers will frequently acknowledge that the health of the land underpins the health of those upon it 'our first patient must be the land itself, if we can heal the land, we will have healed ourselves' (McGowan 2009).

Thus, we should think about planting gullies thickly with shrubs to hold the soil and slow the water flows in rainfall events. Shrubs such as snowberry (*Gaultheria antipoda*) or koromiko (*Hebe strictissima*) provide habitat and can be browsed. Both have veterinary health properties, the former a galactagogue, the latter alleviates scouring. We should line the banks of waterways with healing shrubs and herbs and sow thick resilient pastures. Trees and hedges provide environmental support, carbon sequestration, microclimates, nutrient leaching control, soil erosion and water filtration (Moreno et al. 2018) – healing the land.

When mapping we identify areas of exceptional soil or microclimates, these areas may be excluded from general grazing, designated as horticultural areas or may be planted differently, as species that cannot generally be grown on the farm may grow there. For example, the endemic kawakawa (*Piper excelsum*), helping with pain, infection and wound healing, only grows in niche microclimates below 45 degrees south.

As we study the farm we are aiming for diversity and to optimise health and welfare, but we are also committed to conserving fragile areas such as wetland and to enhancing water quality. Through conservation we introduce more diversity and habitat and can also support cultural values, such as the harvest of traditional medicines and foods, ceremonies and aesthetic values.

Drawing upon local knowledge, tradition, practice and science, we can begin to plan on the maps. We ask questions such as which areas would be suitable for silvopasture, where might hedges be planted, where should riparian margins be fenced, where might denser plantings of high value trees be useful or where should pioneer species be planted with a view to succession as the land rebuilds? Where might we plant 'king' trees that will be left to grow for hundreds of years into the future? As we illustrate the ideas, farmers can make suggestions and correct assumptions.

## 2.4 Animal health

Pasture, containing a mixture of grasses, herbs and forbs, forms part of the natural diet of ruminants. It is rich in minerals, nutrients and allelochemicals. Agroecological diversity ensures that there will always be a range of plant species available, supporting health through either nutrition, immune support, tonic or direct defence as 'each of the estimated 400,000 species of plants on earth makes hundreds of thousands of compounds' (Provenza et al. 2015). Plants in different localities have different arrays of chemicals, and allelochemical arrays change within the plant depending upon season and growth stage; there is a great deal of choice for an animal and a great deal of learning. Specific feedback loops and consequences influence an animal's choice of foodstuff and the homeostatic mechanism balances their intakes in concert with their life stage, which influences nutrient demand or state of wellness.

Numerous researchers have suggested animals can sense illness or parasitism and given the opportunity, and knowledge, will choose species to alleviate the symptoms before the disease manifests (e.g. Hart 2005, Engel 2007, Villalba and Landau 2012). Given the chance, livestock will eat a range of species within one meal and over the course of a grazing day. They will also vary their food category choices (protein, energy, allelochemicals) depending upon their physiological states and requirements.

The grasses, legumes and forbs in a pasture contribute to nutrition and health, ensuring that an animal's potential can be reached. When managed sensitively, pastures will also provide habitat, protect and build soils and contribute to a diverse foragescape. Studies at the Louis Bolk institute have shown that cattle grazing pastures with a diversity of herbs require less antibiotic treatment (Laldi 2012a). Some beneficial herbs found in the pastures included white clover (*Trifolium* spp), dandelion (*Taraxacum officinale*), docks (*Rumex* spp) and plantain (*Plantago* spp), all medicinal species. Pastures also contained nettle (*Urtica dioica*) an antibacterial, chamomile (*Matricaria recutita*) a nervine, red clover (*Trifolium pratense*) an alterative, chickweed (*Stellaria spp*) a demulcent, cow parsley (*Anthriscus sylvestris*) an anti-inflammatory and shepherd's purse (*Capsella bursa-pastoris*) an astringent. Foster (1988) reviewing pasture research, notes that up until the early 20th century it was usual to have seed mixtures containing 15–20 species – true herbal leys. The records of species planted in older pastures emphasise the importance that was attributed to mixed pastures, and the behaviour of livestock today in seeking out and eating 'weeds' show that they are still valuable (Johnson 2015b).

With a medicinal agroecological mindset we aim to fill as many niches as possible. In Europe, hedgerows, although diminished, are a renowned source of biodiversity. In New Zealand, many shelter belts simply comprise a line of macrocarpa (*Cupressus macrocarpa*), which can cause cows to abort if they are hungry and browse too hard (Blackwood 2016). Savvy shelter should start at ground level with forbs and herbs, for example hedge mustard (*Sisymbrium officinale*), lemon balm (*Melissa officinalis*), violet (*Viola* spp), ramsons (*Allium ursinum*), nettle (*Urtica dioica*), meadowsweet (*Filipendula ulmaria*) or raspberry (*Rubus idaeus*). Shrubby trees might include elder (*Sambucus nigra*), crab apple (*Malus sylvestris*), hawthorn (*Crataegus laevigata*) or hazel (*Corylus avellana*) and climbers might include cleavers (*Gallium aparine*), honeysuckle (*Lonicera* spp), hops (*Humulus lupulus*) or ivy (*Hedera helix*). Interspersed trees might be oak (*Quercus* spp), elm (*Ulmus* spp), ash (*Fraxinus excelsior*) or rowan (*Sorbus acuparia*). Thus, animals are encouraged to search and browse and choose from a multitude of species.

Trees are vital members of our farmscapes. Silvopastures are not always lines of trees interspersed with grazing; trees can be planted to suit the terrain and the stock, to flow over the landscape, to aid management and aesthetics and can slowly merge into woodland pasture and then forest.

As we begin to choose which species we might plant, with our aims of an agroecological matrix centring on animal health, we need to ask what health issues are currently prevalent on farm and in the district? What diseases and problems generally affect the species carried on the farm? If our aim is prophylaxis through a broad diet and heightened immunity, then we need to recall the species that stock once chose to eat; for example, broad leaf, kapuka (*Griselinia littoralis*), mahoe (*Melicytus ramiflorus*) and five finger, whauwhau (*Pseudopanax arboreus*) are highly preferred native New Zealand browse, species with a reputation for supporting health (Riley 1994). Ash (*Fraxinus excelsior*), oak (*Quercus* spp), elm (*Ulmus* spp), willow (*Salix* spp), poplar (*Populus* spp), hazel (*Corylus avellana*) and chestnut (*Castanea sativa*) are all trees with a long history as browse (Vandermeulen et al. 2018) and medicine. There are large numbers of herbs with a reputation for healing that can be woven into the landscape with a view of supporting the health of the animals on a particular farm.

Planting choices should also reflect the situation. For example, when stock are yarded there is always the chance of accidental injury. Trees such as willow (*Salix* spp), alder (*Alnus glutinosa*), akeake (*Dodonoea viscose*) and miro (*Prumnopitys ferruginea*) have wound healing activity and provide shade, shelter and browse. Herbs such as plantain (*Plantago* spp), sow thistle (*Sonchus* spp), yarrow (*Achillea millefolium*) and marigold (*Calendula officinalis*) are both edible and promote recovery from injury.

## 2.5   INCORPORATING NATIVE PLANTS IN A NEW ZEALAND CONTEXT

In thinking about medicinal agroecology in New Zealand we can draw on both plants and knowledge from a number of cultures, as by the year 2000 there were more introduced vascular species in New Zealand than native (Brake and Peart 2013). But there has been a loss of native biodiversity on productive lands and an agroecological approach is an ideal way to address the problem. We are not asking or mandating farmers to plant species for the sake of it; we are illustrating all the benefits, not least of which is the health of their livestock and asking them to choose a pathway that meets their needs and the needs of the farm. All medicinal plants have a multiplicity of purpose, supporting the health of livestock and the land, and providing habitat for a myriad of other species both above and below the ground.

There is plenty of evidence of the ingestion of native vegetation by farmed stock, either in publications describing early European farmers living in the country, or in the records of the damage caused by farmed or feral animals in conservation areas (Forsyth et al. 2002). From these records and the knowledge of older farmers we can identify trees and shrubs that are preferentially browsed, such as broadleaf, kapuka, (*Griselinia littoralis*), whiteywood and mahoe (*Melicytus ramiflorus*), which is so palatable it was known to early farmers as cow leaf, five finger whauwhau (*Pseudopanax arboreus*), koromiko (*Veronica salicifolia*) and seven finger pate (*Schefflera digitata*). Many of these also have medicinal qualities.

Guthrie Smith (1907), the owner of a large sheep station in the Hawkes Bay, kept detailed records of station life and describes the changes in the pastures. Of the 21 species of native grass on the farm, six were valued highly as stock feed, with *Trisetum antarcticum* being so palatable 'that it only survived where the sheep couldn't reach it'. Brooking et al. (2011) quote William Travers, an early New Zealand naturalist, as saying 'all stockmen agree on praising the feeding quality of the native grasses.... the secret of the value for feeding purposes of old pastures lies in the fact that they contain a great variety of grasses of varying times of maturity'. Peden (2011) in his book about Mt Peel Station describes many of the species preferentially foraged by sheep and from these and other observations we can draw conclusions about the palatability of many native plants and add them to the list of potential planting candidates. In certain extensive farming situations livestock could perhaps be produced on purely native mixed pastures, but the native New Zealand grasses did not evolve under a grazing regime, and do not compete well with aggressive imported species.

There are limited records of native species used for animal health in New Zealand but those that exist make an excellent foundation for a medicinal planting (Johnson 2015a). For example, karaka (*Corynocarpus laevigatus*) is highly palatable and used to be given to unwell stock. An old Taranaki saying being 'if the cow won't eat karaka, it's time to shoot her'. Koromiko (*Veronica salicifolia*) is useful for scour, and kawakawa (*Piper excelsum*) is a good tonic, a painkiller and an alterative improving waste elimination through the kidneys. It is also useful for ticks. Flax harakeke (*Phormium tenax*) is anthelminthic, anti-fungal and is preferentially eaten at certain times of the year; cattle grasp the leaves and draw them through their mouths leaving the fibres behind.

Many native species are very palatable, some such as maori anise koheriki (*Gingidia montana*) a diuretic and Cook's scurvy grass nau (*Lepidium oleraceum)* rich in Vitamin C, have been eaten almost to extinction.

New Zealand does not have an ethnoveterinary tradition per se but conversations with farmers and healers will elicit suggestions from observations. For example, red matipo (*Myrsine australis*), flax harakeke (*Phormium tenax*), miro (*Prumnopitys ferruginea*), supplejack kareao (*Ripogonum scandens*), fat hen huainga (*Chenopodium album*) and gully fern piupiu (*Pneumatopteris pennigera*) all have a reputation as anti-parasitics. (Johnson 2012).

Parasite management provides an illustration of the benefits of a medicinal agroecological approach. Animals with a broad diet and access to plants with anthelminthic activity when needed have minimal parasite burdens and do not need to be intensively treated with external inputs in the form of chemical anthelminthics. Teague (2017) describes the unforeseen problems of the regular use of ivermectin as a dewormer in intensively managed cattle. Ivermectin is excreted in the dung, which then impacts both invertebrates and soil biota. Mineral recycling and soil building processes are disrupted, which has a flow on effect to plant health and growth. As the dung is no longer quickly broken down, fly populations proliferate, bringing irritation and disease to people and livestock.

A small number of farmers in New Zealand have taken an agroecological approach with diverse pastures, hedges and bush paddocks and are reaping the benefits of healthier stock and high production. These farmers have pastures containing mixtures of herbs, legumes and grasses, availability of browse either as silvopasture or shelter, and in some cases, controlled access to areas of native bush. They also have low to zero veterinary bills.

## 3   FURTHER THOUGHTS

When developing the farmacy there are many balancing decisions to be made. For example, as far as we are aware burdock (*Arctium lappa*) is only relished by donkeys (de Bairacli Levy 1991), but it is a really useful herb to have growing locally for treating chronic joint pain, stomach ailments, mastitis and chronic skin disorders in livestock (de Bairacli levy 1991, Wynn and Fougere 2007). However, the burrs produced by burdock in the autumn can cause the downgrading of sheep fleeces. It is thus a decision to be made by the farmer whether to plant burdock or not – or to manage access by the animals.

Tutu (*Coriaria arborea*) is an excellent plant for poulticing and healing strains and swellings in stock and is valued in traditional medicine, but it is 'the most important toxic plant in New Zealand' (Connor and Fountain 2009). Even honey containing tutin – one of the toxic compounds in tutu – becomes poisonous. On the other hand, native birds love the fruit. Thus, the risk of cultivation has to be weighed.

Farmacies are integral to agroecological practices, but our medicinal plantings have a further role as an ark. Species such as ash (*Fraxinus excelsior*), valuable fodder and providing medicinal benefits, is being decimated in other parts of the world by Ash dieback. *Ulmus procera*, the English elm once valued for fodder and medicine, has disappeared from the English countryside. Both still thrive in New Zealand.

When considering agroecological farmacies we should create 'a dialogue of wisdoms' (Altieri 2015) where there is no standardisation, rather frameworks into which local, knowledge and understandings are woven. With a changing climate the tolerance zones for plants are changing. Agroecology is local, but it is open to freely exchanging knowledge, so as plants move, the wisdom surrounding them can travel with them. The effectiveness of everyone working together equitably is now recognised to produce sustainable solutions that are actually implemented (Utter et al. 2021).

Creating farmacies is a vital part of an agroecological process, involving communities, reducing inputs, creating diversity and promoting the ecological health and resilience of the land, as well as the livestock kept upon it.

And finally, when it comes to the medicinal aspects of agroecology, as Turner (1955) suggests, 'it is always advisable to consult the cow'.

## 4   REFERENCES

Altieri, Miguel A. 2015. *Agroecology: Key Concepts, Principles and Practices*. Penang: Third World Network and SOCLA.

Altieri, Miguel A. 2009. "Agroecology, Small Farms, and Food Sovereignty." *Monthly Review* 61 (3): 102. https://doi.org/10.14452/mr-061-03-2009-07_8.

Alvaro, Mezcua, Luis Revuelta and Joaquín Sánchez de Lollano. 2019. "The origins of Zoopharmacognosy: how humans learned about self medication from animals" *International Journal of Applied Research* 5 (5):73–79 doi.org/10.22271/allresearch

Arsenos, G., I. Kyriazakis, and B. J. Tolkamp. 2000. "Conditioned Feeding Responses of Sheep towards Flavoured Foods Associated with the Administration of Ruminally Degradable and/or Undegradable Protein Sources." *Animal Science* 71 (3): 597–606. https://doi.org/10.1017/s1357729800055429.

Blackwood, Charlie. 2016. "Abortion in Dairy Cows." Accessed February 4, 2022. https://wvc.com.au/abortion-in-dairy-cows/.

Brake, Lucy, and Raewyn Peart. 2013. *Treasuring Our Biodiversity an EDS Guide to the Protection of New Zealand's Indigenous Habitats and Species*. Environmental Defence Society: New Zealand.

Broom, D. M., F. A. Galindo, and E. Murgueitio. 2013. "Sustainable, Efficient Livestock Production with High Biodiversity and Good Welfare for Animals." *Proceedings of the Royal Society B: Biological Sciences* 280 (1771): 20132025. https://doi.org/10.1098/rspb.2013.2025.

Brooking, Tom, Eric Pawson, and Paul Star. 2011. *Seeds of Empire: The Environmental Transformation of New Zealand*. London: I.B. Tauris.

Buhner, Stephen H. 2004. *The Secret Teachings of Plants: The Intelligence of the Heart in the Direct Perception of Nature*. Rochester, Vt.: Bear & Co.

Clark, L., and J. Russell Mason. 1985. "Use of Nest Material as Insecticidal and Anti-Pathogenic Agents by the European Starling." *Oecologia* 67, 2 169–76. https://doi.org/10.1007/bf00384280.

Connor, Henry E, and John S Fountain. 2009. *Plants That Poison: A New Zealand Guide*. Manaaki Whenua Press Lincoln, New Zealand.

Cunningham, Shaun. 2012. Review of *Establishment and Enhancement of Native Biodiversity on Te Putahi Farm, Banks Peninsula, Christchurch*. Nga Pae o te Maramatanga.

De Baïracli-Levy, Juliette 1991. *The Complete Herbal Handbook for Farm and Stable*. London: Faber.

Duncan, Alan J., and Iain J. Gordon. 1999. "Habitat Selection according to the Ability of Animals to Eat, Digest and Detoxify Foods." *Proceedings of the Nutrition Society* 58 (4): 799–805. https://doi.org/10.1017/s0029665199001081.

Eisler, Mark C., Michael R. F. Lee, John F. Tarlton, Graeme B. Martin, John Beddington, Jennifer A. J. Dungait, Henry Greathead, et al. 2014. "Agriculture: Steps to Sustainable Livestock." *Nature* 507 (7490): 32–34. https://doi.org/10.1038/507032a.

Engel, Cindy.2007 "Zoopharmacognosy" in *Veterinary Herbal Medicine* Mosby Elsevier St Louis ISBN-13978-0323-02998-8.

Enioutina, Elena Yu., Lida Teng, Tatyana V. Fateeva, Jessica C.S. Brown, Kathleen M. Job, Valentina V. Bortnikova, Lubov V. Krepkova, Michael I. Gubarev, and Catherine M.T. Sherwin. 2017. "Phytotherapy as an Alternative to Conventional Antimicrobials: Combating Microbial Resistance." *Expert Review of Clinical Pharmacology* 10 (11): 1203–14. https://doi.org/10.1080/17512433.2017.1371591.

FAO. 2018. "The 10 Elements of Agroecology Guiding the Transition to Sustainable Food and Agricultural Systems." www.fao.org/agroecology.

Fajimi, A. K. and A. A. Taiwo. 2005 "Herbal remedies in animal parasitic diseases in Nigeria: a review." *African Journal of Biotechnology* 4, (4) 303-307.

Falvey, Lindsay. 2015. "Food Security: The Contribution of Livestock." *Chiang Mai University Journal of Natural Sciences* 14 (1). https://doi.org/10.12982/cmujns.2015.0074.

Farnsworth, Norman R., Olayiwola Akerele, Audrey S. Bingel, Djaja D. Soejarto, and Zhengang Guo. 1985. "Medicinal Plants in Therapy." *Bulletin of the World Health Organization* 63 (6): 965–81. https://www.ncbi.nlm.nih.gov/pmc/articles/PMC2536466/.

Forsyth, D. M., D. A. Coomes, G. Nugent, and G. M. J. Hall. 2002. "Diet and Diet Preferences of Introduced Ungulates (Order: Artiodactyla) in New Zealand." *New Zealand Journal of Zoology* 29 (4): 323–43. https://doi.org/10.1080/03014223.2002.9518316.

Foster, Lyndall. 1988. "Herbs in Pastures. Development Research in Britain, 1850–1984." *Biological Agriculture & Horticulture* 5 (2): 97–133. https://doi.org/10.1080/01448765.1988.9755134

Franzel, Steven, Sammy Carsan, Ben Lukuyu, Judith Sinja, and Charles Wambugu. 2014. "Fodder Trees for Improving Livestock Productivity and Smallholder Livelihoods in Africa." *Current Opinion in Environmental Sustainability* 6 (February): 98–103. https://doi.org/10.1016/j.cosust.2013.11.008.

Gliessman, Stephen R. 1998. *Agroecology: Ecological Processes in Sustainable Agriculture*. Ann Arbor Press Michigan. ISBN1-57504-043-3

Gradé, J. T., John R. S. Tabuti, and Patrick Van Damme. 2008. "Four Footed Pharmacists: Indications of Self-Medicating Livestock in Karamoja, Uganda." *Economic Botany* 63 (1): 29–42. https://doi.org/10.1007/s12231-008-9058-z.

Gupta, Lokesh, Ghanshyam Tiwari, and Rajeev Garg. 2015. "Documentation of Ethnoveterinary Remedies of Camel Diseases in Rajasthan, India." *Indian Journal of Traditional Knowledge* 14 (3): 447–53.

Guthrie-Smith, Herbert. 1907. "The Grasses of Tutira." *Transactions of the Royal Society of New Zealand* 40, 506–519.

Hart, Benjamin, L. 2005 "The Evolution of Herbal Medicine: Behavioural Perspectives." 2005. *Animal Behaviour* 70 (5): 975–89. https://doi.org/10.1016/j.anbehav.2005.03.005.

Hanson, Herbert C. 1939. "Ecology in Agriculture." *Ecology* 20 (2): 111–17. https://doi.org/10.2307/1930733.

Houzer, Ella, and Ian Scoones. 2021. *"Are Livestock Always Bad for the Planet? Rethinking the Protein Transition and Climate Change Debate,"* Brighton PASTRES https://doi.org/10.19088/steps.2021.003.

Huffman, Michael A., Shunji Gotoh, Linda A. Turner, Miya Hamai, and Kozo Yoshida. 1997. "Seasonal Trends in Intestinal Nematode Infection and Medicinal Plant Use among Chimpanzees in the Mahale Mountains, Tanzania." *Primates* 38 (2): 111–25. https://doi.org/10.1007/bf02382002.

Huffman, Michael. 2002. "Animal Origins of Herbal Medicine." In *Des Sources Du Savoir Aux Médicaments Du Futur*. IRD Éditions, Société française d'ethnopharmacologie.

Johnson, Marion. 2012 *"Adapting the Principles of Te Rongoā into Ecologically and Culturally Sustainable Farm Practice."* http://www.maramatanga.ac.nz

Johnson, Marion. a 2015. "Native Plants for Animal Health." In *He Ahuwhenua Taketake Indigenous Agroecology*, edited by Marion Johnson and Chris Perley. Nga Pae o te Maramatanga. http://www.maramatanga.ac.nz.

Johnson, Marion. b 2015 *"Rongoā Pastures: Healthy Animals, Resilient Farms."* www.maramatanga.co.nz. http://www.maramatanga.co.nz

Juhnke, J., J. Miller, J.O. Hall, F.D. Provenza, and J.J. Villalba. 2012. "Preference for Condensed Tannins by Sheep in Response to Challenge Infection with Haemonchus Contortus." *Veterinary Parasitology* 188 (1-2): 104–14. https://doi.org/10.1016/j.vetpar.2012.02.015.

Kok, A., E.M. de Olde, I.J.M. de Boer, and R. Ripoll-Bosch. 2020. "European Biodiversity Assessments in Livestock Science: A Review of Research Characteristics and Indicators." *Ecological Indicators* 112 (May): 105902. https://doi.org/10.1016/j.ecolind.2019.105902.

Laldi, S. 2012a. *"Herbs in grassland and health of the dairy herd. 1: The potential medicinal value of pasture herbs"*. Louis Bolk Institute. https://edepot.wur.nl/247302.

Lans, Cheryl, Nancy Turner, Tonya Khan, and Gerhard Brauer. 2007. "Ethnoveterinary Medicines Used to Treat Endoparasites and Stomach Problems in Pigs and Pets in British Columbia, Canada." *Veterinary Parasitology* 148 (3-4): 325–40. https://doi.org/10.1016/j.vetpar.2007.06.014.

Marie-Magdeleine, Carine, Steve Ceriac, Dingamgoto Jesse Barde, Nathalie Minatchy, Fred Periacarpin, Frederic Pommier, Brigitte Calif, Lucien Philibert, Jean-Christophe Bambou, and Harry Archimède. 2020. "Evaluation of Nutraceutical Properties of Leucaena Leucocephala Leaf Pellets Fed to Goat Kids Infected with Haemonchus Contortus." *BMC Veterinary Research* 16 (1). https://doi.org/10.1186/s12917-020-02471-8.

Masé Guido. 2017. *"Herbs and Antimicrobial Resistance."* Accessed March 10, 2021. https://www.urbanmoonshine.com/blogs/blog/herbs-and-antimicrobial-resistance.

Mayer, Maria, Christian R. Vogl, Michele Amorena, Matthias Hamburger, and Michael Walkenhorst. 2014. "Treatment of Organic Livestock with Medicinal Plants: A Systematic Review of European Ethnoveterinary Research." *Complementary Medicine Research* 21 (6): 375–86. https://doi.org/10.1159/000370216.

McCorkle, Constance. 1986. "An Introduction to Ethnoveterinary Research and Development." *Journal of Ethnobiology* 6 129–149.

McGowan, Robert. 2009. *Rongoa Maori: A Practical Guide to Traditional Maori Medicine*. Tauranga N.Z.: Robert McGowan.

Meuret, Michel, and Frederick D Provenza. 2014. *The Art & Science of Shepherding: Tapping the Wisdom of French Herders*. Art and Science of Shepherding Austin, Texas: Acres U.S.A. ISBN 978-1-60173-069-5

Moreno, G., S. Aviron, S. Berg, J. Crous-Duran, A. Franca, S. García de Jalón, T. Hartel, 2018. "Agroforestry Systems of High Nature and Cultural Value in Europe: Provision of Commercial Goods and Other Ecosystem Services." *Agroforestry Systems* 92 (4): 877–91. https://doi.org/10.1007/s10457-017-0126-1.

Nitya Sambamurthi Ghotge, Sagari R Ramdas, and Anthra (Organization. 2008. *Plants Used in Animal Care: The Anthra Collective*. Pune, Maharashtra, India: Anthra.

Ogle, Colin 2018. "*Bring Back This Thorny native*." https://www.nzpcn.org.nz/nzpcn/news/conservation-comment-bring-back-this-thorny-native.

PAEPARD. 2021. "Global Assessment of the Contribution of Livestock to Food Security, Sustainable Food Systems, Nutrition and Healthy Diets." https://paepard.blogspot.com/2021/06/global-assessment-of-contribution-of.html.

Peden, Robert. 2011. *Making Sheep Country: Mt Peel Station and the Transformation of the Tussock Lands*. Auckland: Auckland University Press. ISBN 978-1-186940-485-7

Provenza, Frederick D. 1995. "Postingestive Feedback as an Elementary Determinant of Food Preference and Intake in Ruminants." *Journal of Range Management* 48 (1): 2. https://doi.org/10.2307/4002498.

Provenza, Frederick D. 2003. *Foraging Behavior Managing to Survive in a World of Change: Behavioral Principles for Human, Animal, Vegetation, and Ecosystem Management*. Utah: Logan.

Provenza, Frederick D., Michel Meuret, and Pablo Gregorini. 2015. "Our Landscapes, Our Livestock, Ourselves: Restoring Broken Linkages among Plants, Herbivores, and Humans with Diets That Nourish and Satiate." *Appetite* 95 (December): 500–519. https://doi.org/10.1016/j.appet.2015.08.004

Riley, Murdoch, and Brian Enting. 1994. *Māori Healing and Herbal: New Zealand Ethnobotanical Sourcebook*. Viking Sevenseas N.Z.

Rochfort, Simone, Anthony J Parker, and Frank R. Dunshea. 2008. "Plant Bioactives for Ruminant Health and Productivity." *Phytochemistry* 69 (2): 299–322. https://doi.org/10.1016/j.phytochem.2007.08.017.

Sanga, Udita, Frederick D. Provenza, and Juan J. Villalba. 2011. "Transmission of Self-Medicative Behaviour from Mother to Offspring in Sheep." *Animal Behaviour* 82 (2): 219–27. https://doi.org/10.1016/j.anbehav.2011.04.016.

Teague, W. R., S. Apfelbaum, R. Lal, U. P. Kreuter, J. Rowntree, C. A. Davies, R. Conser, 2016. "The Role of Ruminants in Reducing Agriculture's Carbon Footprint in North America." *Journal of Soil and Water Conservation* 71 (2): 156–64. https://doi.org/10.2489/jswc.71.2.156

Teague, Richard W. 2017. "Bridging the Research Management Gap to Restore Ecosystem Function and Social Resilience." In *Global Soil Security. Progress in Soil Science.*, edited by D J Field, C L S Morgan, and A B McBratney. Springer. https://doi.org/10.1007/978-3-319-43394-3_30.

"The Leucaena Network". n.d. https://www.leucaena.net. Accessed January 01, 2022.

Tomich, Thomas P., Sonja Brodt, Howard Ferris, Ryan Galt, William R. Horwath, Ermias Kebreab, Johan H.J. Leveau. 2011. "Agroecology: A Review from a Global-Change Perspective." *Annual Review of Environment and Resources* 36 (1): 193–222. https://doi.org/10.1146/annurev-environ-012110-121302.

Turner, F. Newman. 1955. *Fertility Pastures: Herbal Leys as the Basis of Soil Fertility and Animal Health*. Faber and Faber London

Utter, Alisha, Alissa White, V, Ernesto Mendez and Katlyn Morris. 2021. "Co-creation of knowledge in Agroecology" *Elementa Science of the Anthropocene* 9 (1) https://doi.org/10.1525/elementa.2021.00026

Vandermeulen, Sophie, Carlos Alberto Ramírez-Restrepo, Yves Beckers, Hugues Claessens, and Jérôme Bindelle. 2018. "Agroforestry for Ruminants: A Review of Trees and Shrubs as Fodder in Silvopastoral Temperate and Tropical Production Systems." *Animal Production Science* 58 (5): 767. https://doi.org/10.1071/an16434.

Villalba, J. J., and F. D. Provenza. 2007. "Self-Medication and Homeostatic Behaviour in Herbivores: Learning about the Benefits of Nature's Pharmacy." *Animal* 1 (9): 1360–70. https://doi.org/10.1017/s175173110 7000134.

Villalba, Juan J., Frederick D. Provenza, and Ryan Shaw. 2006. "Sheep Self-Medicate When Challenged with Illness-Inducing Foods." *Animal Behaviour* 71 (5): 1131–39. https://doi.org/10.1016/j.anbe hav.2005.09.012.

Villalba, Juan J., and Serge Y. Landau. 2012. "Host Behavior, Environment and Ability to Self-Medicate." *Small Ruminant Research* 103 (1): 50–59. https://doi.org/10.1016/j.smallrumres.2011.10.018.

Wilkinson, J. M., and M. R. F. Lee. 2017. "Review: Use of Human-Edible Animal Feeds by Ruminant Livestock." *Animal* 12 (8): 1735–43. https://doi.org/10.1017/s175173111700218x.

Wilson, Kathleen. 2003. "Therapeutic Landscapes and First Nations Peoples: An Exploration of Culture, Health and Place." *Health & Place* 9 (2): 83–93. https://doi.org/10.1016/s1353-8292(02)00016-3.

Wynn, Susan G, and Barbara J Fougere. 2007. "Materia Medica." In *Veterinary Herbal Medicine*, edited by Susan G Wynn and Barbara J Fougere. St Louis Missouri: Mosby Elsevier.

# 8 Tropical trees and shrubs for healthy agroecosystems, including animal health and welfare

*Lucero Sarabia, Francisco Solorio, Francisco Galindo,*
*Pedro González, Immo Fiebrig, Carlos A. Sandoval Castro,*
*Felipe Torres and Juan Ku*

## 1  INTRODUCTION

Population growth and the concomitant demand for food have increased pressure on natural resources, putting at risk the health of ecosystems and human health in general (FAO, 2019). As the areas destined for food production increase, so does the use of agrochemicals (pesticides, insecticides, herbicides) and alongside, the amount of antibiotics applied in animal production systems (Kumar et al. 2020; Sun et al. 2018). Undoubtedly, the use of agrochemicals has contributed to significant increases in food production; however, the excessive use of these substances has also caused serious problems to human health and environmental integrity (FAO and ITPS 2017).

Latin America is considered one of the regions with the greatest biodiversity and freshwater resources worldwide. However, the excessive and inappropriate use of antibiotics, as well as agrochemicals, has caused a serious ecological imbalance that puts at risk both the biodiversity and sustainability of soil biota, mainly bacteria, fungi and invertebrates (Giraldo-Pérez et al. 2021). The principal routes of entry of pollutants into agricultural ecosystems are (1) through the deposition of particles from the atmosphere via wind or rainfall, (2) the dragging of pollutants by floods and (3) through agricultural activities themselves. The increasing detriment to the fertility and productivity of agricultural soils is a reality. Furthermore, the contamination of such ecosystems poses serious risks to human health from the consumption of contaminated food grown on soils loaded with residual antibiotics and pesticides. The long-term consequences of soil degradation and subsequent desertification on soil biota threatens global food security (Rodríguez Eugenio et al. 2018).

Fortunately, in tropical regions there is a great diversity of plant species with enormous potential for restoring soils through the generation of better conditions for soil microorganisms. There is a positive correlation between biodiversity and resilience in general, with greater capacity for recycling nutrients and the self-regulation of pests and diseases, as well as greater potential in soil detoxification. Unarguably, soil microorganisms are essential elements in soil health and the proper functioning of agroecosystems (Barrios et al. 2012).

In this sense, it is crucial to seek and promote strategies aimed at soil restoration based on the use of natural resources that restore and maintain the good health of ecosystems. (Abhilash 2021). A great diversity of trees and shrubs have the appropriate characteristics for the restoration of soils contaminated with agrochemicals and pesticides (Karthikeyan et al. 2004). Plants, through their roots, can absorb contaminants from water or soil and transport them to the woody tissues or leaves of multi-purpose tree or shrub species (such as for wood, firewood, etc.), where contaminants can

DOI: 10.1201/9781003146902-10

remain "sequestered" for prolonged periods with subsequent and gradual degradation (Trapp and Legind 2010).

At the same time, tropical agroecosystems – by virtue of their rich diversity of plant species – provide livestock with a significant amount of nutrients, having positive effects on animal health. Indirectly, the use of antibiotics or other agrochemicals, which is usually necessary in industrial live-stock farming systems, gets reduced. As a result, tropical forestry may provide sustainable habitats for rearing livestock thanks to the nutraceutical potential of fodder from tropical tree and shrub species, and this may help to avoid the use of agrochemical contaminants.

## 1.1 Agroecosystems in the Context of Climate Change

Climate change is having a negative impact on livestock production systems and therefore on the wellbeing of millions of small producers globally. The recent floods during the summer of 2020 in Southern Mexico resulted in millions of dollars lost from both damaged crops and perished animals, driving indigenous communities into extreme poverty and hunger. The emissions from animal agriculture, along with emissions arising from the use of synthetic fertilisers, further contribute to increasing the concentrations of both methane and nitrous oxide in the atmosphere. Thus, it is paramount to keep emissions from the agricultural sector as low as possible by adopting agricultural practices that will lead to the mitigation of greenhouse gases (GHG). Silvopastoral systems of cattle production have proven to be one alternative to mitigate emissions of enteric methane under practical farming conditions in the tropics (Ku-Vera et al. 2020a; Ku-Vera et al. 2020b). It is evident that a new approach to food production is required, particularly in developing countries, to provide animal protein for the burgeoning human population. An agroecological approach is necessary to balance the flows of energy, carbon, nitrogen and water in production systems, whilst avoiding soil erosion, loss of biodiversity, water and air pollution and eventually, the loss of food security in a particular country. The production of agricultural commodities has to be matched with income generation and the wellbeing of the rural population; otherwise, no improvements will be achieved in the foreseeable future. The so-called 'circular food system approach' may be envisioned as an appropriate tool in the design of climate-smart livestock production systems (Van Zanten et al. 2019). It proposes an agroecosystemic approach to food production with the ultimate goal of directing all biomass unsuited for human consumption into animal feed. Such a systemic approach to production and the mitigation of GHG in animal agriculture would contribute to compliance with goals as outlined, for example, in the Nationally Determined Contribution (NDC) of Mexico, which commits itself to a reduction of 22% of its GHG emissions by 2030 (SEMARNAT 2020).

## 1.2 Soil Contamination and Risks to Human Health

Increasingly, attention has been paid to water and soil contamination. Most pollutants are caused by anthropogenic activities, such as industrial processes and inappropriate agricultural practices (FAO and PNUMA 2022). Pollution from agricultural activities generally has negative impacts beyond the places of production. For example, pollutants are carried by aquatic and terrestrial ecosystems, and through the food chain.

Recent studies have shown high concentrations of heavy metals (copper, iron and zinc), as well as pesticides and herbicides (such as glyphosate) present in the soil with effects on the blood, liver, brain, kidneys and lungs (Brevik et al. 2021) and an increased risk of developing chronic diseases such as diabetes and cancer. In relation to the soil, the use of pesticides decreases microbial activity; several studies show that the addition of the pesticides significantly decreased microbial activity and the number of soil bacteria, fungi and actinomycetes. It has been proven that microbial activity and the number of microorganisms is inversely proportional to the concentration of pesticides added to the soil (AL-Ani et al. 2019), negatively affecting soil fertility and soil health. Antibiotics used in agricultural systems easily reach the soil through urine and excreta and then may be transported through the environment. Recent studies show that between 40–90% of antibiotics given to animals are excreted and deposited directly into the soil, subsequently contaminating water, crops and the wider environment (Polianciuc et al. 2020). Even more serious is the careless discharge of unused antibiotics into bodies of water or wastewater streams.

### 1.2.1 ANTIBIOTICS: EFFECTS ON ENVIRONMENT AND SOIL

In recent decades, and due to the growing drive to increase the production of food of animal origin, the use of antibiotics has increased considerably, causing a strong impact on soil and in the environment (Koch et al. 2021; Kumar et al. 2005; Boxall 2004). Antibiotics are used in intensive animal production systems with the purpose of increasing production (growth promoting) and to prevent diseases (prophylactic) or treat diseases (therapeutic). Antibiotics are made up of complex molecules with different chemical structures that are often not biodegradable, depending on their mechanisms of action. However, the excessive and thus inappropriate use of antibiotics not only negatively impacts animal health, but also affects environmental health. Antibiotics that are not metabolised sufficiently within the animal itself are excreted through urine and faeces, and are subsequently introduced to the soil with serious effects on public health and ecosystems. Once released, such pharmaceutical residues are further dispersed reaching water bodies or being spread in the atmosphere via soil particles.

There is different evidence indicating that more than 50% of antibiotics such as tetracyclines, erythromycins, lincomycins and tylosin are excreted in urine and manure (Dolliver et al. 2008; Kumar et al. 2005). The excreta are then applied as fertilisers to crops together with contaminated water for irrigation. The highest concentration of antibiotics has been found in farm wastewater (cows, pigs, or poultry) and aquatic environments, thence negatively affecting soil microorganisms (Sosa-Hernandez et al. 2021). The absence of soil microorganisms has serious consequences for the sustainability of agricultural systems. For example, if the quantity and diversity of soil microorganisms is reduced, the efficient use of nitrogen (N) can be seriously affected (Wang and Tiedje 2020). Nitrate salts are very soluble, causing significant losses due to volatilisation, erosion and leaching, especially considering the low adsorption power of most tropical soils. The amount of nitrates leaching into the subsoil depends on the rainfall pattern and the type of soil. Often soils have abundant negatively charged organic and inorganic colloidal particles, which will repel anions, and consequently, these soils readily leach nitrates. However, many tropical soils, if left undamaged by antibiotics, retain a positive charge with a fine texture and, therefore, show a strong retention capacity for nitrates (Pashaei et al. 2022).

## 2 IMPORTANCE OF 'NUTRACEUTICAL' TREES AND SHRUBS IN LATIN AMERICA

Latin America as a region refers to the part of the Americas whose countries share Romance or Latin languages (Spanish, French, Portuguese). Even though the ethnic-political division may be variable (i.e., inclusion or not of Caribbean countries), it is made up of all the countries of the American continent except the United States and Canada (IPPC 2022). This region is one of the most important for global biodiversity, with six of the 17 mega-diverse countries and 11 of the 14 existing terrestrial biomes, as well as half of the world's jungles and forests (OECD 2018). It also has many of the 'hotspots' in biodiversity, a combination of high levels of endemic species but also high biodiversity loss. The situation is complex in this region: on the one hand, agriculture and livestock have been highlighted as the cause biodiversity loss – together with other activities such as mining – and on the other hand, smallholder farmers rear livestock and manage between 30% and 60% of the agricultural land and forestry areas in these countries, and depend on this activity for their livelihood (FAO 2019).

Silvopastures – with native vegetation – make an important contribution to biodiversity forming a natural vegetation not suitable for agricultural cultivation whilst covering approximately 70% of the world's land surface (Holechek et al. 2011). In Latin America, from Mexico to Chile, such vegetation is distributed throughout 33% of the total land area (Coppock et al. 2017), in very diverse climates and ecosystems, and presents a varied proportion of mono- and dicotyledonous herbaceous plants, creeping and climbing plants, native grasses, trees and shrubs. In the case of trees from the tropical region alone, this region can contain between 19,000 and 25,000 species (Slik et al. 2015). Conserving such woody pastures (silvopastures) should currently be a priority in order to store carbon and fight climate change both in rural and urban regions.

One strategy to restore and preserve rangelands is the inclusion and conservation of shrubs and trees native to Latin America, especially legumes. Cattle, sheep and goat production systems can thus contribute to increasing productivity and improving the sustainability of millions of producers (Chará et al. 2019). At the same time, these systems mitigate GHG emissions including methane emissions to the environment (Peters et al. 2013) and provide ecosystem services (Rao et al. 2015).

Shrubs and trees also are very important for inland restoration. In Latin America, soils and agricultural landscapes are becoming increasingly degraded by intensive agriculture and intensive animal husbandry. Managed appropriately in these areas, trees and shrubs can improve the soil organic carbon (SOC) content and thus restore soil fertility. Together with extensively managed livestock, producers can contribute considerably to the recycling of nutrients and improving soil structure.

The use of autochthonous trees and shrubs for animal production has many advantages that have been investigated for more than 40 years (Schultze-Kraft et al. 2018; Murgueitio et al. 2011; Topps 1992; Rao, 1998) including: i) adaptation to the climate and soil of each region, resistance to drought conditions due to the characteristics of their roots, retention of foliage throughout the year by perennial plants; ii) a large proportion (e.g. subfamily *Caesalpinioidae*) act as natural fertilisers in symbiosis with rhizobium-fixing atmospheric nitrogen in the soil (Boddey et al. 2015; Thomas 1995); iii) many contain at least twice the protein of tropical grasses, and their energy content can be considerable (Lüscher et al. 2014); iv) they contain a variety of secondary compounds present in leaves, pods and seeds that can mitigate enteric methane emissions and can have positive effects on animal health (Ku-Vera et al. 2020b; Bhatta et al. 2015; Hoste et al. 2012).

## 2.1   IDENTIFICATION OF NUTRACEUTICALS IN PRACTICE

Food stuffs that both provide nutrients and also provide a preventive and/or therapeutic health benefit to the organism that consumes them are known as nutraceuticals, or as functional foods in the context of human consumption (Andlauer and Fürst 2002).

Transferring the concept of nutraceuticals to animal husbandry, already being applied for example in ruminants, would not only present a benefit to their health but also to production, such as milk, which in turn could have a nutraceutical effect on humans who consume it (Lund and Ahmad 2021; Addrizzo n.d.). It should be kept in mind that nutraceuticals are not (pharmaceutical) medicines nor phytotherapeutics. These would typically be given to treat an ailment over a limited period and thus they would only have a short-term effect. The ongoing, long-term consumption of nutrients together with secondary compounds provides the animal with the nutraceutical effect (Torres-Fajardo et al. 2020; Hoste et al. 2015; Hoste et al. 2008).

There are several review papers on nutraceuticals in ruminant nutrition (e.g. Patel and Katole, 2018) and veterinary medicine (e.g. Gupta et al. 2019). The use of these nutraceutical plants in the feeding of sheep and goats exerts a positive effect on the control of gastrointestinal parasitism

(Hoste et al. 2015) and includes both temperate and tropical climate forages (Hoste et al. 2012; Sandoval-Castro et al. 2012; Niezen et al. 1995). Plants from the extensive *Fabaceae* family (with approximately 20,000 species) have been among the most investigated tree models for the past 30 years, particularly those that contain secondary compounds such as tannins. *Fabaceae* such as *Hedysarum coronarium, Lysiloma latisiliquum, Havardia albicans* or acacias such as *Acacia pennatula, A. molissima, A. mearnsii* have been under investigation as a model for the *in vivo* evaluation of nutraceutical effects against the parasite *Haemonchus contortus* in sheep. *Gymnopodium floribundum,* a member of the *Eudicot* family, was recently added to the list of model plants containing tannins (Méndez-Ortiz et al. 2019). These families are distributed throughout Latin America; therefore, this macro region has great potential as a provider of nutraceutical resources.

*In vitro* evaluation is conducted using plant extracts and parasites or pathogenic organisms under laboratory conditions. Such experiments are very useful in the exploration and identification of plants that may have nutraceutical potential (Castañeda-Ramírez et al. 2018), as well as in the identification of possible mechanisms of action (Castañeda-Ramírez et al. 2017). However, subsequent *in vivo* studies, that is, offering the forages directly to the animals for their consumption and evaluating the effect on their health, are essential to consider a species as a plant of true nutraceutical value (Torres-Fajardo et al. 2020; Villalba and Provenza 2010).

The digestion process in ruminants is extremely complex and dynamic. Diverse chemical reactions occur to degrade forages, the effects of which depend upon on the characteristics of the saliva and its proteins, the effect of microbial and protozoan populations in the rumen and the mixture of secondary compounds from different plants when consumed at the same time. This diversity may give rise to synergies or antagonisms between the components affecting digestion Therefore, it is necessary to avoid oversimplification: if, for example, a plant extract with a good effect *in vitro* is identified, it should not be assumed that it will also have an automatic effect when consumed by the animal (*in vivo* conditions). Similarly, if a plant extract has no effect *in vitro*, it should not be assumed that it will not work *in vivo* (Villalba and Provenza 2010). The effect of biotic and abiotic conditions on the production of secondary plant components (including tannins) is complex (Hove et al. 2003). In turn, the relationship of plants and their secondary compounds with the herbivore's metabolism under grazing of a heterogeneous vegetation is also complex and will depend on the biotype of vegetation as well as the ruminant species (Gordon and Prins 2019) and their grazing patterns. Finally, the relationship between the nematode parasite and its host (ruminants) is also complex and will depend on a large number of biotic and abiotic variables (Hutchings et al. 2006; Vlassoff 1982). Therefore, determining which plant to offer to the animals at what point in the plant's life cycle still requires research. This includes determining how much of the plant, how often and for how long it needs to be consumed to observe a positive effect (preventive or curative) on health. Clearly, the complexity of the processes involved in feeding ruminants through grazing demands an inter-disciplinary approach to identifying nutraceutical resources and designing the management techniques for their use (Torres-Fajardo et al. 2020; Baumont et al. 2000). It requires the multi-disciplinary collaboration between chemists, botanists, agronomists, nutritionists, ethologists, animal health specialists, ecologists and veterinarians.

There are models for the use of processed nutraceuticals on an industrial scale, successful to varying degrees when used in stabled livestock systems or in grazing on pastures that are not as diverse as those of Latin America. For example, in France, hay from the Sainfoin (*Onobrycis viciifolia*) crop (Desrues et al. 2017; Paolini et al. 2005) or Leucaena (*L. leucocephala*) pellets (Marie-Magdeleine et al. 2020) are available; in the United States, a concentrate to formulate rations based on the quebracho tree (*Schinopsis* spp.) is available on the market (e.g., Silvafeed®). While agroecology is about minimising external inputs and about producing inputs on site, these examples nonetheless go to show that the concept of nutraceuticals has already been taken up by industry.

## 3   TREES AND SHRUBS AS NUTRACEUTICALS FOR LIVESTOCK ON THE YUCATAN PENINSULA

A large proportion of the trees and shrubs in Latin America have the essential characteristic for use as a nutraceutical: they are consumed voluntarily by ruminants. There is a significant body of knowledge about such species, identified through various approaches. On the Yucatan Peninsula, extensive production systems, together with inadequate management practices, have had serious repercussions on the ecology, including high rates of deforestation. An area of this over-exploited land has been purchased and is being restored by adopting livestock production systems based on agroecological principles that encourage the managed integration of animals with natural resources (Fig. 1). Thus, the region can still remain an important reserve of plant resources with nutraceutical properties.

Using an ethnobotanical approach Flores and Bautista (2012) report a list of diverse native forage plants from the seasonal tropical forests of the Yucatan Peninsula (Mexico), which are used by 27 Mayan communities. They report 196 species, amongst which are 35 trees and 17 shrubs used to feed different species of domestic animals (birds, pigs, ruminants). Studies on the forage production of various species of trees and shrubs in silvopastoral systems (Zapata-Buenfil et al. 2009) or the content of tannins (Alonso-Díaz et al. 2010) have also contributed to forming such a list of forage plants. Other methods of identifying forage intake include the use of oesophageal cannulas (Pfister and Malecheck, 1986) or by directly observing browsing by cattle (Albores-Moreno et al. 2020), sheep (Jaimez-Rodríguez et al. 2019; González-Pech et al. 2015) and goats

**FIG. 1**   Principles of agroecology drawn on a typical silvopastoral system on restored land in the tropics of the Yucatan Peninsula, modified after Altieri and Nicholls 2000.

(Ventura-Cordero et al. 2019; Franco-Guerra et al. 2014; Ríos and Riley 1985) grazing heterogeneous native vegetation.

There is an excellent opportunity to incorporate species of trees and shrubs with nutraceutical properties within land restoration measures since most of these contribute to the protein requirement of ruminants. In particular, shrub legumes contain more than 10% (some up to 30%) crude protein, which is higher than the 7% considered as a minimum so that the ruminal fermentation process is not deficient in the production of microbial protein (Ku-Vera et al. 2020b). This has made them an excellent option for use as protein banks (e.g., *Leucaena spp.* banks [Brewbaker and Hutton 2019; Pachas et al. 2019]) for ruminant feed; even more so if we consider the protein that is bypassed or sequestered from the rumen due to the effect of the condensed tannins (CT) – protein that would otherwise be broken down by microbes and not be available in the intestine for further absorbtion (Wang et al. 1996). The presence of the so-called secondary compounds, nutraceutical compounds, is intrinsic to the life forms of shrubs and trees. For a long time, such compounds were considered anti-nutritional factors and, in particular, phenolic compounds such as CT were considered to have a negative effect on digestibility in ruminants (Makkar 2003). However today, the role of CT is better understood, and it is known that many of the species of trees and shrubs with tannins also contain a significant proportion of lignin, the component responsible for the reduction of digestibility and not tannins (Méndez-Ortíz et al. 2018). Ruminants when they have freedom of choice whilst browsing heterogeneous vegetation select the youngest plant parts with a lower lignin content, which are therefore more digestible. Available diversity and selection behaviour allow the ruminant to form a medium-to-good quality ration (González-Pech et al. 2015; Agreil et al. 2005) and obtain the benefit of consuming secondary compounds (Ventura-Cordero et al. 2018).

Although trees and shrubs are a very important source of protein for livestock systems, it is still necessary to overcome at least two challenges to extend their use. The first is that their inclusion in the ruminant diet must be balanced with the correct supply of energy of either an adequate pastoral diet or with ingredients such as corn, molasses, citrus silage or cereal grains, which vary in the availability and price of inputs in each region. In agroecological systems these inputs should be available on farm or at least very locally. Achieving balanced rations with a good energy-protein ratio is one of the greatest challenges; for example, it is traditionally considered that cattle require 12 to 14g of crude protein (CP) per megajoule of metabolisable energy, depending on the physiological stage and productivity of the animal (Ku-Vera et al. 2020b). Any positive effect of nutrition on the health of ruminants will be diminished if the diet is not correctly balanced in terms of quality and if not offered in sufficient quantity. The second challenge is to determine the presence of secondary compounds other than phenolic compounds such as terpenoids, saponins and alkaloids that may represent a dose-dependent health risk for ruminants including hepatotoxic effects, photosensitisation and eye injuries (Walelign and Mekruriaw 2016).

### 3.1 TREES AND SHRUBS IN ANIMAL HEALTH

Condensed tannins have further positive effects. They may help prevent bloating (tympany) which is the rapid accumulation of gas due to the fermentation of highly digestible carbohydrates in ruminants (Min et al. 2003). Li et al. (1996) suggested a minimum of 5g CT/kg dry matter (DM) as the necessary concentration in forage to obtain a beneficial reduction in gas or foam production in the rumen. Given the chemical composition of forage from trees and shrubs, of which a large proportion contain condensed tannins, it is very common to observe consumptions higher than those proposed by Li et al. (1996). Goats and sheep were shown to consume a median of 10.3 and 9.2g of CT/kg DM, respectively, in just four hours of grazing in a silvopasture made up of low deciduous forest (González-Pech et al. 2015). Here, in the rainy season, goats and sheep may consume an average of 80 and 18.4g of CT/kg DM, respectively (Ventura-Cordero et al. 2019). The role played by forage trees and shrubs in preventing tympanism where this becomes a problem (e.g., in very young and vigrous pastures) should be recognised as another reason to supplement animal diets with tree forage.

Nutraceutical trees and shrubs with anthelminthic activity, such as *Lysiloma latisiliquum*, help to control parasitism caused by gastrointestinal nematodes (GIN) in ruminants (Martinez-Ortíz-De-Montellano et al. 2010). The anti-parasitic effect is linked in part to the levels of polyphenols such as tannins within the plant. The plant secondary compounds act directly on the biology of the parasite and the host (Athanasiadou et al. 2005). The compounds may make the host gut inhospitable; they may reduce the establishment of the infective larvae that are ingested by the host whilst grazing, and egg laying may be reduced in adult female parasites, or once deposited onto the ground in the dung egg or larval development may be interrupted. This results in a reduction of the numbers of parasites within the host and a less contaminated pasture.

Trees and shrubs also have indirect anthelminthic effects, the first of which is related to the form or architecture of shrubs or trees, which is very different from that presented by the tillers and leaves of grasses. Trees and bushes tend to be higher than grasses, which makes it difficult for infective larvae to move from the soil level to the edible tree and bush foliage, as 80% of these are located in the first 5 cm above the ground (Vlasoff 1982). In this way, the bush format is less favourable for the ingestion of larvae (Hoste et al. 2001) and would be equivalent to a 'clean' GIN forage. A further indirect effect of nutraceuticals on the health of ruminants is an improvement in the immune response of ruminants against GIN infections (Coop and Kyriazakis 1999), due to the bypass protein sequestered by tannins from ruminal fermentation, which then becomes available in the small intestine (Wang et al. 1996). This protein constitutes an additional element that improves the host's immune response. Such an extra quantitative contribution might be used for the formation of immunoglobulins, mucoproteins and cells of the immune system such as leukotrienes of an essentially protein nature (Coop and Holmes 1996). A comprehensive review of such direct and indirect effects is reported by Hoste et al. (2012). The anthelmintic activity of trees and shrubs is not purely mediated by tannins. *In vitro* tests with extracts of *Annona squamosa*, *A. muricata* and *A. reticulata* showed that the anthelmintic effect on the eggs (a decrease in hatching) and larvae (an inhibition of exsheathing) of the abomasal parasite *Haemonchus contortus* continued to be observed after the application of a tannin blocker (Castañeda-Ramírez et al. 2020). In other plant species, such as *Lysiloma latisiliquum*, the *in vitro* anthelmintic effect (reduction of hatching of *H. contortus* eggs) has been shown to be increased by blocking tannins (Hernández-Bolio et al. 2018; Vargas-Magaña et al. 2014).

This means that the interactions of certain secondary compounds contained in trees and shrubs could reduce the anthelmintic effect on the egg stage, but increase the anthelmintic effect on infective larvae L3 (Castañeda-Ramírez et al. 2017). More research will be necessary to understand the interaction of mixtures of secondary compounds, and the possible synergies or antagonisms. However, field evidence shows that goats keep GIN infections in check when consuming mixtures of trees and shrubs such as *Gymnopodium floribundum*, *Leucaena leucocephala*, *Mimosa bahamensis* and *Neomillspaughia emarginata* (Novelo-Chi et al. 2019), which contain CTs, but also other secondary compounds such as saponins, alkaloids and terpenes.

The diverse ration chosen by the ruminants themselves upon grazing heterogeneous vegetation is a natural use of nutraceuticals; ruminants are naturally consuming true cocktails of secondary compounds. For example, sheep and goats when free-grazing rangelands made up of the low deciduous forest of Yucatan (Mexico) consume 35 species of plants in the dry season, of which 17 are shrubs and ten of them contain CT (González-Pech et al. 2015); during the rainy season they consume 61 species of plants, of which 13 are shrubs and seven of them have CT (Ventura-Cordero et al. 2019). However, in both seasons, the ration with four hours of grazing, which is usual in this type of pasture grazing, is insufficient to cover the energy requirement, so to take advantage of the true nutraceutical potential it is necessary to correctly supplement the diet of the animals – the nature of such a supplement will be different for each region of Latin America.

### 3.2 CONSIDERATIONS FOR THE SUSTAINABLE USE OF THE NUTRACEUTICAL RESOURCE

The advantages of the nutraceutical potential of ecosystems, such as low deciduous forests, are encouraging for small ruminant production (Ventura-Cordero et al. 2017; Torres-Acosta et al. 2016). However, the challenges for its successful adoption are closely related to the characteristics of ruminant production systems based on the grazing of native seasonal vegetation. The pattern of rainfall partly determines the seasonal variation in the availability of nutraceutical forage throughout the year. Therefore, it is important to understand the seasonal or monthly calendar of availability (both in terms of quantity and quality) of the trees and shrubs in each region of any biome. For responsible management, it is also necessary to determine the timing and intensity of grazing. Trees and shrubs not only provide foliage, but also pods; for example, the leguminous *Fabaceae* of the low deciduous forest contribute up to 40% of the biomass consumption of grazing sheep and goats in the dry season (González-Pech et al. 2015) and their nutritional value is excellent (Ortíz-Domínguez et al. 2017). Other pods, such as those of *Enterolobium cyclocarpum* when included in balanced diets, improve the overall quality of the diet and the fermentation patterns, reducing the production of methane gas (Ku-Vera et al. 2020b).

Pods of trees and shrubs in grazing areas need to be included in the calculation of the resources that are effectively available to ruminants to avoid underestimating the quality of the diet.

One cannot speak of the nutraceutical effect of trees and shrubs without mentioning that the consumption of such plant species represents an opportunity for ruminants to express self-medication. For example, Ventura-Cordero et al (2018) report that the consumption of plants rich in tannins by Creole goats both avoids an excess of protein in the diet and prevents a GIN infection from being exacerbated. Novelo-Chi et al. (2019) evaluating Creole goats grazing heterogeneous vegetation confirmed these observations.

The consumption of plants rich in tannins helps the goats to maintain low levels of natural infection by GIN in the rainy season. This consumption pattern remains even if the goats have been dewormed, suggesting a preventive self-medication behaviour. Recent evidence suggests that the specific time of consumption during the grazing day is important. Torres-Fajardo et al. (2019a) observed changes in the grazing behaviour of Creole goats, with grazing heterogeneous vegetation depending on the time of day. They reported a constant consumption pattern of shrubs plus herbaceous throughout the day, whilst the consumption of grasses was avoided during the early hours of the morning when the humidity could lead to a greater ingestion of infective larvae. The authors report such behaviour as a "prophylactic self-medication behaviour" aimed at improving the energy:protein balance in the diet, and avoiding the ingestion of infective larvae, thus keeping the levels of GIN infection under control.

Finally, trees and shrubs contribute to the health of grazing animals by allowing them to express the full range of behaviours with which they evolved and adapted to their environment. This is an aspect that is generally overlooked in grazing systems – as mental health is part of animal health, the presence of positive mental states is essential for animal welfare (Mellor and Beausoleil 2015). Thus, trees and shrubs not only nourish the physiological economy of the ruminant, they also nourish their mental state by allowing them to express their innate curiosity, experiencing the set of physical activities that grazing represents – along several kilometres of diverse vegetation – as well as the mental work involved in the selection of species of plants and parts of plants to ingest (the cost-benefit decision is complex; see Torres-Fajardo et al. 2019b). Under such conditions, the animals also have the opportunity to interact socially in an enriched environment. In the case of young animals grazing together with adult animals, this allows them to learn how to form a diet in proportions similar to that of the adults within just one week (Jaimez-Rodríguez et al. 2019).

**FIG. 2**  Agrosilvopastoral systems can protect and restore degraded ecosystems, while providing healthy food to animals and humans alike and securing local livelihoods.

### 3.3  FROM NEGATIVE TO POSITIVE FOOD PRODUCTION SYSTEMS

Based on the problems related to contamination and degradation of the soil, water and the wider environment, it is essential to reverse the negative aspects of the inappropriate use of pesticides, fertilisers and agrochemicals in general. Recently, the effects of climate change, rapid population growth and the demand for food of animal origin have exerted greater pressure on natural resources and in fact, have increased the demand for agrochemicals, fertilisers and antibiotics. The development and uptake of agrosilvopastoral systems (ASPS) characterised by integrating animal production, pastures or crops within woody vegetation is a strategy that could reverse this trend (Murgueitio et al. 2011).

In this sense, ASPS meet all the scientific and traditional criteria of agroecological science necessary to convert current extensive livestock production into more efficient and sustainable livestock systems (FAO 2018), providing empowerment and new opportunities for families in rural areas. ASPS have the potential to immediately transform fragmented livestock landscapes into connected landscapes based on ecosystems that are able to reverse vulnerability and build resilience and adaptation to climate change. In addition, ASPS contribute to the conservation of biodiversity and to improving animal welfare (Broom, et al. 2013).

### 3.4  INTEGRATING TREES AND SHRUBS IN LIVESTOCK PRODUCTION SYSTEMS

In recent decades, the degradation of resources has occurred at an accelerated rate; rainforests have been the most exposed and vulnerable to this degradation, mostly by fire or by clearing open areas for agriculture. Additionally, soils of rangelands where livestock are being kept are rapidly degrading (Castro-Nunez 2021).

Livestock production systems provide diverse forms of livelihoods, with an ability not only to be more food secure but also to build greater resilience to climate change. At the same time, increasing the production of meat and milk will lead to an economic benefit for smallholder farmers. However, small-scale livestock production systems are characterised by their vulnerability due to low profitability. They need better integration with natural resources to reduce environmental impact and to contribute to environmental restoration (FAO 2022).

Currently, this vulnerability of livestock systems is aggravated by the pandemic caused by COVID-19 (FAO and ECLAC 2020). Food costs have increased, and the sale of animals or meat and milk has become difficult, resulting in decreased incomes at family level and the disappearance of thousands of small producers (FAO 2020).

In this sense, there is an urgency to turn current conventional livestock production towards more inclusive and intelligently managed livestock systems. Efforts must be directed at generating environmentally and economically resilient livestock agroecosystems with concomitantly efficient use of natural resources.

For several years, different alternatives for transforming conventional livestock production have been discussed and ASPS seem to be one of the best options (FAO 2018). ASPS are based on agroecological principles that integrate animal production with tree and shrub species at high densities and with pastures in different topological arrangements. ASPS systems have been shown to restore degraded livestock landscapes, increase forage quality and thus productivity, whilst reducing GHG emissions (Broom et al. 2013).

Trees and shrubs further contribute to maintaining diversified agroecosystems as they improve soil organic carbon content (SOC), microclimatic conditions and nutrient cycling. The main advantages of integrating woody species in agroecosystems are detailed below.

### 3.4.1  Productivity

One of the main objectives is to restore degraded pastures, based on the association with mainly fast-growing woody species. Such shrubs and trees produce foliage of excellent quality (high protein content, low fibre content, high digestibility). The combination of shrubs, trees and pastures generally doubles the forage production and the amount of protein/ha compared to conventional extensive pastures. The ASPS allows ranchers to increase the stocking rate/ha, with considerable increases in milk and meat production (25 to 40%) (Sarabia-Salgado et al. 2020). Additionally, this 'sustainable intensification' of forage contributes to reducing the pressure on the demand for new land for cattle ranching.

### 3.4.2  Connectivity

Poorly managed extensive livestock farming has been characterised by the transformation of the geographical space, degrading ecosystems into scattered and isolated landscapes, with a great negative environmental impact and loss of biodiversity. Reconversion to ASPS with the integration of different woody species constitutes one of the most efficient ways to form biological corridors, restore forest fragmentation and reduce isolation between agroecosystems and therefore promote the conservation of biological diversity and increase resilience.

Trees integrated into livestock farming are essential for enriching livestock landscapes and strengthening networks of biological corridors between productive and conservation land. Likewise, the plant biodiversity of paddocks contributes to a greater richness and abundance of wild fauna, since the conditions for their movement, refuge, nesting and feeding are improved. Therefore, the greater diversity and structure of tree cover in livestock agroecosystems brings together a series of functional traits with potential for the development of sustainable livestock production systems with positive effects at both landscape and agroecosystem levels.

### 3.4.3   Mitigation

An extremely important aspect of trees and shrubs in livestock production is related to GHG mitigation. Specifically, ruminants integrated into ASPS are supplemented with woody species that contain CT, saponins, essential oils and unsaturated fatty acids. This helps mitigate enteric methane emission by up to approximately 20%. Some tree legume species carry additional potential to reduce methane emissions (Ku-Vera et al. 2020a). By associating legumes with grasses, protein availability is improved, consumption is increased and so is digestibility. The increased protein content naturally contributes to reducing the enteric methane emissions produced during ruminal fermentation (Gerber et al. 2013). Relative to other global GHG reduction options, enteric methane reduction through increased productivity is the lowest-cost option with direct economic benefits to farmers (FAO and HLPE 2019).

## 4   CHALLENGES AND PRIORITIES TO BE ADDRESSED

The main challenge facing the implementation of an agroecosystem approach (e.g. ASPS) to food production in many developing countries of Latin America is the lack of commitment by the administrative agricultural leadership towards an optimised way forward for agricultural development in the present century (i.e. science based). The long history of colonisation by foreign powers, which led to the 'plantation concept' (monocultures of bananas, cotton, sugar cane, sisal, tobacco, etc.) for agricultural development in many countries, the lack of democratic governments, the political upheavals of the last century and the lack of a consolidated scientific establishment have all contributed to creating a lack of confidence in the support that science offers in the efforts made towards a regenerative agriculture. Under practical farming conditions, foliage, pods and seeds of tropical trees and shrubs containing either CT, essential oils, saponins or starch, as well as by-products such as vegetable oils or crop residues, can be fed to ruminants to mitigate enteric methane emissions under small-farming conditions. ASPS with legumes (e.g. *leucaena*, rain tree), the intensification of grazing management and supplementation with ground pods of *Enterolobium cyclocarpum*, *Samanea saman* or seeds of *Brosimum alicastrum* are readily available options for small-holders to reduce the intensity of methane emissions in ruminants raised under tropical conditions (Fig. 3). Tannin- and saponin-containing tree and shrub legumes are a viable alternative for reducing soil nitrogen (N) mineralisation and N loss. Trees and shrubs rich in tannins and saponins have been shown to decelerate ruminal protein degradation and, therefore, prevent the formation of ammonia. Tannins bind to proteins at a ruminal pH, thus preventing access by microbes, and saponins in turn hamper the activity of microbes at different steps of protein degradation. Similarly at soil level, tannins may limit N-losses, altering the availability of nutrient pools to plants and microbial communities by complexing with proteins and enzymes during decomposition, therefore increasing the ratio of dissolved organic N to mineral N. Overall, diverse landscapes for livestock may decrease GHG emissions and N losses through eliminating the need for N fertilizer, by altering soil C and N cycling and thus maintaining pasture ecosystem services.

The priorities should be to motivate a larger cadre of agricultural scientists and support genuine commitment to science and financial investments in order to close the gaps in knowledge regarding the balance of elements (carbon, nitrogen, water) in agroecosystems in many developing countries of Latin America.

## 5   CONCLUSIONS AND RECOMMENDATIONS

Despite the great efforts made to understand the mechanisms of action between agrochemicals, antibiotics and their interactions with the environment, the current and future impacts of soil contamination on human health and the environment are still uncertain. Soil contamination is the result of the intensification of modern agriculture, based on the excessive use of agrochemicals

Trees and shrubs improve forage quality (dietary digestibility) and result in decreased enteric $CH_4$ emissions.

Forages with a diversity of biochemical composition contribute to reductions in carbon and nitrogen footprints

$CO_2$

$CH_4$

$N_2O$

Diverse diets: Nutrients and plant secondary metabolites (tannins, flavonoids, terpenoids).

Productivity and Enviromental Co-Benefits

Carbon seqestration through soil health

**FIG. 3** Biodiversity of leguminous trees and shrubs for ruminant production can mitigate methane emissions.

and antibiotics that alter the composition of the soil, influence the natural balance of the ecological system and disrupt the self-purification capacity of the soil, as well as the sustainability of food production.

Food production systems that integrate biodiversity have greater potential to restore degraded soils and to contribute to the reduction of contaminants in the environment. The health of ecosystems depends largely on the diversity of tree and shrub species. It is a matter of urgency to promote and adopt strategies that include reforestation practices and the integration of agricultural production systems with natural resources. The quality of water depends on the health of the soil; likewise, the quality of our food depends on the health of agroecosystems. Having overall healthier agroecosystems will allow us to depend less on synthetic medicine, synthetic antibiotics and synthetic agrochemicals.

Land restoration can be more efficient with a combination of woody species and the microbes associated with their roots, rhizodegradation being an effective phyto remedial process. Additionally, tropical regions are home to a great diversity of tropical species with nutraceutical characteristics. Trees and shrubs, mainly legumes, have direct benefits in animal nutrition to reduce the use of agrochemicals and to control parasites or other pathogenic organisms. Likewise, different species of trees and shrubs present a potential for the mitigation of greenhouse gases.

## 6 ACKNOWLEDGMENTS

This manuscript is part of the PAPIIT project for the integration of sustainability indicators in the evaluation and agroecological restoration of the socio-ecological system of livestock farming on the

Yucatan Peninsula supported by the post-doctoral scholarship programme UNAM–DGAPA at the Faculty of Veterinary Medicine – Universidad Nacional Autónoma de México UNAM, Mexico City.

## 7  REFERENCES

Abhilash, Purushothaman Chirakkuzhyil. 2021. "Restoring the Unrestored: Strategies for Restoring Global Land during the UN Decade on Ecosystem Restoration (UN-DER)." *Land* 10 (2): 201.

Addrizzo, John. n.d. "Use of Goat Milk and Goat Meat as Therapeutic Aids in Cardiovascular Diseases." Accessed May 12, 2022. http://goatdocs.ansci.cornell.edu/Resources/GoatArticles/GoatProducts/Goat ProductBenefits.pdf.

Agreil, Cyril, Hervé Fritz and Michel Meuret. 2005. "Maintenance of Daily Intake through Bite Mass Diversity Adjustment in Sheep Grazing on Heterogeneous and Variable Vegetation." *Applied Animal Behaviour Science* 91 (1-2): 35–56.

Albores-Moreno, S., J.A. Alayón-Gamboa, A Morón-Ríos, P. N. Orti-Colin, J. Ventura-Cordero, P. G. González-Pech, G.E. Mendoza-Arroyo, J. C. Ku-Vera, G. Jiménez-Ferrer and A. T. Piñeiro-Vaázquez. 2020. "Influence of the Composition and Diversity of Tree Fodder Grazed on the Selection and Voluntary Intake by Cattle in a Tropical Forest." Agroforestry Systems 94 (5): 1651–64.

Alonso-Díaz, M.A., J.F.J. Torres-Acosta, C.A. Sandoval-Castro and H. Hoste. 2010. "Tannins in Tropical Tree Fodders Fed to Small Ruminants: A Friendly Foe?" *Small Ruminant Research* 89 (2-3): 164–73.

AL-Ani, Mehjin A M, Rawaa M Hmoshi, Ibtiha A Kanaan and Abdullah A Thanoon. 2019. "Effect of Pesticides on Soil Microorganisms." *Journal of Physics: Conference Series* 1294 (September): 072007. https://doi.org/10.1088/1742-6596/1294/7/072007.

Altieri, Miguel and Clara Nicholls. 2000. *Agroecología: teoría y práctica para una agricultura sustentable.* Pnuma, Mexico City.

Andlauer, Wilfried and Peter Fürst. 2002. "Nutraceuticals: A Piece of History, Present Status and Outlook." *Food Research International* 35 (2): 171–76.

Athanasiadou, S., O. Tzamaloukas, I. Kyriazakis, F. Jackson and RL Coop. 2005. "Testing for direct anthelmintic effects of bioactive forages against Trichostrongylus colubriformis in grazing sheep." *Veterinary Parasitology* 127: (3-4): 233–43.

Barrios, Edmund, Gudeta W. Sileshi, Keith Shepherd and Fergus Sinclair. 2012. Agroforestry and Soil Health: Linking Trees, Soil Biota, and Ecosystem Services. In *Soil Ecology and Ecosystem Services.* Edited by Diana H. Wall, 315–30. Oxford, UK: Oxford University Press.

Bhatta, R., M. Saravanan, L. Baruah and C.S. Prasad. 2015. "Effects of Graded Levels of Tannin-Containing Tropical Tree Leaves on in vitro Rumen Fermentation, Total Protozoa and Methane Production." *Journal of Applied Microbiology* 118 (3): 557–64. https://doi.org/10.1111/jam.12723.

Baumont, R, S Prache, Meuret and P Morand-Fehr. 2000. "How forage characteristics influence behaviour and intake in small ruminants: a review." *Livestock Production Science* 64 (1):15–28.

Boddey, R. M., I. N. O. de Carvalho, C. P. Rezende, R. B. Cantarutti, J. M. Pereira, R. Macedo, R. Tarré, B.J.R. Alves and S. Urquiaga. 2015. "The benefit and contribution of legumes and biological N2 fixation to productivity and sustainability of mixed pastures." *Proceedings of the 1st International Conference on Forages in Warm Climates.* Edited by A. R. Evangelista, C. L. S. Avila, D. R. Casagrande, M. A. S. Lara and T. F. Bernardes. Universidad Federal de Lavras, Lavras, MG, Brazil. 103–140.

Boxall, A.B. 2004. "The environmental side effects of medication: How are human and veterinary medicines in soils and water bodies affecting human and environmental health?" *EMBO Reports*, 5(12): 1110–1116.

Brevik, Kristian, Erika M. Bueno, Stephanie McKay, Sean D. Schoville and Yolanda H. Chen. 2021. "Insecticide Exposure Affects Intergenerational Patterns of DNA Methylation in the Colorado Potato Beetle, Leptinotarsa Decemlineata." *Evolutionary Applications*, https://doi.org/10.1111/eva.13153.

Brewbaker James. L., Hutton and E. Mark. 2019. "Leucaena: versatile tropical tree legume." *New agricultural crops* edited by Gary A Ritchie, 207–259. CRC Press, New York.

Broom, D. M., F. A. Galindo and E. Murgueitio. 2013. "Sustainable, Efficient Livestock Production with High Biodiversity and Good Welfare for Animals." *Proceedings of the Royal Society B: Biological Sciences*, 280 (1771): 2013–2025.

Castañeda-Ramírez G.S., Torres-Acosta J.F.J., Mendoza-de-Gives P., Tun-Garrido J., Rosado-Aguilar J. A., Chan-Pérez, J. I., Hernández-Bolio, G. I., Ventura-Cordero, J., Acosta-Viana, K. Y., Jiménez-Coello, M.

2020. "Effects of different extract of three Annona species on egg-hatching processes of Haemonchus contortus." *Journal of Helminthology*. Volume: 94 .

Castañeda-Ramírez, G. S., M. Rodríguez-Labastida, G. I. Ortíz-Ocampo, P. G. González-Pech, J. Ventura-Cordero, R. Bórges-Argáez, J. F. J. Torres-Acosta, C. A. Sandoval-Castro and C. Mathieu. 2018. "An in vitro approach to evaluate the nutraceutical value of plant foliage against *Haemonchus contortus*." *Parasitology Research* 117 (12): 3979–91.

Castañeda-Ramírez, G. S., J. F. J. Torres-Acosta, C. A. Sandoval-Castro, P. G. González-Pech, V. P. Parra-Tabla and C. Mathieu. 2017. "Is there a negative association between the content of CT, total phenols and total tannins of tropical plant extracts and the in vitro anthelmintic activity against *Haemonchus contortus* eggs?" *Parasitology Research* 116 (12): 3341–48

Castro-Nunez, Augusto, Buriticá, Carolina González, Eliza Villarino, Federico Holmann, Lisset Pérez, Martha Del Río, et al. 2021. "The Risk of Unintended Deforestation from Scaling Sustainable Livestock Production Systems." *Conservation Science and Practice* 3 (9).

Chará, J., E. Reyes, P. Peri, J. Otte, E. Arce and F. Schneider. 2019. Silvopastoral Systems and Their Contribution to Improved Resource Use and Sustainable Development Goals (SDG): Evidence from Latin America. Edited by FAO, CIPAV and Agri Benchmark. FAO. https://www.fao.org/3/ca2792en/CA2792EN.pdf.

Coop R.L., and I. Kyriazakis. 1999. "Nutrition-Parasite Interaction." *Veterinary Parasitology*, 84: (3-4): 187–204.

Coop, R. L. and H Holmes. 1996. "Nutrition and parasite interaction." *International Journal of Parasitology*. 26: (8-9) 951–62.

Coppock, D. L., Fernández-Giménez, M., Hiernaux, P., et al. (2017). "Rangeland systems in developing nations: conceptual advances and societal implications." Rangeland Systems. Edited by D. Briske. Springer, Cham. 569–642.

Desrues, Oliver, Irene Mueller-Harvey, Wilbert F. Pellikaan, Heidi L. Enemark and Stig M. Thamsborg. 2017. "Condensed Tannins in the Gastrointestinal Tract of Cattle after Sainfoin (Onobrychis viciifolia) Intake and Their Possible Relationship with Anthelmintic Effects." *Journal of Agricultural and Food Chemistry*, 65 (7): 1420–27.

Dolliver, Holly, Satish Gupta and Sally Noll. 2008. "Antibiotic Degradation during Manure Composting." *Journal of Environment Quality* 37 (3): 1245. https://doi.org/10.2134/jeq2007.0399.

FAO, 2018. *Be the solution to soil pollution*. Global Symposium on soil Pollution. FAO. Rome Italy. https://www.fao.org/3/ca0362en/CA0362EN.pdf. Accessed 26/5/2022.

FAO. 2019. *Cambio climático y seguridad alimentaria y nutricional en América Latina y el Caribe*. FAO. Santiago de Chile. https://www.fao.org/3/CA2902ES/ca2902es.pdf.

FAO. 2020. *El estado mundial de la agricultura y la alimentación 2020. Superar los desafíos relacionados con el agua en la agricultura. Roma*. https://doi.org/10.4060/cb1447es.

FAO. 2022. *Alignment of FAO's work on livestock to the Strategic Framework 2022-31*. COAG:LI/2022/8. Rome, FAO. https://www.fao.org/3/ni075en/ni075en.pdf. Accessed 31 August 2022.

FAO and ECLAC. 2020. Food systems and COVID-19 in Latin America and the Caribbean: food consumption patterns and malnutrition. Bulletin, No. 10, July. https://www.cepal.org/en/publications/45795-food-systems-and-covid-19-latin-america-and-caribbean-ndeg-10-food-consumption.

FAO and HLPE. 2019. *Agroecological and other innovative approaches for sustainable agriculture and food systems that enhance food security and nutrition*. FAO/HLPE Report 14. Rome. https://www.fao.org/3/ca5602en/ca5602en.pdf. Accessed 26/5/2022.

FAO and ITPS. 2017. *Global assessment of the impact of plant protection products on soil functions and soil ecosystems*. Food and Agriculture Organization of the United Nations. Rome, Italy. http://www.fao.org/3/I8168EN/i8168en.pdf.

FAO and PNUMA. 2022. *Evaluación mundial de la contaminación del suelo – Resumen para los formuladores de políticas*. Rome, FAO. https://doi.org/10.4060/cb4827es

Flores, José Salvador and Francisco Bautista. 2012. "El conocimiento de los mayas yucatecos en el manejo del bosque tropical estacional: las plantas forrajeras." *Revista Mexicana de Biodiversidad*, 83(2): 503–518.

Franco-Guerra, F.J.; M. Sánchez, J.C. Camacho, J.E. Hernández, O.A. Villareal, E.L. Rodríguez and O. Marcito. 2014. Consumo de especies arbóreas, arbustivas y sus frutos y herbáceas por cabras en pastoreo trashumante en la mixteca oaxaqueña, México. *Tropical and Subtropical Agroecosystems*. 17: 267–270.

Gerber, P.J., H. Steinfeld, B. Henderson, A. Mottet, C. Opio, J. Dijkman, A. Falcucci and Tempio, G. 2013. *Tackling climate change through livestock: a global assessment of emissions and mitigation opportunities*. Food and Agriculture Organization of the United Nations (FAO), Rome.

Giraldo-Pérez, Pualina, Raw, Victoria, Greven, Marc and M. R. Goddard. 2021. A small effect of conservation agriculture on soil biodiversity that differs between biological kingdoms and geographic locations.
iScience 24(4): 102280.

González-Pech, Pedro Geraldo, Juan Felipe de Jesús Torres-Acosta, Carlos Alfredo Sandoval-Castro and Juan
Tun-Garrido. 2015. "Feeding Behavior of Sheep and Goats in a Deciduous Tropical Forest during the
Dry Season: The Same Menu Consumed Differently." *Small Ruminant Research* 133 (December): 128–
34. https://doi.org/10.1016/j.smallrumres.2015.08.020.

Gordon, I.J.; Prins, H.H.T. 2019. *The Ecology of Browsing and Grazing II*. Cham, Switzerland: Springer Nature
Switzerland Ag.

Gupta, Ramesh C, Ajay Srivastava, and Rajiv Lall. 2019. *Nutraceuticals in Veterinary Medicine*. Cham,
Switzerland: Springer.

Hernández-Bolio, Gloria Ivonne, Karlina García-Sosa, Fabiola Escalante-Erosa, Gloria Sarahi Castañeda-
Ramírez, Enrique Sauri-Duch, Juan Felipe de Jesús Torres-Acosta and Luis Manuel Peña-Rodríguez.
2018. "Effects of Polyphenols Removal Methods on the in Vitro Exsheathment Inhibitory Activity of
Lysiloma latisiliquum Extracts against Haemonchus Contortus Larvae." *Natural Product Research*, 32
(5): 508–13.

Holechek, Jerry, Rex D Pieper and Carlton H Herbel. 2011. *Range Management: Principles and Practices*.
Boston: Prentice Hall.

Hoste, H., H. Leveque and Ph. Dorchies. 2001. "Comparison of nematode infections of the gastrointestinal tract
in Angora and dairy goats in a rangeland environment: relations with the feeding behevaiour". *Veterinary
Parasitology*, 101: 127–135.

Hoste, H., J.F. Torres-Acosta, M.A. Alonso-Díaz, S. Brunet, S., C. Sandoval-Castro and S.H. Adote. 2008.
"Identification and validation of bioactive plants for the control of gastrointestinal nematodes in small
ruminants." *Tropical Biomedicinie*, 25(1 Suppl): 56–72.

Hoste, H., C. Martinez-Ortiz-De-Montellano, F. Manolaraki, S. Brunet, N. Ojeda-Robertos, I. Fourquaux, J.F.J.
Torres-Acosta and C.A. Sandoval-Castro. 2012. "Direct and Indirect Effects of Bioactive Tannin-Rich
Tropical and Temperate Legumes against Nematode Infections." *Veterinary Parasitology* 186 (1-2): 18–
27. https://doi.org/10.1016/j.vetpar.2011.11.042.

Hoste, H., J.F.J. Torres-Acosta, C.A. Sandoval-Castro, I. Mueller-Harvey, S. Sotiraki, H. Louvandini, S.M.
Thamsborg and T.H. Terrill. 2015. "Tannin Containing Legumes as a Model for Nutraceuticals against
Digestive Parasites in Livestock." *Veterinary Parasitology* 212 (1-2): 5–17. https://doi.org/10.1016/j.vet
par.2015.06.026.

Hove, L., L.R. Ndlovu, S. Sibanda. 2003. "The effects of drying temperature on chemical composition and
nutritive value of some tropical fodder shrubs". *Agroferorestry Systems* 59 (3): 231–241.

Hutchings, Michael, R., Johanna, Judge, Iain, J. Gordon, Spiridoula, Athanasiadou and Ilias Kyriazakis. 2006. "Use
of trade-off theory to advance understanding of herbivore-parasite infections". *Mammal Review*, 36(1): 1–16.

IPPC Secretariat. 2022. *2021 IPPC Annual Report – Protecting the world's plant resources from pests*. FAO on
behalf of the Secretariat of the International Plant Protection Convention. Rome. https://doi.org/10.4060/
cb9334en

Jaimez-Rodríguez, P. R., P. G. González-Pech, J. Ventura-Cordero, D. R. B. Brito, L. M. Costa-Júnior, C.
A. Sandoval-Castro and J. F. J. Torres-Acosta. 2019. "The worm burden of tracer kids and lambs
browsing heterogeneous vegetation is influenced by strata harvested and not total dry matter intake or
plant life form". *Tropical Animal Health and Production*. 51(8): 2243–2251. https://doi.org/10.1016/
s0304-4017(01)00510-6.

Karthikeyan, R., Lawrence C. Davis, Larry E. Erickson, Kassim Al-Khatib, Peter A. Kulakow, Philip L.
Barnes, Stacy L. Hutchinson and Asil A. Nurzhanova. 2004. "Potential for Plant-Based Remediation
of Pesticide-Contaminated Soil and Water Using Nontarget Plants such as Trees, Shrubs, and Grasses."
*Critical Reviews in Plant Sciences* 23 (1): 91–101. https://doi.org/10.1080/07352680490273518.

Koch, Niharika, Nazim F. Islam, Songita Sonowal, Ram Prasad and Hemen Sarma. 2021. "Environmental
antibiotics and resistance genes as emerging contaminants: Methods of detection and bioremedi-
ation." *Current Research in Microbial Sciences* 2 (December): 100027. https://doi.org/10.1016/j.crm
icr.2021.100027.

Kumar, K., S. C. Gupta, S. K. Baidoo, Y. Chander and C. J. Rosen. 2005. "Antibiotic uptake by plants from soil
fertilized with animal manure." *Journal of Environment Quality* 34 (6): 2082. https://doi.org/10.2134/
jeq2005.0026.

Kumar, Shashi B., Shanvanth R. Arnipalli and Ouliana Ziouzenkova. 2020. "Antibiotics in Food Chain: The Consequences for Antibiotic Resistance." *Antibiotics* 9 (10): 688. https://doi.org/10.3390/antibiotics 9100688.

Ku-Vera, J. C., O. A. Castelán-Ortega, F. A. Galindo-Maldonado, J. Arango, N. Chirinda, R. Jiménez-Ocampo, S. S. Valencia-Salazar et al. 2020a. "Review: strategies for enteric methane mitigation in cattle fed tropical forages." *Animal* 14 (S3): s453–63. https://doi.org/10.1017/s1751731120001780.

Ku-Vera, Juan Carlos, Rafael Jiménez-Ocampo, Sara Stephanie Valencia-Salazar, María Denisse Montoya-Flores, Isabel Cristina Molina-Botero, Jacobo Arango, Carlos Alfredo Gómez-Bravo, Carlos Fernando Aguilar-Pérez and Francisco Javier Solorio-Sánchez. 2020b. "Role of Secondary Plant Metabolites on Enteric Methane Mitigation in Ruminants." *Frontiers in Veterinary Science* 7 (August). https://doi.org/10.3389/fvets.2020.00584.

Li, Yu-Guang, Greg Tanner and Phil Larkin. 1996. "The DMACA-HCl Protocol and the Threshold Proanthocyanidin Content for Bloat Safety in Forage Legumes." *Journal of the Science of Food and Agriculture* 70 (1): 89–101. https://doi.org/3.0.co;2-n">10.1002/(sici)1097-0010(199601)70:1<89::aid-jsfa470>3.0.co;2-

Lund, Arab and Muhammead, Ahmad. 2021. "Production potential, nutritive value and nutraceutical effects of goat milk." *Journal of Animal Health and Production.* 9:65–71.

Lüscher, A., I. Mueller-Harvey, J. F. Soussana, R. M. Rees and J. L. Peyraud. 2014. "Potential of legume-based grassland-livestock systems in Europe: A review." *Grass and Forage Science* 69 (2): 206–28. https://doi.org/10.1111/gfs.12124.

Makkar, H.P.S. 2003. "Effects and fate of tannins in ruminant animals, adaptation to tannins, and strategies to overcome detrimental effects of feeding tannin-rich feeds." *Small Ruminant Research* 49 (3): 241–56. https://doi.org/10.1016/s0921-4488 (03)00142-1.

Marie-Magdeleine, Carine, Steve Ceriac, Dingamgoto Jesse Barde, Nathalie Minatchy, Fred Periacarpin, Frederic Pommier, Brigitte Calif, Lucien Philibert, Jean-Christophe Bambou and Harry Archimède. 2020. "Evaluation of nutraceutical properties of *Leucaena leucocephala* leaf pellets fed to goat kids infected with *Haemonchus contortus*." *BMC Veterinary Research* 16 (1). https://doi.org/10.1186/s12 917-020-02471-8.

Martínez-Ortíz-de-Montellano, C., J.J. Vargas-Magaña, H.L. Canul-Ku, R. Miranda-Soberanis, C. Capetillo-Leal, C.A. Sandoval-Castro, H. Hoste and J.F.J. Torres-Acosta. 2010. "Effect of a tropical tannin-rich plant *Lysiloma latisiliquum* on adult populations of *Haemonchus contortus* in sheep." *Veterinary Parasitology* 172 (3-4): 283–90. https://doi.org/10.1016/j.vetpar.2010.04.040.

Mellor, D.J. and N.J. Beausoleil. 2015. "Extending the "five domains" model for animal welfare assessment to incorporate positive welfare states." *Animal Welfare*, 24 (3): 241–253.

Méndez-Ortiz, F. A., C. A. Sandoval-Castro, J. Ventura-Cordero, L. A. Sarmiento-Franco, R. H. Santos-Ricalde and J. F. J. Torres-Acosta. 2019. "*Gymnopodium floribundum* fodder as a model for the in vivo evaluation of nutraceutical value against *Haemonchus contortus*." *Tropical Animal Health and Production* 51 (6): 1591–99. https://doi.org/10.1007/s11250-019-01855-9.

Méndez-Ortiz, F. A., C. A. Sandoval-Castro, J. Ventura-Cordero, L. A. Sarmiento-Franco, and J.F.J. Torres-Acosta. 2018. "Condensed tannin intake and sheep performance: A meta-analysis on voluntary intake and live weight change." *Animal Feed Science and Technology* 245 (November): 67–76. https://doi.org/10.1016/j.anifeedsci.2018.09.001.

Min, B.R., T.N. Barry, G.T. Attwood, and W.C. McNabb. 2003. "The effect of condensed tannins on the nutrition and health of ruminants fed fresh temperate forages: A review." Animal Feed Science and Technology 106 (1-4): 3–19. https://doi.org/10.1016/s0377-8401(03)00041-5.

Murgueitio, Enrique, Zoraida Calle, Fernando Uribe, Alicia Calle and Baldomero Solorio. 2011. "Native trees and shrubs for the productive rehabilitation of tropical cattle ranching lands." *Forest Ecology and Management* 261 (10): 1654–63. https://doi.org/10.1016/j.foreco.2010.09.027.

Niezen, J. H., T. S. Waghorn, W. A. G. Charleston, and G. C. Waghorn. 1995. "Growth and Gastrointestinal Nematode Parasitism in Lambs Grazing Either Lucerne (Medicago Sativa) or Sulla (Hedysarum Coronarium) Which Contains Condensed Tannins." *The Journal of Agricultural Science* 125 (2): 281–89. https://doi.org/10.1017/s0021859600084422.

Novelo-Chi, L.K., P.G. González-Pech, J. Ventura-Cordero, J.F.J. Torres-Acosta, C.A. Sandoval-Castro and R. Cámara-Sarmiento. 2019. "Gastrointestinal nematode infection and feeding behaviour of goats in a heterogeneous vegetation: No evidence of therapeutic self-medication." *Behavioural Processes* 162 (May): 7–13. https://doi.org/10.1016/j.beproc.2019.01.006.

OECD. 2018. "Environment at a Glance – OECD Indicators – OECD." Www.oecd.org. Accessed May 24, 2022. https://www.oecd.org/environment/environment-at-a-glance/. https://doi.org/10.1787/ac4b8b89-en.

Ortiz-Domínguez, G., Ventura-Cordero, J., González-Pech, P., Torres-Acosta, J.F., Capetillo-Leal, C.M. and Sandoval-Castro, C.A. 2017. Nutritional value and in vitro digestibility of legume pods from seven trees species present in the tropical deciduous forest. *Tropical and Subtropical Agroecosystems*, 20(3): 505–510.

Pachas, Nahuel A., Alejandro Radrizzani, Enrique Murgueitio, Fernando Uribe, Álvaro Zapata Cadavid, Julián Chará, Tomás E. Ruiz, Eduardo Escalante, Rogerio M. Mauricio and Luis Ramírez-Avilés. 2019. "Establishment and management of Leucaena in Latin America." *Tropical Grasslands – Forrajes Tropicales* 7 (2): 127–32. https://doi.org/10.17138/tgft(7)127-132.

Paolini, V., F. De La Farge, F. Prevot, Ph. Dorchies and H. Hoste. 2005. "Effects of the repeated distribution of sainfoin hay on the resistance and the resilience of goats naturally infected with gastrointestinal nematodes." *Veterinary Parasitology* 127 (3-4): 277–83. https://doi.org/10.1016/j.vetpar.2004.10.015.

Pashaei, Reza, Zahedipour-Sheshglani, P., Dzingelevičienė, R., Sajjad Abbasi and Robert M. Rees. 2022. "Effects of pharmaceuticals on the nitrogen cycle in water and soil: a review." *Environmental Monitoring and Assessment* 194 (105). https://doi.org/10.1007/s10661-022-09754-7

Patel, K.S. and S. Katole. 2018. "Nutraceuticals and Ruminants Nutrition – A Review." *Livestock Research International.* 6(4): 76–85.

Peters, M., M. Herrero, M. Fisher, K.H. Erb, I.M. Rao, G.V. Subbarao, A. Castro, J. Arango, J. Chará, E. Murgueitio, R. van der Hoek, P. Läderach, G. Hyman, J. Tapasco, B. Strassburg, B.K. Paul, A. Rincon, R. Schultze-Kraft, S. Fonte and T. Searchinger. 2013. "Challenges and opportunities for improving eco-efficiency of tropical forage-based systems to mitigate greenhouse gas emissions." *Tropical Grasslands – Forrajes Tropicales* 1:137–148.

Pfister, James A. and John C. Malechek. 1986. "Dietary selection by goats and sheep in a deciduous woodland of Northeastern Brazil." *Journal of Range Management* 39 (1): 24. https://doi.org/10.2307/3899680.

Polianciuc, Svetlana Iuliana, Anca Elena Gurzău, Bela Kiss, Maria Georgia Ştefan and Felicia Loghin. 2020. "Antibiotics in the Environment: Causes and Consequences." *Medicine and Pharmacy Reports*, July. https://doi.org/10.15386/mpr-1742.

*Rangeland Ecology & Management* 72 (6): 946–53. https://doi.org/10.1016/j.rama.2019.08.002.

Rao, IM. 1998. "Root distribution and production in native and introduced pastures in the South American savannas." Edited by Box James E. *Root Demographics and Their Efficiencies in Sustainable Agriculture, Grasslands and Forest Ecosystems: Proceedings of the 5th Symposium of the International Society of Root Research,* Held 14-18 July 1996 at Madren Conference Center, Clemson University, Clemson, South Carolina, USA. Dordrecht: Springer Netherlands p. 19–42.

Rao, I., M. Peters, A. Castro, R. Schultze-Kraft, D. White, M. Fisher, J. Miles et al. 2015. "LivestockPlus – the Sustainable Intensification of Forage-Based Agricultural Systems to Improve Livelihoods and Ecosystem Services in the Tropics." *Tropical Grasslands – Forrajes Tropicales* 3 (2): 59. https://doi.org/10.17138/tgft(3)59-82.

Ríos, G. and J.a. Riley. 1985. "Preliminary studies on the utilization of the natural vegetation in the henequen zone of Yucatán for the production of goats I. Selection and nutritive value of native plants." *Tropical Animal Production* 10:1–10.

Rodríguez Eugenio, Natalie, M J Mclaughlin, Daniel John Pennock. 2018. *Soil Pollution: A Hidden Reality.* Rome: Food and Agriculture Organization of the United Nations and Global Soil Partnership. https://www.fao.org/3/I9183EN/i9183en.pdf. Accessed 24/05/2022.

Sandoval-Castro, C.A., J.F.J. Torres-Acosta, H. Hoste, A.Z.M. Salem and J.I. Chan-Pérez. 2012. "Using plant bioactive materials to control gastrointestinal tract helminths in livestock." *Animal Feed Science and Technology* 176 (1-4): 192–201. https://doi.org/10.1016/j.anifeedsci.2012.07.023.

Sarabia-Salgado, Lucero, Francisco Solorio-Sánchez, Luis Ramírez-Avilés, Bruno José Rodrigues Alves, Juan Ku-Vera, Carlos Aguilar-Pérez, Segundo Urquiaga, and Robert Michael Boddey. 2020. "Increase in Milk Yield from Cows through Improvement of Forage Production Using the N2-Fixing Legume Leucaena Leucocephala in a Silvopastoral System." *Animals: An Open Access Journal from MDPI* 10 (4): 734. https://doi.org/10.3390/ani10040734.

Schultze-Kraft, R., I.M. Rao  M. Peters, R.J. Clements, C. Bai, G. Liu. 2018. "Tropical forage legumes for environmental benefits: An overview." *Tropical Grasslands – Forrajes Tropicales*, 16(1): 1–14.

Sosa-Hernández, J. E., L. I. Rodas-Zuluaga, I. Y. López-Pacheco, E. M. Melchor-Martínez, Z. Aghalari, D. S. Limón, H. M. N. Iqbal and R. Parra-Saldívar, R. 2021. "Sources of antibiotics pollutants in the aquatic environment under SARS-CoV-2 pandemic situation." *Case Studies in Chemical and Environmental Engineering*, 4: 100–127.

SEMARNAT. 2020. *Contribución Determinada a Nivel Nacional – Actualización 2020*. Mexico City: Secretaría de Medio Ambiente y Recursos Naturales. https://www4.unfccc.int/sites/ndcstaging/PublishedDocume nts/Mexico%20First/NDC-Esp-30Dic.pdf.

Slik, J. W. Ferry, Víctor Arroyo-Rodríguez, Shin-Ichiro Aiba, Patricia Alvarez-Loayza, Luciana F. Alves, Peter Ashton, Patricia Balvanera, et al. 2015. "An estimate of the number of tropical tree species." *Proceedings of the National Academy of Sciences* 112 (24): 7472–77. https://doi.org/10.1073/pnas.1423147112.

Sun, S., V. Sidhu, Y. Rong, and Y. Zheng. 2018. "Pesticide pollution in agricultural soils and sustainable remediation methods: a review". *Current Pollution Reports*, 4(3): 240–250.

Thomas, R. J. 1995. "Role of legumes in providing N for sustainable tropical pasture systems." *Plant and Soil* 174 (1-2): 103–18. https://doi.org/10.1007/bf00032243.

Topps, J.H. 1992. "Potential, composition and use of legume shrubs and trees as fodders for livestock in the tropics." *Journal of Agricultural Science*. 118: 1–8.

Torres-Acosta, J.F.J., P.G. González-Pech, G.I. Ortiz-Ocampo, I. Rodríguez-Vivas, J. Tun-Garrido, J. Ventura-Cordero, G.S. Castañeda-Ramírez, G.I. Hernández-Bolio, C.A. Sandoval-Castro, J.I. Chan-Pérez and A. Ortega-Pacheco. 2016. "Revalorizando el uso de la selva baja caducifolia para la producción de rumiantes." *Tropical and Subtropical Agroecosystems*, 19: 73–80.

Torres-Fajardo, R. A., P. G. González-Pech, C. A. Sandoval-Castro, J. Ventura-Cordero and J. F. J. Torres-Acosta. 2019a. "Criollo goats limit their grass intake in the early morning suggesting a prophylactic self-medication behaviour in a heterogeneous vegetation." *Tropical Animal Health and Production* 51 (8): 2473–79. https://doi.org/10.1007/s11250-019-01966-3.

Torres-Fajardo, Rafael Arturo, Jorge Augusto Navarro-Alberto, Javier Ventura-Cordero, Pedro Geraldo González-Pech, Carlos Alfredo Sandoval-Castro, José Israel Chan-Pérez and Juan Felipe de Jesús Torres-Acosta. 2019b. "Intake and selection of goats grazing heterogeneous vegetation: Effect of gastro-intestinal nematodes and condensed tannins."

Torres-Fajardo, Rafael Arturo, Pedro Geraldo González-Pech, Carlos Alfredo Sandoval-Castro and Juan Felipe de Jesús Torres-Acosta. 2020. "Small ruminant production based on rangelands to optimize animal nutrition and health: Building an interdisciplinary approach to evaluate nutraceutical plants." *Animals* 10 (10): 1799. https://doi.org/10.3390/ani10101799.

Trapp, Stefan and Charlotte N. Legind. 2010. "Uptake of organic contaminants from soil into vegetables and fruits." *Dealing with Contaminated Sites*. Edited by Frank A. Swartjes. Springer Dordrecht. 369–408. https://doi.org/10.1007/978-90-481-9757-6_9.

Van Zanten, H.H.E., M.K van Ittersum and I.J.M de Boer. 2019. "The role of farm animals in a circular food system." *Global Food Security* 21 (June): 18–22. https://doi.org/10.1016/j.gfs.2019.06.003.

Vargas-Magaña, J.J., J.F.J. Torres-Acosta, A.J. Aguilar-Caballero, C.A. Sandoval-Castro, H. Hoste and J.I. Chan-Pérez. 2014. "Anthelmintic activity of acetone–water extracts against *Haemonchus contortus* eggs: Interactions between tannins and other plant secondary compounds." *Veterinary Parasitology* 206 (3-4): 322–27. https://doi.org/10.1016/j.vetpar.2014.10.008.

Ventura-Cordero, J., P. G. González-Pech, J. F. J. Torres-Acosta, C. A. Sandoval-Castro and J. Tun-Garrido. 2019. "Sheep and goat browsing a tropical deciduous forest during the rainy season: Why does similar plant species consumption result in different nutrient intake?" *Animal Production Science* 59 (1): 66. https://doi.org/10.1071/an16512.

Ventura-Cordero, J., P.G. González-Pech, P.R. Jaimez-Rodríguez, G.I. Ortíz-Ocampo, C.A. Sandoval-Castro and J.F.J. Torres-Acosta. 2018. "Feed resource selection of Criollo goats artificially infected with *Haemonchus contortus*: nutritional wisdom and prophylactic self-medication." *Animal*. 12(6): 1269–1276.

Ventura-Cordero, J., P.G. González-Pech, P.R. Jaimez-Rodríguez, G.I. Ortiz-Ocampo, C.A. Sandoval-Castro and J.F.J. Torres-Acosta. 2017. "Gastrointestinal nematode infection does not affect selection of tropical foliage by goats in a cafeteria trial." *Tropical Animal Health and Production* 49: 97–104.

Villalba, J.J. and F.D. Provenza. 2010. "Challenges in extrapolating in vitro findings to in vivo evaluation of plant resources. In: *In vitro screening of plant resources for extra nutritional attributes in ruminants: Nuclear and related methodologies*. Edited by Vercoe, P.E., H.P.S. Makkar and A.C. Schlink. FAO/IAEA Springer Edition, 2010. pp 233–242

Vlassoff, A. 1982. "Biology and population dynamics of the free-living stages of gastrointestinal nematodes in sheep." In *Control of Internal Parasites of Sheep*. Lincoln College, Lincoln, UK. pp 11–20.

Walelign, B. and E. Mekuriaw. 2016. "Major toxic plants and their effect on livestock: a review." *Advances in Life Science and Technology*, 45: 1–12.

Wang, Y., G. C. Waghorn, W. C. McNabb, T. N. Barry, M. J. Hedley and I. D. Shelton. 1996. "Effect of condensed tannins in *Lotus corniculatus* upon the digestion of methionine and cysteine in the small intestine of sheep." *The Journal of Agricultural Science* 127 (3): 413–21. https://doi.org/10.1017/s00218 59600078576.

Wang, F. and J.M. Tiedje. 2020. "Antibiotic Resistance in Soil." Edited by Manaia, Célia M, Erica Donner, Ivone Vaz-Moreira and Peiying Hong. *Antibiotic Resistance in the Environment: A Worldwide Overview*. Cham: Springer International Publishing.

Zapata-Buenfil, G., F. Bautista-Zúñiga, and M. Astier-Calderón. 2009. "Caracterización forrajera de un sistema silvopastoril de vegetación secundaria con base en la aptitud de suelo." *Técnica Pecuaria en México* 47(3): 257–270.

# Section III

CASE STUDIES and RESEARCH
METHODS

# 9 Conservation of oligotrophic grassland of high nature value (HNV) through sustainable use of *Arnica montana* in the Apuseni Mountains, Romania

*Florin Păcurar, Albert Reif and Evelyn Ruşdea*

## 1. INTRODUCTION

During the last decades, the globalisation of markets has strongly influenced the prices of agricultural products and their position in the market, leading to changes in land use practices (Foley et al. 2005) and the transformation of landscapes (Fuchs et al. 2014, Fuchs et al. 2013). Fertile, high-yielding land underlies more and more intensification and specialisation for production, performed by agro-technical businesses instead of family farmsteads. Less productive or remote sites underlie abandonment and subsequent natural succession, or active reforestation. This has led to serious regional inequalities in living conditions, associated with an exodus from remote or mountainous rural regions towards urban areas (Figueiredo and Pereira 2011, Foggin 2008, McKinney 2002). In Europe, at least 30% of grasslands had been left abandoned, especially in the mountainous and Mediterranean areas (Peyraud and Peeters 2016). A side effect is that valuable habitats of the traditional cultural landscape become increasingly endangered (Emanuelsson 2009). A significant decline of semi-natural grassland of high nature value (HNV) has been observed in many places in Europe (Vaida et al. 2021, Tokarczyk 2018).

In Eastern European countries in transition, large agricultural and grassland areas were abandoned after 1989 (Peyraud and Peeters 2016). This affected oligotrophic hay meadows and extensively grazed pastures, which belong to the endangered key habitats in Europe. The agro-environment measures supported by the Common Agricultural Policy (CAP) of the European Union are not sufficient to maintain these habitats with their characteristic biodiversity (McGurn et al. 2017). The 2014 CAP reform strengthens low-input grasslands, but it is unlikely that the level of support is sufficient to reverse their overall decline (Luick and Roeder 2016).

There are strategies to counteract this, to maintain traditional landscape structures and to provide adequate livelihood for the people; for example, by designation of protected areas such as biosphere reserves, and establishing related payment opportunities for specific land uses. In some regions of the Carpathians, for example, the CAP measures have improved the management of semi-natural grasslands and their biodiversity (Rotar et al. 2020, Halada et al. 2017).

Another option is to promote further grassland ecosystem services besides the natural production for animal livestock (e.g., landscape beauty for tourism), nectar provision for pollinating insects, or providing medicinal plants. This requires innovative approaches built on ecology, economic potential and interactive social learning processes. All these are important elements in the effort

DOI: 10.1201/9781003146902-12

of maintaining semi-natural mountainous grasslands in a sustainable way in the context of climate change, rural demographic changes and farm abandonment (Darnhofer et al. 2017).

In South-Eastern Europe, especially in Romania, traditional farming systems still exist, which include arable farming, animal husbandry and grassland management. Since 2007, when Romania became a member of the European Union, major structural changes to the economic and societal system were initiated. At the end of 2017, Romania had a population of 19.6 million people, of which 5.5 million were active employees, 9.1 million were pupils, students, pensioners or unemployed and approximately 5 million of the population had left the country to work abroad (Otovescu and Otovescu 2019). Particularly in rural regions receiving less allocation of direct payments for rural development, the exodus has been immense. This has led to a considerable socio-economic marginalisation of landscapes, in correlation to the agrarian production (Galluzzo 2018, Galluzzo 2017).

In Romania, there are still important large areas of oligotrophic grasslands of HNV, which are subject to traditional extensive management (Păcurar et al. 2020, Vîntu et al. 2011), totalling about 2 million ha (out of a total of 4.8 million ha grassland) (MADR 2022, pp 9–10). Our study aims to present the potentials – and difficulties – of further conservation of these habitats through sustainable management, and focuses on the perspective of using medicinal plants to enhance income and profit contribution.

## 2. *ARNICA MONTANA* – A FLAGSHIP SPECIES

*Arnica montana* is a perennial herbaceous plant of the Asteraceae family, which has been used as a herbal medicinal plant for centuries. The underground parts develop rootstocks (rhizomes); the hairy stem reaches 20–60 cm; the leaves are elliptical and lanceolate, opposite distributed and crowded on the base forming a rosette. The yellow-orange florets, with a delicate characteristic fragrance, result from 1–3 up to seven flower heads per stem with a diameter of 5–7 cm. The flowering period in Central Europe is between May and August, depending on geographical distribution and altitude. The morphology differs with respect to the habitat; it grows as a hemicryptophyte in grasslands and pastures, and becomes a tall forb on fallow ground (Titze et al. 2020, Radušiene and Labokas 2007, Schwabe 1990).

### 2.1. *Arnica montana* and its medicinal use

*Arnica montana* has been used for hundreds of years as a medicinal plant against rheumatic pains, skin inflammations, bruises and other complaints. Today it is one of the most commonly used phytotherapeutic (allopathic) and homeopathic medicines. The plant extracts have anti-bacterial, anti-tumour, anti-oxidant, anti-inflammatory, anti-fungal and immunomodulatory effects. In the different parts of the plant, a wide range of chemical compounds (150 therapeutically active substances) can be found, including sesquiterpene lactones and their short-chain carbonic acid esters, flavonoids, carotenoids, essential oils, diterpenes, arnidiol, pyrrolizidine alkaloids, coumarins, phenolic acids, lignans and oligosaccharides (Kriplani et al. 2017, Kos et al. 2005). The flower heads have greater medicinal value and are used as anti-phlogistic, inotropic, anti-biotic, anti-inflammatory, immunomodulatory, anti-platelet, uterotonic, anti-rheumatic and analgesic in febrile conditions (Oberbaum et al. 2005). *Arnica montana* can be used to obtain tinctures, creams, ointments, oils or gels or in the form of wet poultices consisting of a solution that has been valued for curing osteo-arthritis, alopecia and chronic venous insufficiency (Clair 2010).

It has also been reported that decoction, infusion or macerated extracts of *Arnica montana* flowerheads, leaves or aboveground parts of the plant can be used for the treatment of numerous ailments such as bowel ache, cough, contusion, cuts, haematomas, headaches and rheumatism. It also has soothing and healing properties for the hair or skin (Kriplani et al. 2017). In addition, it has been shown that the sesquiterpene lactones from flowers and roots of *Arnica montana* heal inflammations of the human body (Meyer and Straub 2011). In particular, the two components, helenalin and 11α, 13-dihydrohelenalin, are responsible for an increased potential medicinal effect (Lyss et al. 1997, Schröder et al. 1990). They inhibit the NF-κB transcription factor, which is necessary for the transcription of some immune-specific genes. Among other things, these genes are responsible for the synthesis of inflammatory cytokines, which trigger the inflammatory reaction with typical signs such as pain, heat, redness, swelling and loss of function.

## 2.2. *Arnica montana* and its habitat

*Arnica montana* is a temperate European plant species growing in grass- and heathlands, from lowlands near sea level up to alpine regions (Duwe et al. 2017). The species is mainly distributed in Central Europe (Hultén and Fries 1986), reaching to Scandinavia in the north, to Poland, Lithuania and southern Russia in the east, Belgium in the west and the Pyrenees, Portugal and Spain in the south (Meusel and Jäger 1992, Dapper 1987, Hegi 1987). Important arnica populations exist in south-eastern Europe, in Romania, Bulgaria, Serbia and Ukraine (Fig. 1). *Arnica montana* occurs in mountain grasslands on moderately acidic, nutrient-poor soils (Maurice et al. 2012, Michler 2005, Kahmen and Poschlod 2000, Luijten et al. 1996, Sugier et al. 2019), managed in a traditional way (Păcurar et al. 2009, Reif et al. 2008, Reif et al. 2005, Michler et al. 2005). These oligotrophic grasslands are rich in plant diversity and listed in Annex I of the Natura 2000-Program of the European Union as "*Nardus stricta* grasslands" (code R 6230) and "mountain hay meadows" (code R 6520). Their conservation is therefore of high public interest.

*Arnica montana* populations at lower altitudes are more threatened than those at higher altitudes (Duwe et al. 2017, Titze et al. 2020). The main reasons for their decline are: (1) the abandonment of mowing and grazing on low-productivity grasslands, followed by succession towards forest, and (2) agricultural intensification leading to eutrophication with the consequence of significant changes in species composition (Michler et al. 2005, Korneck et al. 1998, Korneck et al. 1996, Fukarek et al. 1978). The flowers are widely collected for domestic medicinal purpose and for commercial trade,

**FIGURE 1** Oligotrophic grassland rich in *Arnica montana* – Apuseni Mountains/Romania – Photo © Arnica System.

which has been recognised as another threat factor (Korneck et al. 1998, Korneck et al. 1996), even though the harvesting impact on *Arnica montana* populations is not clear (Schippmann et al. 2002).

Although the arnica populations have declined significantly on a local level, they are still large on a global level. The decrease in population density has not yet led to a reduction in distribution areas. However, *Arnica montana* is a protected species in Europe, being listed in various categories of Red Lists depending on the country (e.g., as "endangered" in Romania, or "critically endangered" in the Netherlands). The species is listed in Annex V of the Nature 2000-Program – FFH directive (Council Directive 92/43/EEC). In the IUCN Red List, it is assessed as "of least concern" (Bilz et al. 2011).

## 3. *ARNICA MONTANA* – HABITAT PROTECTION THROUGH USE

In the Romanian Carpathians, the characteristic landscape is formed by forest areas and open lands dominated by grasslands. Grasslands are hardly subject to intensification, but many of them, especially in remote areas, are now threatened by abandonment or have already been abandoned (Maruşca 2016). Their continued existence depends on the management applied.

In the Apuseni Mountains, in the upper Arieş Valley, the local people (called 'moţi') have traditionally lived from subsistence production, including grassland management, livestock farming, forest exploitation and craft work as their livelihoods (Păcurar et al. 2014, Auch 2006). Their traditional land use system has created a landscape characterised by its peculiarity and typical biodiversity, which makes it a unique cultural landscape in Romania and Europe (Ruşdea et al. 2005). The open land is dominated by grassland, which is managed as meadow, pasture or in a mixed system. Since about 1995, serious changes in land use have taken place and are still ongoing. Farmers living in mountainous areas with a high proportion of HNV grasslands are most at risk in terms of economic vulnerability (Jitea and Arion 2015). Many people have left the Apuseni Mountains, and marginal-yield grassland sites in particular face abandonment and subsequent succession towards forests, despite the subsidies provided by the agro-environmental schemes of the CAP (through the Agency for Payments and Intervention in Agriculture – APIA). Therefore, other sources of enhancing income are needed to ensure the conservation of traditional landscapes and ecosystems.

A successful example of integrating the agricultural use of oligotrophic mountainous grassland with the harvesting of medicinal plants, in particular *Arnica montana*, with resource management and value-adding can be found in the Apuseni Mountains, Romania (Vaida et al. 2016).

### 3.1 ARNICA SYSTEM: FROM CONCEPT TO PRACTICE

*Arnica montana* is an important plant species from a nature conservation and an economic point of view. In the Apuseni Mountains, it has been harvested from the wild for more than 50 years (Pop and Florescu 2008). Additional income for the local people can be generated by harvesting and processing medicinal plants, and specifically *Arnica montana*. This helps to counteract the emigration mainly of young people towards urban areas and foreign countries and increases the local people´s interest in maintaining their environment, which is traditionally managed oligotrophic grassland. These efforts are supported in the Apuseni Mountains by the company, *Arnica System*.

*Arnica System* is organised in the form of a company and goes back to the German-Romanian interdisciplinary research *Proiect Apuseni*, funded by the German Federal Ministry for Education and Research (BMBF) and located in the village of Ghețari, which belongs to the community of Gârda de Sus, Alba County. Between 2000 and 2004, the rural landscape, the land uses and perspectives for regional development were investigated. These studies provided the base for several trans-disciplinary implementation projects, including ecotourism, rural architecture, improved farming practices and sustainable use of medicinal plants for the benefit of biodiversity conservation and welfare of the people (Ruşdea et al. 2005). From 2004 up to 2007, the collection of medicinal plants was professionalised and developed in the follow-up *Arnica Project*, funded by WWF-UK and the Darwin Foundation. It developed a management plan for the sustainable use and conservation of *Arnica montana* and initiated the co-operation of the Swiss-German company, Weleda (Michler et al. 2006). Until then, *Arnica montana* was collected in an unsystematical and unsustainable way, low quality was produced and a low purchase price was obtained for the harvesters (Kathe 2006, Michler 2005, Michler et al. 2004).

In 2007, a small local company named *Ecoherba* was founded in Gârda. From 2010 onwards, the activity of *Ecoherba* was extended also to the neighbouring communities from the northern and central parts of the Apuseni Mountains. In order to face these new challenges, a new company called *Bioflora Apuseni* was founded in 2010, which expanded the collection of *Arnica montana* to the northern Apuseni Mountains. These two companies have merged and are now working together as *Arnica System*, which implements the developed ideas and grown experiences. The vision of *Arnica System* is to preserve oligotrophic grasslands, often hay meadows, for the cultural landscape, for continued habitat tradition and biodiversity by providing additional income to the local people through the sustainable use and trade of medicinal plants. *Arnica System* has a close co-operation with the main beneficiary of the arnica plant material, the Weleda Company in Schwäbisch Gmünd, Germany – as well as with other beneficiaries; with universities (the University of Agricultural Sciences and Veterinary Medicine Cluj-Napoca, Romania and the Albert-Ludwigs University of Freiburg, Germany); with institutions for the preservation of natural resources (Apuseni Natural Park, Institute for Biological Research Cluj, Romania; National Agency for Environmental Protection, Romania; Commission for Natural Monuments, Romania); and with local administrations (13 communities from the Apuseni Mountains).

The study and activity area has increased since 2000, starting from a small area of 287 ha of grasslands with arnica in the transect area from Gârda de Sus up to the Plateau Gheţari-Poiana Călineasa (Michler 2005); it was then extended to the whole area of the Gârda de Sus community (87 km$^2$), where an area of 550 ha of oligotrophic grasslands with *Arnica montana* was identified (Michler 2007). And finally, the *Arnica System* extends the activity to an area of 13 communities in the central and northern part of the Apuseni Mountains (Gârda de Sus, Arieşeni, Scărişoara, Albac, Horea, Beliş, Călăţele, Mărgău, Săcuieu, Mărişel, Măguri-Răcătău, Băişoara, Râşca), covering 1,470 km$^2$, where the region of oligotrophic grasslands with *Arnica montana* ranges between 4,000 and 5,000 ha. Much of this area is located in the protected area of the Apuseni Natural Park.

From the socio-economic point of view, *Arnica System* today employs four people in permanent positions and 15–20 local people seasonally – for between one and two months during the arnica season – ensuring the stages of processing (checking, sorting, drying and packaging). In addition, the harvesters and collectors are paid according to the amount of arnica flowers collected. In 2015 and 2016, for example, about 550 people were involved in harvesting, most of them women, but recently the number of men has increased, mainly because the timber resources in the region have been plundered.

Another major direction in which *Arnica System* is constantly involved is the management of grassland spaces on its own property or on the property of others (locals, other companies), who want to implement a management system. Through these activities, *Arnica System* contributes to the restoration and reuse of abandoned meadows and provides the resources for the cultivation of the arnica species and at the same time, its maintenance in oligotrophic natural meadows.

The cultivation of *Arnica montana* species can – on the one hand – contribute to reducing the pressure on populations from spontaneous flora and – on the other hand – can be an important means of supplementing the income of the locals (Melero et al. 2012). It would reduce the harvesting pressure of the spontaneous flora because it would stabilise the purchase price and will encourage the locals to harvest from the culture rather than collecting from spontaneous flora.

## 3.2 Sustainable use of *Arnica montana*: Development and implementation

*Arnica System* has developed a model for the conservation of oligotrophic grasslands through the sustainable use of *Arnica montana*. The basic idea of this model is "protection through use" (Neitzke 2015). Given the current general conditions in the Apuseni Mountains, adding economic value to oligotrophic grasslands is essential for their conservation. This is all the more important considering that the use of wood, the main monetary resource of the area, has decreased significantly. The principle of *Arnica System* consists of a series of subsequent activities, including habitat character-isation and mapping, the collecting of arnica and other medicinal plants, as well as processing and selling without intermediaries (Fig. 2). The individual activities of this model are presented in more detail below.

### 3.2.1.  Oligotrophic grasslands with *Arnica montana* and their management

In the Apuseni Mountains, the oligotrophic grasslands below the climatic treeline are the result of historical deforestation (Sângeorzan et al. 2018, Goia 2005, Goia and Borlan 2005, Reif et al. 2005). *Arnica montana* occurs in grassland types within a floristic and edaphic gradient from the nutrient-poor, oligotrophic Nardo-Callunetea (Violo declinatae-Nardetum) to the oligo-mesotrophic Molinio-Arrhenatheretea (*Festuca rubra-Agrostis capillaris*-community) (Gârda 2010, Brinkmann et al. 2009, Michler et al. 2005). *Arnica montana* can be considered a "flagship" species, which is associated with other species like *Polygala vulgaris, Gentianella lutescens, Scorzonera rosea, Hieracium aurantiacum, Viola declinata, Crocus heuffelianus, Gymnadenia conopsea* and *Traunsteinera globosa* (Stoie 2011).

The long-term traditional management of grasslands has created a great diversity of species and habitats (Brinkmann et al. 2009). The small-scale farmers live under difficult working conditions with a high proportion of subsistence production. They traditionally use hay meadows with autumnal after-grazing and permanent pasture near the villages. This system is linked to the grazing on the communal high mountain pasture during summer. Some of the hay meadows are fertilised with manure to increase production. In addition, the *Arnica montana* oligotrophic grasslands are used in this way through a combination of mowing and grazing (hay meadows), or only through grazing (permanent pastures).

Particularly rich in *Arnica montana* and associated species are the oligotrophic and extensively used hay meadows in the neighbourhood of farmsteads. On these grasslands the following main-tenance measures are traditionally carried out: the removal of stones, the levelling of anthills, the removal of woody vegetation, weed control and fertilisation. All maintenance works are done manu-ally, with different tools, and for the application of fertilisers, horse carts are used.

Mowing was traditionally carried out with a scythe, nowadays it is conducted with mowing machines; the start of mowing depends on the weather, but generally takes place between the end of June and the beginning of August (i.e., after the seed formation of *Arnica montana* and many other grassland species is completed). This traditional management requires the cooperation of many people and their horses. Afterwards, in late autumn, grazing with cattle and horses takes place.

Maintaining traditional management is crucial for the existence and conservation of oligotrophic grasslands, since already minor changes can shift the species composition. Only manure from cows and horses with a six-months deposition is used for fertilisation in small quantities. Fertilisation with an amount of 10 t/ha manure per year causes *Arnica montana* and other oligotrophic species to disappear (Bogdan 2012). Heavy permanent overgrazing also leads to species depletion (Reif et al. 2005).

### 3.2.2.  Approvals for harvesting

*Arnica montana* is a European threatened species (Coldea et al. 2003), listed in Annex 5 of the EU-FFH-directive (Council directive 92/43/EEC), and is additionally protected by national laws (Sugier

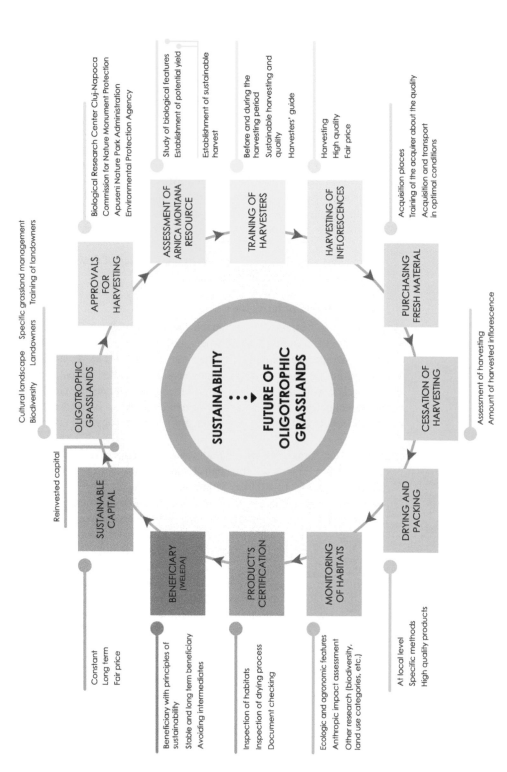

**FIGURE 2** Model for sustainable use of *Arnica montana* in the Apuseni Mountains.

et al. 2018). Its harvesting is prohibited except in certain regions in Spain (Obon et al. 2012) and Romania, where arnica was assessed as vulnerable (Oltean et al. 1994). In Romania, the harvesting of arnica flowers is legal but requires an official authorisation from the state institutions according to the Governmental Decree no. 647 from 26.07.2001. The harvestable quantity of arnica flower heads is assessed by the Institute of Biological Research in Cluj-Napoca, and the evaluating study must be approved by the Commission for Natural Monuments of the Romanian Academy, the Agency for Environmental Protection and the Administration of protected areas, in this case the Apuseni Natural Park.

### 3.2.3. Assessment of the resource

In order to establish the potential and sustainable yield, various biological features and parameters of *Arnica montana* are recorded (see also section 3.2.7. Monitoring and research). This is also important for the development of models for forecasting the harvest quantity. Based on indicators recorded in early spring (April to May), the possible arnica quantity of the coming season can be estimated, which represents an essential basis for the negotiations with customers on the potential quantity of arnica to be supplied.

The monitoring carried out annually reveals information about the resource of flower heads and harvesting methods and management can therefore be optimised.

### 3.2.4. Participatory approach: training of harvesters

Before harvesting, the harvesters have to be trained in sustainable harvesting methods (Williams and Kepe 2008). *Arnica System* works with harvesters, landowners, collectors (purchasers), seasonal workers and permanent employees. Specific training is carried out for each group depending on the job activity required. Training for the harvesters and landowners is conducted annually before the start of the harvesting season. The materials used in the training are the "Harvesting Guidelines Manual" (Michler 2007) and the Guidelines for management of *Arnica montana* habitats (Michler 2007), as well as posters, flyers, leaflets and presentations concerning the biology and ecology of arnica.

Landowners are informed of and encouraged to apply a traditional management regime on arnica grasslands (Păcurar et al. 2008). Of importance are the control and removal of woody vegetation and fertilisation with small manure quantities (6–10 t/ha), which should be applied every 2–3 years. The grasslands must be mown, beginning at the end of June when arnica has mature seeds. The grass should be spread on the ground for drying, which promotes the dispersal of seeds from the herbaceous plants. In autumn, after-grazing is recommended, beginning when the sward is 8–10 cm high, and ending three weeks before the first frost is expected (Păcurar et al. 2009).

The manuals and posters used for training harvesters and collectors (Fig. 3) demand the following measures: "*Distinction between flowers of arnica and similar species, harvesting flower heads only under dry weather conditions, using the textile bags they get from the sourcing team, picking only the full blooming flower heads, picking flower heads without stem, leaving buds, leaving flower heads for seed production, delivering the flower heads in textile bags immediately after picking to the collection points where quality is checked*" (Michler 2007).

### 3.2.5. Harvesting and purchasing of arnica

Harvesting and collecting *Arnica montana* flower heads involves several stages (Fig. 4), playing a decisive role in obtaining a good-quality final product. The locals collect arnica partly from their own land, partly from other landowners' grasslands. To obtain raw material of good quality, the harvesting is done in sunny and dry weather conditions, in the morning after the dew has evaporated. Only fully developed flowers without peduncles are collected and visually checked for quality (e.g., bloomed flowers or buds are not accepted). It is recommended to leave at least one flower head per plant to ensure seed production and further propagation.

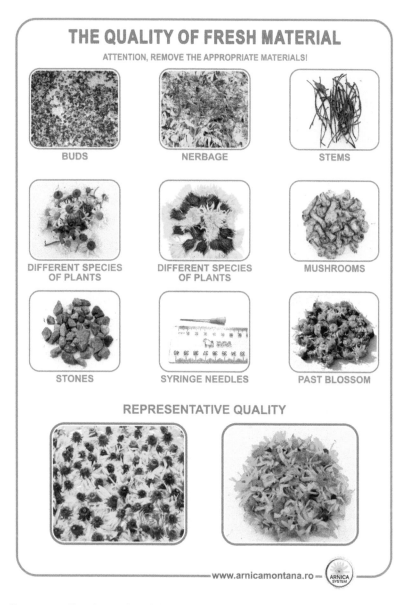

**FIGURE 3**    Poster regarding the quality of *Arnica montana* fresh material used in the trainings.

The fresh material is stored in textile bags that allow air ventilation and keep the bulk of the flower heads cool. The full bags are delivered to nearby collecting points, where the flower heads are once more visually inspected for quality, sorted and weighed. *Arnica System* has 17 acquisition points in the 13 communities mentioned. The fresh material is transported on the same day to the drying facility located in the village of Ghețari. From remote regions, the transport is done by cooling cars, where a constant temperature of 4°C is maintained. Harvesting and collecting under hygienic conditions is the premise for obtaining a dried product of highest quality (Dugalić 2003).

The entire harvesting process lasts between 5 and 18 days (on average, 10.9 days) and starts regionally at different times, depending on altitude and weather conditions. The harvesting activity takes between 4 and 7 hours per day (on average, 5.7 hours/day). In an average family, about two people are engaged in collecting arnica flower heads. From interviews, we know that a harvester

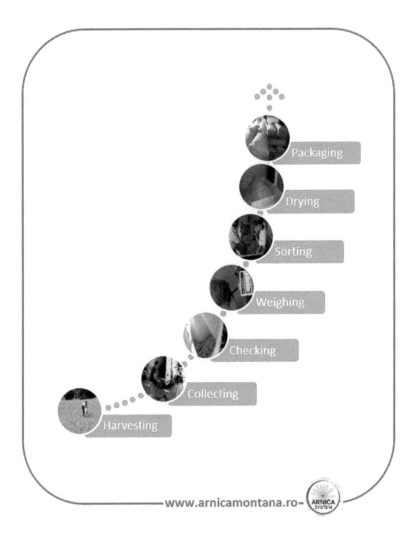

**FIGURE 4**   *Arnica montana* processing activities.

can collect on average 1.1 kg of fresh flower heads per hour, and between 40 and 200 kg of fresh material in a whole season (Fig. 5). The income generated by harvesting lies between 5 and 10% (7% average) of the total annual family income. The income contribution varies from year to year depending on the arnica purchase price (see also chapter 3.2.8. Marketing).

Harvesting also generates additional income for other people besides the harvesters (e.g., collectors, who ensure the acquisition of flower heads at the collecting points). The processing of fresh material in the drying facility also generates an income for the seasonal workers (15–20 people hired for 1–2 months per year), accounting for between 15–20% of the total family annual income. The four permanent employees of the company earn between 50% and 100% of their total family income per year.

Further observations will reveal more information regarding the impact of harvesting and collecting of arnica in the socio-economic context.

### 3.2.6.  Development of processing and drying
The processing of the fresh arnica material guarantees the preservation of volatile oils and lactones without affecting the specific smell of the flower. A specifically designed drying and storage facility

**FIGURE 5**   Harvesting of *Arnica montana* flower heads at Hănăşeşti in the Apuseni Mountains – Photo © Arnica System.

planned by Architecture for Humanity, UK was constructed in 2006 during the *Arnica Project* (Michler 2007). This wooden building was relatively small and designed with the aim of providing simple and practical solutions for a local catchment area, which suffers from poor access to paved roads and a water supply. It included a built-in dryer with shelves for the drying process and a heating system based on firewood. The development of this on-site drying facility with local value-adding was a requirement for establishing business relationships.

With the improvement of transport conditions, the harvesting area and the quantity of harvested flower heads increased and, consequently, a higher drying capacity was needed. To face the new challenges, the existing drying facility was modified in 2011, the building was enlarged and the capacity improved considerably (Table 1). The quality control was enhanced on different levels and the quality of the product has increased significantly (Fig. 6). Under the framework of *Arnica System*, a detailed documentation of the drying process parameters was implemented.

### 3.2.7.  Monitoring and research

Monitoring of the arnica population and its habitats is essential for assessing the long-term effects of harvesting on the arnica flower heads as a resource. The presence of *Arnica montana* and the resource of flower heads on a specific site is the result of a dynamic process of: (1) growing conditions (e.g., soil, micro-climate, water availability); (2) management activities (including absence of management and abandonment); and (3) harvesting activities.

Our goal is to study certain biological features of *Arnica montana* species and to assess the flower heads resource in order to establish the potential and sustainable yield and, respectively, to optimise the harvesting method.

The monitoring focuses on documenting the status of the arnica populations, and of the habitat (species composition and grassland management) and must be repeated annually.

**FIGURE 6** Quality control of fresh arnica flowers heads in the drying facility at Gheţari in the Apuseni Mountains – Photo © Arnica System.

## TABLE 1
### Development of the drying system in Gheţari – Apuseni Mountains.

| Drying process *Arnica project* (2006) | Drying process *Arnica System* (from 2011) |
|---|---|
| Construction of the drying facility in 2006, consisting of three rooms: reception, drying room and storage; Total usable area of 156 m² | Enlargement of the existing drying facility in 2011, consisting now of five rooms: reception, drying, sorting-packing, storage and heating room; Total usable area of 254 m² |
| Reception room of 18 m² for receiving the fresh plant material and weighing. | Enlarged reception room to 35 m², additionally space for sorting and careful quality control on special tables. |
| Drying room with a single drying tunnel (10 m long, 6 m wide, 3 m high) provided with fixed shelves, where the drying racks with frames were placed, the distance between frames being 12–15 cm; a corridor of 2 m width in the middle was left for handling the drying frames; the heating generators were inside the drying room; Total drying surface on the frames is 240 m². | Drying room has been divided into three drying tunnels (each 10 m long, 2 m wide, 2 m high); every tunnel contains nine mobile trolleys with 27 drying frames each, the distance between frames being 3 cm; one-way workflow: the trolleys enter the drying room from the reception side and leave on the opposite side towards the sorting and packing room; the heating generators are in a separate room; Total drying surface on the frames is 1,458 m². |
| During the drying process the workers are in contact too frequently with arnica flower heads (shaking material on the frames and manual packaging). | The arnica material is placed on the mobile trolleys and no further manual handling is required and therefore proper hygiene is guaranteed. |

**TABLE 1 (Continued)**
**Development of the drying system in Gheţari – Apuseni Mountains.**

| Drying process *Arnica project* (2006) | Drying process *Arnica System* (from 2011) |
|---|---|
| Weak ventilation inside the drying room: the four ventilators have a performance of up to 256 m³/hour each, resulting in a drying capacity of 1,024 m³/hour related to 180 m³ dryer volum (but in connection to a small drying surface of 240 m² on the frames). | Ventilation has been increased using ventilators with high performance of 8,000 m³/hour per tunnel, resulting in a drying capacity of 24,000 m³/hour related to 120 m³ dryer volum (but in connection to a bigger drying surface of 1,458 m² on the frames). |
| Lack of moisture release due to the poor performance of the dryer. | Humidity is better removed with additional 9 fans of 2,500 m³/hour each. |
| Absence of a quality control point after drying. | With the new quality check after drying (in the sorting-packing room), the product's quality has increased significantly. |
| Absence of a dry material sorting-packaging room. | Packing directly in large paper bags in the sorting-packing room by using a large funnel. |
| The dried and packed material is manually transported through the reception room (on a narrow staircase) to the storage room upstairs. | The dried and packed material is transported by means of an elevator directly from the sorting-packing room to the storage room upstairs. |
| Heating system with two small heat generators (with a heating power of 16 kWh each) could provide a temperature of 20–35 °C. | Two new more powerful heat generators were introduced (with a heating power of 93 kWh each) enable the necessary temperature of 40–44 °C and 45% air humidity. |
| High consumption of firewood for heating; e.g. in 2008 12 m³ of beech wood were used for drying 3,100 kg fresh arnica flower heads, meaning 4 m³ wood needed for 1,000 kg of dried arnica. | Reduced firewood consumption; e.g. in 2017 30 m³ of beech wood were consumed for drying 33,600 kg of fresh material, meaning 0,89 m³ wood for 1,000 kg dried arnica. |
| Controlling of temperature and air humidity was done manually and was time-consuming. | Temperature and air humidity are measured automatically every 15 minutes; the drying system is regulated according to these data. |
| Drying duration of a batch took between 4-6 days (Morea and Michler 2008). | Drying duration of a batch is considerably reduced, ranging from 16 to 20 hours. |
| Drying capacity was limited: in 2008 the maximum drying capacity was 3,100 kg fresh arnica although more fresh flower heads could be available. | Drying capacity has increased considerably; e.g. in 2017, 33,600 kg fresh arnica flower heads were dried. |

Currently, the monitoring activities are only carried out in the perimeter of the Gârda de Sus community. The entire area where monitoring is performed covers 183.5 ha, representing 33.3% out of the total area of arnica habitats (550 ha) in the Gârda community. In the future, this activity will be extended to the areas of the other communities, but this will require new and more efficient monitoring methods.

Biological parameters like number of rosettes, number of flowering stems per rosette, number of flower heads per stem, as well as agronomic parameters are recorded (i.e., the number of stems harvested totally, or harvested partially, or remaining unharvested). The methodology was initiated during the *Apuseni Project* (Michler 2005) and the *Arnica Project* (Michler 2007). *Arnica System* adapted the methodology after 2007; the parameters used for monitoring being the relation between harvest potential and harvest rate. This provides basic information on the density of arnica flower heads and the population size in the study area.

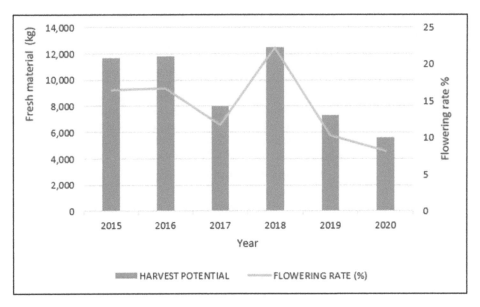

**FIGURE 7**   Flowering rate and harvest potential of *Arnica montana* for Gârda de Sus community

Annual monitoring reports are prepared providing important information for future harvesting planning, including the development of harvested arnica populations.

The flowering rate is one of the most important characteristics, having a decisive role in planning the harvest. Therefore, the number of flowering and the number of non-flowering rosettes is counted in transects on selected sites. The flowering rate is determined as the number of flowering rosettes divided by the total number of rosettes (sum of flowering and non-flowering rosettes) (Michler 2007).

However, the flowering intensity of arnica is subject to large fluctuations between different years (Fig. 7). For example, the flowering rate in the area of the Gârda de Sus community varied greatly in the period between 2015 and 2020, registering the lowest flowering rate in 2020 (8.12%), and the highest in 2018 (22.24%), depending on weather conditions. From our experience, flowering appears to be most intensive after a snow-rich winter period and under wet weather conditions during the vegetation period.

The total harvest potential, meaning the maximum amount of flower heads that could be harvested (calculated on the basis of our monitoring data), varied accordingly, with a minimum value of 5,578 kg (in 2020) and a maximum of 12,457 kg (in 2018) of calculated green mass flower heads (Fig. 7).

Harvesting does not mean picking all flower heads of a population; on the contrary, it is recommended to harvest a maximum of half of the flower heads on the stem and to leave the rest for seed production (Michler 2007). The harvest rate is represented by the difference between the number of harvested flower heads and the total number of flower heads expressed as a percentage (Fig. 8). Within the area of the Gârda de Sus community, the highest harvest rate was recorded in 2019 (64%), and the lowest values were registered in 2015 and 2020 (47%). The results show that a large part of flower heads remain unharvested, contributing to species conservation, being an effect of the annually applied training for harvesters, collectors and landowners as part of the participatory approach.

In several studies, it is claimed that harvesting is a threat to the arnica populations (Korneck et al. 1998). Therefore, about 50% of the arnica flower heads remain unharvested to keep the population in good condition. This is debatable because these claims are based on observations, and not on

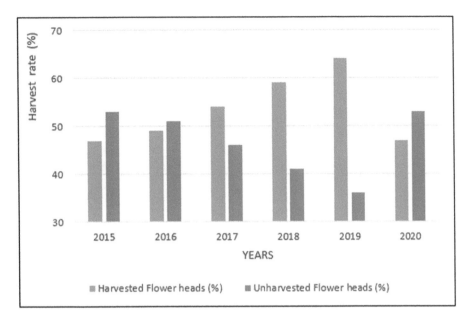

**FIGURE 8**   Harvest rates of *Arnica montana* species for the community of Gârda de Sus.

experiments. In mountainous regions, *Arnica montana* reproduces more vegetatively and less by sexual reproduction (Maurice et al. 2012).

From our experience, it is not overharvesting but habitat loss that is a much more pressing issue. Populations can be threatened very quickly due to several influences, such as nitrogen enrichment from the atmosphere (Dupré et al. 2010), which promotes the vitality of grasses as a superior competitor to *Arnica montana* (Sugier et al. 2018, Maurice et al. 2012). Other threats are permanent overgrazing or abandonment and successional spreading of *Vaccinium myrtillus* (Mardari et al. 2015).

### 3.2.8   Marketing

The final product, the dried *Arnica montana* flower heads, are packed in paper bags and prepared with all valid export permits to be sold on the international market. The product offered is certified organic and of the highest quality, which guarantees continuity in business relations with customers. Large quantities of the dried arnica flower heads are sold to the Weleda company from Schwäbisch Gmünd, Germany. Weleda, the main beneficiary, is open to supporting *Arnica System*'s vision and the sustainable use of *Arnica montana* grasslands.

Before establishing the cooperation with Weleda, the harvesters of the region received a price of 0.50 €/kg, for example for fresh material from different buyers in 2002 (Michler 2005). In the meantime prices have risen. The fair and favourable price offered by Weleda allowed a continuous increase in the purchase price of fresh arnica flower heads (Fig. 9). In 2015, the purchase price for fresh material was 2.28 €/kg and reached the amount of 6.61 €/kg in 2020. In 2021, we faced an unexpected situation: the price for fresh arnica flower heads had doubled because of a particularly high demand on the market and, consequently, high competition between many buyers and commercial traders. The contribution of arnica harvesting to the annual family income of harvesters was accordingly higher.

The *Arnica montana* supply chain is buyer-driven; the main and constant beneficiary is the company, Weleda. In the meantime, *Arnica System* also established cooperations with other trading partners and in addition the range of products has been diversified. The products are sold directly to the beneficiary companies without any intermediate purchaser.

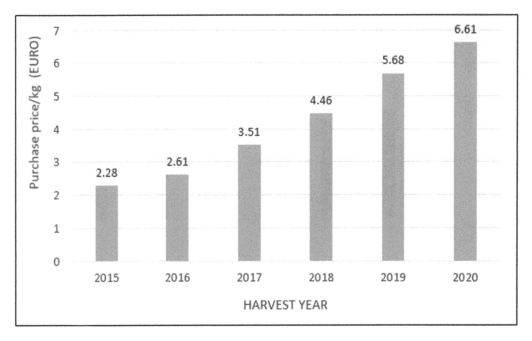

**FIGURE 9**   Evolution of purchase price for fresh *Arnica montana* flower heads.

In order to encourage sustainable use and create an incentive to continue the management of oligotrophic grasslands, it is important to pay a higher and fair price for fresh arnica flower heads. This higher market price entails the promise of high quality, organic certification and sustainable harvesting and could be a stable contribution to annual family incomes.

### 3.2.9.   Product certification

Certificates support the certified product economically and contribute to a social and environmental balance (Bhattacharyya et al. 2009). Certification of medicinal and aromatic plants (MAP) supports the enterprise by reducing the risk of recalls and rejections, by enhancing the confidence of buyers and by assuring compliance with all legal requriements (Kala 2015). Such certification of plant resources collected from the wild entails an independent assessment to guarantee sustainable management. If a company imports natural products as raw materials, they should be sure to refer to a reliable supply chain in terms of harvesting, cultivation and importation for the seamless supply of authorised medicines (Ahn 2017). Nowadays, certification of medicinal and aromatic plants is becoming more important, because both traders and consumers demand certified goods of high quality.

The quality of the dried material from the Apuseni Mountains is checked by the beneficiary Weleda. The quality is corresponding to the International Specification Raw Material (Dried Drug) *Arnica montana*, Flos sicc., organic (NOC), conform with the European Pharmacopoeia, monograph 1391 on *Arnicae flos*, the International Specification on Herbal Drugs and the National Testing Instruction (current versions).

For the organic certification of its products, *Arnica System* co-operates with the control and certification organisation, ECOINSPECT SRL, from Cluj-Napoca (identification code RO-ECO-008 according to RENAR), which is accredited and acknowledged by the Ministry for Agriculture and Rural Development (www.ecoinspect.ro). The certification includes the control of all processing documentation in the office and a field visit to verify the status of habitats and other economic or social processes. Certification of the final organic product – dried arnica flower heads – consists of the following steps: certification of suitability of the arnica habitat and the harvesting yield; and certification of the harvesting method, of collecting, transport, sorting, drying, packing and storage of arnica flower heads, including marketing. For each of these activities, *Arnica System* provides the ECOINSPECT Certification Company with all necessary documentation. As a result of the certification process, a control report is issued and the "Certificate of Conformity" with annexes is released.

### 3.2.10. Reinvested capital

A special characteristic of *Arnica System* is the reinvestment of capital for ecological, social and economic issues. The profit from the system varied between 11.1% and 28.1% of the turnover, with a decreasing trend (Fig. 10). The reason for this decrease is that the purchase price offered to local harvesters and collectors is constantly increasing due to business competition between different traders on the arnica market. Every year, nearly the entire profit, sometimes even more than that, is reinvested in the region – and not skimmed off by shareholders as in typical corporate business models – to promote the sustainable use of the species *Arnica montana*.

From an ecological point of view, the capital is invested in research activities (monitoring and evaluation of the resource and the habitat); in scientific experiments (impact of harvesting on arnica populations); in activities for the maintenance of grassland management, in restoration and the use of abandoned grasslands; in the cultivation of arnica and/or in training the landowners in the sustainable management of their grasslands. Most recently, *Arnica System* co-financed the acquisition of two drones for monitoring the arnica habitats and the sustainable harvesting of flower heads. The corresponding methodology is being developed in an ongoing implementation project funded by the DBU (Deutsche Bundesstiftung Umwelt) from Germany.

From a social perspective, the capital was invested in the creation of seasonal (15–20 people) and permanent (four people) job positions. Employment of local people counteracts emigration and offers employment opportunities for the locals, including young adults.

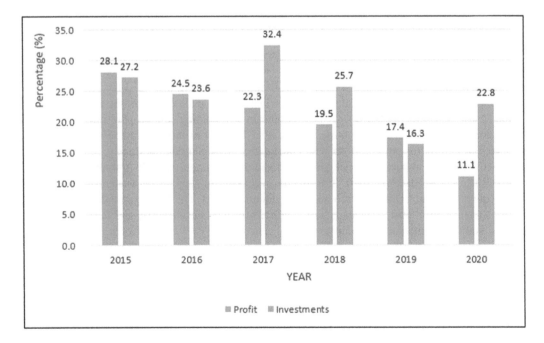

**FIGURE 10**   'Arnica System' annual level of profit and investments (expressed in % of the turnover).

From an economic point of view, *Arnica System* has invested in facilities and infrastructure; in the establishment of 17 seasonal collecting points distributed in the region; in the purchase of refrigerated transport vehicles; in off-road vehicles for transport in remote areas; in the aquisition of a warehouse with controlled temperature including a cooling chamber; in the modernisation of the drying and packaging system; and in equipment for grassland management (professional mulcher and mowing machine). Another issue concerns providing resources for the cultivation of additional arnica, which will reduce the harvesting pressure on collecting from the wild.

### 3.3. Perspectives

Sustainable production must be based on resource assessment, habitat monitoring, habitat management monitoring, monitoring of harvesting guidelines and quotas and quality control at different levels ensuring high quality and thus a fair price. Harvesters of arnica flower heads and landowners of oligotrophic grasslands with arnica are integrated into the process through the participatory approach, with annual training on sustainable harvesting and sustainable management of the grassland. The business is clearly linked to the revenue and profit generated. If the revenues of the arnica business do not cover the costs and no profit is generated with sustainable use of the arnica, the locals would be inclined to concentrate solely on collecting arnica and give up the management of habitats. If the resource declines due to habitat loss, abandonment or reforestation, the remaining resource will be overexploited (Michler 2007).

Increasing the purchase price of fresh *Arnica montana* flower heads is not a promising solution for the reason that many local people would not follow the principles of sustainable harvest. Instead, emphasis should be given on supporting the management of the still-existing arnica habitats. Some authors recommend the cultivation of *Arnica montana* to reduce the over-use of flower heads resources (Sand 2015, Pljevljakušić et al. 2014, Sugier et al. 2013). The risk of this strategy is that local farmers would try to cultivate arnica and stop harvesting from the wild and consequently, the interest in the management of HNV grassland would decrease. Instead, a combined system (conservation of oligotrophic grasslands and crop compensation) regulated through contractual conditions could be a sustainable solution for the future.

Management of oligotrophic grasslands is essential for the conservation of cultural landscapes and their biodiversity. However, this is under severe pressure, as payments provided by CAP compensation are not sufficient to preserve the oligotrophic grasslands. Additional income for rural households must be generated by the harvesting and selling of arnica flower heads, whilst the added economic value of processing them into a local product contributes to the maintenance of the HNV grassland.

Increasing the intrinsic interest of local people in maintaining oligotrophic grasslands is the key factor for the conservation of these mountainous landscapes in the future (Fig. 11). This has also been described for other regions with endangered grassland communities (e.g., from British grasslands where local people are acknowledged and play an important role in traditional management) (Blakesley & Buckley 2016). In Romania, successful models for the sustainable use of natural resources are rare (Drăgulănescu and Drăgulănescu 2013). The business model developed by *Arnica System* over almost 20 years is based on transparency, sustainability and fairness and takes into account environmental, economic and social concerns.

## 4.   CONCLUSIONS

The presence of oligotrophic grasslands and the harvesting of *Arnica montana* from the wild is, and will continue to be, important for the Apuseni Mountains region. The creation of local revenue and value adding is an important premise for the stability and balance of the area. The sustainable use of oligotrophic grasslands in the Apuseni Mountains contributes to the living standard of the people and to the preservation of the grasslands and their traditional management techniques.

**FIGURE 11** Traditional cultural landscape with oligotrophic grassland (Ocoale Plateau in the Apuseni Mountains) – Photo © Arnica System.

*Arnica System* has developed a model for the maintainance of HNV oligotrophic grasslands and conservation of traditional cultural landscapes in the Apuseni Mountains through the sustainable production and trade of the flagship species *Arnica montana*. The model shows how incentives and capacities can be created for the conservation of species-rich, traditionally managed habitats and landscapes containing medicinal plants.

Principles and lessons learnt from this case study can be transferred to other regions (e.g., to arnica habitats on other sites in Romania or in other European countries), or to other medicinal plant species in Europe and the rest of the world.

## REFERENCES

Ahn, Kyungseop. 2017. "The Worldwide Trend of Using Botanical Drugs and Strategies for Developing Global Drugs." *BMB Reports* 50 (3): 111–16. https://doi.org/10.5483/bmbrep.2017.50.3.221.

Auch, Eckhard. 2006. *Überlebensstrategien waldnutzender Familienwirtschaften im Apuseni-Gebirge, Rumänien: Sustainable Livelihoods Analyse und Handlungsempfehlungen.* Freiburg im Breisgau: Schriftenreihe des Instituts für Forstökonomie der Universität Freiburg, 27, 345 pp., ISBN: 978-3980673686.

Bhattacharyya, Rajasri, Aparna Asokan, Prodyut Bhattacharya, and Ram Prasad. 2009. "The Potential of Certification for Conservation and Management of Wild MAP Resources." *Biodiversity and Conservation* 18 (13): 3441–51. https://doi.org/10.1007/s10531-009-9653-z.

Bilz Melanie, Shelagh Kell, Nigel Maxted, and Richard Lansdown. 2011. "European Red List of Vascular Plants." Luxembourg: Publications Office of the European Union. 144 pp., https://doi.org/10.2779/8515.

Blakesley, David, and Peter Buckley. 2016. *Grassland Restoration and Management.* Exeter: Pelagic Publishing., 272 pp., ISBN: 978-1907807800.

Bogdan, Anca. 2012: "Cercetări privind folosirea şi menţinerea pajiştilor montane cu low-input". Teză de doctorat, USAMV Cluj-Napoca.

Brinkmann Katja, Florin Păcurar, Ioan Rotar, Evelyn Ruşdea, Eckhard Auch, and Albert Reif. 2009. "The Grasslands of the Apuseni Mountains, Romania." In *Grasslands in Europe of High Nature Value*, 226–37. KNNV Publishing. https://doi.org/10.1163/9789004278103_026.

Clair, Sandra. 2010. "Arnica: A Proven First Aid Remedy for Injuries and Accidents." *JN Zeal. Assoc. Med. Herbal*, pp. 12–13.

Coldea Gheorghe, Ion Sârbu, Vasile Cristea, Anca Sârbu, Gaftil Negrean, Adrian Oprea, Ion Cristurean, and Gheorghe Popescu. 2003. *Ghid pentru identificarea importantelor arii de protecţie şi conservare a plantelor din România*. Ed. Alo, Bucureşti, 113 pp., ISBN 973-86364-0-x.

Dapper, Heinrich. 1987. *Liste der Arzneipflanzen Mitteleuropas = Check-List of the Medicinal Plants of Central Europe*. Berlin Innova-Verlag, 73 pp., ISBN: 3-925096-01-9.

Darnhofer Ika, Markus Schermer, Melanie Steinbacher, Marine Gabillet, and Karoline Daugstad. 2017. "Preserving Permanent Mountain Grasslands in Western Europe: Why Are Promising Approaches Not Implemented More Widely?" *Land Use Policy* (68): 306–15. https://doi.org/10.1016/j.landuse pol.2017.08.005.

Drăgulănescu, Irina-Virginia and Natalia Drăgulănescu. 2013: "Unele teorii de durabilitate ecologică." *Revista Română de Statistică*, 12, 3–13. https://www.revistadestatistica.ro/wp-content/uploads/2014/04/RRS_12_2013_A1_ro.pdf

Dugalić, Sretenka. 2003. "The Treatment of Medical Plants: Opportunity and Feasability of Domestic Industry." *Hemijska Industrija* 57 (4): 171–80. https://doi.org/10.2298/hemind0304171d.

Duprè, Cecilia, Carly Joanne Stevens, Traute Ranke, Albert Bleeker, Cord Peppler-Lisbach, David J. G. Gowing, Nancy B. Dise, Edu Dorland, Roland Bobbink, and Martin Diekmann. 2010. "Changes in Species Richness and Composition in European Acidic Grasslands over the Past 70 Years: The Contribution of Cumulative Atmospheric Nitrogen Deposition." *Global Change Biology* 16 (1): 344–57. https://doi.org/10.1111/j.1365-2486.2009.01982.x.

Duwe, Virginia K., Ludo A.H. Muller., Thomas Borsch and Sascha A. Ismail. 2017. "Pervasive Genetic Differentiation among Central European Populations of the Threatened *Arnica montana* L. And Genetic Erosion at Lower Elevations." *Perspectives in Plant Ecology, Evolution and Systematics* 27: 45–56. https://doi.org/10.1016/j.ppees.2017.02.003

Emanuelsson, Urban. 2009. *The Rural Landscapes of Europe: How Man Has Shaped European Nature*. The Swedish Research Council Formas, Uppsala, 384 pp. ISBN: 978-9154060351

Figueiredo, Joana, and Henrique M. Pereira. 2011. "Regime Shifts in a Socio-Ecological Model of Farmland Abandonment." *Landscape Ecology* 26 (5): 737–49. https://doi.org/10.1007/s10980-011-9605-3.

Foggin, J. Marc. 2008. "Depopulating the Tibetan Grasslands." *Mountain Research and Development* 28 (1): 26–31. https://doi.org/10.1659/mrd.0972.

Foley, Jonathan, Ruth De Fries, Gregory P. Asner, Carol Barford, Gordon Bonan, Stephen R. Carpenter, Stuart F. Chapin, Michael T. Coe1, Gretchen C. Daily, Holly K. Gibbs, Joseph H. Helkowski, Tracey Holloway, Erica A. Howard, Christopher J. Kucharik, Chad Monfreda, Jonathan A. Patz, Colin I. Prentice, Navin Ramankutty, and Peter K. Snyder 2005: "Global Consequences of Land Use." *Science*, 309: 570–74. https://doi.org/10.1126/science.1111772.

Fuchs, Richard, Martin Herold, Peter H.Verburg, and Jan G. P. W. Clevers. 2013. "A High-Resolution and Harmonized Model Approach for Reconstructing and Analysing Historic Land Changes in Europe." *Biogeosciences* 10 (3): 1543–59. https://doi.org/10.5194/bg-10-1543-2013.

Fuchs, Richard, Herold Martin, Verburg Peter H., Jan G.P.W. Clevers, and Jonas Eberle. 2014. "Gross Changes in Reconstructions of Historic Land Cover/Use for Europe between 1900 and 2010." *Global Change Biology* 21 (1): 299–313. https://doi.org/10.1111/gcb.12714.

Fukarek, Franz, Hans Dieter Knapp, Stephan Rauschert, and Erich Weinart. 1978. "Karten der Pflanzenverbreitung in der DDR – 1. Serie." *Hercynia – Ökologie Und Umwelt in Mitteleuropa* 15 (3): 231–234. https://pub lic.bibliothek.uni-halle.de/hercynia/article/view/1080.

Galluzzo, Nicola. 2017. "The Development of Agritourism in Romania and Role of Financial Subsides Allocated under the Common Agricultural Policy." *Geographia Polonica* 90 (2): 25–39. https://doi.org/10.7163/gpol.0087.

Galluzzo Nicola. 2018. "Impact of the Common Agricultural Policy Payments Towards Romanian Farms", *Bulgarian. Journal Agricultural Sciences*. 24 (2): 199–205. https://www.agrojournal.org/24/02-04.pdf

Gârda, Nicoleta. 2010. "Studiul unor elemente de landşaft montan (cu privire specială asupra ecosistemelor de pajişti din comuna Gârda de Sus, Munţii Apuseni), Teză de doctorat USAMV, Cluj- Napoca.

Goia, Augustin. 2005. "Die traditionelle Lebensweise". In: Ruşdea Evelyn, Reif Albert, Povară Ioan, and Werner Konold „Perspektiven für eine traditionelle Kulturlandschaft in Osteuropa. Ergebnisse eines inter-und transdisziplinären, partizipativen Forschungsprojektes in Osteuropa". Culterra, (Schriftenreihe des Inst. Landespflege, Univ. Freiburg) 34: 115–19.

Goia, Augustin and Zoltan Borlan. 2005. "Siedlungsgeschichte der Dörfer im Motzenland". In: Ruşdea Evelyn, Reif Albert, Povară Ioan, and Werner Konold „Perspektiven für eine traditionelle Kulturlandschaft in Osteuropa. Ergebnisse eines inter-und transdisziplinären, partizipativen Forschungsprojektes in Osteuropa". Culterra, (Schriftenreihe des Inst. Landespflege, Univ. Freiburg) 34: 109–14.

Halada, Ľuboš, David Stanislav, Juraj Hreško, Alexandra Klimantová, Andrej Bača, Tomáš Rusňák, Miroslav Buraľ and Ľuboš Vadel. 2017. "Changes in Grassland Management and Plant Diversity in a Marginal Region of the Carpathian Mts. In 1999–2015." Science of the Total Environment 609: 896–905. https://doi.org/10.1016/j.scitotenv.2017.07.066.

Hegi, Gustav. 1987. Illustrierte Flora von Mitteleuropa. VI/4, Pteridophyta – Spermatophyta, 2nd ed., Berlin, Germany, Verlag Paul Parey, 704-710, ISBN: 978-3489860204

Hultén, Eric, and Magnus Fries. 1986. Atlas of North European Vascular Plants: North of the Tropic of Cancer. Königstein: Koeltz Botanical Books, 1188 pp., ISBN: 978-3874292634

Jitea, Mugurel I., and Felix H. Arion. 2015. "The Role of Agri-Environment Schemes in Farm Economic Sustainability from High Natural Value Transylvanian Areas." Environmental Engineering and Management Journal (EEMJ), 14 (4): 943–53. https://doi.org/10.30638/eemj.2015.105.

Kahmen, Stefanie, and Peter Poschlod. 2000. "Population Size, Plant Performance, and Genetic Variation in the Rare Plant Arnica montana L. in the Rhön, Germany." Basic and Applied Ecology 1 (1): 43–51. https://doi.org/10.1078/1439-1791-00007.

Kala, Chandra Prakash. 2015. "Medicinal and Aromatic Plants: Boon for Enterprise Development." Journal of Applied Research on Medicinal and Aromatic Plants 2 (4): 134–39. https://doi.org/10.1016/j.jarmap.2015.05.002.

Kathe, Wolfgang. 2006. Conservation of Eastern-European Medicinal Plants: Arnica montana in Romania. In: Medicinal and Aromatic Plants: Agricultural, Commercial, Ecological, Legal, Pharmacological and Social Aspects". Bogers, Robert J.; Craker, Lyle E.; Lange, Dagmar (Eds.), 203–11. Springer, Netherlands, https://library.wur.nl/ojs/index.php/frontis/article/download/1233/805

Korneck, Dieter, Martin Schnittler, Frank Klingenstein, Gerhard Ludwig, Melanie Takla, Udo Bohn, and Rudolf May. 1998. "Warum verarmt unsere Flora? Auswertung der Roten Liste der Farn-und Blütenpflanzen Deutschlands." In "Ursachen des Artenrückgangs von Wildpflanzen und Möglichkeiten zur Erhaltung der Artenvielfalt." Schriftenreihe für Vegetationskunde, 29, Bad Godesberg: Bundesamt für Naturschutz, 299–444.

Korneck, Dieter, Martin Schnittler, and Immo Vollmer. 1996. "Rote Liste der Farn- und Blütenpflanzen (Peridophyta et Spermatophyta) Deutschlands.", Schriftenreihe für Vegetationskunde, 28, Bonn-Bad Godesberg. Bundesamt Für Naturschutz, 21–187.

Kos, Olha, Maja T. Lindenmeyer, Aurelia Tubaro, Silvio Sosa, and Irmgard Merfort. 2005. "New Sesquiterpene Lactones from Arnica Tincture Prepared from Fresh Flowerheads of Arnica montana." Planta Medica 71 (11): 1044–52. https://doi.org/10.1055/s-2005-871284.

Kriplani, Priyanka, Kumar Guarve, and Uttam S. Baghael. 2017. "Arnica montana L. – a Plant of Healing: Review." Journal of Pharmacy and Pharmacology 69 (8): 925–45. https://doi.org/10.1111/jphp.12724.

Luick, Rainer and Norbert Roeder. 2016. "The First Pillar of the New CAP – Implications for Low Input Grasslands." In The Multiple Roles of Grassland in the European Bioeconomy, 603–5. Trondheim, Norway: Organising Committee of the 26th General Meeting of the European Grassland Federation, NIBIO.https://www.europeangrassland.org/fileadmin/documents/Infos/Printed_Matter/Proceedings/EGF2016.pdf

Luijten, Sheila H., Gerard B. Oostermeijer, Nico C. van Leeuwen, and Hans C. M. den Nijs. 1996. "Reproductive Success and Clonal Genetic Structure of the Rare Arnica montana (Compositae) in The Netherlands." Plant Systematics and Evolution 201 (1-4): 15–30. https://doi.org/10.1007/bf00989049.

Lyss, Guido, Thomas J. Schmidt, Irmgard Merfort, and Heike L. Pahl. 1997. "Helenalin, an Anti-Inflammatory Sesquiterpene Lactone from Arnica, Selectively Inhibits Transcription Factor NF-KB." Biological Chemistry 378. 951–61, https://doi.org/10.1515/bchm.1997.378.9.951.

MADR (Ministerul Agriculturii şi Dezvoltării Rurale), 2022 – *Ghid informativ pentru beneficiarii Măsurii 10-agro-mediu şi climă din Programul naţional de dezvoltare rurală* (PNDR) 2014-2020, Anexa nr. 1 la Ordinul MADR din 2022. https://apia.org.ro/wp-content/uploads/2022/03/Ghid-M10-FINAL.pdf

Mardari, Constantin, Doina Dănilă, Ciprian Bîrsan, Tiberius Balaeş, Camelia Ştefanache and Călin Tănase. 2015. "Plant Communities with *Arnica montana* in Natural Habitats from the Central Region of Romanian Eastern Carpathians." *Journal of Plant Development* 22, 95–105. https://plant-journal.uaic.ro/docs/2015/10.pdf.

Maruşca Teodor, 2016. *Praticultura pe înţelesul tuturor.* Editura Profesional – Agromedia, Bucharest, 263 pp., ISBN 978-973-0-23325-4.

Maurice, Tiphaine, Guy Colling, Serge Muller, and Diethart Matthies. 2012. "Habitat Characteristics, Stage Structure and Reproduction of Colline and Montane Populations of the Threatened Species *Arnica montana.*" *Plant Ecology* 213 (5): 831–42. https://doi.org/10.1007/s11258-012-0045-1.

McGurn, Patrick, A. Browne, G. NíChonghaile, L. Duignan, J. Moran, Daire ÓHuallacháin, and J. A. Finn. 2017. "Semi-Natural Grasslands on the Aran Islands, Ireland: Ecologically Rich, Economically Poor." In *Grassland Resources for Extensive Farming Systems in Marginal Lands: Major Drivers and Future Scenarios. Proceedings of the 19th Symposium of the European Grassland Federation, Alghero, Italy, 7-10 May 2017*, edited by Claudio Porqueddu, Antonello Franca, Giampiero Lombardi, Giovanni Molle, Giovanni Peratoner, and Alan Hopkins, 197–99. Sassary, Italy: CNR-ISPAAM. https://www.europeangrassland.org/fileadmin/documents/Infos/Printed_Matter/Proceedings/EGF2017.pdf

McKinney, Michael 2002. "Urbanization, Biodiversity, and Conservation.: The Impacts Of Urbanization On Nattive Species Are Poorly Studied, but Educating a Highly Urbanized Human Population about These Impacts Can Greatly Improve Species Conservation In All Ecosystems" BioScience 52 (10): 883–90. https://doi.org/10.1641/0006-3568(2002)052[0883:ubac]2.0.co;2

Melero, Roser, Mónica Fanlo, Eva Moré, Irene Vázquez, and Roser Cristóbal. 2012. "Sustainable Production of *Arnica montana* in the Catalan Pyrenees (NE Spain): Wild Harvesting or Cultivation?" In *Ith International Symposium on Medicinal, Aromatic and Nutraceutical Plants from Mountainous Areas (MAP-Mountain 2011), Acta Horticulturae* 955: 225–30. https://doi.org/10.17660/actahortic.2012.955.32.

Meusel, Hermann, and Eckehart J. Jäger. 1992. *Vergleichende Chorologie der Zentraleuropäischen Flora.* Jena: Gustav Fischer Verlag, 688 pp., ISBN: 978-3334004111.

Meyer, Frank and Michael Straub. 2011: *Die magischen 11 der heilenden Pflanzen.* Gräfe und Unzer, München, 192 pp., ISBN: 978-3833823268.

Michler, Barbara. 2007: *Management Plan for Project Conservation of Eastern European Medicinal Plants: Arnica montana in Romania, Case Study Gârda de Sus.* 83 pp., http://biolaya.com/wp-content/uploads/Managenmentplan-Arnica-montana-Apuseni-Romania-1.pdf.

Michler, Barbara. 2005: "Leitprojekt Heilpflanzen". In: Ruşdea Evelyn, Reif Albert, Povară Ioan, and Werner Konold „Perspektiven für eine traditionelle Kulturlandschaft in Osteuropa. Ergebnisse eines inter-und transdisziplinären, partizipativen Forschungsprojektes in Osteuropa". *Culterra*, (Schriftenreihe des Inst. Landespflege, Univ. Freiburg) 34: 378-80. https://www.landespflege.uni-freiburg.de/ressourcen/pub/2005%20-%20Endbericht%20Proiect%20Apuseni.pdf.

Michler, Barbara, Ioan Rotar, and Florin Păcurar. 2006. "Biodiversity and Conservation of Medicinal Plants: A Case Study in the Apuseni Mountains in Romania." *Bulletin of University of Agricultural Sciences and Veterinary Medicine* 62: 86–7. http://journals.usamvcluj.ro/index.php/agriculture/article/download/1716/1685

Michler, Barbara, Ioan Rotar, Florin Păcurar, and Andrei Stoie. 2005. "*Arnica montana*, an Endangered Species and a Traditional Medicinal Plant: The Biodiversity and Productivity of Its Typical Grasslands Habitats." In *Integrating Efficient Grassland Farming and Biodiversity*, Proceedings of EGF, Grassland Science in Europe, Estonia,10, 336–39.

Michler, Barbara, Wolfgang Kathe, Susanne Schmitt and Ioan Rotar. 2004. "Conservation of Eastern European Medicinal Plants: *Arnica montana* in Romania." *Bulletin USAMV-CN, Seria Agricultura* 60: 228–30.

Morea, Adriana, and Barbara Michler. 2008. "Drying of the *Arnica montana* Flower Heads in Apuseni Mountains, County Garda." *Bulletin of University of Agricultural Sciences and Veterinary Medicine Cluj-Napoca. Agriculture* 62. http://journals.usamvcluj.ro/index.php/agriculture/article/download/1645/1614.

Neitzke, Mechthild. 2015. "Heilpflanzendiversität in den Acker-Ökosystemen Nordrhein-Westfalens." *Natur in NRW*, 4, 32–36, Landesamt für Natur, Umwelt und Verbraucherschutz Nordrhein-Westfalen (LANUV) https://www.lanuv.nrw.de/fileadmin/lanuvpubl/5_natur_in_nrw/50036_Natur_in_NRW_4_2015.pdf.

Oberbaum, Menachem, Narine Galoyan, Liat Lerner-Geva, Shepherd Roee Singer, Sorina Grisaru, David Shashar, and Arnon Samueloff. 2005. "The Effect of the Homeopathic Remedies *Arnica montana* and *Bellis perennis* on Mild Postpartum Bleeding-a Randomized, Double-Blind, Placebo-Controlled Study-Preliminary Results." *Complementary Therapies in Medicine* 13 (2): 87–90. https://doi.org/10.1016/j.ctim.2005.03.006.

Obón, Concepción, Diego Rivera, Alonso Verde, José Fajardo, Arturo Valdés, Francisco Alcaraz, and Ana Maria Carvalho. 2012. "Árnica: A Multivariate Analysis of the Botany and Ethnopharmacology of a Medicinal Plant Complex in the Iberian Peninsula and the Balearic Islands." *Journal of Ethnopharmacology* 144 (1): 44–56. https://doi.org/10.1016/j.jep.2012.08.024.

Oltean, Miruna, Gavril Negrean, A. Popescu, Nicolae Roman, Gheorghe Dihoru, Vasile Sanda, and Simona Mihăilescu. 1994. "*Lista Roşie a plantelor superioare din România.*" In: *Studii, sinteze, documentaţii de ecologie*, Institutul de Biologie, Academia Română, 1, 52.pp. https://www.researchgate.net/publication/318654406_Lista_rosie_a_plantelor_superioare_din_Romania

Otovescu, Cristina, and Adrian Otovescu. 2019. "The Depopulation of Romania – Is It an Irreversible Process?" *Revista de cercetare şi intervenţie socială* 65, 6, 1: 370–88. https://doi.org/10.33788/rcis.65.23.

Păcurar, Florin, Ioan Rotar, and Nicoleta Gârda. 2008. "Elemente de dezvoltare durabilă în managementul tradiţional al pajiştilor cu *Arnica montana.*" *Hameiul şi plantele medicinale* XVI, 1-2 (31–32): 251–54.

Păcurar, Florin, Ioan Rotar, Nicoleta Gârda, and Adriana Morea. 2009. "The Management of Oligotrophic Grasslands and the Approach of New Improvement Methods." *Transylvanian Review of Systematical and Ecological Research,* 7 The Arieş River Basin: 59–68. https://magazines.ulbsibiu.ro/trser/trser7/59-68.pdf.

Păcurar, Florin, Ioan Rotar, Albert Reif, Roxana Vidican, Vlad Stoian, Stefanie M. Gärtner, and Robert B. Allen. 2014. "Impact of Climate on Vegetation Change in a Mountain Grassland-Succession and Fluctuation." *Notulae Botanicae Horti Agrobotanici Cluj-Napoca* 42: 347–56. https://doi.org/10.15835/nbha4229578.

Păcurar, Florin, Ágnes Balazsi, Ioan Rotar, Ioana Vaida, Albert Reif, Roxana Vidican, Evelyn Ruşdea, Vlad Stoian, and Dragomir Sângeorzan. 2020. "Technologies Used for Maintaining Oligotrophic Grasslands and Their Biodiversity in a Mountain Landscape." *Romanian Biotechnological Letters* 25, 1: 1128–35. https://doi.org/10.25083/rbl/25.1/1128.1135.

Peyraud, Jean-Louis, and Alain Peeters. 2016. "The Role of Grassland Based Production System in the Protein Security." In *The Multiple Roles of Grassland in the European Bioeconomy. Proceedings of the 26th General Meeting of the European Grassland Federation, Trondheim, Norway, 4-8 September 2016*, 29–43. Wageningen: NIBIO. https://www.europeangrassland.org/fileadmin/documents/Infos/Printed_Matter/Proceedings/EGF2016.pdf.

Pljevljakušić, Dejan, Teodora Janković, Slavica Jelačić, Miroslav Novaković, Nebojša Menković, Damir Beatović, and Zora Dajić-Stevanović. 2014. "Morphological and Chemical Characterization of *Arnica montana* L. Under Different Cultivation Models." *Industrial Crops and Products* 52: 233–44. https://doi.org/10.1016/j.indcrop.2013.10.035.

Pop, Oliviu G, and Florentina Florescu. 2008. *Ameninţări potenţiale, Recomandări de management şi monitorizare*. Editura Universităţii Transilvania Braşov., 84 pp., ISBN: 978-9735983499, https://www.protectiamediului.ro/wp-content/uploads/2019/10/Publication.Amenintari.Management.Pajisti.Ro_.pdf.

Radušienė, Jolita, and Juozas. Labokas. 2007. "Population Performance of *Arnica montana* L. In Different Habitats." In *Crop Wild Relative Conservation and Use*, Maxted, N., Ford-Lloyd, B., Kell, S., Iriondo, J., Dulloo, E., Turok, J., (Eds.); CAB International: Wallingford, UK, 386–91, ISBN 978-1-84593-099-8. https://www.cabi.org/environmentalimpact/ebook/20093268686.

Reif, Albert. Gheorghe Coldea, and Georg Harth. 2005: "Pflanzengesellschaften des Offenlandes und der Wälder." In: Ruşdea Evelyn, Reif Albert, Povară Ioan, and Werner Konold „Perspektiven für eine traditionelle Kulturlandschaft in Osteuropa. Ergebnisse eines inter-und transdisziplinären, partizipativen Forschungsprojektes in Osteuropa". *Culterra*, (Schriftenreihe des Inst. Landespflege, Univ. Freiburg) 34: 78–87.

Reif, Albert, Evelyn Ruşdea, Florin Păcurar, Ioan Rotar, Katja Brinkmann, Eckhard Auch, Augustin Goia, and Josef Bühler. 2008. "A Traditional Cultural Landscape in Transformation." *Mountain Research and Development* 28 (1): 18–22. https://doi.org/10.1659/mrd.0806.

Rotar, Ioan, Ioana Vaida, and Florin Păcurar. 2020. "Species with Indicative Values for the Management of the Mountain Grasslands." *Romanian Agricultural Research,* Nardi Fundulea, 37: 189–96. https://www.incda-fundulea.ro/rar/nr37/rar37.22.pdf

Ruşdea, Evelyn, Albert Reif, Ioan Povară, and Werner Konold. 2005. "Perspektiven für eine traditionelle Kulturlandschaft in Osteuropa – Ergebnisse eines inter- und transdisziplinären, partizipativen Forschungsprojektes im Apuseni-Gebirge in Rumänien. " *Culterra,* (Schriftenreihe des Inst. Landespflege, Univ. Freiburg) 34. 401 pp. ISBN: 978-3933390219, https://www.landespflege.uni-freiburg.de/ressourcen/pub/2005%20-%20Endbericht%20Proiect%20Apuseni.pdf.

Sand Sava, Camelia. 2015. "*Arnica montana* L. As a Medicinal Crop Species." *Scientific Papers Series-Management, Economic Engineering in Agriculture and Rural Development* 15 (4): 303–7. http://managementjournal.usamv.ro/pdf/vol.15_4/Art45.pdf

Sângeorzan, Dragomir, Ioan Rotar, Florin Păcurar, Ioana Vaida, Alina Şuteu, and Valeria Deac. 2018. "The Definition of Oligotrophic Grasslands." *Romanian Journal of Grassland and Forage Crops* 17: 33–41. https://sropaj.ro/documente/ro/revista/articole/RJGFC-17-2018_art-5.pdf.

Schippmann, Uwe, Danna J. Leaman, and Anthony B. Cunningham. 2002. "Impact of Cultivation and Gathering of Medicinal Plants on Biodiversity: Global Trends and Issues." In *Biodiversity and the Ecosystem Approach in Agriculture, Forestry and Fisheries*, 142–67. Rome, Italy: Food and Agriculture Organization of the United Nations (FAO). https://www.researchgate.net/publication/265157471_Impact_of_Cultivation_and_Gathering_of_Medicinal_Plants_on_Biodiversity_Global_Trends_and_Issues

Schröder, H., Wolfgang Lösche, Hans Strobach, Walter Leven, Günter Willuhn, Uwe Till, and Karsten Schrör. 1990. "Helenalin and 11α,13-Dihydrohelenalin, Two Constituents from *Arnica montana* L., Inhibit Human Platelet Function via Thiol-Dependent Pathways." *Thrombosis Research* 57 (6): 839–45. https://doi.org/10.1016/0049-3848(90)90151-2

Schwabe, Angelika. 1990. "Syndynamische Prozesse in Borstgrasrasen: Reaktionsmuster von Brachen nach erneuter Rinderbeweidung und Lebensrhythmus von *Arnica montana L*" *Carolinea* , 48, Karlsruhe, 45–68. https://www.researchgate.net/publication/272686300_Syndymanische_Prozesse_in_Borstgrasrasen_Reaktionsmuster_von_Brachen_nach_erneuter_Rinderbeweidung_und_Lebensrhythmus_von_Arnica_montana_L

Stoie, Andrei. 2011. "Cercetări asupra ecosistemelor de pajişti cu *Arnica montana* în Bazinul Superior al Arieşului." Teză de Doctorat, USAMV, Cluj- Napoca.

Sugier, Danuta, Piotr Sugier, and Urszula Gawlik-Dziki. 2013. "Propagation and Introduction of *Arnica montana* L. into Cultivation: A Step to Reduce the Pressure on Endangered and High-Valued Medicinal Plant Species." *The Scientific World Journal* 2013: 1–11. https://doi.org/10.1155/2013/414363.

Sugier, Piotr, Aleksander Kołos, Dan Wołkowycki, Danuta Sugier, Andrzej Plak, and Oleg Sozinov. 2018. "Evaluation of Species Inter-Relations and Soil Conditions in *Arnica montana* L. Habitats: A Step towards Active Protection of Endangered and High-Valued Medicinal Plant Species in NE Poland." *Acta Societatis Botanicorum Poloniae* 87 (3):1–16. https://doi.org/10.5586/asbp.3592.

Sugier, Piotr, Danuta Sugier, Oleg Sozinov, Aleksander Kołos, Dan Wołkowycki, Andrzej Plak, and Olha Budnyk. 2019. "Characteristics of Plant Communities, Population Features, and Edaphic Conditions of *Arnica montana* L. Populations in Pine Forests of Mid-Eastern Europe." *Acta Societatis Botanicorum Poloniae* 88 (4):1–13. https://doi.org/10.5586/asbp.3640.

Titze, Andreas, Claudia Hepting, Verena Hollmann, Lilith Jeske, Ilona Leyer, Sascha Liepelt, Annika Peters, and WeiseJörg. 2020. *Wilde Arnika – Ein Leitfaden für die Praxis.* Botanischer Garten der Philipps-Universität Marburg: Arnikahessen, Botanischer Garten der Philipps-Universität Marburg, 229 pp. ISBN: 978-3-8185-0561-5. https://www.uni-marburg.de/de/botgart/forschung/doppelseiten_20-4mb.pdf.

Tokarczyk, Natalia. 2018. "Challenges for the Conservation of Semi-Natural Grasslands in Mountainous National Parks – Case Studies from the Polish Carpathians." *Carpathian Journal of Earth and Environmental Sciences* 13 (1): 187–98. https://doi.org/10.26471/cjees/2018/013/017.

Vaida, Ioana, Florin Păcurar, Ioan Rotar, Liviu Tomoş, and Vlad Stoian. 2021. "Changes in Diversity due to Long-Term Management in a High Natural Value Grassland." *Plants* 10 (4): 739. 1–20. https://doi.org/10.3390/plants10040739.

Vaida, Ioana, Ioan Rotar, Florin Păcurar, Roxana Vidican, Anca Pleşa, Anamaria Mălinaş, and Vlad Stoian. 2016. "Impact on the Abandonment of Semi-Natural Grasslands from Apuseni Mountains." *Bulletin of University of Agricultural Sciences and Veterinary Medicine Cluj-Napoca. Agriculture* 73 (2): 323–31. https://doi.org/10.15835/buasvmcn-agr:12417.

Vîntu, Vasile, Costel Samuil, Ioan Rotar, Alexandru Moisuc, and Iosif Razec. 2011. "Influence of the Management on the Phytocoenotic Biodiversity of Some Romanian Representative Grassland Types." *Notulae Botanicae Horti Agrobotanici Cluj-Napoca* 39 (1): 119–25. https://doi.org/10.15835/nbha 3915867

Williams, Samantha, and Thembela Kepe. 2008. "Discordant Harvest: Debating the Harvesting and Commercialization of Wild Buchu (*Agathosma betulina*) in Elandskloof, South Africa." *Mountain Research and Development* 28 (1): 58–64. https://doi.org/10.1659/mrd.0813

# 10 Ethno-veterinary Science and Practices as an alternative to antibiotics for certain veterinary diseases

*M N Balakrishnan Nair and Natesan Punniamurthy*

## 1. INTRODUCTION

India is the largest milk producing country in the world and produced about 194,800 thousand tonnes of milk in 2020. Livestock keeping is a full-time livelihood occupation for many farmers in India. About 70% of rural indigent people depend on livestock as a critical source of income (Nair and Unnikrishnan 2010). Marginal and small farmers hold half of the cattle and buffaloes. Livestock rearing is also used as an additional source of income for the people engaged in agriculture. The National Dairy Development Board (NDDB) was instrumental in augmenting the growth of the dairy sector in this country through targeted programs like Operation Flood I, II and III. Women accounted for 93% of total employment in dairy production. A woman devotes 3.5 hours per day for animal husbandry activities like feeding and animal healthcare-related activities performed at home.

Cross-breeding with exotic breeds was introduced to India mainly to enhance production of milk. This led to the loss of local breeds which have resistance to many diseases. High incidence of diseases in cross-bred animals and the indiscriminate use of antibiotics in dairy animals caused high antibiotic residues in animal products like milk and meat (Asif et al. 2020; Moudgil et al. 2019; Nair 2019; Kumaraswamy et al. 2018; Vishnuraj et al. 2016; Moharana et al. 2015; Gaurav et al. 2014; Dinki and Balcha 2013; Ram et al. 2000). Antimicrobial resistance (AMR) is a worldwide problem affecting both human and animal health (Jeena et al. 2020; Prajwal et al. 2017; Priyanka et al. 2017; Nisha 2008). About 90% of the antibiotics used end up in the environment, affecting the quality of water, soil and biodiversity. During the past 5 years, the annual rate of use of antibiotics in India has risen by 6–7% (Thomas et al. 2015). It is also predicted that the global consumption of antimicrobials will go up to 67% from 2010 to 2030 and it is estimated that by 2050 the antimicrobial resistance (AMR) will cause 10 million deaths per year (Mutua et al. 2020; Klein et al. 2018; Shallcross et al. 2015; Thomas et al. 2015; Review on Antimicrobial Resistance 2014; Nisha 2008). As the antibiotics find their way through the food chain, many countries in the world are looking for safer herbal alternatives to antibiotics. Presently, implementation of regulations and policy focus on reducing and controlling the use of antibiotics in India is poor. A high priority need in the livestock sector is to find safe and cost-effective herbal remedies from medicinal plants to replace to a feasible extent high-cost antibiotics and other chemical veterinary drugs.

Prevalence of mastitis continues to remain the most challenging disease and it is increasing among those cattle with higher production of milk. In India, the average prevalence of mastitis in the 1960s to the early 1990s was not more than 30%. In the last three decades the prevalence increased to even more than 60%. In 2012 alone, the economic loss due to mastitis was about 71655.1 million Indian Rupees (986.65 million USD) per annum (National Dairy Research Institute 2012).

DOI: 10.1201/9781003146902-13

## 2. A HISTORICAL PERSPECTIVE OF ETHNO-VETERINARY MEDICINE

The Indian systems of medicine can be broadly divided into codified systems and non-codified (oral) folk medicine. The codified systems are based on the theory of physiological functioning, disease etiology and clinical practices. The non-codified or folk traditions are as old as humankind and have a symbolic relationship with the codified system. They are dynamic, innovative, evolving and spread across 4639 ethnic communities all over India. They are self-perpetuating, transmitted orally without support from any agency or institution.

A historical perspective of evolution of traditional veterinary medicine to take care of the health of domesticated animals was reported earlier (Sagari and Nitya 2004, Nair and Unnikrishnan 2010). Codified veterinary knowledge exists in the form of medical texts and manuscripts on various aspects of veterinary care and has a documented history of around 5000 years. The veterinary and animal husbandry practices are mentioned in *Rigveda* and *Atharvaveda*.

Texts and manuscripts on various aspects of veterinary care like *Asvasatra* (1800 BCE), *Hastyayurveda* (1000 BCE), *Asva Vaidyaka, Garuda Purana, Asvayur Veda Sarasindu, Sahadeva Pasu Vaidya Sastramu* (Telugu), *Mattu Vaidya Bodhini* (Tamil) *Pashuvaidya Mattuvagadam* (Tamil), *Matsyapurana, Garudapurana, Agnipurana, Brahmanandapurana, Lingapuranah* and *Arthasatra,* which are available for research in some libraries in India (e.g. Saraswathi Mahal Library, also called Thanjavur Maharaja Serfoji's Saraswathi Mahal Library located in Thanjavur (Tanjore), Tamil Nadu; Bhandarkar Oriental Research Institute (BORI), Pune, Maharashtra; Sampurnanand Sanskrit Viswavidyalaya (University), Varanasi, Uttar Pradesh or Lal Chand Research Library at DAV College, Chandigarh, India and outside India (e.g. The British Library, London, UK, shelf mark Or 4481 (Rig Veda)).

## 3. CURRENT ETHNO-VETERINARY PRACTICES

Rich ethno-veterinary health traditions still prevail in India. The ethno-veterinary practices (EVP) existing in the villages form an integral part of village life and play an important social, religious (see Figures 1 A–D) and economic role. These traditions are based on location and ethnic community-specific health related knowledge, practices, beliefs, lifestyles, food habits, customs and skills related to health care and management of livestock (McCorkle et al. 1999; Nair et al. 2017a; Nair and Unnikrishnan 2010).

The local healers and some knowledgeable farmers use the locally available medicinal plants for treatment of animals. The codified medical traditions share a similar worldview as that of oral folk traditions (Unnikrishnan and Darshan Shankar 2006). They also contain sophisticated clinical theory apart from practical therapies (Nair and Unnikrishnan 2010; Unnikrishnan and Darshan Shankar 2006). A perusal of the literature on EVP shows there are many reports of medicinal plants used for animal health. A list of the plants used for animal health conditions prepared from a few selected publications is presented in Table 1. One hundred and fourteen community-based animal health practices for 18 health conditions were published earlier (Nair and Unnikrishnan 2010). For an increasing number of marginalised farmers cost-effective treatment and timely intervention will be of priority in the coming years. Therefore, urgent revival of these traditional veterinary practices has high priority in light of the constraints imposed by Western medicine. Sixty-one clinical conditions in cattle were prioritized based on a participatory rural appraisal (PRA, for identifying disease conditions) and matrix ranking methodology (for subsequent prioritizing) from 24 locations in the country (Nair et al 2017a). Out of these prioritized conditions, treatment for eight clinical conditions such as mastitis (see Table 2), foot and mouth disease (see Table 3), foot lesions (see Table 4), diarrhoea (see Table 5), udder pox (see Table 6), external parasites (see Table 7), fever (see Table 8) and repeat breeding syndrome (see Table 9) are presented for reference.

**FIGURE 1A–D** Some EVP have a religious and social role for the people in India as shown in Figures A – D.
**FIGURE 1 A** A member of "Kole Basava" nomad community taking a heavily decorated bull house to house as part of their folklore tradition for giving blessing to be people. These bulls (sometimes cows) are trained to obey simple commands. People attribute spiritual value to these bulls and pray to the bull for blessings. In return for the blessing people give fruits, fodder and money to the owner.

**FIGURE 1B** A cow and a bull perform a small episode based on the story from '*Ramayana*' an ancient epic. The owners also use musical instruments as part of the command given to the animals to perform.

**FIGURE 1C AND D**  A '*Gau Pooja*' performed in a house at Perunthurai, Tamil Nadu, India. This pooja is done on the last Friday of every Tamil month, which is based on the moon calendar. It is mainly performed for prosperity and creating awareness about protecting cows.

## TABLE 1

Medicinal plants and preparations thereof used for animal health conditions listed from a selection of publications. Sources: Ghosh 1999 (Pos. 1-8); Harsha et al. 2005 (Pos. 9-32); Lakshmi Narayana and Rao 2013 (Pos. 33-64); Mallik et al. 2012 (Pos. 64-81); Sharma, et al. 2014 (Pos. 82-100); FMD: Foot and Mouth Disease.

| No | Botanical name | Parts used | Uses |
|---|---|---|---|
| 1 | *Achyranthes aspera* L. | root | Retention of placenta |
| 2 | *Adina cordifolia* (Roxb.) | bark + rhizome | Hair loss in the skin |
|  | *Curcuma longa* L. | equal parts | |
| 3 | *Albizia lebbeck* (L.) Benth. | rotten petiole | Swelling on the throat |
| 4 | *Anthocephalus chinensis* Hassk. Lam. | leaf juice + molasses | Colic pain |
| 5 | *Bryophyllum pinnatum* (Lam.) Oken. | tender leaves + mustard oil | Poisonous insect bites |
| 6 | *Coccinia grandis* (L.) Voigt | root | Typhoid fever |
|  | *Nerium indicum* L. | root | |
|  | *Borassus flabellifer* L. | tender leaves | |
|  | *Nigella sativa* L. | fruits | |
| 7 | *Curcuma longa* L. | rhizome | Leach biting |
| 8 | *Tamerindus indica* L. | leaves | Higher milk yield |
|  | *Ameranthus spinosus* L. | leaves | |
| 9 | *Alangium salvifolium* (L.f.) | bark | Fever, intestinal disorders, madness |
| 10 | *Alseodaphne semecarpifolia* Nees | bark | Rinderpest & dysentery in cattle |
| 11 | *Aspwragus recemosa* Willd | root | Arthritis in cattle |
| 12 | *Capsicum frutescens* L. | fruits | Haemorrhagic septicaemia |
|  | *Musa paradisiaca* L. | stem | |
| 13 | *Careya arborea* Roxb. | leaves | Dislocated bones |
|  | *Hibiscus rosa-sinensis* L. | leaves | |
|  | *Crossandra undulaefolia* (L.) Nees | leaves | |
|  | *Randia dumentorum* (Retz.) Poir. | not specified | |
| 14 | *Cryptolepis buchanani* Roemer & Schultes | leaves | Snake bite in cattle |
| 15 | *Cucurbita maxima* Duchesne ex Lam. | paste of fruit stalk | Dengu fever in cattle |
| 16 | *Curcuma longa* L. | rhizome | FMD |
|  | *Coriandrum sativum* L. | whole plants + ground nut oil | |
| 17 | *Dendrothoe falcate* (L.f.) Ettingsh. | whole plant | Rinderpest |
|  | *Piper nigram* L. | fruits | |
|  | *Trachyspermum ammi* (L.) Sprague | fruits | |
|  | *Vernonia anthelmintica* (L.) willd. | leaves | |
| 18 | Elephanopus scaber L. | whole plants | Increase lactation in cattle |
| 19 | *Erventania heyneana* | not specified | Poisonous bite |
| 20 | *Oroxylum indicum* (L.) Kurz | bark | Dysentery, diarrhoea |
|  | *Trachyspermum ammi* (L.) Sprague | fruit | |
| 21 | *Ficus glomerata* L. | bark | Tympanites |
| 22 | *Gymnema sylvestre* R. Br. | bark | Fever in cattle |
|  | *Alstonia scholaris* (L.) R. Br. | not specified | |
|  | *Ervantamia heyneane* Wall. Ex T. Cooke | not specified | |
|  | *Anacardium occidentale* L. | not specified | |
|  | *Piper nigrum* L. | Fruit not specified | |
| 23 | *Machillus macrantha* (Nees) Kosterm | bark | Fever in cattle |
|  | *Cinnamomum wightii* Meisuer | | |
| 24 | *Musadra frontosa* L. | root | Poisonous bite |

*(continued)*

**TABLE 1 (Continued)**
**Medicinal plants and preparations thereof used for animal health conditions listed from a selection of publications. Sources: Ghosh 1999 (Pos. 1-8); Harsha et al. 2005 (Pos. 9-32); Lakshmi Narayana and Rao 2013 (Pos. 33-64); Mallik et al. 2012 (Pos. 64-81); Sharma, et al. 2014 (Pos. 82-100); FMD: Foot and Mouth Disease.**

| No | Botanical name | Parts used | Uses |
|---|---|---|---|
| 25 | *Musa paradisiaca* L. | root and young leaves | Reduce heat in cattle |
| 26 | *Nothopodytes foeitida* (Wight) Sleumer | root | Poisonous bite |
| 27 | *Oroxylum indicum* (L.) Kurz | not specified | Paralysis |
| | *Terminalia paniculata* Roth | root | |
| | *Adhatoda vasica* Nees | root | |
| | *Trachispermum ammi* (L.) Sprague | fruit | |
| | *Piper nigram* L. | fruits | |
| 28 | *Pathos scandens* L. | whole plants | Increase lactation |
| 29 | *Spondias mangifera* (L. f.) Kurz. | bark | Dysentery in cattle |
| 30 | *Syzygium carophyllatum* (L.) Alston. | bark | Tympanites |
| | *Annona squamosa* L. | bark | |
| 31 | *Vitex negundo* L. | leaf juice | Poisonous bites |
| | *Leucas aspera* (Willd.) Link. | | |
| 32 | *Zanthoxylum alatum* DC. | bark | Corneal opacity |
| 33 | *Abrus precatorius* L. | root | Wounds |
| 34 | *Alangium salvifolium* (L.f.) Wangerin | leaves | Opacity of cornea |
| 35 | *Aloe vera* (L.) Burm.f. | leaves | Septicaemias in cattle |
| 36 | *Annona squamosa* L. | leaves | Maggot wounds |
| 37 | *Azima tetracantha* Lam. | branches | FMD |
| 38 | *Butea monosperma* (Lam.) Taub. | seed | Intestinal worms |
| 39 | *Carissa spinarum* L. | root | Wound cattle |
| 40 | *Cassia fistula* L. | bark | Conjunctivitis in cattle |
| | *Ocimum tenuiflorum* L. | leaves | |
| | *Piper nigraum* L. | fruits | |
| 41 | *Chloroxylon swietenia* (Roxb.) DC. | leaves | Wounds and ulcers in cattle |
| | *Curcuma longa* L. | rhizome | |
| 42 | *Cissus trilobata* Lam. | leaves | Fibrous tissue formation at neck |
| 43 | *Cleistanthus collinus* (Roxb.) Benth. ex Hook.f | bark | Sores |
| 44 | *Cochlospermum religiosum* (L.) Alston | fibres from stem | Ticks |
| 45 | *Corallocarpus epigaeus* (Rottl.) C.B.Clark | tuber | Colic pain |
| 46 | *Cryptolepis buchanani* Roemer & Schultes | leaves | Enhance milk production, galactagogue |
| 47 | *Dodonaea viscosa* (L.) Jacq. | leaves | Fractured bone |
| 48 | *Elephantopus scaber* L. | leaves | Loose motion |
| 49 | *Grewia tiliaefolia* Vahl. | root bark | Dislocated joints |
| 50 | *Leonotis nepetaefolia* (L.) R. Br. | root | Mastitis |
| 51 | *Litsea glutinosa* (Lour.) C.B.Rob. | bark | Bone fractures in cattle |
| 52 | *Manilkara hexandra* (Roxb.)Dubard | bark | Throat diseases |
| | *Cissus quadrangularis* L. | stem | |
| 53 | *Martynia annua* L. | leaves | Sores in cattle |
| 54 | *Oroxylum indicum* (L.) Kurz. | root bark | Wounds |
| 55 | *Pergularia daemia* (Forssk.) Chiov. | leaves | Muscular pain |
| | *Calotropis procera* (Aiton) W.T. | | |
| 56 | *Plumbago zeylanica* L. | juice | Wounds |
| 57 | *Pterolobium hexapetalum* (Roth) Santapu &Wagh | leaves | Dyspepsia |
| 58 | *Ricinus communis* L. | root paste | Maggot wounds |
| 58 | *Rubia cordifolia* L. | bark | Post-natal diseases |

**TABLE 1 (Continued)**
**Medicinal plants and preparations thereof used for animal health conditions listed from a selection of publications. Sources: Ghosh 1999 (Pos. 1-8); Harsha et al. 2005 (Pos. 9-32); Lakshmi Narayana and Rao 2013 (Pos. 33-64); Mallik et al. 2012 (Pos. 64-81); Sharma, et al. 2014 (Pos. 82-100); FMD: Foot and Mouth Disease.**

| No | Botanical name | Parts used | Uses |
|----|----------------|------------|------|
| 59 | *Strychnos potatorum* L.f. | seed | Eye infections |
| 60 | *Tinospora cordifolia* (Tunb.) Miers | stem | FMD |
|    | *Shorea robusta* Gaertn. f | resin | |
| 61 | *Toddalia asiatica* (L.) Lam. | root | Galactagogue |
| 62 | *Trichosanthes tricuspidata* Lour. | tubers | Bloat |
|    | *Maerua oblongifoli* (Forssk.) A. Rich. | | |
| 64 | *Xanthium strumarium* L. | plant juice | Swelling of the glands in cattle |
| 65 | *Abelmoschus esculentus* (L.) Moench | root | Blocked urination |
| 66 | *Acacia catechu* (L.f.) Willd. | wood | Wounds |
| 67 | *Acalypha indica* L. | leaves | Scabies |
| 68 | *Achyranthes aspera* L. | root | Parturition and bronchitis |
| 69 | *Aegle marmelos* (L.) Correa | fruits | Internal fever |
| 70 | *Alangium salviifolium* (L.f.) Wangerin | root | Snake bites |
| 71 | *Albizia lebbeck* (L.) Benth. | bark | Wound of rat bite |
| 72 | *Allium cepa* L. | bulb | Cough |
| 73 | *Andrographis paniculata* (Burm. f.) Wall. ex Nees | leaves | Fever, FMD |
| 74 | *Annona squamosa* L. | leaves | Maggot wounds |
| 75 | *Atylosia scarabaeoides* (L.) Benth. | leaves | Diarrhoea |
| 76 | *Azadirachta indica* A. Juss. | leaves and fruits | Constipation, internal fever, removal of ectoparasites, killing of intestinal worms |
| 77 | *Bambusa arundifolia* Roxb. | leaves | Blood dysentery |
| 78 | *Bauhinia racemosa* Lam. | leaf | Redness of eye |
| 79 | *Bombax ceiba* L. | bark | Dislocated bones |
| 80 | *Borassus flabellifer* L. | inflorescence | Clearance of uterus |
| 81 | *Brassica campestris* L. | seed | Cough and cold |
| 82 | *Achyranthes aspera* L. | root with turmeric | Shivering in cattle |
| 83 | *Bauhinia vahlii* Wight and Arn. | leaves | Cold in buffalo |
| 84 | *Bauhinia veriegata* L. | leaves | Ensure the successful conception in cow |
| 85 | *Calotropis procera* (Ait) R.Br. | leaves | Throat problem |
| 86 | *Cassia fistula* L. | fruits | Stomach ache |
| 87 | *Colebrookia oppositifolia* Smith | leaves | Cough in cows |
| 88 | *Dendrocalamus strictus* Nees. | leaves | Dysentery |
| 89 | *Eruca sativa* Lamk | seeds | Increase lactation |
| 90 | *Euphorbia geniculata* Ort. ex Boiss | whole plants | Increase lactation in cattle |
| 91 | *Grewia oppositifolia* Roxb. | leaves | Fodder |
| 92 | *Hordeum vulgare* L. | leaves & seeds | Increase lactation |
| 93 | *Murraya koenigii* (L.) Spreng | leaves | Gastric troubles |
| 94 | *Ocimum sanctum* L. | leaves | Cold & cough |
| 95 | *Phyllanthus emblica* Linn. | Not specified | |
| 96 | *Syzygium cumini* (Linn.) Skeels | leaves | Worms |
| 97 | *Thalictrum foliolosum* DC | leaves | Conjunctivitis in cattle |
| 98 | *Trifolium alexandrinum* L | not specified | Increase lactation in cattle |
| 99 | *Woodfordia fruticosa* Kurz. Dhavi | flower | Grinding of teeth in cattle |
| 100 | *Mangifera indica* L. | bark | Stoppage of chewing the cud |

## TABLE 2

Protocol for treatment of mastitis. Method of preparation and application: "Cut *Aloe vera* leaves into small pieces, add turmeric and calcium hydroxide to it. Grind the ingredients into a fine paste. Divide this paste into 10 parts for applying 10 times. Wash the udder well with water. Remove the milk completely from the udder. Take one part of the paste and add 200 ml water to dilute it and apply it all over the udder. After one hour wash the udder with water, remove the milk completely and take one portion of the paste, dilute it and apply as described earlier. Repeat this application for 10 times a day. Continue to apply for 5 days. Subclinical, acute and chronic mastitis can be cured using this formulation. Add two pieces of *Cissus quadrangularis* with the above formulation in the case of chronic mastitis and apply for at least three weeks."

| No. | Botanical name | Common name | Parts used |
|-----|----------------|-------------|------------|
| 1 | *Aloe vera* | Aloe | leaves |
| 2 | *Curcuma longa* | Turmeric | rhizome |
| 3 | --- | Slaked lime | [powder] |
| 4 | *Cissus quadrangularis* | Veldt grape, winged treebine | stem |

## TABLE 3

Protocol for treatment of Foot and Mouth Disease (FMD). Method of preparation and application: "Soak cumin, fenugreek and pepper in water for one hour. Grind them well along with turmeric and garlic. Add 120 g of jaggery and mix them well along with one grated coconut. This represents one dose for large animals and four doses for goats. Slowly feed the animal little by little the whole preparation. Repeat this three times a day. Prepare fresh formulation every time. This treatment can be continued till the animal is recovered completely."

| No. | Botanical name | Common name | Parts used |
|-----|----------------|-------------|------------|
| 1 | *Allium sativum* | Garlic | bulbs |
| 2 | *Cocos nucifera* | Coconut | fruit |
| 3 | *Cuminum cyminum* | Cumin | seed |
| 4 | *Curcuma longa* | Turmeric | rhizome |
| 5 | *Piper nigrum* | Black pepper | seed |
| 6 | *Trigonella foenum-graecum* | Fenugreek | seed |
| 7 | --- | Jaggery* | --- |

*'Jaggery' is a concentrated product of cane juice and sometimes of date or palm sap without a separation of the molasses and crystals. It is a traditional 'non-centrifugal' cane sugar consumed in the Indian Subcontinent and Southeast Asia.

## 4. DOCUMENTATION AND RAPID ASSESSMENT OF EVP

A process of participatory documentation and a rapid assessment methodology was developed to establish safety and efficacy of these EVP (Raneesh et al. 2008; Nair and Raneesh 2011). The flowchart of the process of documentation and rapid assessment of EVP is given in Figure 2. There is an inherent relation between the classical textual knowledge (such as in Ayurveda or Mrugayurveda, which is legal knowledge in India) and the folk knowledge. This is highlighted in the classical texts of Ayurveda in a number of locations. Due to this inherent relationship and similarity in the worldview, Ayurveda or Mrugayurveda is used as a tool for studying the folk/

## TABLE 4

Protocol for treatment of foot lesion. Methods preparation and application: "Grind the ingredients and add them to the oil. Warm the oil and cool it. Clean the hoof with a dry cloth and apply the oil on the hoof, legs and wherever there is ulceration as many times as possible. Repeat this till the ulcers are healed. Oil mixed with camphor is applied for one day to kill any maggots if present in the wound and then the above oil is used for healing the ulcers (this is only if any maggot is present; not added in the above table for clarity). This oil can also be used for wounds."

| No. | Botanical name | Common name | Parts used |
|---|---|---|---|
| 1 | *Acalypha indica* | Indian Acalypha | leaves |
| 2 | *Ocimum sanctum* | Sacred basil | leaves |
| 3 | *Lawsonia inermis* | Henna | leaves |
| 4 | *Curcuma longa* | Turmeric | rhizome |
| 5 | *Azadirachta indica* | Neem | leaves |
| 6 | *Allium sativum* | Garlic | bulbs |
| 7 | *Cocos nucifera* | Coconut | oil |

Note: If there is swelling on the leg, sesame oil must be used instead of coconut oil.
Time of administration: Any time
Quantity: Sufficent
Number of times: As many times as possible
Duration of treatment: Till the condition is cured
Specification: The wound can be dressed with clean cloth if needed.

Mode of application: External

## TABLE 5

Protocol for treatment of diarrhorea. Method of preparation and administration: "Cumin, asafoetida, poppy seeds, black pepper and fenugreek are fried till they become dark like charcoal. Remove them from the fire, cool and powder them. Grind shallot, garlic and curry leaves into a paste. Add the prepared powder and jaggery. Mix them well and make into small balls. Touch the ball on salt (crystal, not the free-flowing powder) and rub it on the tongue of the cattle or calf.

| No. | Name of the ingredients | Common name | Parts used |
|---|---|---|---|
| 1 | *Allium cepa* var. *aggregatum* | Shallot | bulbs |
| 2 | *Allium sativum* | Garlic | bulbs |
| 3 | *Cuminum cyminum* | Cumin | seeds |
| 4 | *Curcuma longa* | Turmeric | rhizome |
| 5 | *Ferula assa-foetida* | Asafoetida | resin |
| 6 | *Papaver somniferum* | Poppy seeds | seeds |
| 7 | *Piper nigrum* | Black pepper | fruit |
| 8 | *Murraya koenigii* | Curry leaves | leaves |
| 9 | *Trigonella foenum-graecum* | Fenugreek | seeds |
| 10 | --- | Jaggery* | --- |

Time of administration: No specification.
Quantity: Up to 200 g.
Number of times: 3 times.
Duration: 3 days

Mode of administration: Oral

**TABLE 6**
Protocol for the treatment of udder pox. Method of preparation and application: "Soak cumin in water for an hour. Grind all the ingredients into a fine paste and mix well with butter using a spoon. Apply this paste on the affected udder every two hours till the udder pox is cured."

| No. | Botanical name | Common name | Parts used |
|---|---|---|---|
| 1 | *Ocimum basilicum* | Sweet basil | leaves |
| 2 | *Curcuma longa* | Turmeric | powder/rhizome |
| 3 | *Cuminum cyminum* | Cumin | seeds |
| 4 | *Allium sativum* | Garlic | bulbs |
| 5 | --- | Butter | --- |

**TABLE 7**
Protocol for treatment of external parasites (ticks). Method of preparation and administration: "Grind the ingredients and mix well in one liter of water. Filter it through a fine cloth. Spray this formulation on the animal or apply externally against the hair using a cloth every alternate day till there is no tick on the animal."

| No. | Botanical name | Common name | Parts used |
|---|---|---|---|
| 1 | *Acorus calamus* | Sweet flag | rhizhome |
| 2 | *Allium sativum* | Garlic | bulbs |
| 3 | *Azadirachta indica* | Neem | leaves |
| 4 | *Lantana camera* | Common lantana | leaves |
| 5 | *Ocimum sanctum* | Sweet basil | leaves |
| 6 | *Curcuma longa* | Turmeric | rhizome |
| 7 | *Zingiber officinalis* | Ginger | rhizome |
| 8 | *Cissus quadrangularis* | Veldt grape | stem |

**TABLE 8**
Protocol for the treatment of fever/ephemeral fever. Method of preparation and application: "Cumin, pepper and coriander are soaked in water for 15 minutes. Blend and mix all ingredients to form a paste. Administer orally in small portions in the morning and evening. Prepare fresh paste every time."

| No. | Botanical name | Common name | Parts used |
|---|---|---|---|
| 1 | *Allium cepa var. aggregatum* | Shallot | bulb |
| 2 | *Allium sativum* | Garlic | bulb |
| 3 | *Andrographis paniculata* | Kariyat | areal parts |
| 4 | *Azadiracta indica* | Neem | leaves |
| 5 | *Coriandrum sativum* | Coriander | fruits |
| 6 | *Cuminum cyminum* | Cumin | fruits |
| 7 | *Curcuma longa* | Turmeric | rhizhome |
| 8 | *Cinnamomum tamala* | Indian Bay Leaf | leaves |
| 9 | *Ocimum basilicum* | Sweet basil | leaves |
| 10 | *Ocimum sanctum* | Sacred basil | leaves |
| 11 | *Piper betel* | Betel leaf | leaves |
| 12 | *Piper nigrum* | Black pepper | fruits |
| 13 | --- | Jaggery | --- |

**TABLE 9**
Treatment for repeat breeding syndrome in cows. Method of preparation and administration: "In order to remove the uterine infections, feed the cow with one radish per day for 3 to 5 days after applying salt on it. Feed the cow with 4 handfuls of *Aloe Vera* once a day for 4 days. Next, feed with 4 handfuls of *Moringa oleifera* leaves once a day for 4 days. Subsequently, feed with 4 handfuls of *Cissus quadrangularis* once a day for 4 days. Lastly feed the animal a paste of with 4 handfuls of *Murrya koenigii* leaves made into a paste along with turmeric. If the cow does not conceive, this treatment can be repeated once again."

| No. | Botanical name | Common name | Parts used |
|---|---|---|---|
| 1 | *Raphanus sativus* | Raddish | radish |
| 2 | *Alove vera* | Aloe | leaves |
| 3 | *Moringa oleifera* | Drum stick tree | leaves |
| 4 | *Cissus quadrangularis* | Velt grape | stem |
| 5 | *Murraya koenigii* | Curry leaf | leaves |

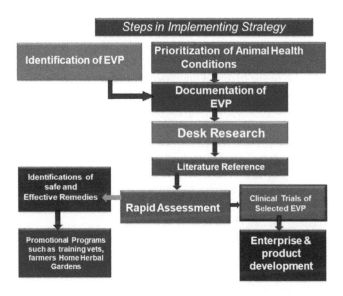

**FIGURE 2**

ethno-veterinary knowledge (Unnikrishnan and Darshan Shankar 2006) to confirm EVP safety and efficacy.

Documentation of EVP from 24 locations in 10 states were done and rapidly assessed using Ayurveda or Siddha as a base and established that 353 formulations out of 441 are safe and efficacious (Nair et al. 2017a). This assessment is an essentially participatory method of assessment, by triangulation of local evidence with Ayurveda, Siddha and Western pharmacology. It involves the community, *vaidyas* (healers) and medical practitioners from various systems of medicine, pharmacologists, botanists and the facilitators like NGOs and People's Organizations (Raneesh et al. 2008). The prioritized list of health conditions/illness and the causes and symptoms are presented to the panel of pharmacologists drawn from Ayurveda and veterinary medicine. The panel discusses with the community and folk healers, and seeks clarifications if required. The members of the panel then finalize their respective rapid assessment based on their knowledge systems using the standard forms (Appendix 1–4). After the panel has filled in their data sheets, the following steps are undertaken:

1. Treatment or practice followed in the community with all the details as given in the EVP data sheet is presented by the folk ethno-veterinary healers including their understanding of cause of health conditions and illness.
2. Short-listing of formula having consensus from the community and all the experts of the Indian Systems of Medicine and modern veterinary medicine.
3. Short-listing the number of formulae having consensus from the community, Ayurveda and Siddha experts.
4. Formulae suggested for modification or additions.
5. Formulae to be discouraged/encouraged as per consensus of the participants.

## 4.1.  Process of assessment of mastitis is given below as an example

### 4.1.1.  Etiology

Folk understanding of etiology of mastitis:

- Incomplete milking from the udder and presence of residual milk in the udder
- Trauma or insect bite on the udder
- Unhygienic condition of the byre and improper feeding methods
- Evil eye

Ayurveda understanding of etiology of mastitis (*Sthanavidhradi*):

- Consumption of stale, very dry, or fermented food, wrong food habits and wrong regimen
- Inadequate sleep
- Staying confined to the stable, without having proper exercise
- Incomplete milking
- Not properly feeding the calf (baby)
- Injury (external or internal) to the udder
- Unhygienic conditions

*Western veterinary science understanding of mastitis*

- Management: milking machine, technique, milkers
- Environment: global warming, bad ventilation, overcrowding, feed, housing, hygiene
- Host: age: breed, immunity, milk yield, milking interval, injury, lactation stage, disease
- Microorganisms: virulence, number & type of drug resistance, poor veterinary care, methicillin resistant *Staphylococcus aureus* (MRSA)

### 4.1.2.  Clinical features

*Folk understanding of clinical features of mastitis*

- First Stage: Salty taste of milk
- Second Stage: Curdled milk

- Third Stage: Hard udder, less milk yield
- Others: The outer ear (pinna) will be thickened; fever; "animal doesn't take food and looks dull"

*AYURVEDA[1] UNDERSTANDING OF CLINICAL FEATURES OF MASTITIS*
There are three stages of this disease:

- *Samavastha*: When the imbalance of *doshas* has just lodged itself in the udder – heaviness in the udder, slight pain, anorexia, etc.
- *Pachyamanavastha*: When the imbalance of *doshas* starts getting expressed – tenderness in the udder, swelling, fever (slight), reddish or blackening of the udder.
- *Pakvavastha*: When the imbalance of *doshas* is fully expressed – extreme pain and tenderness, swelling, fever, hard and warm to touch; milk is often yellow or curdled, has blood tinges at times and also suspended particles can be seen in the milk of the affected mammary gland (Kumar et al. 2018b).

*WESTERN VETERINARY SCIENCE UNDERSTANDING OF CLINICAL FEATURES OF MASTITIS*

- Invasion of microorganisms (bacterial/viral/fungal (yeast) origin) into the udder and its multi-plication *in-situ* and gradual destruction of the udder alveoli.
- Three phases in the development of mastitis have been described:
  - Invasive phase: Bacteria were able to enter the teat orifice and are present in the teat canal as well as cisterns.
  - Infection phase: The organisms are able to overcome the animal´s immune system and thus multiply.
  - Inflammatory phase: The organism has invaded the entire udder.

### 4.1.3.  Assessment of the folk herbal formulation for mastitis using Ayurveda principles

Mastitis can be compared with *sthanavidhradi* as described in Ayurveda, which is a disease of *pitta* origin. The drugs used in this formulation are potent *pitta shamaka* (pacifies *pitta* humour). This formulation consists of *Aloe vera*, *Curcuma longa* and calcium hydroxide, which have the properties of *Agni deepana* (digestive), *Amapachana* (carminative), *Krimighna* (to reduce microbial load), *Vranashodaka* (wound cleanser), *Vranaropaka* (promotes wound healing), *Shothahara* (anti-inflammatory) and *Srotoshodaka* (channel cleanser). Such formulation is considered good in combatting mastitis (Kumar et al. 2018b)

### 4.1.4.  Western pharmacology assessment of such formulation

There are several publications in western biomedicine on the antimicrobial and anti-inflammatory properties of *Aloe vera* (e.g. Sahu et al. 2013; Vázquez et al. 1996).Regarding *Curcuma longa* extracts, there is literature on the safety and anti-inflammatory activity of curcumin, one of the components of *C. longa* (Nita Chainani-Wu 2003; Betül Kocaadam and Nevin Şanlier 2017). Curcumin has been reported to be a safer and effective supplement for osteoarthritis (OA) patients in a meta-analysis. For example, it is being recommended to use *Curcuma longa* extract and curcumin supplements in

---

1    Ayurveda medicine is based on the idea that the world is made up of five elements — *aakash* (space), *jala* (water), *prithvi* (earth), *teja* (fire) and *vayu* (air). A combination of each element results in three humors, or *doshas*, known as *vata, pitta* and *kapha*. Every person is said to have a unique ratio of each *dosha*, usually with one standing out more than the others. The three *doshas* – *vata, pitta* and *kapha* – are responsible for maintaining homeostasis. Disease, on the other hand, occurs when there is an imbalance between these *doshas*.

OA patients for more than 12 weeks (Zeng et al. 2021). Calcium oxide (Na-ngam et al. 2004) has been known as an antibacterial for a long time and is currently being tested too.

### 4.1.5.  Validation of the herbal formulation for mastitis

*In vitro* antimicrobial activity of the extracts of the herbal formulation against mastitis showed inhibitory activity against *Escherichia coli* and *Streptococcus aureus* (Punniamurthy et al. 2017 a). Clinical study using a traditional formulation against mastitis showed that somatic cell counts (SCC), electrical conductivity (EC) and pH of the milk become normal within 6 days, indicating cure of mastitis (Nair et al. 2017 b). An *in-silico* approach was used to find the effect of the herbal preparation against the infection. The bioactive compounds were tested for their effect against the target proteins of *S. aureus* using molecular docking studies (Punniamurthy et al. 2017 b). It has been shown that traditional medicine can be used during dry periods to reduce the incidence of mastitis (Kumar et al. 2018a).

A microbiome study of the milk from mastitis-affected udders before and after treatment with ethno-veterinary herbal formulations indicates that there is reduction of the abundance of the 'mastitis causing microbes' to minimum after 6 days of treatment, thus indicating cure of mastitis (Santhosha et al. 2021)

Three hundred and seventy-five grams of *Dhanwandharam Kashaya* powder (available in herbal shops) was boiled with 15 litres of water on moderate fire and reduced to 5 litres. The instructions then were to: "Mix 250 ml of this decoction with feed and administer to the cows orally twice daily for 30 days." The treatment controls postpartum complications including retention of placenta (Suresh et al. 2018).

## 5.  MAINSTREAMING EVP

The steps of mainstreaming EVP are given in Figure 2. One of the major challenges for the mainstreaming of EVP is the lopsided western science-based veterinary services provided in India. The veterinary education and curriculum development are controlled by the Veterinary Council of India. The inclusion of EVP as part of the curriculum in veterinary education was proposed by the Government of India recently. The Tamil Nadu Veterinary and Animal Science University (TANUVAS) and the University of Trans-Disciplinary Health Sciences and Technology (TDU) initiated a unique post-graduate diploma course on EVP for field veterinarians. Subsequently, TDU created a facility for a certificate course and MSc. and Ph.D. programmes at the Center of Ethno-Veterinary Science and Practice at TDU. The initial intervention on training of veterinarians was supported by ETC Netherlands for training 150 veterinarians from Karnataka, Kerala and Tamil Nadu to use EVP for reducing antibiotic use and drug residue in milk. One hundred and fifty-eight veterinarians from Kerala, Karnataka and Tamil Nadu were trained to use such EVP. After the training, an intervention impact analysis indicated reduction of antibiotic residue in the milk up to 49% in one year (Nair et al. 2017b). The southern states of India had an outbreak of foot and mouth disease (FMD) during that period. Ethno-veterinary medicine could probably save thousands of cows and also be able to prevent fresh infections of FMD using EVP based herbal formulation. It is also important to note that the Government of Kerala, Department of Animal Husbandry, issued a notification to include EVP as part of the protocol to manage FMD in the year 2013–2014 .

Training veterinarians both from government and dairy unions was the strategy used to mainstream EVP. The veterinarians are responsible for the health care of the animals and are trained in veterinary sciences. It is assumed that they can influence farmers to use EVP. An intervention impact analysis related to our studies and done through a baseline and end line survey conducted among 220 farmers using a format with 1 to 10 scales and personal interview indicates that farmers are not aware of the benefits of EVP and the challenges of antibiotic residues in milk and associated AMR issues. It is also important to equip farmers with simple herbal formulations that can be easily prepared at home and use them as the first response to any disease conditions. Furthermore, these

**TABLE 10**
**Feedback from various milk societies from NDDB through INAPH and ABBOTT Private Ltd. of the efficacy of EVPs for 20 clinical conditions in cattle. NDDP/INAPH: Information Network for Animal Productivity & Health, an application that facilitates capturing of real time reliable data on breeding, nutrition and health services delivered at farmer's doorstep. It helps to assess and monitor progress of the projects; see also https://www.nddb.coop/resources/inaph; Abbott India, also see: https://abbottnutrition.com/**

| S No | Ailment | Total treated cases | Total clinical recovery | % Clinical recovery |
|------|---------|--------------------|-----------------------|--------------------|
| 1 | Fever | 113172 | 94583 | 83.6 |
| 2 | Diarrhoea | 110046 | 93658 | 85.2 |
| 3 | Acute Mastitis | 104475 | 82878 | 79.3 |
| 4 | Chronic mastitis | 52791 | 41502 | 78.6 |
| 5 | Indigestion | 27358 | 22961 | 83.9 |
| 6 | Sub-clinical Mastitis | 23986 | 19780 | 82.5 |
| 7 | Anoestrus | 17617 | 13132 | 74.5 |
| 8 | Blood in milk | 15718 | 13269 | 84.4 |
| 9 | Repeat breeder | 13262 | 9017 | 68.0 |
| 10 | Deworming | 11916 | 10690 | 89.7 |
| 11 | Udder oedema | 9567 | 7993 | 83.5 |
| 12 | Wound | 6534 | 5339 | 81.7 |
| 13 | Retention of placenta | 5744 | 4094 | 71.3 |
| 14 | Bloat | 5220 | 3959 | 75.8 |
| 15 | Ectoparasites/ticks | 4164 | 3444 | 82.7 |
| 16 | Teat obstruction | 4030 | 2714 | 67.3 |
| 17 | Endometritis | 3770 | 3056 | 81.1 |
| 18 | Agalactia | 2721 | 2048 | 75.3 |
| 19 | Downer | 2720 | 1801 | 66.2 |
| 20 | Wart | 2573 | 1802 | 70.0 |
| 21 | Lumpy Skin Disease | 2258 | 1693 | 75.0 |
| 22 | Swelling/ Joint Pains | 1913 | 1424 | 74.4 |
| 23 | Prolapse | 1543 | 1052 | 68.2 |
| 24 | Poisoning (unknown origin) | 647 | 448 | 69.2 |
| **Total EVM Treatment** | | 543745 | 442337 | 81.4 |

formulations can be used to prevent the diseases in the first place and also save substantial expenses for health management of their livestock.

One thousand seven hundred fifty veterinarians were trained from various organizations including the National Dairy Development Board (NDDB), AMUL (an Indian dairy cooperative society, based at Anand in the Indian state of Gujarat) and Karnataka Milk Federation (KMF); 30 milk unions from 14 states in India[2]; veterinarians from government of Sikkim and The Lala Lajpat Rai University of Veterinary and Animal Sciences LUVAS Haryana, Abbott India Ltd. and BAIF Development Research Foundation accorded to use EVP for 24 clinical conditions in cattle. Thirty thousand farmers and 552 Village resource persons were also trained to use EVP. See Table 10 for cure rates of each condition treated with EVP. An intervention impact study undertaken with the support from the Department of Science and Technology, Government of India encompassing 220

2   Fourteen states in India engaged in EVP training for veterinarians: Andra Pradesh, Assam, Delhi, Gujarat, Hriyana, Karnataka, Kerala, Maharashtra, Punjab, Sikkim, Tamil Nadu, Telangana, Utter Pradesh and West Bangal.

farmers (80 control group and 140 intervention group) from 11 milk societies in Karnataka, Kerala and Tamil Nadu indicated that milk samples from 123 (87.86%) out of 140 farmers showed no detectable antibiotic residue. There was a reduction in the incidence of mastitis 83.3%, enteritis 63.6%, repeat breeding 96% and cowpox 100%. These low-cost herbal formulations helped farmers save on average 75% of health expenditure for their livestock (average expenditure for one episode of clinical condition for conventional medicine was Rs. 1713 and for herbal formulation Rs. 425). The amount saved was Rs. 1288 – (1 USD equivalent to Rs.73.52 on 01/12/2020; Nair et al. 2022).

## 6. CONCLUSION

Promotion of farmer-oriented herbal EVP is an important way to encourage sustainable organic dairy farming by immediate application in-field, thus reducing dependence on antibiotics, other chemical (synthetic) veterinary drugs and hormones. This would help eliminate/reduce antibiotic and other chemical residues in animal products and reduce associated AMR. EVP will also lead to safer and more cost-effective dairy products as well as enhanced health of livestock. In the long run it can promote animal products from organic production and increase the concept of One Health (i.e. the wellness of humans, animals and the environment).

Adopting ethno-veterinary science and practices to combat infectious and other clinical conditions in livestock has been identified and tested as a key alternative in reducing the use of antibiotic and other veterinary drugs. It is also indicated that the EVP formulations are cost effective and could be prepared and used by the farmers themselves whilst being extremely helpful to prevent and manage the healthcare of their cattle.

## APPENDIX 1

### ETHNOVETERINARY PRACTICES DATA SHEET

| Local name of the prioritized condition: | | | | |
|---|---|---|---|---|
| Description (cause/stages) of the condition | | | | |
| Ingredient name | Botanical Name | Part used | Proportion used | Purification/ Remarks |
| | | | | |
| Preparation of the medicine: | | | | |
| Dosage & Administration | | | | |
| How much? | | | | |
| How many times? | | | | |
| How long? | | | | |
| Pathya/Apathya (Food & Regime advice) | | | | |
| Any other remarks (contra indications, special precaution etc) | | | | |

| Reports of Successful Community Use | | |
|---|---|---|
| Specify the number of people/farmers who reported using the remedy OR | Specify the number of villages, the remedy was reported from OR | Specify the number of Talukas/Districts the remedy was reported from |
| | | |

## APPENDIX 2

### AYURVEDIC/MRUGAYURVEDA DATA SHEET
(For individual resource references)

| Local/common name of the plant: | |
|---|---|
| 1 | Scientific name |
| 2 | *Rasa* (taste) |
| 3 | *Guna* (qualities possessed by matters which are incapable of independent actions) |
| 4 | *Veerya* (the potency of a substance immediatelyafter ingestion) |
| 5 | *Vipakam* (the post-digestion state of a substance) |
| 6 | *Prabhava* (unique biological activity of a substance) |
| 7 | *Karma* (Action) |

References from classical texts:

| Disease/condition & stage | Text | Chapter | *Sthaana* (Location) | Mode of Administration | Other ingredients | Remarks |
|---|---|---|---|---|---|---|
| | | | | | | |

## APPENDIX 3

### PHARMACOLOGY DATA SHEET

| Resource name | |
|---|---|
| Botanical/scientific name: (including family name) | |
| Part | |
| Ethno-veterinary uses | |
| Active constituents | |
| Biological activity | |

| Clinical reports | |
|---|---|
| Remarks | |
| Reference | |

## APPENDIX 4
TREATMENT PROTOCOL FOR FMD

Report Incidence of FMD cases/symptoms similar to that of FMD immediately to SADEC (0471 3256288) and District Epidemiologist so as to facilitate steps for proper sample collection and containment vaccination.

## MANAGEMENT

- All infected animals should be isolated from the rest of the herd.
- Suckling calf of ailing cows shall be weaned and given boiled milk.
- Milker should disinfect himself thoroughly before and after each milking.
- Disinfect the area with 4% Sodium Carbonate (washing soda) morning and evening.
- Do not allow grazing and bathing in common places.
- Do not buy or sell animals.
- Avoid fodder/straw from affected area.
- Prevent visitors from affected area and provide foot bath at farm entrance.
- Protect animals from the sun and give them plenty of water.
- Provide soft feed such as green soft lush grass.
- Do containment vaccination
- Also simultaneously vaccinate for *Hemorrhagic septicemia* where ever it is prone.

## TREATMENT

Since there is no specific treatment for FMD, supportive therapy shall be done to alleviate animal suffering and prevent secondary infection/complications.

Give antibiotics wherever necessary to prevent secondary infection. *Hemorrhagic septicemia, Anaplasma and Theileria* are reported as a complication from many places.

Avoid oral administration of antibiotics.

Steroids and non-steroidal anti-inflammatory drugs should be avoided. Antipyretics such as Sodium Salicylate and Paracetamol can be used.

## MOUTH AND UDDER LESIONS

Apply Melboracis on mouth lesions OR make boric acid powder in the form of a paste preferably with plantain (Robesta) and apply over lesions. Avoid excessive handling of lesions in mouth. Repeat until the wounds heal.

Boric acid can be applied to the udder lesions. Regularly milk (full hand only) the affected animals to avoid mastitis/complications.

## HOOF LESIONS

Apply non-irritant antiseptic and fly repellent dressing.

Herbal non-irritant anti-inflammatory topical application can be used.

Traditional practices are being followed with various degrees of success rates, Eg: Lugol's Iodine and Phenol combination, Ethno veterinary medicines etc.

## ETHNO-VETERINARY MEDICINES

Herbal mixture using ingredients like cumin, garlic, pepper, turmeric, fenugreek, jaggery and coconut gratings and applying it on the ulcerated and blistered gums of the infected animal, thrice a day in small quantities is useful. A mixture of Kuppaimeni (*Acalypha indica*), garlic, turmeric, pepper, neem leaves and gingelly oil can be used as foot dressing. (Field-tested and certified medicine by Ethno Veterinary Medicine Centre attached to TANUVAS).

**Reference web sites:** www.pdfmd.ernet.in/; www.tanuvas.ac.in; www.oie.int/;

## 7. ACKNOWLEDGMENTS

We are grateful to the healers for sharing their knowledge and also thank NDDB, Abbot Private Ltd. and the farmers for accepting EVP and sharing data.

## 8. REFERENCES

Asif, M.H., C. Latha, K. Vrinda Menon, Deepa Jolly, N. Suresh Nair, and C. Sunanda. 2020. "The occurrence of antibiotic residues in pooled raw cow milk samples of Palakkad, Kerala." *Journal of Veterinary and Animal Science* 51 (1): 34–39.

Betül Kocaadam and Nevin Şanlier. 2017. "Curcumin, an active component of turmeric (*Curcuma longa*), and its effects on health." *Critical Reviews in Food Science and Nutrition*, 57:13, 2889–2895, DOI: 10.1080/10408398.2015.1077195

Dinki, N. and E. Balcha. 2013. "Detection of antibiotic residues and determination of microbial quality of milk from milk collection canters." *Advances of Animal and Veterinary Sciences* 1(3): 80–83.

Gaurav, Abhishek, J. P. S. Gill, R. S. Aulakh, and J. S. Bedi. 2014. "ELISA Based Monitoring and Analysis of Tetracycline Residues in Cattle Milk in Various Districts of Punjab." *Veterinary World* 7 (1): 26–29. https://doi.org/10.14202/vetworld.2014.26-29.

Ghosh, A. 1999. "Herbal veterinary medicine from the tribal areas of Bankura district, West Bengal." *Journal of Economic and Taxonomic Botany* 23: 557–560.

Harsha, V. H., V. Shripathi, and G. R. Hegde. 2005. "Ethno-veterinary practices in Uttara Kannada district of Karnataka." *Indian Journal of Traditional Knowledge 4:* 253–258.

Jeena, S., N. Venkateswaramurthy, and R Sambathkumar. 2020. "Antibiotic Residues in Milk Products: Impacts on Human Health." *Research Journal of Pharmacology and Pharmacodynamics* 12 (1): 15. https://doi.org/10.5958/2321-5836.2020.00004.x.

Klein, Eili Y., Thomas P. Van Boeckel, Elena M. Martinez, Suraj Pant, Sumanth Gandra, Simon A. Levin, Herman Goossens and Ramanan Laxminarayan. 2018. "Global Increase and Geographic Convergence in Antibiotic Consumption between 2000 and 2015." *Proceedings of the National Academy of Sciences* 115 (15): E3463–70. https://doi.org/10.1073/pnas.1717295115.

Kumar, S. K., P. M. Deepa, N. Punnimurthy, and M.N.B. Nair. 2018a. "Prevention of mastitis in cattle during dry period using herbal formulation." *Research & Reviews Journal of Veterinary Sciences* 4 (1). https://www.rroij.com/open-access/prevention-of-mastitis-in-cattle-during-dry-period-using-herbalformulation.pdf. Accessed 14 March 2022.

Kumar Seethakempanahalli Kempanna, Balakrishna Nair Mannoor Narayanan, Natesan Punniamurthy, Girish Kumar Venkateshappa 2018b. "Ayurveda Understanding of Mastitis in Dairy Animals." *Journal of Ayurveda Medical Sciences* Apr-Jun 3(2): 349–52.

Kumaraswamy, N. P., C. Latha, Vinda, K. Menon, C. Sethulakshmi, and K. A. Mary. 2018. "Detection of antibiotic residue in raw cow milk in Thrissur, India." *The Pharma Innovation Journal* 8: 452–454.

Lakshmi Narayana, V, and G. M. N. Rao. 2013. "Traditional veterinary medicinal practices in Srikakulam district of Andhra Pradesh." *Asian Journal of Experimental Biological Sciences* 4: 476–479.

Mallik, Bikram K., Tribhuban Panda, and Rabindra N. Padhy. 2012. "Ethnoveterinary Practices of Aborigine Tribes in Odisha, India." *Asian Pacific Journal of Tropical Biomedicine* 2 (3): S1520–25. https://doi.org/ 10.1016/s2221-1691(12)60447-x.

McCorkle, Constance M., D.V. Rangnekar and Evelyn Mathias. 1999. "Introduction: Whence and whither ER & D." In: *Ethnoveterinary Medicine: Alternatives for livestock development*. Vol.1: Selected papers. Proceedings of an international conference held in Pune, India on 4-6 1997, 1-12. https://www.vetwork. org.uk/pune11.htm#introduction. Accessed 9 March 2022.

Moharana, B., P.K. Venkatesh, S.P. Preetha, and S. Selvasubramanian. 2015. "Quantification of enrofloxacin residues in milk sample using RP-HPLC." *World Journal of Pharmacy and Pharmaceuticals Sciences* 4: 1443.

Moudgil, Pallavi, Jasbir Singh Bedi, Rabinder Singh Aulakh and Jatinder Paul Singh Gill. 2019. "Antibiotic Residues and Mycotoxins in Raw Milk in Punjab (India): A Rising Concern for Food Safety." *Journal of Food Science and Technology* 56 (11): 5146–51. https://doi.org/10.1007/s13197-019-03963-8.

Mutua, Florence, Garima Sharma, Delia Grace, Samiran Bandyopadhyay, Bibek Shome, and Johanna Lindahl. 2020. "A Review of Animal Health and Drug Use Practices in India, and Their Possible Link to Antimicrobial Resistance." *Antimicrobial Resistance & Infection Control* 9 (1). https://doi.org/10.1186/ s13756-020-00760-3.

Na-ngam N, Angkititakul S, Noimay P, Thamlikitkul V. 2004. "The effect of quicklime (calcium oxide) as an inhibitor of Burkholderia pseudomallei." *Transactions of the Royal Society of Tropical Medicine and Hygiene* 98(6):337–41. doi: 10.1016/j.trstmh.2003.10.003. PMID: 15099988.

Nair, M.N.B. and P.M. Unnikrishnan. 2010. "Revitalizing Ethno-veterinary medical traditions – A perspective from India." In *Ethno-veterinary Botanical Medicine-Herbal Medicine for Animal Health*. Ed. Katerere, D. R. and D. Luseba, 95–124. CRC-Press/Taylor & Francis Group, USA.

Nair, M N B and Raneesh 2011. "Documentation and Assessment of Ethno-veterinary Practices – A strategy for mainstreaming local health traditions for sustainable veterinary care" in India in *Testing and Validation of Indigenous knowledge, the COMPAS experience from South India*. (Edts) A V Balasubramanian, K Vijayalakshmi, Shylaja R Rao and R Abarna Thooyavathy. CKS Chennai pp 60–94

Nair, M. N. B., N. Punniamurthy, and S. K. Kumar. 2017a. "Ethno-veterinary practices and the associated medicinal plants from 24 locations in 10 states of India." *Journal of Veterinary Science* 3 (2): 16–25.

Nair, M. N. B., N. Punniamurthy, P. Mekala, N. Ramakrishnan, and Kumar, S.K. 2017b. "Ethno-veterinary Formulation for Treatment of Bovine Mastitis." *Journal of Veterinary Science* S1: 25–29.

Nair, M. N. B. 2019. "Ethno-Veterinary Sciences and Practices for reducing the use of Antimicrobial and Other Veterinary Drugs in Veterinary Practices". EC *Veterinary Science* RCO.01: 16–17.

Nair M N B., Punniamurthy, N. and Kumar, S. K. 2022. "Reduction of antibiotic residue in milk through the use of cost effective ethno-veterinary practices (EVP) for cattle health." *The Pharma Innovation Journal*; SP-11 (7): 181–189.

National Dairy Research Institute. 2012. Director's desk. NDRI *News Letter of Dairy Science and Technology*. Deemed University. April –June 2012, Vol. 17: No 1 https://www.yumpu.com/en/document/view/36618 957/directors-national-dairy-research-institute

Nisha, A. 2008. "Antibiotic Residues – a Global Health Hazard." *Veterinary World* 2 (2): 375. https://doi.org/ 10.5455/vetworld.2008.375-377.

Nita Chainani-Wu 2003. "Safety and anti-inflammatory activity of curcumin: a component of tumeric (*Curcuma longa*)" *Journal of Alternative and Complementary Mededicine* 2, Feb;9(1):161–8. doi: 10.1089/ 107555303321223035.

Prajwal, S, V Vasudevan, T Sathu, and Kuleswan Pame. 2017. "Antibiotic residues in food animals: Causes and health effects." *The Pharma Innovation Journal* 6 (12): 1–04. https://www.thepharmajournal.com/archi ves/2017/vol6issue12/PartA/6-11-134-234.pdf. Accessed Dec, 15, 2020

Priyanka, Sumitra Panigrahi, Maninder Singh Sheoran, and Subha Ganguly. 2017. "Antibiotic residues in milk – a serious public health hazard. Review Article." *Journal of Environment and Life Sciences* 2 (4): 99–102.

Punniamurthy, N., N. Ramakrishnan, M.N.B. Nair, and S. Vijayaraghavan. 2017 a. "*In-vitro* Antimicrobial Activity of Ethno-veterinary Herbal Preparation for Mastitis." *Dairy and Veterinary Science Journal* 3(2), 555607. DOI: 10.19080/JDVS.2017.03.555607002

Punniamurthy, N., P. L Sujatha., S. P. Preetha, and Ramakrishnan, N. 2017 b. "Analysis of the mechanism of action by molecular docking studies of one ethno-veterinary herbal preparation used in bovine mastitis." *International Journal of Applied and Natural Sciences* 6 (5): 23–30. ISSN(P): 2319-4014; ISSN(E): 2319-4022

Ram, C., M.K. Bhavadasan, and G.V. Vijaya. 2000. "Antibiotic residues in milk." *Indian Journal of Dairy and Biosciences* 11(5): 151–154.

Raneesh, S., A. Hafeel, G. Hariramamurthi, and P. M. Unnikrishnan. 2008. "Documentation and participatory rapid assessment of ethno-veterinary practices." *Indian Journal of Traditional knowledge* 7(2): 360–364.

Review on Antimicrobial Resistance. 2014. "Antimicrobial Resistance. Tackling a crisis for the health and wealth of Nations." HM Government/Wellcome Trust. https://amr-review.org/sites/default/files/ AMR%20Review%20Paper%20-%20Tackling%20a%20crisis%20for%20the%20health%20and%20 wealth%20of%20nations_1.pdf. Accessed 9 March 2022.

Sagari, R.R. and Nitya, S.G. 2004. *Ethno-veterinary research in India: an annotated bibliography.* Mudra, 383 Narayan Peth, Pune, India.

Sahu, Pankaj K., Deen Dayal Giri, Ritu Singh, Priyanka Pandey, Sharmistha Gupta, Atul Kumar Shrivastava, Ajay Kumar, and Kapil Dev Pandey. 2013. "Therapeutic and Medicinal Uses of Aloe Vera: A Review." *Pharmacology & Pharmacy* 04 (08): 599–610. https://doi.org/10.4236/pp.2013.48086.

Santhosha Hegde, Pavithra Narendran, Malali Gowda and M N Balakrishnan Nair. 2021. "Metagenomic Profiling of Bovine Milk from Mastitis Infected Udder of the Cows before and after Treatment with Ethno-Veterinary Practice (EVP)". *EC Veterinary Science* 6.7.

Shallcross, Laura J., Simon J. Howard, Tom Fowler, and Sally C. Davies. 2015. "Tackling the Threat of Antimicrobial Resistance: From Policy to Sustainable Action." *Philosophical Transactions of the Royal Society B: Biological Sciences* 370 (1670): 20140082. https://doi.org/10.1098/rstb.2014.0082.

Sharma, A., V. K. Santvan, P. Sharma, and S. Chandel. 2014. "Ethno-veterinary Practices in Jawalamukhi, Himachal Pradesh, India." *International Research Journal of Biological Sciences* 3(10): 6–12. http:// updatepublishing.com/journal/index.php/ripb/article/view/2614. Accessed 11 March 2022

Suresh, B., N. Punniamurthy, and M.N.B. Nair. 2018. "*Dhanwantharam Kashayam* for Preventing Post-Partum Complications in Cross-Bred Cows." *Dairy and Veterinary Science Journal* 5(4): 555666. DOI: 10.19080/JDVS.2018.05.555666004

Thomas, P., Van Boeckel, Charles Brower, et.al. 2015. "Global trends in antimicrobial use in food animals." *Proceedings of the National Academy of Sciences* 112 (18): 5649–5654. https://doi.org/10.1073/ pnas.1503141112

Unnikrishnan, P. M. and Darshan Shankar. 2006. "Ethno-veterinary medicine in Ayurvedic perspective." *In Proceedings of National conference on contemporary relevance of ethno-veterinary medical traditions of India,* ed. Nair M N B, 28–34. FRLHT, Bangalore India.

Vázquez, Beatriz, Guillermo Avila, David Segura, Bruno Escalante. 1996. "Anti-inflammatory activity of extracts from *Aloe vera* gel." *Journal of Ethnopharmacology.* Volume 55, Issue 1, 69–75. https://doi.org/ 10.1016/S0378-8741(96)01476-6.

Vishnuraj, M.R., G. Kandeepan, K.H. Rao, S. Chand, and V. Kumbhar. 2016. "Occurrence, Public Health Hazards and Detection Methods of Antibiotic Residues in Foods of Animal Origin: A Comprehensive Review." Edited by Pedro González-Redondo. *Cogent Food & Agriculture* 2 (1). 123–126. https://doi. org/10.1080/23311932.2016.1235458.

Zeng L, Yu G, Hao W, Yang K, Chen H. 2021. "The efficacy and safety of *Curcuma longa* extract and curcumin supplements on osteoarthritis: a systematic review and meta-analysis." *Biosciences Reports.* Jun 25;41(6):BSR20210817. doi: 10.1042/BSR20210817. PMID: 34017975; PMCID: PMC8202067

# 11 Reverse Pharmacognosy: Traditional Knowledge Guided Assessment of Medicinal Plant Quality and Efficacy

*Padma Venkatasubramanian, Subramani Paranthaman Balasubramani and Subrahmanya Kumar Kukkupuni*

## 1. INTRODUCTION

Pharmacognostic methods are used by official Pharmacopoeias that publish monographs on the quality standards of plant drugs. These standards must be adhered to by the manufacturers to assure the quality of every batch of plant-based products manufactured by the industry. Pharmacognosy (*pharmakon*-drug and *gnosis*-knowledge) is a branch of pharmacology that deals with the composition, use and development of medicinal substances of biological origin, especially plants. The American Society of Pharmacognosy defines the scope of Pharmacognosy as 'the study of physical, chemical, biochemical, and biological properties of drugs, drug substances or potential drugs or drug substances of natural origin as well as the search for new drugs from natural materials' ("What Is Pharmacognosy?" 1998).

Physical, organoleptic, botanical, microscopical, phytochemical and microbiological methods have been typically used in Pharmacopoeias to characterise and authenticate medicinal plant raw drugs, powders and extracts. Bioactivity-related phytochemical markers are available for well-researched plant raw drugs. For example, the terpenoid saponins, ginsenosides, are used as bioactivity markers for Ginseng (*Panax ginseng* C.A. Meyer), a popular herbal tonic used in Traditional Chinese Medicine (TCM). Several *Panax* spp. are used as Ginseng, all having different levels of ginsenosides, reflecting different grades (Shahrajabian, Sun, and Cheng 2019).

There are no standards or methods stipulated by pharmacopoeias to control the batch-to-batch variability in the bioactivity of plant drugs; instead, the biological activity is assumed if the quality of the drug meets the pharmacopeial standards of identity, purity, and safety. Here, the bioactivity refers to the expected effect (beneficial and adverse) of the herb on living tissue. For example, morphological, anatomical and chemical methods and standards have been used to characterize plant raw drugs, powders or extracts in the latest edition of British Pharmacopoeia (British Pharmacopoeia Commission 2019), but there are no bioassays or standards indicated to assure the batch-to-batch bioactivity. Emerging Systems Biology approaches, Big data analytics and untargeted metabolomics provide new hope to assess the quality of medicinal plants in a holistic manner (Khoomrung et al. 2017; Fitzgerald, Heinrich, and Booker 2020). DNA and omics-based markers have been explored

DOI: 10.1201/9781003146902-14

to authenticate the quality of plant drugs during the past decade or so, but they have yet to become part of mainstream pharmacopoeial standards (Devaiah, Balasubramani, and Venkatasubramanian 2011; Efferth and Greten 2012).

Medicinal plants have been a part and parcel of ancient health traditions (HTs) across the world including Traditional Chinese Medicine (TCM), Indian Systems of Medicine (ISM), the ancient Greek medicine and other systems. The philosophies, principles and science behind HTs were formalized as codified systems over time involving trial and error, experimentation, empiricism and theorisation (i.e. they were documented in a systematic way) (Lemonnier et al. 2017). The local oral (non-codified) traditions and codified HTs continue to informally co-exist and grow in countries like China, India and other South Asian traditions, mainly since they are an integral part of food and cultural practices (Payyappallimana 2010). However, in formal healthcare structures, this vast knowledge system has either been replaced by mainstream modern biomedical systems, like in western countries, or side-lined as alternative/complementary medicine, as seen in South Asian countries (Wujastyk and Smith 2008). Globalization and economy-driven trends have led to rapid erosion of HTs and marginalization of its practitioners. This indicates the loss of vital wisdom regarding medicinal plants and usage practices.

Despite enormous development in conventional biomedicine, the popularity of HTs and medicinal plant usage for curative and health promotive purposes is on the rise, mainly because they are perceived as safe by the user, thus increasing the demand for herbal drugs. Around 80% of the global population still use plant-based medicines for healthcare (Willcox and Bodeker 2004). About 88% of WHO member states have acknowledged the use of Traditional & Complementary Medicines (WHO 2019). Around 33% of drug molecules used over the past 30 years in modern medicine have been directly or indirectly derived from natural products (Newman and Cragg 2016), such as artemisinin, an antimalarial drug derived from a plant called *Artemisia annua* L., mentioned in TCM. HTs have significantly contributed to the development of the dietary supplement and nutraceutical industry as well. Medicinal plants like *Phyllanthus emblica* L., *Punica granatum* L., *Zingiber officinale* Roscoe and *Piper longum* L. that are used in functional foods/nutraceutical products are an inseparable part of HTs (Table 1).

Medicinal plants are an important part of the *Materia Medica* of ancient HTs and have largely been local ecosystem based (Payyappallimana 2008). *Materia Medica* is the collected body of knowledge about the therapeutic properties of substances used in a medical system. The *Materia Medica* of HTs comprise of information on materials of plant/animal/mineral origin. For example, *Vidanga* (*Embelia ribes* Burm. f.), a medicinal species endemic to the Western Ghats and North-Eastern parts of India, is mentioned in *Charaka Samhita*, an Ayurveda treatise (*c.* 1500 BCE) of India (Sastry 1997). Through the ages, the HTs have also been dynamic in integrating exotic species into the *Materia Medica*. For example, *Aloe vera* L. (Burm.), a native of Africa, was added to the *Materia Medica* of Ayurveda *c.* 10th Century CE, as *Kumari* (= *that which helps girls*) (Kumar and Yadav 2014). It is important to note that codified HTs such as Ayurveda follow a systematic methodology to assess and classify drugs.

A raw drug (plant/part) is included in *Materia Medica* only after its properties and therapeutic utility are well established. They are assessed per *traditional* protocols and parameters of quality assessment and drug classification that are very different from those used in modern pharmacognosy or pharmacology. A dedicated branch of Ayurveda called *Dravyaguna Sastra* (*dravya-guna-sastra* = *drug-properties-science*) describes the taste (*rasa*), properties (*guna*), potency (*virya*), biotransformation (*vipaka*) and overall bioactivity (*karma*) of drugs to classify the organoleptic and pharmacodynamic properties of the drug (Chunekar 2004). It also provides information on the dose, dosage forms, indications and contraindications (Mishra 2007).

Efforts at revitalizing HTs have primarily focused on materials, like medicinal plants and new chemical entities, rather than on the foundations of HTs *per se* (i.e., the principles/science/practices

## TABLE 1
## Plant derived drug molecules.

| Drug molecule | Source plants, examples | Therapeutic application | References |
|---|---|---|---|
| Artemisinin | *Artemisia annua* L. (annual mugwort) | Anti-malarial | Su and Miller 2015 |
| Salicylic acid (Aspirin) | *Salix alba* L. (white willow), *Spirea spp.* (wintergreens), and *Betula spp.* (birch) | Treatment of pain, inflammation and fever. COX inhibitor | Miner and Hoffhines 2007 |
| Atropine, hyoscy-amine, scopolamine | *Atropa belladonna* L. (belladonna), *Datura metel* L., *D. stramonium* L. (jimson weed), *Hyoscyamus niger* L. (hen-bane), *Mandragora officinarum* L. (European mandrake) | Anticholinergics | Bradley and Elkes 1953 |
| Colchicine | *Colchicum autumnale* L. (autumn crocus) | Antigout, pain, anti-inflammatory | Graham and Roberts 1953 |
| Cocaine | *Erythroxylum coca* Lam. (coca leaves) | Local anaesthetic | Middleton and Kirkpatrick 1993 |
| Digoxin | *Digitalis lanata* Ehrh. (Foxglove) | Heart failure, and cardiac arrhythmias | Gheorghiade et al. 2004 |
| Diosgenin, hecogenin and stigmasterol | *Dioscorea* spp. | Oral contraceptives and other steroid drugs and hormones | Jesus et al. 2016 |
| Opiates, morphine and codeine | *Papaver somniferum* L. (Opium poppy) | Management of moderate to severe pain and cough | Akbar 2020 |
| Paclitaxel (Taxol) | *Taxus brevifolia* Nutt. (Pacific Yew) | Chemotherapeutic agent | Wall and Wani 1995 |
| Physostigmine | *Physostigma venenosum* Balf. (Calabar bean) | Cholinergic | Scheindlin 2010 |
| Pilocarpine | *Pilocarpus jaborandi* Holmes (jaborandi) and related species | Cholinergic | Sneader 2005 |
| Quinine | *Cinchona calisaya* Wedd., *C. pubescens* and *C. officinalis* | Malaria and babesiosis medication | Wilcox 2004 |
| Reserpine | *Rauvolfia serpentina* (L.) Benth ex Kurz (East Indian snakeroot) | Antihypertensive, psychotropic | López-Muñoz et al. 2004 |
| Vincristine and Vinblastine | *Catharanthus roseus* (L.) G. Don (Madagascar periwinkle plant) | Chemotherapeutic agents | Duffin 2002 |
| d-Tubocurarine (Curare alkaloid) | *Strychnos toxifera* R.H. Schomb. ex Lindl., *Chondrodendron tomentosum* Ruiz & Pavon | Skeletal muscle relaxant | Bevan 1994 |

are not adequately investigated) (Valiathan 2016). Modern Science and Technology (S&T) paradigms have been used to standardize the quality and safety of medicinal plants and to generate evidence for HT's efficacy. While there is little doubt that modern S&T-based methods and instrumentation are the need of the hour to standardise medicinal plants and for scaling up, it is also a fact that in the process of 'contemporising' HTs, the context of the knowledge system from where it originated has either been distorted or represented only partially (Mazzocchi 2006). It is important to keep the context of knowledge and wisdom intact, to obtain the best benefits from the materials. For example, the corrosive seeds of *Semecarpus anacardium* L. f. (*Bhallataka,* marking nut) are converted to a health-promoting tonic (*rasayana*) in Ayurveda through traditional processing methods (Rangasamy et al. 2012). Without the traditional wisdom, the marking nut will either be left aside because of its corrosive nature or else injure people if used improperly. Wisdom on how not to use a plant drug is equally important when it comes to safety. Therefore, traditional knowledge (TK)-guided standards of quality, safety and efficacy would be relevant and beneficial (Shankar, Payyappallimana, and Venkatasubramanian 2007).

This chapter introduces *Reverse Pharmacognosy* as a methodology that builds on TK to develop scientific standards that are pertinent to batch-to-batch quality and efficacy of plant drugs. It provides examples of research done in this area from the Indian Systems of Medicine.

## 2. HTs AND MEDICINAL PLANTS IN INDIA

The WHO defines Traditional Medicine as "the sum total of the knowledge, skills and practices based on the theories, beliefs and experiences indigenous to different cultures, whether explicable or not, used in the maintenance of health, as well as in the prevention, diagnosis, improvement or treatment of physical and mental illnesses". The terms complementary/alternative/non-conventional medicine are used interchangeably with traditional medicine (WHO 2000).

---

"Scientific methodology without philosophy is like seeing the trees but missing the forest" – Albert Einstein

---

Indigenous worldviews recognize and emphasize the human-nature and human-human interconnectedness and interdependence. Medicinal plant usage in HTs is inextricably dependent on the ecosystem and the cultural context. HTs are mostly passed on from generation to generation orally. However, many countries also have codified systems of medicine that have evolved systematic principles and practices that are documented in treatises. For example, Ayurveda, Siddha, Unani and Sowa-rigpa are codified HTs of India. Ayurveda has ancient palm leaf medical manuscripts that date back to 1500 BCE. The methods used to analyse medicinal plants in HTs are holistic and qualitative, and seem more subjective as opposed to the scientific/biomedical methods used in modern medicine that are by design more objective whilst reductionist (Mazzocchi 2006). *Cognitive mining* of the knowledge for validation using modern scientific methods alone, without considering the wisdom behind it, can distort applications and/or threaten the traditional knowledge base (Nakashima and Roué 2002, 5:1–11). Any scientific methodology, however precise, if following a single epistemological dimension without integration of the philosophy behind HTs or without being open to include methodologies from other knowledge systems, becomes restrictive in its own growth.

---

*"There is no plant on this universe without medicinal property"*
*- Charaka Samhita, an ancient Ayurvedic treatise (c. 1500 BCE)*

---

There are > 300,000 known vascular plants on this earth while several thousands are still being discovered every year ("Catalogue of Life" n.d.). About one-tenth of them have an established use as medicinal plants known to modern science (Salmerón-Manzano, Garrido-Cardenas, and Manzano-Agugliaro 2020). HTs like Ayurveda, Traditional Chinese Medicine (TCM), Tibetan medicine, Siddha and Unani use thousands of plants as medicines, and the local communities have knowledge of a greater number. More than 2,400 species of medicinal plants are used by the codified systems of medicine, whereas non-codified oral traditions use >5000 species of medicinal plants in the Indian subcontinent. About 85% of these are sourced from the wild (Ved and Goraya 2008) (Fig. 1).

---

*"Go to the forest dwellers and shepherds if you wish to know about a medicinal plant"*
*- Charaka Samhita, an ancient Ayurvedic treatise (c. 1500 BCE) (Sastry 1997)*

---

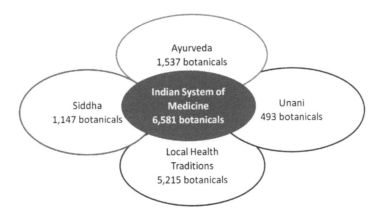

**FIGURE 1**    Medicinal plants used in Indian HTs (adapted from Ved and Goraya 2008; Ved et al. 2016). Several plant species are common across the HTs.

## 3.   MEDICINAL PLANT DEMAND

During the year 2000, the international trade of medicinal plants and their products was US$60 billion, which has been growing at an average annual rate of 7% and experts predict it to be around US$5 trillion by the year 2050 (WHO 2009). However, these figures are likely to be an underestimate due to the Covid-19 pandemic.

The whole world was witness to the best healthcare systems being overwhelmed by the Covid-19 pandemic. Lack of evidence-based biomedical interventions against the SARS-Cov-2 virus brought into focus the importance of immunity and self-care. Millions of people started using alternative healing systems for 'boosting' immunity and improving wellness. The price of several kitchen herbs including turmeric, ginger and cumin sky-rocketed in 2020 since they were being used as home remedies (Decker 2020). The sale of TM products (including Ayurveda, TCM and Siddha) also increased during 2020 as people were looking for any solution to protect them against the disease. The TM market in India has been predicted to grow by 16% from USD 4.5 Bn in 2017 to USD 15 Bn by 2026 (Lakshmanan and Aggarwal, 2020). With the rise in demand come the issues of supply of good quality plant material. The herbal sector is rampant with plant drug substitution and adulteration (Goraya and Ved 2017). Assuring good quality of plant materials is fundamental to better health of consumers and thriving business.

## 4.   TRADITIONAL QUALITY STANDARDS (TQS)

We define *Traditional Quality Standards* (TQS) as the parameters and standards used by HTs to assess the quality of medicinal plants as well as the recommended protocols of collection and use of plant drugs. Many TQS found in classical texts may appear cryptic and obscure to scientific scrutiny but nevertheless hold important details about the plant. For example, Turmeric (*Curcuma longa* L.) is commonly called *Haridra* in Sanskrit, but is also referred to as *Nisha* or *Rajani*, meaning *night*, suggesting the preferred time of harvest/usage to be night, for better therapeutic actions (Shankar, Payyappallimana, and Venkatasubramanian 2007). However, some TQS are clear and elaborate. A striking set of generic recommendations on the best seasons of plant collection have been mentioned in the ancient Ayurvedic text, *Charaka Samhita* (Sastry 1997). A few examples are provided in Table 2.

**TABLE 2**
**Examples of Traditional Quality Standards of Medicinal Plants.**

| TQS Category | Sanskrit Name | Botanical Name | Part used | TQS details | References |
|---|---|---|---|---|---|
| **Generic TQS Guidelines for seasonal collection of plants/ plant-parts** | NA | NA | Stem bark | Collect during late autumn | Sastry 1997 |
| | NA | NA | Stem | Collect during spring | |
| | NA | NA | Roots & tubers | Harvest during early summer or late winter | |
| | NA | NA | Heartwood | Collect during winter or spring | |
| | NA | NA | Oleo-gum-resin | Collect during winter – spring | |
| | NA | NA | Whole plant (usually herbs) | Collect during late autumn | |
| | NA | NA | Flowers & Fruits | Collect during natural flowering/fruiting seasons | |
| **Identity** | *Vidanga* | *Embelia ribes* Burm.f. | Seeds | Seeds with a pattern/picture (*Chitratandula*) | Chunekar 2004 |
| | *Guduci* | *Tinospora cordifolia* (Thunb.) Miers | Mature stem | Wheel shaped appearance of cut stem (*Chakrangi* appreciable in thick, mature stem) Moon shaped seeds (*Chandrahaasini*) | |
| | *Daruharidra* | *Berberis aristata* DC | Mature stem | Dark yellow coloured woody stem (*Daruharidra*) | |
| | *Vidari* | *Pueraria tuberosa* (Roxb. ex Willd.) DC. | Mature tuber | Underground tubers resemble ash gourd (*Bhookoosmandi*) | |
| | *Kshiravidari* | *Ipomoea mauritiana* Jacq. | Mature tuber | Plant with lengthy aerial part and with abundant white exudation (*Dheerghakanda & Bahukshira*) | |
| **Substitutes** | *Ativisha* | *Aconitum heterophyllum* Wall. ex Royle | Roots | Can be substituted with *Musta* Tubers of *Cyperus rotundus* L. | Mishra 2007 |
| | *Daruharidra* | *Berberis aristata* DC. | Stem | Can be substituted with *Haridra khanda* Rhizomes of turmeric | |
| | *Yashtimadhu* | *Glycyrrhiza glabra* L. | Roots | Can be substituted with *Dhataki pushpa* Flowers of *Woodfordia fruticosa* (L.) Kurz. | |
| **Processing** | *Pippali* | *Piper longum* L. | Dried fruit | Dried fruits of *Pippali* is processed as milk decoction | Sastry 1997 |

**TABLE 2 (Continued)**
**Examples of Traditional Quality Standards of Medicinal Plants.**

| TQS Category | Sanskrit Name | Botanical Name | Part used | TQS details | References |
|---|---|---|---|---|---|
| | Haridra | Curcuma longa L. | Fresh/dry Rhizome | Harvest, process and consume during night | Shankar, Payyappallimana, and Venkata-subramanian 2007 |
| | Gunja | Abrus precatorius L. | Dry seeds | Always remove husk, crush, boil in milk, dry and consume to reduce toxicity | Sharma 2006 |
| **Storage** | Madanaphala | Randia dumetorum (Retz.) Lam. | Seeds | Wash and wrap with dry Kusa (Desmostachya bipinnata [L.] Stapf) grass and cover with fresh cow dung; store for 8 days in the heap of barley husk/black gram/rice/horse gram/green gram | Sastry 1997 |
| | Vidanga | Embelia ribes Burm.f. | Seeds | Store Vidanga for a year | |
| | Shali | Oryza sativa L. | Seeds | Store rice for at least 6 months | |

NA- Not applicable

Codified HTs like Ayurveda have sophisticated methods to examine the quality, safety and efficacy of materials. Such methods and practices (*vyavahara*) are not modern technology based but based on human pharmacodynamics, backed by fundamental principles (*tatva*) and theories (*shastra*). The science of Ayurveda is built on the fundamental principles of *panchamahabhuta*; i.e., five basic elements namely earth (*pṛthvi*), water (*ap/jala*), fire (*teja*), air *(vayu)* and space (*akasha*). Per the Ayurvedic principle, the pharmacological action of materials is determined by the predominance and combination of the five elements to form the *tridosha*[1], the three humours (*vata, pitta* and *kapha*). The *tridosha* theory of Ayurveda is fundamental to the practical application of interventions for healing since individuals, diseases/health conditions and healing materials are all classified according to the three *doshas*. For example, black gram (*Vigna mungo* (L.) Hepper) is a *kapha* predominant material that is normally considered a good and nutritious pulse for all individuals during the winter season. However, the same would aggravate *kapha dosha* in individuals with compromised digestive capacity (*~agni*), and cause lethargy and drowsiness. Excessive intake

---

1    *Tridoshas* are three bodily humours, namely *vata, pitta* and *kapha*. *Doshas* have specific quality and actions on physiology. When the doshas are in equilibrium, they bring about health and wellbeing while imbalance leads to disease. *Vata dosha* is predominantly made up of basic elements air and space and is responsible for initiation of biological processes, movements in the body, respiration, manifestation of natural urges, normal functioning of sense organs, mental functions and maintenance of the tissues. *Pitta dosha* is made up of fire and water components and is responsible for digestion and metabolism, maintenance of body temperature, vision, hunger and thirst, complexion, intelligence and courage in an individual. *Kapha dosha* is made up of earth and water and gives stability, brings about compactness and provides lubrication in tissues and firmness of the joints. It provides resilience and makes the body withstand physical, mental and pathological challenges (Sastry 1997).

**FIGURE 2**   Fresh (left) and dried rhizome (right) of *Curcuma longa* L. (*Haridra*).

**TABLE 3**
*Rasapanchaka* **of turmeric – rich Traditional Quality Standards (TQS) in Ayurvedic**
*Materia Medica.*

| | |
|---|---|
| **Ayurvedic/Sanskrit Name** | *Haridra* |
| **Botanical identity** | Rhizomes of *Curcuma longa* L. (Fam. Zingiberaceae) |
| **Important Synonyms in Sanskrit and their meaning** | *Haridra* (yellow coloured); *Rajani, Nisha, Nishi* (night); *Dirgharanga* (stable colour); *Krimighna* (antimicrobial); *Mangala* (auspicious) |
| *Rasapanchaka* (functional profile – as per Ayurveda) | |
| *Rasa* (taste) | *Tikta* (bitter) and *katu* (pungent) |
| *Guna* (properties) | *Ruksha* (causing dryness) |
| *Virya* (potency) | *Ushna* (hot) |
| *Vipaka* (post digestive effect) | *Katu* (pungent) |
| *Dosha-karma* (pharmacological actions – on *doshas*) | *Kapha-pitta hara* (alleviates *kapha* and *pitta doshas*) |
| *Roga-karma* (uses in specific diseases) | *Krimighna* (antiseptic), *vranaropana* (wound healing) *kushtaghna* (useful to treat skin diseases), *vishaghna* (treats poisonous conditions), *twachya & varnya* (helps to bring skin health & complexion), *prameha nashaka* (useful to treat diabetes) |
| *Yoga* (selected poly-herbal formulations prepared with turmeric as a major ingredient) | *Haridrakhanda, Nishamalaki, Rajanyadi Churna* |
| TQS* | *Collection:* at night (*Nisha, Rajani*) Processing: milk decoction (*ksheerapaka*) |

\* TQS recommendations about a plant are not provided as standard operating procedures but need to be gathered/documented from texts and traditional practices.

can lead to the diseases of *kapha*, like obesity and lipidemia (*medoroga*) (Samagandi, Jagriti, and Kumar 2013).

Per Ayurveda, materials are assessed using five parameters called *Rasapanchaka* (Ayurvedic principles of drug-action), which include 6 tastes (*rasa*), 20 properties (*guṇa*), 2 potencies (*virya*), 3 postdigestive effects (*vipaka*) and unanticipated potency (*prabhava*). These five factors contribute to the pharmacological actions (*karma*) (Dhyani 1994; Chunekar 2004). *Karma* or the bioactivity of a medicinal herb is classified as *doshakarma* (at the level of *doshas, i.e., vata, pitta* and *kapha*) and *dhatukarma* (at the level of *dhatu*, the tissue system) and act on specific diseases (*roga- karma*). *Materia Medica* of Ayurvedic medicinal plants lists the qualities under each of the five parameters (*Rasapanchaka*) and the overall *dosha* and *dhatu karma* (Sharma 2006). Illustration of *Rasapanchaka* of a well-known plant drug *Curcuma longa* L. (*Haridra*, Fig. 2) as mentioned in Ayurvedic literature is given in Table 3. *Rasapanchaka in toto* provide insight into the therapeutic potential of the medicinal plant drug.

In Ayurveda, *Rasa* of plant drugs is assessed through sensory analysis, mainly through taste and smell and the immediate effects, while the other *Rasapanchaka* parameters like *vipaka* and *virya* are analysed based on the pharmacodynamic (physiological) effect of the materials.

Discernible organoleptic/sensorial attributes like colour (*rupa*), taste (*rasa*), smell (*gandha*), tactile sensation (*sparsha*), sound (*shabda*), etc., are also used in Ayurveda as quality assessment parameters, based on which the therapeutic activities (*karma*) of the plant are also interpreted (Table 4).

Protocols on ideal ways of collection or cultivation and harvest of medicinal plants, processing and storage are an essential part of TQS. However, these recommendations do not figure as Standard Operating Procedures (SOPs) in a single section in classical texts but are found scattered. Some of these practices are alive among traditional health practitioners even today, especially the elderly. Documenting this valuable knowledge on medicinal plants, creating a fully referenced database on TQS and managing it are critical for the future of medicinal plant applications, whose uses stem from TK.

## TABLE 4
### Five sensorial parameters used in TQS.

| TQS Parameter | Example | Interpretations per Ayurvedic principles |
|---|---|---|
| Colour (*rupa*) | Rhizomes of turmeric (*Curcuma longa* L.) | Dark yellow in colour. Signifies the involvement of fire element (*agni mahabhuta*) and hot potency (*ushna veerya*) Hot potency helps it to alleviate aggravated *kapha dosha* |
| Taste (*rasa*) | Fruit of pomegranate (*Punica granatum* L.) | Sweet and astringent taste. Implies earth (*prithvi*), water (*jala*) and air (*vayu*) predominance The presence of contrasting qualities of sweet and astringency makes it a fruit suitable to all for daily consumption, without affecting any *dosha* |
| Smell (*gandha*) | Flowers of jasmine (*Jasminum officinale* L.) | Mild and refreshing fragrance of jasmine flowers indicates its calming nature, due to the involvement of earth (*prithvi*) and water (*jala*) elements |
| Tactile sensation (*sparsha*) | Leaves of coral jasmine (*Nyctanthes arbor-tristis* L.) | Leaves of coral jasmine (*Parijata*) are rough (*khara*) to touch indicating the dominance of air (*vayu*) and space (*akasha*), therefore prompting the use in management of *kapha* disorders (with water and earth element predominance) |
| Sound (*shabda*) | Long pods of Indian trumpet (*Oroxylum indicum* [L.] Kurz.) | Fruit pods hang and dance with a sound (*tuntuka*). It is a TQS indicator of maturity of the fruits. Sound is related with air (*vayu*). It is interesting to note that, *O. indicum* is one of the drugs of choice to treat *Vata dosha*. |

## 5.  REVERSE PHARMACOGNOSY

At best the current pharmacopoeial and modern plant quality standards address species authenticity, purity and potency and not necessarily the batch-to-batch efficacy. This is where integrating TQS about the plants would help. We define 'Reverse Pharmacognosy' as a novel TK-guided strategy to standardize medicinal plant quality to assure batch-to-batch consistency in bioactivity. Reverse Pharamcognosy of a plant drug of HT origin is an integrated approach that uses modern scientific quality standardization techniques, on medicinal plants that are identified, collected, processed and stored per traditional recommendations (Venkat, 2009).

The following guidelines can be adopted in the Reverse Pharmacognosy approach (Fig 3):

### 5.1  DOCUMENTATION OF TQS

TK about medicinal identity (species, part), collection (best time/season/region) and best processing (single plant/formulation/methods), storage practices and usage lie scattered in classical texts of HTs and as living traditions among a few traditional practitioners and healers. Documenting and creating a database of this TQS is key for *Reverse Pharmacognosy*. Documentation of TK is not a plug-and-play process but implies building an environment of mutual trust and rapport between the TK holders and scientists. It is crucial to obtain the consent of the stakeholders and agree to acknowledge and credit the TK source appropriately (see Figure 3 representing the proposed hierarchy of the TQS).

> *"You will find in the lower animals' mechanisms and adaptations of exquisite beauty and the most surprising character..." (Krogh, 1929)*

*Prior Informed Consent*: While information from classical, published HT texts may be archived with citations in databases, documenting unwritten oral traditions from healers, elderly practitioners and communities requires a sensitive, due diligent and ethical process of obtaining prior informed consent from the knowledge providers. International and national laws on accessing materials

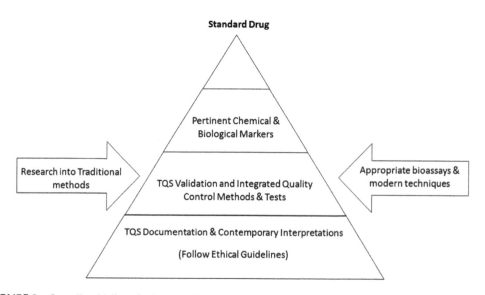

**FIGURE 3**   Overall guidelines for Reverse Pharmacognosy.

and knowledge would be applicable (Abbott 2014). The purpose of documentation of TQS is normally to enhance quality standards of medicinal plants and products and research. However, the documenter(s) need to be aware that the new knowledge obtained could lead to new products of commercial potential. The intellectual property of the source of knowledge should therefore be properly acknowledged and benefits agreed upon beforehand.

The Nagoya protocol for Access and Benefit Sharing with local communities became a part of the Convention of Biodiversity in 2010, in recognition that the materials and TK were the sovereign properties of the land and the peoples. Abbott (2014) provides a comprehensive overview of the benefits and risks involved in documenting TK and a checklist to be considered. The World Intellectual Property Organization (WIPO) has brought out a generic toolkit that helps one to be sensitive to the importance of TK and steps to be taken while documenting the same (WIPO 2017). Creating electronic databases of TQS is yet another process, requiring a clear policy on the users, access, intellectual property, benefits and exploitation. TQS databases are one way of documenting and acknowledging prior existing knowledge and thus protecting the intellectual rights. However, they can also be misused.

So far there have been no concerted and comprehensive efforts to document TQS globally or in India, other than limited information in the herbal pharmacopoeias and formularies and a few research publications that have addressed TQS (NMPB, n.d.).

## 5.2 TRANSDISCIPLINARY RESEARCH

Several modern scientific instrumental techniques are available to characterize the botanical, physical, chemical and biological properties of a plant drug. The organoleptic characteristics, like appearance, smell, taste, etc., are checked by experienced herb collectors and floor mangers in industries using sensory perception. It is imperative that the natural and cultural contexts of TQS be understood as well and interpreted before selecting the tool(s) and methods of quality assessment (Fig. 4). Improper interpretations and non-traditional extrapolations of HTs have led to several adverse events. For example, *Ephedra sinica* L. (*Ma Huang*) is a well-known plant drug used in TCM that contains a highly potent alkaloid called ephedrine. It is traditionally prescribed by qualified practitioners as medicine for treating cold, cough, wheezing and fever. However, in the early 2000s in the United States the herb and the potent ephedrine were promoted as dietary supplements for weight loss as well as for increasing energy and athletic performance (United States Pharmacopeial Convention 2015). These new and non-traditional claims of use were supported by limited scientific studies. In 2004 the plant drug and the alkaloid had to be banned in the United States because of adverse effects of seizure, heart attack and even sudden death ("Ephedra" n.d.). This kind of misuse of traditional knowledge not only brings disrepute to HTs but can also be dangerous and unsafe to

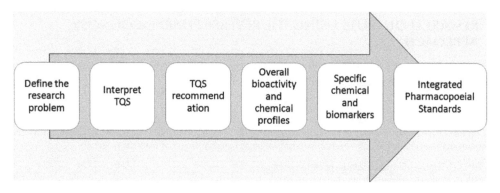

**FIGURE 4**  TQS Research Strategy.

humans. TQS has evolved over centuries of practice and is time-tested. *Ma Huang* is not TCM's chosen plant for energy drinks; other plants like Ginseng are recommended instead. Ignoring such recommendations can be fatal.

The authors' research group adopted the following strategy (Fig. 4 and sections 5.2.1–5.2.4) to research TQS.

### 5.2.1   Contemporary Interpretation

Define the research problem and specify the TQS about a plant drug that needs to be corroborated. Accurately interpret TQS through a healthy dialogue with practitioners and knowledge holders.

### 5.2.2   Apply TQS recommendation

Collect/process/store plants per TQS recommendations and also randomly. Test the overall biological and chemical differences in quality: To quickly learn the validity of the TQS holistic bioassays, holistic bioassays on small organisms can be performed. Chromatography[2] (HPTLC/HPLC/LC-MS/GC-MS) is a good technique to compare samples collected/processed per TQS and otherwise. Any established whole organism that is sensitive to the small changes in the plant drug can be used. Since testing on humans or bigger animal models have ethical implications, invertebrates and small organisms that are simple and amenable to rapid throughput in a basic laboratory can be used. Common yeast (*Saccharomyces cerevisiae*), brine shrimp (*Artemia salina*), fruit fly (*Drosophila melanogaster*), roundworm (*Caenorhabditis elegans*), hydra (*Hydra attenuata*), planaria (*Dugesia japonica*), zebra fish (*Danio rario*) and African clawed frog (*Xenopus laevis*) are well-accepted and commonly used surrogate organisms used in R&D labs globally for several applications ranging from environmental pollution monitoring to teratology and drug discovery (Dietrich et al. 2020; Giacomotto and Ségalat 2010). These organisms are sensitive and react to even small changes in the medium/environment in which they are grown. Thus study the veracity of the TQS at the gross level.

### 5.2.3   Develop specific bio- and chemical markers

A plant drug is usually used for multiple purposes in HTs. However, there may be 1–2 predominant bioactivities that are well known. Select an appropriate pharmacological model to corroborate the TQS recommendation. Similarly, identify specific phytochemical markers that are linked to bioactivity. Metabolomics and hyphenated technologies (e.g., LC-MS) are useful tools for the analysis of thousands of molecules in plants in one go (Salem et al. 2020).

### 5.2.4   Integrate with Pharmacopoeial Standards

Identify specific morphological, organoleptic, chemical and biological properties of collected/processed plant material per TQS.

## 6.   RESEARCH OUTPUTS USING THE REVERSE PHARMACOGNOSY APPROACH

In this section we share a few research studies that have used the Reverse Pharmacognosy approach to strengthen quality standards of medicinal plants of TMs.

---

2   Chromatography techniques are used to separate, identify and quantify chemical components present in a mixture. The commonly used chromatography tools in standardisation and quality control of plant drugs are:

  HPTLC: High-performance thin-layer chromatography
  HPLC: High-performance liquid chromatography
  LC-MS: Liquid chromatography–mass spectrometry
  GC-MS: Gas chromatography–mass spectrometry

## 6.1    *Authentic Botanical Identity*

Correlation of local names with correct botanical entities is in itself a challenge (Payyappallimana 2008). One can obtain clues from the several synonyms used to describe the identity of a plant drug in HTs. For example, there is an issue of the authentic botanical identity of *Vidanga,* a top-traded plant drug in India. *Embelia ribes* Burm. f., *Embelia tsjeriam-cottam* (Roem. & Schult.) A. DC., *Maesa indica* (Roxb.) A. DC. and *Myrsine africana* L. are all traded as *Vidanga* in the raw drug markets. Which is the authentic species and which an adulterant/substitute? This issue was resolved by digging deep into the classical HT texts and looking for references that describe the characteristics of *Vidanga.* Synonyms such as *chitratandula* (seeds with pattern) and *kapaali* (seeds resembling bowl) match for both *E. ribes* and *E. tsjeriam-cottam* (Fig 5) and not for the other two species. Molecular markers were developed to distinguish the species (Devaiah and Venkatasubramanian, 2008). Venkatasubramanian et al. (2013) demonstrated that both *E. ribes* as well as *E. tsjeriam-cottam* contain embelin, the phytochemical marker, and possess good anthelmintic properties (Venkatasubramanian et al. 2013). Similarly, *Coscinium fenestratum* (Gaertn.) Colebr. and *Berberis aristata* DC (Figure 5) are used as *Daruharidra* by the Ayurvedic industry in south and north India, respectively, while the Ayurvedic Pharmacopoeia of India only recognises the latter (The Ayurvedic Pharmacopoeia of India, 1999). Can *C. fenestratum* be used as *Daruharidra*? Research indicates that *C. fenestratum* is equally active pharmacologically. The aqueous and methanolic extracts of *B. aristata* and *C. fenestratum* demonstrated better anti-inflammatory activity on carrageenan-induced paw oedema model in Wistar rats, when compared

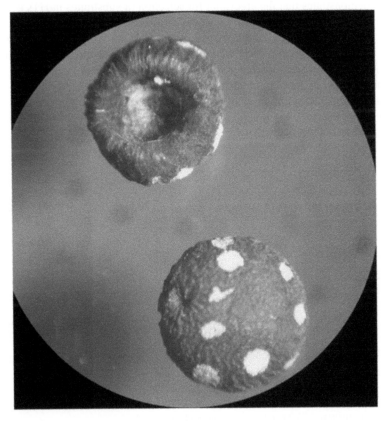

**FIGURE 5**    Seeds of *Vidanga* resembling a bowl with pattern: the features correlate with the etymology of two Sanskrit names of this plant, *chitratandula* (seeds with pattern) and *kapali* (resembling a bowl).

to indomethacin, a standard allopathic drug (Tamilselvi, Venkatasubramanian, and Vasanthi 2014). However, it is worth mentioning that the occurrence of *C. fenestratum,* a species of the Western Ghats of India, is rapidly declining in the wild and is mentioned as 'critically endangered' in the Convention on International Trade in Endangered Species (CITES) list of wild flora and fauna (Dutta and Jain, 2000).

## 6.2 SUBSTITUTION

Several medicinal plant drugs mentioned in HTs are endangered and others in short supply today, requiring legitimate substitutes. Ayurvedic texts recommend the use of substitutes (*Abhava pratinidhi dravya*) with similar pharmacological properties when a plant drug is unavailable. Also, understanding the Ayurvedic rationale behind selection of substitutes can lead to the discovery of new plant drugs, new functions for existing drugs and new ways of drug discovery (Venkatasubramanian, Kumar, and Nair 2010). Systematic studies on substitutes during recent decades have helped validate the use of a few of these as indicated in the Ayurvedic literature and those in traditional/industrial practice. The rationale behind the selection of substitutes by HTs is not the phytochemical match between the rare plant and the substitute, but the principles of *Rasapanchaka* (Ayurvedic pharmacology) and the pharmacological similarity as shown in Table 5 comparing *Ativisha* with its Ayurveda substitute *Musta* (cited from Kukkupuni 2015). Tubers of *Ativiṣha*, a rare and endangered plant drug *Aconitum heterophyllum* Wall. ex. Royle, and the Ayurvedic substitute *Musta* (*Cyperus rotundus* L. tubers) are similar in terms of the *panchamahabhuta, rasa* and *guna*. Only the *virya* of the drugs were different. However, it is important to note that the *doshakarma* and *dhatukarma* were identical. This demonstrates that the algorithm for selection of substitutes in HTs is internally consistent with its own scientific principles. They also show significant similarities in terms of HPLC profile and main pharmacological activities. *A. heterophyllum* (400 mg/kg body weight) and its substitute *C. rotundus* (800 mg/kg body weight) significantly controlled pyrexia induced by yeast in rats in about two hours after administration compared to the blank control ($p<0.01$). Standard control for the study was paracetamol 150 mg/kg body weight. Higher dose of the substitute was necessary to bring about a similar level of pharmacological activity to that of the original drug. It is worth mentioning that in living practice, the dose of the substitute is doubled in instances when *Ativisha* is unavailable (Fig 6 a & b).

---

**TABLE 5**
Ayurvedic drug characteristics of *Ativisha (Aconitum heterophyllum* Wall. ex Royle) and *Musta (Cyperus rotundus* L.).

| Properties | Similarities | Differences | |
|---|---|---|---|
| | | Ativisha | Musta |
| *Panchamahabhuta* predominance | *Agni* (fire), *vayu* (air), *akasha* (space) | | |
| *Rasa* (taste) | *Katu* (pungent), *tikta* (bitter) | | *Kashaya* (astringent) |
| *Guṇa* (properties) | *Laghu* (light), *ruksha* (dry) | | |
| *Virya* (potency) | | *Ushna* (hot) | *Shita* (cold) |
| *Vipaka* (taste after digestion) | *Katu* (pungent) | | |
| *Dosakarma* (action on doshas) | *Kapha-pittahara* (reduces *kapha* and *pitta* doshas) | | |
| *Dhatukarma* (action on tissues) | *Dipana-pachana* (enhances digestive-fire), *grahi* (prevents water loss), *lekhaniya* (scrapes off excess tissues) | | |

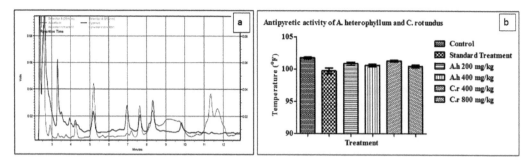

**FIGURE 6**   (a) Comparative HPLC profile of water extracts of *Aconitum heterophyllum* and *Cyperus rotundus;* (b) Effect of crude powders of *A. heterophyllum* and *C. rotundus* on pyrexia in rats induced with fever.

Limited natural distribution and yet a high demand for medicinal plant species have pushed several of them into the endangered species list, demanding immediate conservation strategies. Cultivation of medicinal plant species has been promoted by various government and non-government agencies, but only for a limited number of medicinal plant species used by the codified systems of medicine in India ("Prioritised List of Medicinal Plants for Cultivation" 2016). Agro-technology for the cultivation of several rare species like *Aconitum heterophyllum* is still in its infancy. Substitute species acceptable per TM logical framework and demonstrated through modern scientific research may offer a solution.

## 6.3   BEST COLLECTION PRACTICES

The best time to collect the rhizomes of turmeric (*Curcuma longa* L.) 'at night' is a recommendation found in Ayurveda texts and reflected in local practices. Research indicates that the total curcuminoids as well as the bioactivity of turmeric rhizomes collected at night were significantly higher than those collected during the day (as demonstrated by the $LD_{50}$ value on brine shrimps; Shankar, Payyappallimana, and Venkatasubramanian 2007).

Also, for example, Ayurveda recommends that the stems of *Tinospora cordifolia* (Willd.) Miers. need to be harvested when they attain the thickness of a 'thumb' (*angushtha pramana*). One kilogram of each thin (1.0–1.5 cm diameter), medium (1.6–2.0 cm diameter) and thick (2.1–2.5 cm diameter) stem of *T. cordifolia* were collected from the same plant. Stem was washed, cut into small pieces, crushed and soaked with potable water. Then, the entire mass was macerated, the liquid was filtered and the sediment white starch was completely dried. The yield of the stem starch/polysaccharides (*satva*) was found to be higher in stems collected according to recommendation, when compared to that in stems that were too thin or thick (Sharma et al. 2012). Mature stem of *T. cordifolia* were collected during three different seasons, and the starch (*satva*) was extracted. The starch was subjected to physico-chemical and phyto-chemical screening. The alkaloid content was maximum in stem samples collected during spring season and polysaccharides during winter (Sharma et al. 2013). However, such information is not integrated in *The Ayurvedic Pharmacopoeia of India* (1999).

## 6.4   PROCESSING FOR EFFICACY ENHANCEMENT

*Piper longum* L. (*Pippali*, long pepper) is a popular herb in Indian HTs and is used as a bioavailability enhancer in several polyherbal formulations. As a single herb, the fruit powder is processed in milk for use in certain pulmonary conditions (*svasa, kasa*). The bioactivity of milk decoction was found to be 27 times that of an aqueous decoction (Preethi Sudha, Ugru, and Venkatasubramanian 2004).

## 6.5 Processing for Detoxification

Toxicology (*Agada tantra*), a separate discipline of study in Ayurveda, explains the detoxification (*shodhana*) processes of several toxic plants. Maurya et al. (2015) present the traditional purification processes for some of the toxic plants used in Ayurveda including *Strychnos nux-vomica* L. (*Kupeelu*), *Aconitum ferox* Wall. ex Ser. (*Vatsanabh*), *Datura metel* L. (*Dhattura*) and *Abrus precatorius* L. (*Gunja*). Research studies have demonstrated that traditional processing methods make the toxic plants safe and efficacious. Processing in *Kanji* (rice gruel) per TQS reduced the toxic alkaloid content of strychnine and brucine up to 60%, in *Nux vomica* seeds (Mitra, Shukla, and Acharya 2011). The seeds of *Semecarpus anacardium* L. f. (*Bhallataka*) or the Marking Nut is an excellent tonic (*rasayana*) and possesses anti-arthritic, anti-cancer and analgesic properties, provided it is purified before use; otherwise, it is extremely corrosive and toxic. Soaking the seeds in cow's urine, milk and rubbing them in brick powder reduced the causticity and converted the toxic urushiol to non-toxic anacardol (Sahu and Tiwari 2020).

> A good physician is one who is knowledgeable about the medicinal plant's properties and skilfully manipulates them to treat diseases.
> - *Charaka Samhita, an ancient Ayurvedic treatise (c.* 1500 BCE) (Sastry 1997)

## 6.6 Usage

Iron Deficiency Anemia (IDA) is a major public health condition affecting ~2 billion people worldwide. Pomegranate (*Punica granatum* L.) is one of the *rasayana* (tonic/dietary supplement) plants mentioned in Ayurveda for managing *pandu*, akin to IDA. Rasayana is a dedicated branch of Ayurveda that deals with methods to increase vitality and delay aging through the use of diet, herbal supplements and other lifestyle practices. The effect of pomegranate juice (PJ) on reversing the 'IDA like' condition simulated in the common yeast (*Saccharomyces cerevisiae*) was studied. Culturing iron deficient (ID) cells in the presence of 10% PJ-supplemented medium (IDP) improved iron status by at least 7-fold and reversed mitochondrial degeneration induced by iron deficiency. Percentage of healthy reticulate mitochondria in IDP cells was >30% higher than that in the ID cells grown in iron deficient medium (IDD) and at least 14% more than that in ID cells grown in 10% PJ-equivalent iron substituted media. Interestingly, PJ substitution improved the functional ferrous ($Fe^{2+}$) form as well as the bio-assimilated heme form of iron, but not the ferric ($Fe^{3+}$) storage form in ID cells (Balasubramani et al. 2014). Yeast is a useful whole organism model to rapidly validate nutritional supplement claims of HTs. Pomegranate's potential role as a nutritional supplement in IDA management and as a hematinic is worthy of further research.

Pomegranate juice when tested on *Drosophila melanogaster* (fruit fly) as model extended the lifespan by up to 19% as compared to control (no PJ). Whereas the lifespan of flies fed with resveratrol[3] (used as a positive control) was enhanced by 11% when compared to the control. A two-fold enhancement in fecundity, improved resistance to oxidative stress ($H_2O_2$ and paraquat induced) and to *Candida albicans* infection were observed in PJ-fed flies. Further, the flies in the PJ-fed group were found to be physically more active when compared to the flies in the control and resveratrol groups. This was tested using the climbing assay (Balasubramani et al. 2014).

Enhancing the health span while extending the lifespan have been the trending foci of biomedical research. Research into the principles and practices of HTs can provide new insights and creative applications for leading a healthy life.

---

3 Resveratrol is a phenolic compound from the skin of grapes and a well-known lifespan enhancer. This was used as a positive control.

## 7.  DISCUSSION AND CONCLUSIONS

Quality, safety and efficacy of medicinal plant drugs depend on identity (plant species/genetic composition/part/age), collection (stage/time/season/region), processing, storage and usage. Prior knowledge about these aspects exists among classical texts and living practices of HTs, whereas this kind of knowledge is neither common to modern biomedicine nor part of pharmacopoeial methods and standards. Non-contextual standardization of medicinal plants used in HTs does not do justice to the full potential of medicinal plant bioactivity.

Apart from the beneficial phytocompounds a plant may also have several toxic compounds, the effect of which needs to be neutralized before usage. The good qualities of most medicinal plants also start waning over storage and the yield may vary regionally, seasonally and diurnally. Therefore, it is important to fully understand the seasonality and factors influencing the medicinal plants. The processing methods applied on a plant drug can also change its properties and biological action. The common tea (*Camelia sinensis*), which is probably the most popular global beverage, is one example into which heavy R&D investments have been made to learn every little detail about the plant in terms of when to collect, how to process, etc. Plucking the leaves at the 4-leaf stage, and particular ways of fermentation are crucial for a good, flavourful cup of tea. The total funds received for all R&D in Tea in India was ~INR 230 million in 2017–18 (*Research and Development Statistics 2019–20* 2020), whereas the budget for the Pharmacopoeia Commission for Indian Medicine and Homeopathy, the apex body that sets standards of all the drugs (~250000) of HTs in India, was only INR 103 million (AYUSH, 2021). Medicinal plants that are used to treat diseases, and promote health, should at least be given *substantial investments*, if not *more* than that for tea!

*Reverse Pharmacognosy* evolved as a novel approach to develop better and more pertinent standards by integrating TQS of plant drugs with modern ones. However, the parameters and standards of quality assessment in HTs are very different from those used in modern scientific parlance. Quality control was manageable in ancient times since a physician who was familiar with TQS used to collect, process and administer the plant drug to patients. However, the TQS may be inadequate to handle the modern-day challenges of large-scale manufacturing of herbal drugs, adulteration, substitution and contamination. Hence, researching the TQS and integrating them appropriately with modern protocols and standards of medicinal plant usage can enhance the efficacy, quality and safety.

Modern as well as TQS have their own pros and cons (Table 6). The full potential of medicinal plants and products can be obtained by adopting the *Reverse Pharmacognosy* approach. The TQS

---

## TABLE 6
### Comparison of Modern vs. Traditional Quality Standards (TQS).

| Modern Quality Standards | Traditional Quality Standards (TQS) |
| --- | --- |
| • Pharmacognosy and Pharmacology are separate disciplines with regard to medicinal plant quality control (QC) | • Pharmacognosy and Pharmacology were inseparable part of a drug's quality and assessed as such (*Dravyaguna*) |
| • Bioactivity is not a part of routine QC | • Bioactivity guides QC methods |
| • Mostly instrument-dependent, including objective protocols and standards | • Human perception and experience based; appear subjective, generic and scientifically un-validated |
| • SOPs and Pharmacopeial standards exist | • TQS scattered in classical texts and in living traditions. SOPs do not exist in a systematic manner |
| • Modern physician is the prescriber of medicine. He/she is not required to know about the quality or preparation procedures of drugs | • A traditional practitioner knows about the plant drug, quality, preparation and administration |
| • Industry-friendly | • Applicability and scalability need to be researched; currently not industry-friendly |
| • Scope for enhancement of quality standards through Reverse Pharmacognosy approach | • Need for documentation and R&D to arrive at objective standards |

provides the background wisdom about medicinal plant collection, processing and storage methods; the modern pharmacopoeial methods standardize the quality, and the bioassays checks the overall batch-to-batch bioactivity.

Using whole organism-based models and non-target based comprehensive technologies like metabolomics in medicinal plants batch-to-batch quality assurance of bioactivity is possible. Brine shrimps, fruit flies, *C. elegans* and yeasts are simple, whole organisms and ideal for use in QC laboratories. These models can also be used for studying the mode of action of the plants and drug research.

Research on TQS is at a very nascent stage with unexplored scientific studies on the recommendations (Table 7). This provides opportunities for concerted, coordinated efforts from multi-disciplinary

**TABLE 7**
**Unexplored Reverse Pharmacognosy approach for Quality Standardization of Medicinal Plants used in HTs – Turmeric (*Curcuma longa* L.) as an example.**

| Parameter | Traditional Recommendations (codified and living traditions) | WHO Monograph (WHO Health Systems Library 1999) *Curcuma longa* L. Quality Standards |
|---|---|---|
| Identity | • Yellow rhizomes<br>• Dark yellow rhizomes preferred*<br>• Pharmacodynamic Properties (*Rasapanchaka*)*<br>• Taste (*Rasa*) – pungent (*Katu*), bitter (*Tikta*)<br>• Properties (*Guna*) – dry (*Ruksha*)<br>• Potency (*Virya*) – hot (*Usna*)<br>• Post-digestive effect (*Vipaka*) – Pungent (*Katu*) | • Botanical identity – *Curcuma longa* L. (Zingiberaceae family)<br>• Organoleptic-aromatic odour, bitter taste. When chewed colours saliva yellow<br>• Macroscopic and microscopic (anatomical) description<br>• Physico-chemical identity – ash values and extractives; essential oil (not less than 4%) and curcuminoids (not less than 3%)<br>• Purity tests for contaminants and adulterants |
| Collection/Harvest | • Mature rhizomes*<br>• During early winter*<br>• At night-preferred for medicinal use* | • None provided |
| Processing | • Fresh rhizomes preferred for medicinal use*<br>• Boil and dry rhizomes for storage/powdering purposes | • None provided |
| Dosage form | • Dosage form – milk decoction*<br>• Several polyherbal formulations | • Powder/dried rhizome |
| Pharmacological action | A few examples<br>• Antimicrobial (*Krimighna*),<br>• Cures skin ailments (*Kushtaghna*),<br>• Improves complexion (*Varnya*),<br>• Improves wound healing (*Vrana ropana*),<br>• Reduces pain (*Ruja hara*),<br>• Reduces frequency and turbidity of urination (*Pramehanasaka*),<br>• Pacifies *kapha* and pitta (*Kaphapitta shamaka*) | • Some of the traditional and folk use listed<br>• A few pharmacology studies cited |

\* The highlighted traditional recommendations, if researched, can add value to quality standards and medicinal value of the plant.

teams, including traditional practitioners, floor managers, language experts, taxonomists and modern scientists (chemists, pharmacologists, statisticians etc.). Engaging with the herbal industry is equally important to prioritise the plant drugs to work on and the challenges being faced.

Documentation and digitalization of TQS have to be done with care, taking into consideration the laws of the land regarding access and benefit sharing of medicinal plants. Some of the rare, endangered or vulnerable species are forbidden from being accessed and the *Reverse Pharmacognosy* approach can help find substitutes for species of conservation concern. It can open up TK-based algorithms for new drug discovery. Working on HTs is not only exciting for scientists but also rewarding as it generates new knowledge. However, it is important to acknowledge the prior knowledge of HTs and follow ethical processes while using it.

## REFERENCES

Abbott, Ryan. 2014. *Documenting Traditional Medical Knowledge*. WIPO: Geneva, Switzerland. https://www.wipo.int/export/sites/www/tk/en/resources/pdf/medical_tk.pdf.

Akbar, Shahid. 2020. "*Papaver somniferum* L. (Papaveraceae)." In *Handbook of 200 Medicinal Plants*, 1377–83. Cham: Springer. https://doi.org/10.1007/978-3-030-16807-0_142.

AYUSH. 2021. *Notes on Demands for Grants, 2021-2022*. New Delhi: Ministry of Ayurveda, Yoga and Naturopathy, Unani, Siddha and Homoeopathy (AYUSH). https://www.indiabudget.gov.in/doc/eb/sbe4.pdf.

Balasubramani, Subramani Paranthaman, Jayaram Mohan, Arunita Chatterjee, Esha Patnaik, Subrahmanya Kumar Kukkupuni, Upendra Nongthomba, and Padmavathy Venkatasubramanian. 2014. "Pomegranate Juice Enhances Healthy Lifespan in Drosophila Melanogaster: An Exploratory Study." *Frontiers in Public Health* 2 (December). https://doi.org/10.3389/fpubh.2014.00245.

Bevan, David R. 1994. "Newer Neuromuscular Blocking Agents." *Pharmacology & Toxicology* 74 (1): 3–9. https://doi.org/10.1111/j.1600-0773.1994.tb01065.x.

Bradley, PB, and J Elkes. 1953. "The Effect of Atropine, Hyoscyamine, Physostigmine and Neostigmine on the Electrical Activity of the Brain of the Conscious Cat." *The Journal of Physiology* 120 (1-2). https://pubmed.ncbi.nlm.nih.gov/13062251/.

British Pharmacopoeia Commission. 2019. *British Pharmacopoeia 2020 [Complete Editionprint + Download + Online Access]*. S.L.: Tso.

"Catalogue of Life." n.d. Catalogueoflife. Accessed March 15, 2021. https://www.catalogueoflife.org/col.

Chunekar, Krishna Chand. 2004. *Bhavaprakasa Nighantu of Bhavamisra*. Varanasi: Chaukhambha Bharati Academy.

Decker, Kimberly J. 2020. "Growth in a Global Pandemic." *Nutritional Outlook*, September 2020. https://cdn.sanity.io/files/0vv8moc6/nutrioutlook/5a31705aa76b473e39c89f9896191bb2f70fc228.pdf.

Devaiah, K., and Padma Venkatasubramanian. 2008. "Genetic Characterization and Authentication of Embelia Ribes Using RAPD-PCR and SCAR Marker." *Planta Medica* 74 (2): 194–96. https://doi.org/10.1055/s-2008-1034279.

Devaiah, Kambiranda, Subramani Paranthaman Balasubramani, and Padma Venkatasubramanian. 2011. "Development of Randomly Amplified Polymorphic DNA Based SCAR Marker for Identification OfIpomoea MauritianaJacq (Convolvulaceae)." *Evidence-Based Complementary and Alternative Medicine* 2011: 1–6. https://doi.org/10.1093/ecam/neq023.

Dhyani, CS. 1994. *Rasapanchaka*. Varanasi: Krishnadas Academy.

Dietrich, Michael R., Rachel A. Ankeny, Nathan Crowe, Sara Green, and Sabina Leonelli. 2020. "How to choose your research organism." *Studies in History and Philosophy of Science Part C: Studies in History and Philosophy of Biological and Biomedical Sciences* 80 (April): 101227. https://doi.org/10.1016/j.shpsc.2019.101227.

Duffin, Jacalyn. 2002. "Poisoning the Spindle: Serendipity and Discovery of the Anti-Tumor Properties of the Vinca Alkaloids." *Pharmacy in History* 44 (2): 64–76. https://pubmed.ncbi.nlm.nih.gov/12240681/.

Dutta, Ritwick, Pushp Jain. 2000. *CITES Listed Medicinal Plants of India: An Identification Manual*. New Delhi, India: Traffic-India https://www.traffic.org/site/assets/files/9623/cites-listed-medicinal-plants-of-india.pdf

Efferth, Thomas, and Henry Johannes Greten. 2012. "Quality Control for Medicinal Plants." *Medicinal & Aromatic Plants* 01 (07). https://doi.org/10.4172/2167-0412.1000e131.

"Ephedra." n.d. NCCIH. National Center for Complementary and Integrative Health. Accessed April 12, 2021. https://www.nccih.nih.gov/health/ephedra.

Fitzgerald, Martin, Michael Heinrich, and Anthony Booker. 2020. "Medicinal Plant Analysis: A Historical and Regional Discussion of Emergent Complex Techniques." *Frontiers in Pharmacology* 10 (January). https://doi.org/10.3389/fphar.2019.01480.

Gheorghiade, Mihai, Kirkwood F. Adams, and Wilson S. Colucci. 2004. "Digoxin in the Management of Cardiovascular Disorders." *Circulation* 109 (24): 2959–64. https://doi.org/10.1161/01.cir.0000132482.95686.87.

Giacomotto, Jean, and Laurent Ségalat. 2010. "High-Throughput Screening and Small Animal Models, Where Are We?" *British Journal of Pharmacology* 160 (2): 204–16. https://doi.org/10.1111/j.1476-5381.2010.00725.x.

Goraya, G.S, and D.K Ved. 2017. "Medicinal Plants in India: An Assessment of Their Demand and Supply." National Medicinal Plants Board, Ministry of AYUSH, Government of India, New Delhi and Indian Council of Forestry Research & Education, Dehradun.

Graham, W., and J. B. Roberts. 1953. "Intravenous Colchicine in the Management of Gouty Arthritis." *Annals of the Rheumatic Diseases* 12 (1): 16–19. https://doi.org/10.1136/ard.12.1.16.

Jesus, Mafalda, Ana P. J. Martins, Eugenia Gallardo, and Samuel Silvestre. 2016. "Diosgenin: Recent Highlights on Pharmacology and Analytical Methodology." *Journal of Analytical Methods in Chemistry* 2016: 1–16. https://doi.org/10.1155/2016/4156293.

Khoomrung, Sakda, Kwanjeera Wanichthanarak, Intawat Nookaew, Onusa Thamsermsang, Patcharamon Seubnooch, Tawee Laohapand, and Pravit Akaraserenont. 2017. "Metabolomics and Integrative Omics for the Development of Thai Traditional Medicine." *Frontiers in Pharmacology* 8 (July). https://doi.org/10.3389/fphar.2017.00474.

Kukkupuni, Subrahmanya Kumar. 2015. "Study of Abhava and Abhava Pratinidhi Dravyas." Thesis. http://shodhganga.inflibnet.ac.in:8080/jspui/handle/10603/53233#.

Kumar, Sandeep, and J.P Yadav. 2014. "Journal of Medicinal Plant Research Ethnobotanical and Pharmacological Properties of Aloe Vera: A Review." Https://Academicjournals.org/Journal/JMPR/Article-Full-Text-Pdf/3A183ED49602 8 (48): 1387–98. https://doi.org/10.5897/JMPR2014.5336x.

Lakshmanan, Remya, and Aarushi Aggarwal. 2020. "Invigorating Ayurveda in the times of COVID-19: India's Position and Investment Opportunities" www.investindia.gov.in. July 9, 2020. https://www.investindia.gov.in/siru/invigorating-ayurveda-times-covid-19-indias-position-and-investment-opportunities.

Lemonnier, Nathanaël, Guang-Biao Zhou, Bhavana Prasher, Mitali Mukerji, Zhu Chen, Samir K. Brahmachari, Denis Noble, Charles Auffray, and Michael Sagner. 2017. "Traditional Knowledge-Based Medicine." *Progress in Preventive Medicine* 2 (7): e0011. https://doi.org/10.1097/pp9.0000000000000011.

López-Muñoz, F., V. S. Bhatara, C. Alamo, and E. Cuenca. 2004. "Historical Approach to Reserpine Discovery and Its Introduction in Psychiatry." *Actas Espanolas de Psiquiatria* 32 (6): 387–95. https://pubmed.ncbi.nlm.nih.gov/15529229/.

Maurya, Santosh Kumar, Ankit Seth, Damiki Laloo, Narendra Kumar Singh, Anil Kumar Singh, and Dev Nath Singh Gautam. 2015. "Śodhana: An Ayurvedic Process for Detoxification and Modification of Therapeutic Activities of Poisonous Medicinal Plants." *Ancient Science of Life* 34 (4): 188–97. https://doi.org/10.4103/0257-7941.160862.

Mazzocchi, Fulvio. 2006. "Western Science and Traditional Knowledge: Despite Their Variations, Different Forms of Knowledge Can Learn from Each Other." *EMBO Reports* 7 (5): 463–66. https://doi.org/10.1038/sj.embor.7400693.

Middleton, Robert M., and Michael B. Kirkpatrick. 1993. "Clinical Use of Cocaine." *Drug Safety* 9 (3): 212–17. https://doi.org/10.2165/00002018-199309030-00006.

Miner, Jonathan, and Adam Hoffhines. 2007. "The Discovery of Aspirin's Antithrombotic Effects." *Texas Heart Institute Journal* 34 (2): 179–86. https://www.ncbi.nlm.nih.gov/pmc/articles/PMC1894700/.

Mishra, Siddhinandan. 2007. *Bhaishajya Ratnavali*. Varanasi: Chaukhamba Surbharati Prakashan.

Mitra, Swarnendu, VJ Shukla, and Rabinarayan Acharya. 2011. "Effect of Shodhana (Processing) on Kupeelu (Strychnos Nux-Vomica Linn.) with Special Reference to Strychnine and Brucine Content." *AYU* 32 (3): 402–7. https://doi.org/10.4103/0974-8520.93923.

Nakashima, Douglas, and Marie Roué. 2002. *Encyclopedia of Global Environmental Change: Social and Economic Dimensions of Global Environmental Change*. Edited by Peter Timmerman and Ted Munn. Vol. 5. Chichester: John Wiley & Sons, Ltd. http://www.unesco.org/new/fileadmin/MULTIMEDIA/HQ/SC/pdf/sc_LINKS-art%20EGEC.pdf.

Newman, David J., and Gordon M. Cragg. 2016. "Natural Products as Sources of New Drugs from 1981 to 2014." *Journal of Natural Products* 79 (3): 629–61. https://doi.org/10.1021/acs.jnatprod.5b01055.

NMPB. n.d. "Good Field Collection Practices Standard for Medicinal Plants – Requirements." https://www.nmpb.nic.in/sites/default/files/publications/Good_Field_Collection_Practicies_GFCPs_Standard_for_Medicinal_Plants.pdf.

Payyappallimana, Unnikrishnan. 2008. "Ayurvedic Pharmacopoeia Databases in the Context of the Revitalization of Traditional Medicine." In *Modern and Global Ayurveda: Pluralism and Paradigms*, edited by Dagmar Wujastyk and Frederick M Smith, 139–55. Albany: State University of New York.

Payyappallimana, Unnikrishnan. 2010. "Role of Traditional Medicine in Primary Health Care: An Overview of Perspectives and Challenges." *Yokohama Journal of Social Sciences* 14 (6): 58–77. https://ynu.repo.nii.ac.jp/?action=repository_action_common_download&item_id=3027&item_no=1&attribute_id=20&file_no=1 orhttps://www.researchgate.net/publication/284697212_Role_of_Traditional_Medicine_in_Primary_Health_Care_An_Overview_of_Perspectives_and_Challenges

Preethi Sudha, VB, Geeta G Ugru, and Padma Venkatasubramanian. 2004. "Bioactivity of Traditional Preparation of Piper Longum L. (Piperaceae)." *Journal of Tropical Medicinal Plants* 5 (2): 179–82.

"Prioritised List of Medicinal Plants for Cultivation." 2016. Nmpb.nic.in. National Medicinal Plants Board, Government of India. 2016. https://www.nmpb.nic.in/content/prioritised-list-medicinal-plants-cultivation.

Rangasamy, Ilanchezhian, Rabinarayan Acharya, Roshy C Joseph, and Vinay J Shukla. 2012. "Impact of Ayurvedic Shodhana (Purificatory Procedures) on Bhallataka Fruits (Semecarpus Anacardium Linn.) by Measuring the Anacardol Content." *Global Journal of Research on Medicinal Plants & Indigenous Medicine* 1 (7): 286–94.

*Research and Development Statistics 2019-20*. 2020. New Delhi: Government of India, Ministry of Science & Technology, Department of Science & Technology. https://dst.gov.in/sites/default/files/Research%20and%20Deveopment%20Statistics%202019-20_0.pdf.

Sahu, Pratap Kumar, and Prashant Tiwari. 2020. "Impact of Shodhana on Semecarpus Anacardium Nuts." *Www.intechopen.com*, November 5, 2020. https://www.intechopen.com/online-first/impact-of-shodhana-on-semecarpus-anacardium-nuts?jwsource=cl.

Salem, Mohamed A., Leonardo Perez de Souza, Ahmed Serag, Alisdair R. Fernie, Mohamed A. Farag, Shahira M. Ezzat, and Saleh Alseekh. 2020. "Metabolomics in the Context of Plant Natural Products Research: From Sample Preparation to Metabolite Analysis." *Metabolites* 10 (1): 37. https://doi.org/10.3390/metabo10010037.

Salmerón-Manzano, Esther, Jose Antonio Garrido-Cardenas, and Francisco Manzano-Agugliaro. 2020. "Worldwide Research Trends on Medicinal Plants." *International Journal of Environmental Research and Public Health* 17 (10): 3376. https://doi.org/10.3390/ijerph17103376.

Samagandi, Kashinath, Samagandi Sharma Jagriti, and Sharma Kamlesh Kumar. 2013. "Proficient Modulation of Comestibles: An Aesthetic Choice for Obese." *International Journal of Research in Ayurveda and Pharmacy* 4 (2): 279–83. https://doi.org/10.7897/2277-4343.04240.

Sastry, Kasinatha. 1997. *Caraka Samhita of Agnivesa with Cakrapanidatta Tika*. 1st ed. Vol. 2. Varanasi: Chaukhambha Sanskrit Sansthan.

Scheindlin, S. 2010. "Episodes in the Story of Physostigmine." *Molecular Interventions* 10 (1): 4–10. https://doi.org/10.1124/mi.10.1.1.

Shahrajabian, Mohamad, Wenli Sun, and Qi Cheng. 2019. "Journal of Medicinal Plants Research a Review of Ginseng Species in Different Regions as a Multipurpose Herb in Traditional Chinese Medicine, Modern Herbology and Pharmacological Science." *Journal of Medicinal Plants Research* 13 (10): 213–26. https://doi.org/10.5897/JMPR2019.6731.

Shankar, Darshan, Unnikrishnan Payyappallimana, and Padma Venkatasubramanian. 2007. "Need to Develop Inter-Cultural Standards for Quality, Safety and Efficacy of Traditional Indian Systems of Medicine." *Current Science* 92 (11): 1499–1505.

Sharma, Priyavrat. 2006. *Dravyaguna Vijnana*. Vol. 1. Varanasi: Chowkhambha Bharati Academy.

Sharma, Rohit, C.R Harisha, Ruknuddin Galib, B.J Patgiri, and P.K Prajapati. 2012. "Quantitative Estimation of Satva Extracted from Different Stem Sizes of Guduchi (Tinospora Cordifolia (Willd.) Miers." *Journal of Pharmaceutical and Scientific Innovation* 1 (1): 38–40.

Sharma, Rohit, Galib R, PK Prajapati, and Hetal Amin. 2013. "Seasonal Variations in Physicochemical Profiles of Guduchi Satva (Starchy Substance from Tinospora Cordifolia [Willd.] Miers)." *Journal of Ayurveda and Integrative Medicine* 4 (4): 193. https://doi.org/10.4103/0975-9476.123685.

Sneader, Walter. 2005. *Drug Discovery: A History*. Hoboken, N.J.: John Wiley & Sons.

Su, Xin-Zhuan, and Louis H. Miller. 2015. "The Discovery of Artemisinin and the Nobel Prize in Physiology or Medicine." *Science China Life Sciences* 58 (11): 1175–79. https://doi.org/10.1007/s11 427-015-4948-7.

Tamilselvi, S, Padma Venkatasubramanian, and N S Vasanthi. 2014. "Physico Chemical Characterization and Anti-Inflammatory Activity of Stem Extracts of *Berberis aristata* DC and *Cosinium fenestratum* Linn in Carrageenan Induced Wistar Rats." *Pharmacognosy Journal* 6 (4): 72–77. https://doi.org/10.5530/pj.2014.4.11.

*The Ayurvedic Pharmacopoeia of India*. 1999. Vol. 2. New Delhi: Government of India, Ministry of Health and Family Welfare, Department of Indian Systems of Medicine & Homoeopathy.

United States Pharmacopeial Convention. 2015. *Dietary Supplements Compendium*. Rockville, Md: United States Pharmacopeial Convention. https://www.usp.org/sites/default/files/usp/document/products-servi ces/products/2015-dsc-vol-1-table-of-contents-ref-standard-index.pdf.

Valiathan, M. S. 2016. "Ayurvedic Biology: The First Decade." *Proceedings of the Indian National Science Academy* 82 (1). https://doi.org/10.16943/ptinsa/2016/v82i1/48376.

Ved, D K, and G S Goraya. 2008. *Demand and Supply of Medicinal Plants in India*. Dehra Dun: Bishen Singh Mahendra Pal Singh.

Ved, D.K, Suma Tagadur Sureshchandra, Vijay Barve, Vijay Srinivas, Sathya Sangeetha, K Ravikumar, R Kartikeyan, et al. 2016. "Environmental Information System (ENVIS) Centre on Medicinal Plants." Envis.frlht.org. 2016. http://envis.frlht.org/.

Venkat, Padma. 2009. "Reverse Pharmacognosy: A Novel Strategy to Standardise ISM Drugs – CRISM." Yumpu.com. April 6, 2009. https://www.yumpu.com/en/document/read/35594143/reverse-pharmacogn osy-a-novel-strategy-to-standardize-crism.

Venkatasubramanian, Padma, Ashwini Godbole, R Vidyashankar, and Gina R. Kuruvilla. 2013. "Evaluation of Traditional Anthelmintic Herbs as Substitutes for the Endangered *Embelia ribes*, Using Caenorhabditis Elegans Model." *Current Science* 105 (11): 1593–98.

Venkatasubramanian, Padma, Subrahmanya K Kumar, and Venugopalan S.N Nair. 2010. "*Cyperus rotundus*, a substitute for *Aconitum heterophyllum*: Studies on the Ayurvedic concept of *Abhava Pratinidhi Dravya* (Drug Substitution)." *Journal of Ayurveda and Integrative Medicine* 1 (1): 33. https://doi.org/10.4103/0975-9476.59825.

Wall, M. E., and M. C. Wani. 1995. "Camptothecin and Taxol: Discovery to Clinic--Thirteenth Bruce F. Cain Memorial Award Lecture." *Cancer Research* 55 (4): 753–60. https://pubmed.ncbi.nlm.nih.gov/7850785/.

"What Is Pharmacognosy?" 1998. Web.archive.org. The American Society of Pharmacognosy. December 1, 1998. https://web.archive.org/web/19981201075709/http://www.phcog.org/definition.html.

WHO. 2000. *General Guidelines for Methodologies on Research and Evaluation of Traditional Medicine*. https://apps.who.int/iris/bitstream/handle/10665/66783/WHO_EDM_TRM_2000.1.pdf.

WHO. 2002. *WHO Policy Perspectives on Medicines — Traditional Medicine – Growing Needs and Potential*. Geneva: World Health Organization. https://apps.who.int/iris/bitstream/handle/10665/67294/WHO_E DM_2002.4.pdf?sequence=1.

WHO. 2009. *The Use of Herbal Medicines in Primary Health Care*.: World Health Organization, Regional Office for South-East Asia. https://apps.who.int/iris/bitstream/handle/10665/206476/B4260.pdf?seque nce=1&isAllowed=y.

WHO. 2019. *WHO Global Report on Traditional and Complementary Medicine, 2019*. Geneva, Switzerland: World Health Organization. https://www.who.int/traditional-complementary-integrative-medicine/WhoGlobalReportOnTraditionalAndComplementaryMedicine2019.pdf?ua=1.

WHO Health Systems Library. 1999. "WHO Monographs on Selected Medicinal Plants – Volume 1: Rhizoma Curcumae Longae." Digicollection.org. WHO. 1999. http://digicollection.org/hss/en/d/Js2200e/14.html.

Willcox, Merlin L, and Gerard Bodeker. 2004. "Traditional Herbal Medicines for Malaria." *BMJ* 329 (7475): 1156–59. https://doi.org/10.1136/bmj.329.7475.1156.

WIPO. 2017. *Documenting Traditional Knowledge – a Toolkit.* Geneva: World Intellectual Property Organization (WIPO). https://www.wipo.int/edocs/pubdocs/en/wipo_pub_1049.pdf.

Wujastyk, Dagmar, and Frederick M Smith, eds. 2008. "Plural Medicine and East-West Dialogue." In *Modern and Global Ayurveda: Pluralism and Paradigms*, 29–41. Albany: State University Of New York Press.

# 12 Technical Report: Effect-Directed Analysis (EDA) by High-Performance Thin-Layer Chromatography (HPTLC) – principles of the method to detect harmful and beneficial substances in plant material and the environment

*Christel Weins*

## 1  MOTIVATION

Effect-Directed Analysis (EDA) is fundamentally different in its approach as opposed to other analytical methodologies. EDA does both: it looks for a known molecule of a specific structure and at the same time it detects the effects in a complex mixture of components. The preceding chromatographic method as a separation technique is HPTLC (High Performance Thin-Layer Chromatography). It separates and concentrates groups of components according to their polarity. EDA follows and allows the assessment of toxicities or beneficial effects of unspecified/unknown molecules onto specific biological model systems (e.g., biological toxicity tests, enzyme inhibition tests). Such active ingredients, or so-called 'effective agents' (EAs), are substances that cause biological effects in living organisms. EAs comprise components of a substance that may undergo chemical change (transformation) by biotic or abiotic degradation, like thermal oxidation or photo oxidation. Such degradation products may cause biologically very bio-effective products. One example is the abiotic oxidation of parathion (insecticide E 605, also known as 'thiophos') to paraoxon, which is much more toxic than the genuine compound.

Exposure to an EA in turn initiates a characteristic biological response that is associated with the onset or exacerbation of a wide spectrum of physical illnesses. Given that EAs are capable of fitting onto an appropriate receptor of the biological system, the biological response can be, for example, growth inhibition, cell proliferation or enzyme induction. Furthermore, EAs can even provoke mutation causing cell death. In general, EAs can show effects at very low concentrations; for example, natural hormones acting at extremely low serum concentrations. The range of estrogenic hormones

DOI: 10.1201/9781003146902-15

**TABLE 1**
**Significance of instrumental analysis and biotests for detecting EAs.**

| Trace Analysis (Instrumental analysis: LC-MS, LC-MS MS, GC-MS, TLC)[a] | Biotests (*in vivo* and *in vitro* tests: e.g., fish test, bioluminescence, genotoxicity, antibiotic effects on an agar plate…) |
|---|---|
| • bioeffective substances in a sample are selectively enriched <br> • analytical separation <br> • the appropriate substances are identified using selected reference substances using physical detection methods <br> • identified toxins can be quantified | • determination of bioactivity of bioeffect substances using test organisms in un-processed water samples or other matrices after a simple sample preparation <br> • discussion of synergetic or masking effects <br> • no identification and quantification of only one single substance in an unknown sample |

a   Liquid Chromatography Mass Spectrometry (LC-MS/MS) is an exceedingly sensitive and specific analytical technique that can precisely determine the identities and concentration of compounds within your sample;

  Gas Chromatography and Mass Spectrometry (GC-MS) to analyse, for example, volatile organic (synthetic) compounds in drinking water;

  High-Performance Thin-Layer Chromatography (HPTLC) and Thin-Layer Chromatography (TLC) separates different substances in a planar chromatographic system.

in adult females is 0.5–9 pg/ml (Vandenberg et al. 2012). To detect, identify and quantify EAs at such very low concentrations and within different matrices[1] is a substantial challenge for any analytical procedure

## 2   ANALYTICAL CHALLENGE

The investigation of samples via EDA such as water, soil, food and plants to detect active or toxic components (i.e., beneficial or harmful) presents a challenge for any analytical technique.

In conventional instrumental analysis instead, it is only possible to detect those substances that are actively sought by the analyst. The analysis of individual substances does not detect unknown substances or metabolites having adverse (toxic) effects or beneficial effects in an environmental sample (Weins 2010). Table 1 shows the main differences between conventional instrumental trace analysis and biotests.

EDA is based on the idea that the biological effect of the total sample or an extract on selected target systems (e.g., cell cultures, microorganisms, enzymes) is determined as a sum of effects. In the case of conspicuous effects, a specific search is subsequently performed to identify the responsible substance(s) (Brack 2003).

Thus, the purpose of EDA is to bridge the gap between cause and effect, that is, primarily not to identify a given analyte but to identify a class of compounds with a defined bioactivity. This is the basis of screening tests for substances with special properties (e.g., fungicides) and for screening tests on new compounds with a particular biological activity. This approach is of interest in the search for new compounds that show a particular biological activity and for investigating samples containing substances whose identity is unknown or unavailable as reference standards.

EDA is used to detect and identify known and unknown bio-effective compounds such as pharmacologically active substances or contaminants from samples having biological activity at

---

1   'Matrix' refers to the components of a sample other than the analyte of interest.

trace concentrations within the 0.1–200 ng/kg range and thus is as sensitive as conventional instrumental analytics.

As such, EDA involves a coupling of two different analytical methods as mentioned in Table 1. On the one hand, an instrumental analysis, using conventional (trace) analytical methods, is used for the determination of selected organic (synthetic) bioactive substances and, on the other hand, the physical/chemical assessment followed by a biological or biochemical activity test, thus, allowing a direct activity-dependent evaluation to be made after chemical/physical characterisation. EDA can be summarised as a method designed to integrate biological or biochemical tools for the detection of potentially adverse effects, physico-chemical fractionation procedures and chemical-analytical methods for structure elucidation and toxicant quantification.

The selected method should be as universal as possible, in order to detect unknown contaminants with biological effects in the sample being analysed. It is not the selectivity with respect to individual substances that is of importance, but rather the detection of all, or at least as many of the bio-effective substances as possible.

## 2.1 HPTLC: HIGH-PERFORMANCE THIN-LAYER CHROMATOGRAPHY

HPTLC is the High-Performance version of Thin-Layer Chromatography (TLC) and a state-of-the-art technique (e.g., for plant analysis). HPTLC offers significantly shorter developing times, lower solvent consumptions and improved resolution as compared to TLC. Highly reproducible results and traceable records are achieved through a standardised methodology and the use of suitable instruments (typically controlled by software) for all steps of the analysis. A system suitability test is used to qualify results.

The stationary phase is an HPTLC glass plate or aluminum sheet coated with a uniform thin layer (typically 200 micron) of porous particles (2–10 µm) with an average particle size of 5 µm. The layer typically consists of silica gel with a pore size of 60 Ångstroms, a polymeric binder and a so-called fluorescence indicator ($F_{254}$). The standard format of the plate is 20×10 cm (HPTLC Association 2022).

The apparatus for HPTLC consists of:

- A device suitable for the application of samples as bands providing control of dimension and position of the application as well as applied sample volume
- A suitable chromatographic chamber (typically a twin trough chamber) providing control of saturation and developing distance
- A device suitable for controlling the activity of the stationary phase via relative humidity
- A device suitable for reproducible drying of the developed plate
- Suitable devices for reagent transfer and heating as part of the derivatisation procedure
- A device suitable for electronic documentation of chromatograms under UV 254 nm, UV 366 nm and white light
- For quantitative determinations, a densitometer or image evaluation software

## 2.2 CHEMICAL/PHYSICAL PROCEDURE OF INSTRUMENTAL ANALYSIS

TLC, or HPTLC, is a planar separation technique and one of the best-known methods used in the analysis of bio-effective compounds in different and complex matrices. The separation of the compounds is carried out in so-called 'normal phase chromatography', which is characterised by a polar solid phase-like silica gel ($SiO_2$) and a non-polar mobile phase-like ethyl acetate. The components of a mixture are separated chromatographically according to their polarity, which means that genuine substances can easily be separated from their oxygenated metabolites. The first step of identification can be determined by the retention factor (Rf-value) in the chromatogram and the second step of identification is carried out by *in situ* reflectance measurement at seven wavelengths (200–320 nm). That means the individual substances are subjected to preliminary identification on the basis of their position in the chromatogram and the reflectance spectrum.

NOTE: In this case the detection limit of the substances is generally dependent on the absorption coefficients of the substance or of its derivative being tested.

## 2.3   Effect Test (directly on the plate, *in situ*)

The effect test consists of the detection of the active substance on the same chromatogram by coupling the chromatogram with a biological/biochemical test.

Test organisms, such as mould spores, yeast cells – such as genetically modified yeast cells (*Saccharomyces cerevisiae*) – bacteria or cell organelles – such as chloroplasts – in a suitable nutrient medium are applied to the chromatogram. The biological signal is used to localise bio-effective substances on the chromatogram. In addition, the use of organisms or sub-organisms (e.g., enzymes) allows the detection of EAs.

NOTE: In this case the sensitivity depends on the effectiveness of the substance concerned on this defined test system.

EDA in TLC couples two different methods (see Figure 1). On the one hand, the TLC plate is used as the stationary phase in a chromatographic system, pre-washed and activated at 100°C, optimised for the chromatographic development. On the other hand, this stationary phase, contaminated with components of the mobile phase and bio-effective compounds, should act as the underlayer for living organisms. For biological assays especially, the apparent pH of the sorbent must be taken into account. This is why strongly acidic solvents, such as those containing formic or acetic acid should be avoided in the mobile phase. Toxic solvents like toluene should not be used (Weins 2010).

A key issue for the identification of bio-effective compounds is the establishment of suitable biotests using organism and biochemical tests. The biological or biochemical signal, such as inhibition or stimulation of growth, inhibition or stimulation of luminescence, or inhibition of photosynthesis, is used to localise bio-effective substances on the chromatogram (see Fig. 2).

Documentation can be carried out by means of a flatbed scanner or a video camera and the inhibition can be reported quantitatively in defined toxicity units. The sensitivity here depends on the toxicity of the substance with respect to each test system.

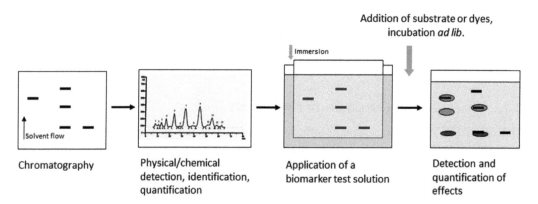

**FIG. 1**   Principle of the procedure of EDA using thin-layer chromatography (TLC) combined with a biological/biochemical detection direct on the plate (*in situ*).

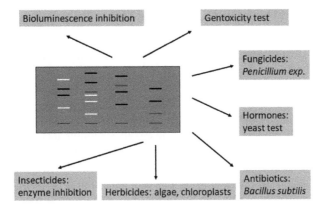

**FIG. 2**  Examples for *in situ* toxicological tests on HPTLC plates.

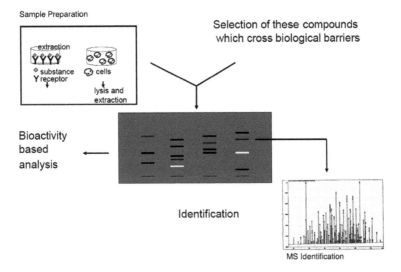

**FIG. 3**  Analytical concept of effect-directed analysis (EDA).

## 2.4  IDENTIFICATION OF UNKNOWN SUBSTANCES BY MASS SPECTROMETRY (MS)

Final identification of compounds can be achieved through LC-MS[2] or GC[3]-MS analysis of TLC fractions exhibiting activity. The coupling of planar chromatography with direct analysis in real time time-of-flight mass spectrometry (DART-TOF-MS)[4] was shown for the first time in 2007 (Morlock and Ueda 2007). Cutting the plate within a track led to substance zones positioned on the plate edge, which were directly introduced into the DART gas stream. Mass signals were obtained instantaneously within seconds. Detectability was shown in the very low ng-range per zone on the example of isopropylthioxanthone (Morlock and Ueda 2007).

In Figure 3, the analytical concept of EDA with TLC in combination with biotests is demonstrated.

2  LC-MS: Liquid Chromatography followed by Mass Spectrometry
3  GC: Gas Chromatography
4  Direct Analysis in Real Time (DART) coupled with a time-of-flight (TOF) mass spectrometer is an emerging technique that yields highly definitive screening data leading to the identity of controlled substances present in a sample.

## 3   EXAMPLES FOR EFFECT DIRECTED ANALYSIS

This chapter will demonstrate some examples of how to detect different EAs by different biological and biochemical test systems according to their specific biological activity.

### 3.1   ENZYME TESTS ON THE HPTLC PLATE

Effective analysis generally follows these steps: sample application, (HP)TLC separation, evaporation of the mobile phase, dipping the plate into an enzyme solution, incubation (which means time for building the enzyme/agent complex), spraying with or dipping into a substrate solution, stopping the enzyme reaction by heat or drying and plate evaluation. The inhibitory effect is determined by the reduction of enzymatic substrate turnover. Enzyme inhibition is revealed by bands of different colour than the background, depending on the substrate.

#### 3.1.1   Acetylcholinesterase inhibition test

The inhibition of cholinesterase has long been recognised as a biochemical method for the detection of the enzyme-inhibiting effects of organophosphate esters and carbamates. Acetylcholinesterase (AChE) inhibitors inhibit the cholinesterase enzyme from breaking down Acetylcholin (Ach), increasing both the level and duration of the neurotransmitter action (Čolović et al. 2013). The inhibition of cholinesterase is the result of an irreversible phosphorylation or carbamylation of the serine OH groups in the active centre of the enzyme. The organophosphorus pesticides and carbamate insecticides inhibit the cholinesterase to very different degrees. Most irreversible inhibitors inhibit the enzymatic reaction completely, frequently by formation of a covalent bond if their concentration exceeds that of the reacting groups in the enzyme (Weins 2010).

Acetylcholinesterase is able to hydrolyse various esters. In former studies, Weins and Jork used 1-naphthyl acetate as substrate in combination with fast blue salt B (see Fig. 3). In Table 2 the procedure is described in detail (Weins 2006).

Figure 4 shows the inhibition of cholinesterase caused by different organophophates after chromatography.

In 2018, Weiß could show that indoxylacetate could act as substrate and responding dye in one single step which lowered the detection limit to 0.5 ng of paraoxon-methyl (Weiß 2018).

To validate this method, it is mandatory to define an exact procedure. Weiß (2018) recommends the parameters listed in Table 3.

---

**TABLE 2**
**Procedure of the AChE inhibition test using 1-Napthylacetate as substrate and blue salt B as dye (Weins 2006).**

- Dissolve 11 mg cholinesterase (50 U/mg) in 180 ml of 0.05 M pH 7.8 tris(hydroxymethyl)-aminomethane-HCl buffer and add 100 mg of bovine serum albumin to dilute the enzyme and improve its stability on the plate.
- The diazonium cation reagents, (250 mg 1-napthylacetate) in 100 ml absolute ethanol and 400 mg fast blue salt B are dissolved in 160 ml $H_2O$.
- Dip the dried plate in the enzyme solution for 2 s and incubate for 30 minutes at 37°C in a chamber at 90% relative humidity.
- Dip the plate for 2 s in a freshly made solution of the diazonium cation reagent. After about 3 min, colourless spots appear on a violet background.
- Stop the enzymatic turnover of the substrate by drying the plate.
- The result can be evaluated at 533 nm or flatbed scanner.

NOTE: The enzyme solution can be stored at 4°C for about three weeks.

---

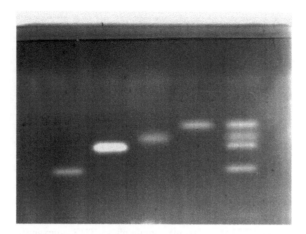

**FIG. 4** HPTLC–cholinesterase assays with 1-naphthyl acetate as substrate coupled with fast blue salt B. From left to right: paraoxon-methyl (0.4ng/zone), paraoxon-ethyl (2ng/zone), naled (0.4ng/zone) and dichlorvos (2 ng/zone). Stationary phase HPTLC KG 60 F254 (10 x 10 cm), mobile phase, THF, n-hexane, (7+25, v/v). The migration distance is 5 cm and the migration time 15 min. For documentation a flatbed scanner was used (Weins 2006).

## TABLE 3
## Optimised parameters for the AChE inhibition test using indoxylacetate as substrate (Weiß 2018).

| Parameters | Amount |
|---|---|
| Activity of the enzyme [Units/ml] | 2,5 |
| Time of incubation for the reaction of the inhibitor with the enzyme [min] | 5 |
| Spraying reagent on the plate [µg/mm²] | 0,030 |
| Time of incubation for the turnover of the substrate [min] | 5 |

Based on the high sensitivity of the procedure, even low traces of impurities and metabolites can be detected in reference substances. Figure 5 shows the detection of several inhibitors of cholinesterase in a solution of methiocarb as reference. The test shows the presence of the metabolites and impurities.

### 3.2 APPLICATION OF PHOTOBACTERIA

Dipping a TLC plate in photobacterial solution is an elegant way of detecting toxic substances directly on the plate. Photobacteria emit light via luciferase enzymes using long-chained aldehydes as a substrate. Their bioluminescence depends directly on the metabolic status of the cell (see Figure 6). If there is a certain concentration of these cells in the medium, this sets free an autoinducer, [N-(ß-hydroxybutyryl)homoserin-lactone], which incites the bacteria to luminosity. This is the case when the (slightly shaken) bacteria have been kept incubated at room temperature for 24 to 40 hours at ambient temperature. The luminous emission lasts for a maximum of about 20 hours (Weins 2010).

The procedure is described in Table 4.

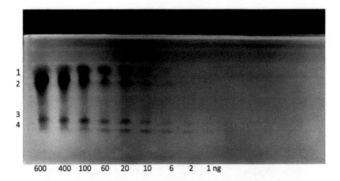

**FIG. 5** HPTLC–cholinesterase assays with 1-naphthyl acetate as substrate coupled with fast blue salt B. (1) methiocarb, (2) methiocarb sulfoxide, (3) methiocarb phenolsulfoxide, (4) methiocarb phenolsulfon. Stationary phase HPTLC KG 60 F254 (10 x 10 cm), mobile phase: THF, n-hexane, (7+25, v/v). The migration distance was 5 cm and the migration time 15 min. For documentation a flatbed scanner was used (Weins 2006).

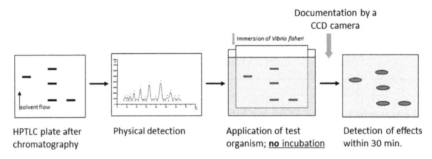

**FIG. 6** Principle of the procedure of detecting luminescence inhibitors applying *Vibrio fischeri* as photobacteria directly on the plate (*in situ*).

---

## TABLE. 4
## Procedure for the preparation of the dipping solution of photobacteria such as *Vibrio fischeri* (Weins 2006).

Nutrition broth (calculated for 2 L):
60.0g NaCl
12.2g $NaH_2PO_4*H_2O$
4.2 g $K_2HPO_4$
0.4 g $MgSO_4*7H_2O$
1.0 g $(NH_4)_2HPO_4$
3.4 ml glycerine (87%)
10.0g peptone from casein
1.0g yeast extract

- Make up the volume to 2 L and adjust the pH to 7.0±0.2. The nutrient medium is heated at 121°C for 20 min and then stored in a refrigerator. The solution should have a pale-yellow colour.
- For reactivation, thaw 5 ml of the bacteria (kept as cold as possible) in a reactivation solution*.
- Leave the mixture to stand for 15 minutes and then pour into 220 ml of nutrient medium.
- Shake this mixture at room temperature for 25 to 30 hours, after which it will be ready for use.
- Dip the TLC plate into the bacteria suspension for 3 seconds and then gently wipe off the dipping solution with a wiper.
- Place a clean glass plate on top of the layer and with a light-sensitive camera measure the luminescence for 2 to 30 minutes.

*NOTE: Reactivate freeze-dried luminous bacteria as recommended by the manufacturer.

## 3.3  APPLICATION OF BACTERIUM *BACILLUS SUBTILIS*

The bacterium *Bacillus subtilis* (ATCC 6633) is a suitable organism to detect growth inhibition substances like antibiotic active compounds. After the chromatographical development, *Bacillus subtilis* can be used as the indicator organism in the activity analysis that follows. The growth of the test organism on the thin-layer plate is inhibited by antibiotic active compounds and the zones of inhibition are detected by means of a bacterial vitality test, where the bacterial lawn on the thin-layer chromatogram is sprayed with an MTT tetrazolium salt (Móricz 2008).

The size of the inhibition zones is determined by the sample amount applied to the layer and by the specific activity of the substance. Figure 7 shows the separation of different amounts of chloroamphenicol, oxytetracycline and lasalocide. After separation the plate was dipped in a *Bacillus subtilis* suspension and after incubating for 16 hours was stained with 3-(4,5-dimethylthiazol-2-yl)-2,5-diphenyltetrazolium bromide (MTT). The stained plate was scanned by a simple flatbed scanner. Inhibition zones show white areas on a blue-red background. Dehydrogenase activity (which indicates living bacteria) is detected by an oxidation of the light-yellow MTT to a blue-red formazone dye.

The procedure is described in Table 5.

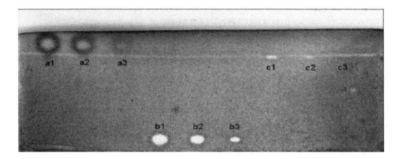

**FIG. 7**   Separation of different amounts of chloroamphenicol (a1 – a3), oxytetracycline (b1 – b3) and lasalocide (c1 – c3; at 25, 15 and 5 ng each). The plate was dipped post-chromatically in a *Bacillus subtilis* suspension and after incubating for 16 hours stained with 3-(4,5-dimethylthiazol-2-yl)-2,5-diphenyltetrazolium bromide (MTT).

---

## TABLE 5
### Procedure of detection of antibiotics post chromatography.

The nutrient solution consists of:
- 1000 ml of meat broth (boiled minced meat) with 10 g of peptone
- 3 g of NaCl and 2 g of $Na_2HPO_4$ – all mixed together to a pH optimum of 5.5–8.5

NOTE:   The nutrient solution must be sterilized by autoclaved at 121°C for 20 min before use. The bacterial suspension is added under sterile conditions to the nutrient.

- The dried TLC plate is dipped into the bacteria suspension
- Time of incubation is 19 hours at 23–35°C in a humid chamber (Choma 2004; R Eymann 2001).
- For the reagent solution dissolve 20 mg of triphenyltetrazolium chloride in 10 ml of water.
- After incubation the plate is sprayed with the dye solution. Yellow inhibition zones appear on a blue-violet background after 5–30 minutes.

NOTE:   Most important is that the plate should not contain solvent residues, leftover from the mobile phase. In some cases it could be necessary to redevelop the plate with pentane after the separation to wash out mobile phase residues.

**TABLE 6**
**Procedure of detection of fungicides post chromatography**

Nutrient solution contains in 1 L of sterilised water:

- $7g$ $KH_2PO_4$, $3g$ $Na_2HPO_4 * 7H_2O$
- $4g$ $KNO_3$, $1g$ $MgSO_4 * 7H_2O$
- $1g$ NaCl

Mix 10 ml of this solution with 60 ml of a sterilized 30% glucose solution in 0.1% Tween 20 solution. The suspension should contain 107–109 cells per ml.

The dried plate is dipped into this suspension for 3 s and then incubated in a humid chamber for 1–3 hours at 25°C.

The chamber should be draped with a piece of wet filter paper to keep the atmosphere moist. Penicillin spores will turn green and inhibition zones can be detected as white zones.

NOTE:    When using TLC plates containing a fluorescent indicator the inhibition effect can be documented by illumination with a light source of 254 nm. The inhibition spots are visible as bright zones due to excitation of the phosphorescence indicator in the stationary phase.

## 3.4    APPLICATION OF *PENICILLIUM* AND YEAST CELLS

To detect and quantify the effect of anti-fungal agents, the TLC plate can be used as a layer for suitable test organisms like *Penicillium* strains or *Candida albicans*. Normal household baking yeast can be used for this purpose as well. The yeast strain *Rhodotorula rubra* is a useful test organism that imparts a red colour to the layer when growing successfully. In the presence of an anti-fungal agent, a white inhibition zone is produced whose size depends on the amount of fungicide and on its specific activity (Weins 2006). The exact procedure is described in Table 6.

## 3.5    APPLICATION OF MODIFIED YEAST CELLS (*SACCHAROMYCES CEREVISIAE*)

The Yeast Estrogen Screen (YES) is a reporter gene assay that can be used for the measurement of the activation of the estrogen receptor (ER) in the presence of an estrogenic sample. The human ER is heterologously expressed in the yeast cell under control of a copper-dependent promoter. The ER belongs to the family of nuclear hormone receptors. If agonists of the ER enter the yeast cell, they bind to the ER protein and thus induce its conformational change. As a consequence, two receptor proteins form a receptor dimer that is translocated in the nucleous of the yeast cell. This activation of the ER is measured by the induction of the reporter gene lacZ, which encodes the enzyme ß-galacosidase. The lacZ is fused to an estrogen-dependent promoter that contains estrogen responsive elements (ERE). The ER-dimer binds to the promoter and by this activates the expression of the ß-galacosidase. Finally, the activity of the ß-galacosidase as a measure for the estrogenic potential of the sample is determined using an appropriate substrate that is cleaved to a coloured or fluorescent reaction product (Routledge and Sumpter 1996).

YES was introduced as a new bio-autographic detection method for HPTLC analysis. In 2004, Müller, Dausend and Weins selected the YES screen from the then existing *in vitro* bioassays for estrogenic compounds because the yeast cells are more suitable for cultivation on HPTLC plates than other test organisms or organelles. Recombinant yeast cultures can be grown directly on HPTLC silica gel plates, where in the presence of estrogenic substances the enzyme ß-galactosidase is produced. Chlorophenol red-ß-D-galactopyranoside (CPRG) and 4-methylumbelliferyl ß-D-galactopyranoside (MUG) are used as enzyme substrates (100 mg per 100 ml of growth medium) (Müller et al. 2004).

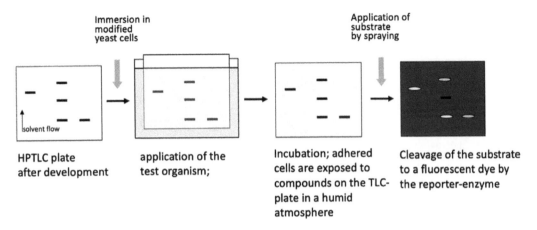

**FIG. 8**   Detection of Hormones. Complex workflow with test organism and enzyme reaction.

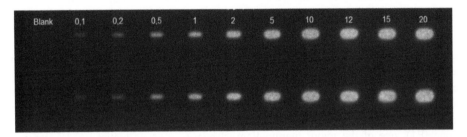

**FIG. 9**   Detection and estimation of the LLOQ of 17-β-estradiol (E2), by p-YES test, 17b-estradiol LLOQ ~ 0.1 pg/band (image provided by Sebastian Buchinger 2018, German Federal Institute of Hydrology, Koblenz, Germany).

The performance of the test as the combination of planar chromatography and YES test (p-YES) is shown in Figure 8.

An ISO guideline for the planar YES assay is under preparation by the DIN working group NA 119-01-03-05-13 AK Planar Yeast Estrogen Screen (p-YES). The p-YES Test is performed in three steps:

**Step 1**: immobilisation of the sample on a TLC plate (silica gel), granting an option to separate the estrogenically effective substances from its matrix during a chromatographic procedure (clean-up).

**Step 2**: immobilisation of the test organism on the solvent-freed silica gel layer and establishing good conditions for cell cultivation with contact of the cells with the investigated substances. The solvent freed TLC plate is dipped for about 2 seconds into a 24-hour-old culture of yeast-estrogenic-cells and incubated for nearly 3 hours at 32°C and 90–95% humidity.

**Step 3**: after the incubation and induction of the ß-galactosidase by the presence of estrogenically effective substances, a metabolic turnover of an appropriate substrate will prove the built enzyme. A fluorgenic substrate (4-methylumbelliferyl ß-D-galactopyranosid (MUG)) is sprayed on to the plate. Within 15–30 minutes, this substrate is cleft by ß-galactosidase to galactose and the fluorescent 4-methylumbelliferon (4MU) at 32°C. The fluorescent product can be evaluated by video-documentation or by an HPTLC scanner using the fluorescent mode of the instrument to quantify the effect. Figure 9 shows the Lower Limit of Quantification (LLOQ) of 17-β-estradiol (E2).

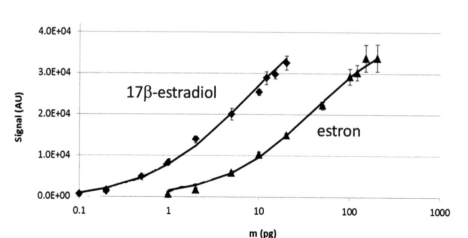

**FIG. 10** Determination of the LLOQ of 17-β-estradiol (E2) and estrone (E1) by a TLC scanner of the p-YES test (image provided by Sebastian Buchinger 2018, German Federal Institute of Hydrology, Koblenz, Germany).

17-β-estradiol is also commonly used as a component of the oral contraceptive pill and in hormone replacement therapy (HRT).

The results of the p-YES tests can be quantified as well by a densitometric evaluation using a TLC scanner. By this evaluation method, the dosis/effect relation of 17-β-estradiol (E2) and estrone (E1) could be shown (see Figure 10).

Estrone (E1), one of the major mammalian estrogens, is an aromatised C-18 steroid with a 3-hydroxyl group and a 17-ketone, while the LLOQ of estradiol (E2) is 0.1 pg/band on the TLC plate and the LLOQ of estrone is 1.0/band. For 17β-estradiol (E2), 'fish' were the most sensitive taxa and 'chronic reproductive effects' were the most sensitive endpoint. Using the SSD methodology, Cladwell et al. (2012) derived a PNEC (predicted-no-effect concentration) of 2 ng/L for E2. The authors could show that induction studies to determine the relative potency of the steroid estrogens to induce vitellogenin (VTG) derive PNECs of 6 and 60 ng/L for estrone (E1) (Cladwell et al. 2012). If we take into account that the ranges of estrogenic hormones in adult females is 0.5–9 pg/ml, which show effect in women, we can find that the p-YES test is suitable for determining single estrogen derivatives in the concentration range and traces within their effect level. Quantification is performed by using known estrogenic effective substances like 17 α-ethinylestradiol (EE2) or 17β-estradiol (E2) as reference substances. The content of unknown substances can be expressed by an estrogenic effective unit (EEQ) using a dose-response curve generated with the reference substances mentioned above (Buchinger et al. 2013).

### 3.5.1 Advantages of the p-YES test

In contrast to the *in vitro* YES test, the duration of the test could be diminished from around 18 hours to around 3 hours. The present limit of detection (LOD) is 0.5–2.0 pg/application spot.

This method is applicable to all kinds of matrices such as:

- drinking water and raw water
- surface water
- wastewater
- wastewater, highly contaminated with bacteria and micro-organisms
- wastewater with a high content of salts
- aqueous extracts and leachates

- eluates of sediments (fresh water)
- pore water
- aqueous solutions of single substances or of chemical mixtures
- water with a high turbidity
- sunscreens
- extractions of medical plants

## 3.6   DETECTION OF ESTROGENIC AND ANDROGENIC COMPOUNDS

EDA, in combination with HPTLC chromatography, allows the multi-parallel detection of diverse endocrine disruption activities. New *Saccharomyces cerevisiae*-based bioreporter strains could be constructed, responding to compounds with either estrogenic or androgenic activity, by the expression of green (EGFP), red (mRuby) or blue (mTagBFP2) fluorescent proteins. The applicability of the system could be demonstrated by separating influent samples of wastewater treatment plants, and simultaneously quantifying estrogenic and androgenic activities of their components (Moscovici et al. 2020.) This method and procedure will be described in detail in Chapter 14

## 3.7   DETECTION OF CARCINOGENIC AGENTS WITH **HPTLC**

There are several *in vitro* tests that have been used for three decades to predict the detection and assessment of DNA-damaging chemicals in the environmental genotoxicity field (Oda 2016). One of these is the *umu*[5] *in vitro* test, which has been widely used (see DIN EN ISO 38415-T3 1996-12).

There are two publications where the umu test was selected for determining potential genotoxic substances with HPTLC (Weiss 2021). Since the umu test necessitates a long incubation period (2 hours), one challenge on the plate is to minimise the band broadening caused by diffusion. According to Baumann et al., Weiss used a complex combination of calcium alginate solidification and the medical gauze (Baumann et al. 2003). Thus, he succeeded in using the umu test on the HPLC plate and detecting the effective genotoxic substance 4-nitroquinoline-N-oxide (4-NQO) known from DIN 38415-3 on the HPLC plate. The detection limit for 4-NQO is equivalent to an application amount of 3 ng.

In 2019, Shakibai et al. presented an innovative technological platform for monitoring the direct genotoxicity of individual components in complex environmental samples, based on bioluminescent *Escherichia coli* genotoxicity bioreporters, sprayed onto the surface of an HPTLC plate. These sensor strains harbour plasmid-borne fusions of selected gene promoters of the *E. coli* SOS DNA repair system to the *Photorhabdus luminescens* luxABCDE gene cassette, and mark by increased luminescence the presence of potentially DNA-damaging sample components separated on the plate. The authors showed an 'on plate' quantifiable dose-dependent response to several model genotoxicants (without metabolic activation). A specific application of this test will be described in Chapter 14.

## 4   CONCLUSION

It has been demonstrated that HPTLC is particularly eligible for Effect-Directed Analysis (EDA). The main advantages and results of this method are listed below:

1. The detection limit depends on the effectiveness of the substance.
2. The more effective a chemical in a defined test environment is, the lower the detection limit.

---

5   The umu test is based on the abilities of DNA-damaging agents to induce expression of the *umuC* gene responsible for SOS mutagenesis induced by radiation or chemical agents in *E. coli* [6]. The *umuC* gene is regulated by the *lexA* and *recA* genes of bacterial SOS response.

Using sample separation by planar chromatography offers several advantages for this purpose:

- separated components are preserved on the (HP)TLC plate where they can be analysed using a combination of physical, chemical and biological detection methods;
- separation of compounds from a wide polarity range (e.g., alkylphenols and alkylphenol ethoxylates), as well as a focusing effect, can be achieved by using special development techniques for HPTLC such as automated multiple development (AMD);
- up to 20 samples can be analysed simultaneously; and
- the HPTLC plate can be used as a growth surface for the bioassay test organism.

It has to be noted that it is not possible to keep the required metabolising system upright on the HPTLC plate to successfully detect the indirectly acting substances, which only have a genotoxic effect after activation by the metabolic enzymes.

## 5   REFERENCES

Baumann, Urs, Caroline Brunner, Ernst Pletscher, and Nicole Tobler. 2003. "Biologische Detektionsverfahren in Der Dünnschichtchromatographie." *Umweltwissenschaften und Schadstoff-Forschung* 15 (3): 163–67. https://doi.org/10.1065/uwsf2001.12.080.

Brack, Werner. 2003. Effect-directed analysis: A promising tool for the identification of organic toxicants in complex mixtures? *Analytical and Bioanalytical Chemistry*. 377. 397–407. 10.1007/s00216-003-2139-z. Accessed April 6, 2021. https://www.researchgate.net/publication/10624752_Effect-directed_analysis_A_promising_tool_for_the_identification_of_organic_toxicants_in_complex_mixtures

Buchinger, Sebastian, Denise Spira, Kathrin Bröder, Michael Schlüsener, Thomas Ternes, and Georg Reifferscheid. 2013. "Direct Coupling of Thin-Layer Chromatography with a Bioassay for the Detection of Estrogenic Compounds: Applications for Effect-Directed Analysis." *Analytical Chemistry* 85 (15): 7248–56. https://doi.org/10.1021/ac4010925.

Caldwell, Daniel J., Frank Mastrocco, Paul D. Anderson, Reinhard Länge, and John P. Sumpter. 2012. "Predicted-No-Effect Concentrations for the Steroid Estrogens Estron, 17β-Estradiol, Estriol, and 17α-Ethinylestradiol." *Environmental Toxicology and Chemistry* 31 (6): 1396–1406. https://doi.org/10.1002/etc.1825

Choma, I. M. et al. "Semiquantitative estimation of enrofloxacin and ciprofloxacin by thin-layer chromatography – direct bioautography". *Journal of Liquid Chromatography & Related Technologies*. 27 (2004), 2071–2085.

Colovic, Mirjana B., Danijela Z. Krstic, Tamara D. Lazarevic-Pasti, Aleksandra M. Bondzic, and Vesna M. Vasic. 2013. "Acetylcholinesterase Inhibitors: Pharmacology and Toxicology." *Current Neuropharmacology* 11 (3): 315–35. Accessed April 6, 2021 https://www.ncbi.nlm.nih.gov/pmc/articles/PMC3648782/

Eymann R., W. Fischer, H.E. Hauck and Ch. Weins. 2001. "Nachweis von Antibiotika in Futtermitteln durch wirkungsbezogene Analytik." *Fleischwirtschaft* 8 (2001), 95–96

HPTLC Association. 2022. "What is HPTLC." https://www.hptlc-association.org/about/what_is_hptlc.cfm. Accessed April 18, 2022

Liu J, Fu K, Wang Y, Wu C, Li F, Shi L, Ge Y, Zhou L. 2017. "Detection of Diverse N-Acyl-Homoserine Lactones in Vibrio alginolyticus and Regulation of Biofilm Formation by N-(3-Oxodecanoyl) Homoserine Lactone In vitro. " *Front Microbiol*. 2017 Jun 16; 8:1097. doi: 10.3389/fmicb.2017.01097. PMID: 28670299; PMCID: PMC5472671.

Móricz, Ágnes, Nóra Adányi, Erzsébet Horváth, Péter Ott, and Ernő Tyihák. 2008. "Applicability of the BioArena System to Investigation of the Mechanisms of Biological Effects." *Journal of Planar Chromatography – Modern TLC* 21 (6): 417–22. https://doi.org/10.1556/jpc.21.2008.6.4.

Morlock, Gertrud, and Yoshihisa Ueda. 2007. "New Coupling of Planar Chromatography with Direct Analysis in Real Time Mass Spectrometry." *Journal of Chromatography* A 1143 (1): 243–51. Accessed April 6, 2021, https://doi.org/10.1016/j.chroma.2006.12.056.

Moscovici, Liat, Carolin Riegraf, Nidaa Abu-Rmailah, Hadas Atias, Dror Shakibai, Sebastian Buchinger, Georg Reifferscheid, and Shimshon Belkin. 2020. "Yeast-Based Fluorescent Sensors for the Simultaneous

Detection of Estrogenic and Androgenic Compounds, Coupled with High-Performance Thin-Layer Chromatography." *Biosensors* 10 (11). https://doi.org/10.3390/bios10110169.

Müller, M. B., C. Dausend, Ch. Weins, and F.H. Frimmel. 2004. "A New Bioautographic Screening Method for the Detection of Estrogenic Compounds." *Chromatographia*, 60(3-4). https://doi.org/10.1365/s10 337-004-0315-8

Oda, Yoshimitsu. 2016. "Development and Progress for Three Decades in Umu Test Systems." *Genes and Environment* 38 (1). https://doi.org/10.1186/s41021-016-0054-8.

Routledge, Edwin J., and John P. Sumpter. 1996. "Estrogenic Activity of Surfactants and Some of Their Degradation Products Assessed Using a Recombinant Yeast Screen." *Environmental Toxicology and Chemistry* 15 (3): 241–48. https://doi.org/10.1002/etc.5620150303.

Shakibai, Dror, Carolin Riegraf, Liat Moscovici, Georg Reifferscheid, Sebastian Buchinger, and Shimshon Belkin. 2019. "Coupling High-Performance Thin-Layer Chromatography with Bacterial Genotoxicity Bioreporters." *Environmental Science & Technology* 53 (11): 6410–19. https://doi.org/10.1021/acs. est.9b00921.

Vandenberg, Laura N., Theo Colborn, Tyrone B. Hayes, Jerrold J. Heindel, David R. Jacobs, Duk-Hee Lee, Toshi Shioda, et al. 2012. "Hormones and Endocrine-Disrupting Chemicals: Low-Dose Effects and Nonmonotonic Dose Responses." *Endocrine Reviews* 33 (3): 378–455. Accessed April 03, 2021 https:// www.ncbi.nlm.nih.gov/pmc/articles/PMC3365860/.

Weins, Christel, "Möglichkeiten und Grenzen der wirkungsbezogenen Analytik mit der Hochleistungs – Dünnschichtchromatographie", Dissertation dated May 02, 2006 University of Basel, Accessed April 04, 2021 https://edoc.unibas.ch/411/1/DissB_7500.pdf

Weins, Christel. 2010. "Bioeffective-linked Analysis in modern HPTLC ", in *Quantitative Thin-Layer Chromatography*, 201–29, edited by Spangenberg, Bernd, Colin F. Poole, and Christel Weins. 2010. https://doi.org/10.1007/978-3-642-10729-0_8.

Weins, C., and H. Jork. 1996. "Toxicological Evaluation of Harmful Substances by *In Situ* Enzymatic and Biological Detection in High-Performance Thin-Layer Chromatography." *Journal of Chromatography. A* 750 (1-2): 403–7. https://doi.org/10.1016/0021-9673(96)00601-2.

Weiss, Stefan. n.d. "Einsatz Der Planarchromatographie mit wirkungsbezogener Detektion zur Untersuchung von Wässern. Dissertation." Accessed April 6, 2021. https://pub-data.leuphana.de/frontdoor/deliver/ index/docId/893/file/Dissertation_StefanWeiss2018.pdf.

# 13 Technical Report: Use of HPTLC to identify medicinal plants according to the EurPh and USP followed by detection of unknown bio-effective substances in plant drugs using effect-directed analysis (EDA)

*Christel Weins*

## 1 MOTIVATION

Seventy to 95% of the population still use traditional medicine (TM) for primary healthcare in Africa, Asia, Latin America and the Middle East. Some 100 million people are believed to use traditional, complementary or herbal medicine in the European Union (EU) alone – as much as 90% of the population in some countries. In 2012, global sales of Chinese herbal medicine reached US$ 83 billion –an increase of more than 20% compared to 2011.

Amid the COVID-19 crisis, the global market for Herbal Supplements and Remedies estimated at US$64.5 Billion in the year 2020, is projected to reach a revised size of US$119.9 Billion by 2027, growing at a CAGR of 9.3% over the period 2020-2027. Multi-Herbs, one of the segments analyzed in the report, is projected to record a 10.3% CAGR (Compound Annual Growth Rate) and reach US$68.5 Billion by the end of the analysis period. (Herbal Supplements: Global Strategic Business Report 2022)

The use of herbal medicinal products and supplements has increased tremendously over the past three decades with no fewer than 80% of people worldwide relying on them for some part of primary healthcare. The use of herbal remedies has also been widely embraced in many developed countries with complementary and alternative medicines (CAMs) now becoming mainstream in the UK and the rest of Europe, as well as in North America and Australia (Ekor 2014).

Meanwhile, modern medicine is desperately short of new treatments. Drugs take years to get through the research and development pipeline at enormous cost. And rising drug resistance, partly caused by misuse of medicines, has rendered several antibiotics and other life-saving drugs ineffective. Thus, scientists and pharmaceutical companies are increasingly searching TM for new drug sources. The best known is *artemisinin*, used to treat malaria. Ethnobotanical and other studies are now seeking other new anti-malarials. For example, a team at the University of Cape Town,

DOI: 10.1201/9781003146902-16

South Africa, has identified a compound that could evolve into the first single-dose cure for malaria. And researchers found healers using 28 plants to manage malaria in a single district of Zimbabwe (Ngarivhume 2014). TM could uncover new active compounds or validate treatments used as a first-line response against uncomplicated malaria (Graz et al. 2011).

With tens of thousands of plant species on earth, we are endowed with an enormous wealth of medicinal remedies from Mother Nature. Natural products and their derivatives represent more than 50% of all the drugs in modern therapeutics. Due to the low success rate and huge capital investment needed, the research and development of conventional drugs is very costly and difficult. Over the past few decades, researchers have focused on drug discovery from herbal medicines or botanical sources, an important group of CAM therapy (Pan et al. 2013).

Today, approximately 80% of anti-microbial, cardiovascular, immunosuppressive and anti-cancer drugs are of plant origin; their sales exceeded US$65 billion in 2003 (Gordaliza 2009). It is widely accepted that more than 80% of drug substances are either directly derived from natural products or developed from a natural compound (Maridass and de Britto 2008).

Although therapies involving these agents have shown promising potential, with the efficacy of a good number of herbal products clearly established, many of them remain untested and their use is either poorly monitored or not even monitored at all. It has been observed that most of the problems associated with the use of traditional and herbal medicines arise mainly from the classification of many of these products as foods or dietary supplements in some countries (Ekor 2014; see also Chapter 3 for reference).

The integration of herbal medicine into modern medical practices, including cancer treatments, must take into account the inter-related issues of quality, safety and efficacy. Quality is the paramount issue because it can affect the efficacy and/or safety of the herbal products being used. Current product quality ranges from very high to very low due to intrinsic, extrinsic and regulatory factors. Intrinsically, species differences, organ specificity, diurnal and seasonal variations can affect the qualitative and quantitative accumulation of active chemical constituents in the source medicinal plants (Fong 1970).

To describe and ensure the quality of herbal drugs, a suite of appropriate tests is recommended by regulatory agencies and organisations. Such tests, as well as specifications for compliance, are described in pharmacopoeial or other quality monographs. They include verification of identity and purity as well as determination of the amount of the active substance(s) or marker(s). In order to perform all tests, different analytical techniques and expertise are needed, and together with additional experiments (e.g., test for pesticides, mycotoxins, etc.), the overall costs of quality testing can increase dramatically (Frommenwiler et al. 2018).

## 2 IDENTIFICATION OF HERBAL MATERIALS BY HIGH-PERFORMANCE THIN-LAYER CHROMATOGRAPHY

Identification of medicinal herbs can be achieved by the application of multiple techniques, including macroscopic and microscopic descriptions, DNA analysis and chemical means. Chemical identification typically employs chromatographic or spectroscopic procedures to achieve the identification by fingerprint comparison against that of a reference standard, monograph description or a reference chromatogram. High-performance thin-layer chromatography (HPTLC) is the most advanced version of thin-layer chromatography (TLC) (HPTLC-Association 2012).

### 2.1 Use of HPTLC by the American Herbal Pharmacopoeia and the European Pharmacopoeia

TLC characterisations are among the key identity tests in most pharmacopoeial monographs. Pharmacopoeial standards are typically used by industry as a basis for meeting QC (Qualitiy Control) requirements and current good manufacturing practices (cGMPs). TLC is a relatively low-cost, highly versatile tool for developing specifications for raw materials, as well as for the various preparations for which pharmacopoeial standards are created. In addition to its use in the development of identity tests, TLC is a valuable tool for screening plant samples that pharmacopoeias must review in the development of monographs and botanical reference materials (BRMs). Specifically, HPTLC is the ideal TLC technique for these purposes because of its increased accuracy, reproducibility and ability to document the results, compared with standard TLC. As a result of this, HPTLC technologies are also the most appropriate TLC technique for conformity with GMPs (Upton 2019). Furthermore, the Inspection Guidelines of the European Medicines Agency say: "Identification tests should be specific for the herbal preparation, and optimally should be discriminatory with regard to substitutes/adulterants that are likely to occur. Identification solely by chromatographic retention time, for example, is not regarded as being specific; however, a combination of chromatographic tests (e.g., HPLC and TLC-densitometry) or a combination of tests into a single procedure, such as HPLC/UV-diode array, HPLC/MS, or GC/MS may be acceptable" (EMEA 2006).

HPTLC is a state-of-the art technique for plant analysis. It features significantly shorter developing times, lower solvent consumptions and improved resolution. Highly reproducible results and traceable records are achieved through a standardised methodology and the use of suitable instruments (typically controlled by software) for all steps of the analysis. A system suitability test is used to qualify results (HPTLC Association 2021).

The stationary phase is an HPTLC glass plate or aluminum sheet coated with a uniform thin layer (typically 200 micron) of porous particles (2 – 10 μm) with an average particle size of 5 μm. The layer typically consists of silica gel with a pore size of 60 Ångstroms, a polymeric binder and a so-called fluorescence indicator ($F_{254}$). The standard format of the plate is 20×10 cm (HPTLC Association 2021).

A Standard Operation Procedure (SOP) for HPTLC was created by Meier & Spriano (2012). The method is also documented in the U.S. Pharmacopeia (USP General Chapter <1064>, European Pharmacopoeia 7.0) and European Pharmacopeia (PhEur 2013). USP and PhEur establish written (documentary) and physical (reference) standards for medicines, food ingredients, dietary supplement products and ingredients. These standards are used by regulatory agencies and manufacturers to help to ensure that these products are of the appropriate identity, as well as strength, quality, purity and consistency. Table 1 shows the differences between a TLC method with the use of HPTLC in the identification of herbal materials.

**TABLE 1**
**The difference between TLC and HPTLC for the identification of herbal materials.**

|  | TLC | HPTLC |
|---|---|---|
| **Principle** | Planar chromatography |  |
| **Primary focus** | Simplicity, low cost | Reproducibility, separation power |
| **Process** | Flexible, no rules | Highly standardised methodology/optimised parameters |
| **Methods** | Only a few parameters defined | Well defined and validated |
| **Flexibility** | Very high | None for validated methods |
|  |  | High between methods |
| **Target** | Rapid, preliminary results | Reliable analytical results |
| **Data structure** | Simple chromatograms/ photographs | Traceable digital images/scan data qualified by system suitability test (SST) on each plate, cGMP compliant reporting |
| **Samples/references** | Side by side on the plate | On the same or on different plates |
| **Plate** | TLC any format | HPTLC 20x10 cm |
| **Instrumentation** | None to simple | Simple to sophisticated |
| **Cost** | Very low | Medium to high compared to TLC; low per sample compared to HPLC |

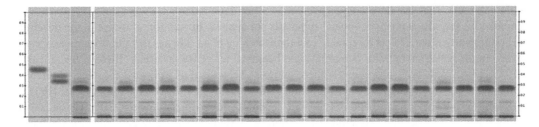

**FIG. 1**   The same chromatography at 20 different days, stationary phase silica gel 60 F $_{254,}$ detection UV light 254 nm (image provided by Eike Reich, CAMAG Switzerland, 2018).

As HPTLC is widely used to monitor the quality of produce of botanical origin on an increasingly globalised level, it seemed important to thoroughly standardise the analytical procedures involved.

In the USP General Chapter <1064> (n.d.) 'Identification of articles of botanical origin by High-Performance Thin-Layer Chromatography', the method is described and documented with pictures. The result of this description is that standardised methodology and validated methods give reproducible results from plate to plate, day to day and lab to lab (see Fig. 1 and Fig. 2). Figure 2 shows convincingly that following the SOP the results of chromatograms are comparable at any time – in this example after four years.

The PhEur (2013) Chapter 2.2.46 includes general information about all chromatographic separation techniques, system suitability definitions/requirements and chromatographic condition adjustments also known as allowable or allowed adjustments.

The extent to which the various parameters of a chromatographic test may be adjusted to satisfy the system suitability criteria, without fundamentally modifying the methods, are listed separately for all chromatographic methods including the TLC methods. In Figure 3, a System Suitability Test (SST) is introduced to check the quality and reproducibility of chromatography.

HPTLC is the method of choice for the identification of plant material in many pharmacopoeias, such as described in the PhEur (2013) Chapter 2.2.27. If combined with a suitable reference material for comparison, HPTLC can provide information beyond identification and thus may simplify quality control (Frommenwiler et al. 2018).

## 3   EFFECT DIRECTED ANALYSIS IN PLANT DRUGS

Plants are an abundant natural source of potential new medicines. Finding new medicines in plant drugs is a challenge. Phytotherapy gains great popularity in modern phytopharmacology. Today, as mentioned above, approximately 80% of anti-microbial, cardiovascular, immunosuppressive and anti-cancer drugs are of plant origin (Pan et al. 2013). This phenomenon is related to the wide and powerful healing properties, including supporting immune, nervous and digestive systems activity. Finding novel antibiotic substances from medicinal plants or substances for the treatment of neurodegenerative disorders, particularly Alzheimer's disease, are on the agenda of many pharmaceutical companies. Also, many women, for example, concerned about the health risks of the synthetic hormones used in conventional hormone therapy, are looking for natural alternatives. Plants can be important sources of many compounds with potential pharmaceutical applications.

Extraction of these matrices[1] is one of the ways of identifying the presence of active agents like inhibitory active substances against enzymes whose high activity leads to serious human diseases including cancer, Parkinson's or Crohn's disease. Besides the inhibitory effects against enzymes,

---

1   The active agent has to be freed from all disturbing substances of the cell matrix. Furthermore, for a better analysis the active agent can be enriched in the extract.

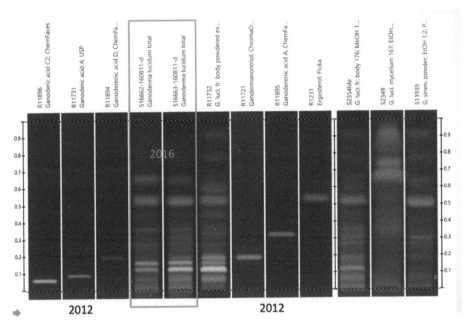

**FIG. 2** The chromatogram of *Ganoderma in 2012 and 2016,* stationary phase silica gel 60 $F_{254,}$ derivatisation: sulfuric acid reagent[a], detection UV light 366 nm (image provided by Eike Reich, CAMAG Switzerland, 2018).

[a] Sulfuric acid reagent preparation: 20 ml sulfuric acid is mixed with 180 ml ice-cooled ethanol, use: dip (e.g., time 0, speed 5), heat 100°C for 3 min (International Association for the Advancement of High Performance Thin Layer Chromatography 2016).

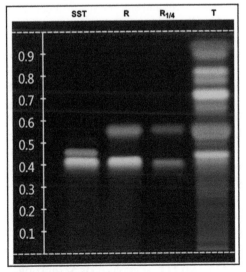

**FIG. 3** System Suitability Test (SST) in HPTLC (provided by Eike Reich, CAMAG Switzerland, 2018)

inhibitory effects against bacteria or fungi can be detected too. The isolation and purification of such inhibitors is time consuming and entails expensive steps in the analysis of the crude extract. Effect-directed chromatography with TLC/HPTLC can combine the separation of the extract on a thin layer with its subsequent biological analysis (Móricza et al. 2021).

As described previously in Chapter 12, bio-effective components of medicinal plants can be analysed by effect-directed analysis (EDA), i.e., characterised, identified and quantified. Unknown effects can be detected using HPTLC, which can be easily combined (hyphenated) with biological assays performed directly on a TLC plate that uses effect-directed detection. Using this method as an example, acetylcholinesterase (AChE) inhibitors can be found. Using different bacteria or fungi, it is possible to detect antibiotics or anti-fungal substances on the same chromatogram. Phytoestrogens are estrogens that occur in some plants. EDA is able to detect the presence and the amount of activity in herbs using transgene yeast cells. This procedure can be finalised by identifying the effective substances by mass spectrometry from the chromatogram (Morlock and Ueda 2007).

Legerskáa and her co-authors (2020) could show that effect-directed methods have been developed for several classes of enzymes including oxidoreductases, hydrolases and isomerases, and that there is a potential for developing functional methods for other classes of enzymes. This chapter summarises known effect-directed methods and their applications for determining the presence of enzyme inhibitors in extracts, and compares the effectiveness of different methodo-logical approaches (Legerskáa et al. 2020; Móricza et al. 2020). EDA with HPTLC represents a new strategy to detect active agents in a short time. Three examples will be presented below.

## 3.1 Active agents in giant goldenrod root extracts

Giant goldenrod (*Solidago gigantea* Ait.) is widespread across Europe and is a serious invader of abandoned fields, forest edges and river banks (Jakobs, Weber, and Edwards 2004).

Goldenrod is also a medicinal plant and is listed in the PhEur (2013) as *Solidaginis* herba (the whole or cut dried flowering aerial part of either *S. gigantea* Ait. and/or *S. canadensis* L.) used to treat disorders of the urinary tract, prostate and kidney. The extract of goldenrod was shown to display favourable anti-bacterial (Kołodziej 2011), anti-fungal (Webster et al. 2008), insecticidal (Benelli et al. 2018) and anti-obesity (Wang et al. 2017) activities that can be attributed to its essential oil, phenolics, saponins and diterpenes (Móricza et al. 2020).

Móricza and her team screened giant goldenrod root extract for bioactive compounds by HPTLC, coupled with EDA, and found anti-bacterial effects (*Bacillus subtilis* F1276, *B. subtilis* subsp. *spizizenii*, *Aliivibrio fischeri* and *Xanthomonas euvesicatoria*), anti-fungal effects (*Fusarium avenaceum*) and enzyme inhibition of acetyl- and butyrylcholinesterases (AChE and BChE), α- and β-glucosidases and α-amylase assays (Móricza et al. 2021).

The result is that the HPTLC chromatograms of the *S. gigantea* root extract showed five zones of AChE inhibition and five zones of BChE inhibition, *B. subtilis* subsp. *Spizizenii* causes two inhibition zones, *A. fischeri* results in six zones of luminescence inhibition. α-Glucosidase, β-glucosidase and α-amylase show three zones of inhibition. The growth and vitality of the fungus was proved by staining with iodo-nitrophenyl tetrazolium chloride (INT) and showed five inhibitory zones with different and separated polarities of the inhibitors.

The results show the anti-fungal, anti-bacterial and enzyme-inhibiting activity of multipotent isolates, which showed potential as lead compounds especially for various infectious plant diseases.

## 3.2 EDA and TLC screening of *Schisandra chinensis* fruits

The fruit of *Schisandra chinensis* (Chinese magnolia vine) is well-known in Traditional Chinese Medicine and is gaining great popularity in modern phytopharmacology. This phenomenon is related to the wide and powerful healing properties, including supporting immune, nervous and digestive systems activity. *S. chinensis* is also known for its adaptogenic properties, which can support the treatment of neuro-degenerative disorders, particularly Alzheimer's disease. (Sobstyl et al. 2020).

In this paper, EDA for acetylcholinesterase (AChE) inhibition and *Bacillus subtilis* growth inhibition was performed, followed by the micro-preparative separation of fractions, which were subsequently subjected to LC-MS tentative identification. The result was that EDA followed by MS analysis of *S. chinensis* fruit revealed components with biological activity, especially anti-bacterials and inhibitors of AChE.

### 3.2.1 Detection and identification of acetylcholinesterase inhibitors in *Annona cherimola* Mill.

Different extracts of *Annona cherimola* Mill. and the phytochemical constituents isolated from several parts of the plant exhibit pharmacological activities such as neuro-protective anti-Alzheimer's disease effect (Kazman et al. 2020). A paper from Chile describes the use of HPTLC-MS to detect and identify novel acetylcholinesterase inhibitors in cherimoya fruit (Galarce-Bustos et al. 2019). EDA analysis by planar chromatography-bioassay-MS was applied to detect and identify AChE inhibitors in pulp, peel and cherimoya seed. The bioassay was optimised, establishing the following conditions: enzymatic solution (1.0 U ml$^{-1}$), 1-naphtyl acetate substrate (1.5 mg ml$^{-1}$) and Fast Blue B salt (1.0 mg ml$^{-1}$). TLC-MS interface was used to directly elute the active zones into a mass spectrometer or to a micro-vial for further off-line studies.

### 3.3 TOXIC CONTAMINATIONS IN PHARMACEUTICAL HERBS AND HERBAL FOOD

Quality is a mandatory requirement of materials in order to accomplish the pharmaceutical 'Good Manufacturing Practices' (GMP). Today, the presence of pesticides in animal and vegetal commodities is a topic of public concern for the potential health hazards derived from them . The presence of pesticide residues in animal or vegetal raw materials can originate in agricultural practices, environmental contamination or through cross-contamination (Pérez-Parada et al. 2011). Active substances that are used in the manufacture of plant protection products are often impure in a characteristic way. These impurities can come from starting materials or may arise as a result of the production process or during storage. The material that comes directly from the production process is referred to as a technical active substance. When an application for authorisation is made, information on impurities must be submitted for each manufacturing site because the impurity profile is influenced by the manufacturing process and the manufacturing plant. Regulation EU No 283/2013 (EU 2014) defines impurities as being significant if they are present in the technical active substance in quantities of 1 g/kg or more. Impurities of toxicological and/or ecotoxicological or environmental concern are referred to as relevant impurities. They must be declared, even at contents less than 1 g/kg. The approval of an active substance is achieved by an Implementing Regulation, which states the minimum purity of the technical material as well as maximum limits for impurities. These are therefore official provisions that are to be considered for the authorisation of a plant protection product (BVL[2] 2021).

*Paeoniae Radix* Alba, *Chaenomelis Fructus* and *Moutan Cortex* represent three medicinal components of Chinese traditional medicine. In 2019, these plants were subjected to toxicological analysis to investigate possible pesticide contamination. Exposure using a point estimate model identified 47 residues that were simultaneously validated by the QuEChERS-UPLC-MS/MS[3] method. Of the 313 samples tested, 94.57% contained pesticide residues, with concentrations ranging from 0.10 to 1199.84 µg kg$^{-1}$, of which >83.17% contained 4–15 different residues. Carbendazim was the

---

2   The Federal Office of Consumer Protection and Food Safety (BVL) fulfils many tasks in the area of food safety. The aim of the BVL is to make the communication of risks more transparent and to manage risks before they turn into crises.

3   QuEChERS = a Quick, Easy, Cheap, Efficient, Rugged, Safe method for sample preparation, QuEChERS was created to facilitate the rapid screening of large numbers of food and agricultural samples for pesticide residues by Looser et al. 2006. UPLC = Ultra Performance Liquid Chromatography.

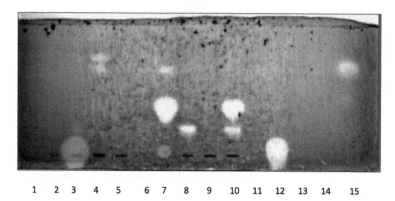

**FIG. 4** EDA with HPTLC demonstrates the presence of fungicides in 50 μL extracts of different plant-derived food samples post chromatographically: (1) control strawberries, (2–5, 9) strawberries, (6) control raisins, (7, 8, 10) raisins, (11–15) fungicide standards (30 ng fenpropathrine, 10 ng imazalile, 30 ng mercaptodimethur, and 10 ng procymidome per zone, exposure of the plate to 254 nm after 15 hours incubation time (Weins 2006).

most frequently detected pesticide (>85%), and procymidone, pendimethalin and phoxim were also abundant (median concentration = 15.33–623.12 μg kg$^{-1}$) (Xiao et al. 2019).

Furthermore, a Polish study of a total of 104 samples of herbal material (herbage of thyme, savory, sage, rock rose, marjoram, horsetail, oregano, basil; seeds of flax; roots of liquorice, valerian and lovage, flowers of coneflower and camomile and fruits of fennel and caraway) were analysed for the content of 250 pesticides. Residues of 16 pesticides were identified in 72.1% of the analysed herbal samples. In 11 of the analysed samples of thyme herbage and in one sample of basil herbage, concentrations exceeding the maximum allowable levels were demonstrated. Residues of the identified substances were detected most frequently in samples of thyme (66.34%), compared to the other groups of analysed herbal material, where the percentage share of samples containing the compounds sought was at the level of approximately 20% (Kowalska 2020),

As described in Chapter 12, the use of instrumental analysis for the identification of pesticides and their toxic metabolites requires the aid of selected reference substances. There exists a fundamental challenge in the selection of relevant reference substances, given that it is only possible to detect those substances that are actively sought by the analyst. Therefore, the analysis of individual substances does not detect unknown substances or metabolites and where the reference substance is lacking (Weins 2011).

Figure 4 shows the presence of fungicides in crude extracts of different plant-derived food samples in a screening test. The residues are fractionated by HPTLC according to their polarity. The effect-directed identification was performed with penicillium spec. The white spots show the inhibitory effects towards the growth rate of penicillium spec. The visualisation was carried out at exposure of the plate to 254 nm after 15 hours incubation time.

In Figure 5, EDA with HPTLC of methanolic extracts of salad and carrots inhibitory detected effects of acetylcholinesterase. The method is described in detail in Chapter 12.
Such screening methods allow the exclusion of negative samples very quickly, identify possible substances as contaminants and allow an idea of the intensity of the effect in correlation with the defined amount of a known active agent.

### 3.3.1  TLC – a validation method in low- and middle-income countries

With a relatively high degree of separation efficiency, target substances and unwanted impurities can be separated and then detected by different staining and visualisation methods. A quantitative scanning is also available in addition to an EDA. In this way, the unknown or related compounds in

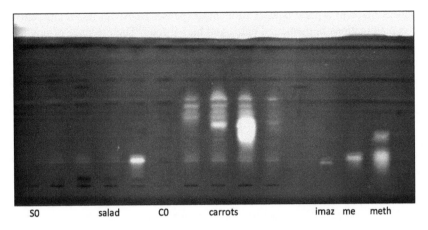

**FIG. 5**  Presence of esterase inhibitors in different plant food samples, S0 control salad, C0 control carrots, imaz: imazalile 10 ng, me: mercaptodimethur 30 ng, meth: methiocarb 10 ng (Weins 2006).

herbs or drugs can be screened quickly and analysed at a relatively low cost. For institutions and in countries with low budgets, especially, a TLC method can be combined with image analysis using a smartphone, 3D-printed light box and open-source ImageJ software. The paper of Sowers et al. (2022) describes that spots produced on the TLC plates were subsequently photographed using a smartphone camera and quantified using ImageJ's image analysis software. The pixel data collected from each plate's standard spots were compared to the data generated from its sample spots. Data sets collected across multiple TLC plates and numerous days of method performance were evaluated to assess linearity, accuracy, precision, specificity and robustness of the analytical method. Across the range of 75–125% of the target concentration, the method was found to have linearity of standard spots (with R2 generally greater than 0.99), overall accuracy of 101.0% (4.1% RSD[4]), repeatability pooled standard deviation of 2.44%, intermediate precision pooled standard deviation of 3.68% and observed demonstration of specificity and robustness. In low- and middle-income countries (LMICs), quality screening of pharmaceutical products can be challenging when testing resources are expensive, difficult to procure or complex to utilise (Sowers et al. 2022).

## 4   CONCLUSION

- HPTLC is very suitable for documenting and identifying herbal drugs according to the PhEur and UPS;
- HPTLC ensures quality of tests from plate to plate, day to day, lab to lab, year to year;
- HPTLC fingerprints allow the differentiation of chemical traits of plants of the same phenotype;
- EDA with HPTLC enables pharmaceutical companies to generate a so-called bio-profiling of the plant extracts, testing various bacterial strains and enzymes by HPTLC-EDA to identify the fingerprint of the active agents;
- This procedure is very useful especially for detecting new antibiotics. This strategy is less time- and cost-consuming than conventional methods because, besides the result of the effectiveness of the extract, you get an impression of the polarity of the agent. This is important with regards to the pharmacological characteristics of the substance. Furthermore, combining EDA with mass spectrometric methods allows the identification of bio-active substances;
- Due to the fact that there are thousands of unknown harmful contaminants and their toxic transformation products, the development of an analytical procedure for each group of

---

4    RSD= relative standard deviation

substances involves a great deal of time and expense. EDA can be an alternative to screen toxic contaminations in medicinal herbs and herbal food. The presence of unknown harmful agents can only be detected by their biological effect, so that the combination of planar chromatography and effect-directed detection has become of great importance;

- TLC methods are a suitable validation method in low- and middle-income countries.

# 5  REFERENCES

Benelli, Giovanni, Roman Pavela, Kevin Cianfaglione, David U. Nagy, Angelo Canale, and Filippo Maggi. 2018. "Evaluation of Two Invasive Plant Invaders in Europe (Solidago Canadensis and Solidago Gigantea) as Possible Sources of Botanical Insecticides." Journal of Pest Science 92 (2): 805–21. https://doi.org/10.1007/s10340-018-1034-5.

BVL 2021, Active Substances, https://www.bvl.bund.de/EN/Tasks/04_Plant_protection_products/01_ppp_tasks/08_ProductChemistry/02_ppp_actsubst_physchem/ppp_actsubst_physchem_node.html, 28.05.2021

Ekor, Martins. "The Growing Use of Herbal Medicines: Issues Relating to Adverse Reactions and Challenges in Monitoring Safety." Frontiers in pharmacology. Frontiers Media S.A., January 10, 2014. https://www.ncbi.nlm.nih.gov/pmc/articles/PMC3887317/.

EMEA. 2006. https://www.ema.europa.eu/en/documents/scientific-guideline/guideline-specifications-test-procedures-acceptance-criteria-herbal-substances-herbal-preparations/traditional-herbal-medicinal-products-revision-1_en.pdf. Accessed April 15, 2022.

EU. 2014. "*Commission Regulation (EU) No 283/2013 of 1 March 2013 setting out the data requirements for active substances, in accordance with Regulation (EC) No 1107/2009 of the European Parliament and of the Council concerning the placing of plant protection products on the market.*" Retrieved May 6, 2022, from https://eur-lex.europa.eu/eli/reg/2013/283/oj.

European Pharmacopoeia 7.0, Chapter 2.2.27, Thin-layer chromatography, https://www.drugfuture.com/Pharmacopoeia/EP7/DATA/20227E.PDF

Fong, Harry H.S. "Integration of Herbal Medicine into Modern Medical Practices: Issues and Prospects." Mendeley, January 1, 1970. https://www.mendeley.com/catalogue/4a7a4117-a1d3-3e48-8292-275f6a4cc689/.

Frommenwiler, Débora, Jonghwan Kim, Chang-Soo Yook, Thi Tran, Salvador Cañigueral, and Eike Reich. 2018. "Comprehensive HPTLC Fingerprinting for Quality Control of a Herbal Drug – the Case of Angelica Gigas Root." *Planta Medica* 84 (06/07): 465–74. https://doi.org/10.1055/a-0575-4425.

Galarce-Bustos, Oscar, Jessy Pavón, Karem Henríquez-Aedo, and Mario Aranda. 2019. "Detection and Identification of Acetylcholinesterase Inhibitors in Annona Cherimola Mill. By Effect-Directed Analysis Using Thin-Layer Chromatography-Bioassay-Mass Spectrometry." Phytochemical Analysis 30 (6): 679–86. https://doi.org/10.1002/pca.2843.

Gordaliza, M. "Terpenyl-purines from the sea," Marine Drugs, vol. 7, no. 4, pp. 833–849, 2009, https://www.mdpi.com/1660-3397/7/4/833/htm

Graz, Bertrand, Andrew Y Kitua, and Hamisi M Malebo. "To What Extent Can Traditional Medicine Contribute a Complementary or Alternative Solution to Malaria Control Programmes? – Malaria Journal." BioMed Central. BioMed Central, March 15, 2011. https://malariajournal.biomedcentral.com/articles/10.1186/1475-2875-10-S1-S6.

Herbal Supplements: Global Strategic Business Report 2022, https://www.researchandmarkets.com/reports/338400/herbal_supplements_and_remedies_global_market

HPTLC Association. 2012. Identification of herbal materials by High-Performance Thin–Layer Chromatography (HPTLC), published in Pharmacopoeial Forum USP PF 403 (2014) and published in USP 38 – NF 33 as General Chapter <1064> Identification of Articles of Botanical Origin by High-Performance Thin-layer Chromatography (2015);

HPTLC Association. 2016. SOP Ganoderma fruiting body, www.hptlc-association.org

HPTLC Association. 2021. International Association for the Advancement of High-Performance Thin-Layer Chromatography, https://www.hptlc-association.org/about/what_is_hptlc.cfm, Accessed 28.05. 2021

Jakobs, Gabi, Ewald Weber, and Peter J. Edwards. 2004. "Introduced Plants of the Invasive Solidago Gigantea (Asteraceae) Are Larger and Grow Denser than Conspecifics in the Native Range." *Diversity and Distributions* 10 (1): 11–19. https://doi.org/10.1111/j.1472-4642.2004.00052.x.

Kazman, Bassam S. M. Al, Joanna E. Harnett, and Jane R. Hanrahan. 2020. "The Phytochemical Constituents and Pharmacological Activities of Annona Atemoya: A Systematic Review." *Pharmaceuticals* 13 (10): 269. https://doi.org/10.3390/ph13100269.

Kowalska, Grażyna. 2020. "Pesticide Residues in Some Polish Herbs." *Agriculture* 10 (5): 154. https://doi.org/10.3390/agriculture10050154.

Kołodziej, Barbara. 2011. "Antibacterial and Antimutagenic Activity of Extracts Aboveground Parts of Three Solidago Species: Solidago virgaurea L., Solidago canadensis L. and Solidago gigantea Ait." *Journal of Medicinal Plants Research* 5 (31). https://doi.org/10.5897/jmpr11.1098.

Legerskáa Barbora, Daniela Chmelováa, Miroslav Ondrejovica and Stanislav Miertušab. 2020. "The TLC-Bioautography as a Tool for Rapid Enzyme Inhibitors detection – A Review". *Critical Reviews in Analytical Chemistry*, https://doi.org/10.1080/10408347.2020.1797467

Looser, Nadja, Drazen Kostelac, Ellen Scherbaum, Michelangelo Anastassiades, and Hubert Zipper. 2006. "Pesticide Residues in Strawberries Sampled from the Market of the Federal State of Baden-Württemberg in the Period between 2002 and 2005." Journal Für Verbraucherschutz Und Lebensmittelsicherheit 1 (2): 135–41. https://doi.org/10.1007/s00003-006-0022-5.

Maridass M. and A. John de Britto. 2008. "Origins of plant derived medicines". *Ethnobotanical Leaflets*, vol.12, pp. 373–387.

Meier, B., and A. Spirano. 2012. Review of *Standard Operating Procedure for HPTLC*. Www.hptlc-Association. org. International Association for the Advancement of High-Performance Thin-Layer Chromatography. December 19, 2012. https://www.hptlc-association.org/media/PM5SU520/SOP_for_HPTLC_ZHAW_v3_26.7.2018.pdf. Accessed 4 May 2022.

Móricza, Ágnes M., Dániel Krüzselyi, Péter G. Ott, Zsófia Garádi, Szabolcs Béni, Gertrud E. Morlock and József Bakonyi, "Bioactive Clerodane Diterpenes of Giant Goldenrod (Solidago gigantea Ait.) Root Extract." 2021. *Journal of Chromatography A* 1635 (January): 461727. https://doi.org/10.1016/j.chroma.2020.461727.

Morlock, Gertrud, and Yoshihisa Ueda. 2007. "New Coupling of Planar Chromatography with Direct Analysis in Real Time Mass Spectrometry." *Journal of Chromatography A* 1143 (1): 243–51. Accessed April 6, 2021, https://doi.org/10.1016/j.chroma.2006.12.056

Ngarivhume, Talkmore, Charlotte I.E.A. van't Klooster, Joop T.V.M. de Jong, and Jan H. Van der Westhuizen. "Medicinal Plants Used by Traditional Healers for the Treatment of Malaria in the Chipinge District in Zimbabwe." Journal of Ethnopharmacology. Elsevier, November 18, 2014. https://www.sciencedirect.com/science/article/pii/S0378874114007946.

Pérez-Parada Andrés, Marcos Colazzo, Natalia Besil, Eduardo Dellacassa, Verónica Cesio, Horacio Heinzen and Amadeo R. Fernández-Alba: Pesticide Residues in Natural Products with Pharmaceutical Use: Occurrence, Advances and Perspectives, in. Pesticides in the Modern World: Trends in Pesticides Analysis. Google Books. Edited by Margarita Stoytcheva 2011; BoD – Books on Demand. https://books.google.de/books?hl=de&lr=&id=i-GdDwAAQBAJ&oi=fnd&pg=PA357&dq=pesticide+residues+in+pharmaceuticals&ots=0axg20tETA&sig=RndyGIneMcFuuzR-sjFBVTTWfwI#v=onepage&q=pesticide%20residues%20in%20pharmaceuticals&f=false.

Pan, Si-Yuan, Shu-Feng Zhou, Si-Hua Gao, Zhi-Ling Yu, Shuo-Feng Zhang, Min-Ke Tang, Jian-Ning Sun, et al. 2013. "New Perspectives on How to Discover Drugs from Herbal Medicines: CAM's Outstanding Contribution to Modern Therapeutics." Evidence-Based Complementary and Alternative Medicine 2013: 1–25. https://doi.org/10.1155/2013/627375.

PhEur. 2013. *European Pharmacopoeia : Published in Accordance with the Convention on the Elaboration of a European Pharmacopoeia (European Treaty Series No. 50).* Strasbourg: Council Of Europe.

Sobstyl Ewelina, Agnieszka Szopa, Halina Ekiert, Sebastian Gnat, Rafał TypekaIren, and Maria Choma "Effect Directed Analysis and TLC Screening of Schisandra Chinensis Fruits." 2020. Journal of Chromatography A 1618 (May): 460942. https://doi.org/10.1016/j.chroma.2020.460942.

Sowers, Mary Elizabeth, Ryan Ambrose, Ed Bethea, Christopher Harmon, and David Jenkins. "Quantitative Thin-Layer Chromatography for the Determination of Medroxyprogesterone Acetate Using a Smartphone and Open-Source Image Analysis." Journal of Chromatography A. Elsevier, March 9, 2022. https://www.sciencedirect.com/science/article/pii/S0021967322001406.

Upton, Roy T. "Use of High-Performance Thin-Layer Chromatography by the American Herbal Pharmacopoeia." OUP Academic. Oxford University Press, November 27, 2019. https://academic.oup.com/jaoac/article/93/5/1349/5655759?login=false.

USP General Chapter <1064>. N.d. Identification of articles of botanical origin by High-Performance Thin-Layer Chromatography, https://hmc.usp.org/about/general-chapters

Wang, Zhiqiang, Ju Hee Kim, Young Soo Jang, Chea Ha Kim, Jae-Yong Lee, and Soon Sung Lim. 2017. "Anti-Obesity Effect of Solidago Virgaurea Var. Gigantea Extract through Regulation of Adipogenesis and Lipogenesis Pathways in High-Fat Diet-Induced Obese Mice (C57BL/6N)." Food & Nutrition Research, February. https://foodandnutritionresearch.net/index.php/fnr/article/view/1144.

Webster D., P. Taschereau, R.J. Belland, C. Sand, R.P. Rennie, "Antifungal Activity of Medicinal Plant Extracts; Preliminary Screening Studies." 2008. Journal of Ethnopharmacology 115 (1): 140–46. https://doi.org/10.1016/j.jep.2007.09.014.

Weins, Christel, "Möglichkeiten und Grenzen der wirkungsbezogenen Analytik mit der Hochleistungs – Dünnschichtchromatographie", Dissertation vom May 02, 2006 University of Basel, Accessed April 04, 2021 https://edoc.unibas.ch/411/1/DissB_7500.pdf

Xiao Jinjing, Xing Xu, Fan Wang, Jinjuan Ma, Min Liao, Yanhong Shi, Qingkui Fang, Haiqun Cao, Analysis of exposure to pesticide residues from Traditional Chinese Medicine, Journal of Hazardous Materials, Volume 365, 2019, Pages 857–867, ISSN 0304-3894, https://doi.org/10.1016/j.jhazmat.2018.11.075. (https://www.sciencedirect.com/science/article/pii/S030438941831104X)

# 14 Technical Report: Detection of harmful substances in water and sediments by Effect-Directed Analysis (EDA) as part of health, safety and environmental risk assessment

*Christel Weins*

## 1 MOTIVATION

Synthetic[1] contaminants like pesticides, including herbicides and insecticides, pharmaceutical substances, including hormones and their metabolites, and transformation products can cause harmful impacts on the environment. Trace levels of synthetic contaminant residues present in water but also in soil, air and sometimes in food may result in harmful effects for human and environmental health (Kookana et al. 1998).

Over time, synthetic contaminants dissolve and accumulate in lipophilic phases of animals (e.g., in fatty tissue) and plants (e.g., within essential oil glands). The net enrichment of contaminants in organisms relative to that in the environment is described by the process of bio-accumulation. Furthermore, an organism can modify the absorbed mixture of any contaminant. Some chemicals are retained and accumulate. Others that are more water soluble or degradable are eliminated from the body, resulting in no net accumulation there. In the food web, animals thus show very different bio-accumulation of various chemicals, both in levels and in the relative composition (Mandal et al. 2016). In 2003, the World Health Organization put its guard on Persistent Organic Pollutants (POPs) of anthropogenic origin that resist degradation and accumulate in the food chain. They can be transported over long distances in the atmosphere, resulting in widespread distribution across the earth, including regions where they have never been used. Owing to their toxicity, they can pose a threat to humans and the environment (WHO 2003).

Consequently, the Environmental Risk Assessment (ERA) for human and veterinary pharmaceuticals was accepted by the European legislation with two EC Directives 2004/27/EC and 2004/28/EC laws. Today, the ERA is conducted based on the EMA[2] guidelines (Sangion and

---

1  'Synthetic' means anthropogenic. It is often referred to as 'organic' because of its relation to 'organic chemistry' originating from 'petrochemistry' (historically opposed to 'inorganic chemistry'). The connotation is not that of 'organic' used in certain types of agriculture, where processes are intended to be more natural or close to nature, precisely devoid of harmful anthropogenic substances.

2  European Medicines Agency

Gramatica, 2016). An ERA refers to the acute toxic risk occurring in the aquatic environment. This risk is calculated based on the ratio between the predicted environmental concentration (PEC) of the compounds, and the highest predicted no-effect concentration (PNEC) of these compounds. A PEC:PNEC ratio <0.1 is considered an insignificant risk; 0.1–1: low risk; 1–10: moderate risk; and >10: high risk. An ERA assesses the likelihood of the organic pollutants causing harm to the environment. This includes describing potential hazards and impacts before taking precautions to reduce the risks (Migrator 2011). As an example of how to carry out an ERA, Migrator proposes five key steps:

- identify any hazards, i.e., possible sources of harm
- describe the harm they might cause
- evaluate the risk of occurrence and identify precautions
- record the results of the assessment and implement precautions
- review the assessment at regular intervals

The emphasis of efforts towards rational environmental protection – in the sense of 'sustainable development' – lies in the careful exploitation of natural resources and the protection of these natural resources. Prophylactic, active, environmental protection means risk management with the aim of protecting people from environmental risks as well as the recognition and limitation of environmental risks. Reliable data concerning the environmental interaction of contaminants forms the basis of an estimate of environmental risks. An estimate of risk requires – on the one hand – a knowledge of the effect of the individual contaminant on the environment and of the cumulative effect in the ecosystem – and on the other hand – the degradation behaviour and the effect of metabolites (degradants).

The most important concern of risk analysis is to identify the effect of damage by quantity (risk identification) followed by the identification of toxic substances or groups of substances. Risk identification should encompass all risks and rapidly lead to precise results that can be evaluated. Risk analysis demands evaluation of the results using independent methods (i.e., obtained using physico-chemical, biological, biochemical and molecular biological methods alike). Within risk analysis, it is necessary to quantify both damaging effects and establish threshold levels in order to determine an 'acceptable risk'.

## 1.1 PROBLEMS WITH THE CURRENT ANALYTICAL CONCEPT REGARDING WATER BODIES

Analytical methods for single substances cannot possibly be applied to a palette of more than 30,000 relevant chemicals and their degradation products at a time. History of environmental analysis reveals that the analysis of single substances has generally underestimated the number of primary contaminants and, hence, it is not a suitable instrument for risk management, in the sense of rational environmental protection.

Targeting the most relevant organic micro-pollutants (OMP) in routine analysis appears difficult due to the formation of transformation products of unknown concentration or toxicity. Performance assessment of water purification, for example, is still based upon limited target data. To assess micro-pollutant elimination, persistence or formation during different seasons, considering local redox conditions, travel distances and total component number in the river, non-target analysis features are the methods of choice.

Effect-directed analysis (EDA) with thin-layer chromatography (TLC) is proven to be valuable for a more comprehensive assessment of unknown OMP (Oberleitner et al. 2020).

## 1.2 Requirements of the method

In order to be able to detect organic pollutants having biological-toxic effects present in the environmental sample, it is necessary that the method selected should be as universal as possible. However, it is not the selectivity with respect to individual substances that is of importance for the method. The goal for the analyst is rather the detection of all or at least as many as possible of the organic pollutants present in the environmental sample.

Hence, the choice of sample preparation is of decisive importance for this procedure: whether selective sample preparation shall be used to detect one specific active substance in the environmental sample, or whether universal sample preparation shall be used to detect as many active substances as possible. A further parameter, the enrichment factor, depends on the toxicity of the specific active substances. It is possible to detect very small traces of highly toxic substances by means of activity analysis, so that enrichment may not have to be carried out under these circumstances.

## 1.3 Principle of the method

EDA with TLC involves a coupling of two different methods. First, a chromatographic separation is used for the determination of selected organic pollutants. This is followed by the physical/chemical assessment and a biological/biochemical toxicity test, thus allowing a direct activity-dependent evaluation to be made after the chemical/physical characterisation.

Specific enzyme inhibition tests on the thin-layer plate or test procedures involving organisms that use inhibition of bacterial growth, inhibition of bacterial luminescence or inhibition of the growth of a yeast strain as the signal serve to detect the presence of toxicologically relevant pollutants. The procedure is described in detail in Chapter 12.

# 2 DETECTION OF UNKNOWN HARMFUL SUBSTANCES IN WATER AND MARINE SEDIMENT BY EDA

The amount of environmental pollution by active agents is increasing, because of their intensive use and release into the environment. Additionally, active agents present in the environment can be accumulated in living organisms; this in turn can have a negative effect on the biotic elements of ecosystems (e.g., toxicity and disturbance of endocrine equilibrium).

Toxic chemicals, as considered here, are individual chemicals or mixtures of chemicals and their by-products that originate from human activities. These are generally toxic chemicals that have not yet been identified in the assessment but may be adversely affecting organisms.

Such chemicals may be unknown because they have not been measured or measurement is difficult (e.g., due to episodic occurrence, unique chemistry or low concentrations) (EPA 2022).

Micro-contamination by incompletely purified wastewater is widespread in the environment. The contaminants are chemicals and by-products with a wide polarity range. In order to separate, detect and identify such chemicals within a mixture of unknown toxic substances with a wide polarity range, a gradient separation method must be used.

## 2.1 AMD to detect unknown toxic substances by a gradient separation

AMD stands for Automated Multiple Development. A multiple development is the simplest method of performing a TLC-gradient separation. The separation is carried out using either short or maximum development distances. As shown in Figure 1, an AMD chromatogram is the result of several chromatographic runs. Usually, the number of runs lies between 10 and 40. The migration distance of the solvent front in each single run is longer for a constant increment (const.) than the migration

Number of steps

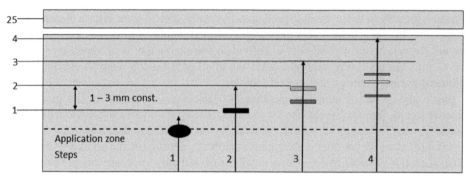

**FIG. 1**   Principle of a separation of unknown substances with a wide range of polarity by AMD (according to Burger 1984).

distance in the previous run. Typical increments of migration distances reach from 1 to 3 mm. After each development, the layer is dried and the whole process is repeated once again using another solvent with another polarity from very polar (e.g., methanol) to non-polar (e.g., hexane). AMD is suitable for separating unknown substances in a crude extract with a wide polarity range and is thus well suited to environmental screening as well as separating complex samples, for example, in food analysis (Burger 1995).

The next examples will show the harmful effect caused by wastewater effluents into a river basin.

## 2.2   DETECTION OF FUNGICIDAL ACTIVE SUBSTANCES IN SURFACE WATER SAMPLES

Water samples from 10 different sites in Luxembourg, France, Rhineland-Palatinate/Germany and Saarland/Germany were investigated according to fungicidal active substances in surface water. As described above, in the analysis of individual substances, it is not possible to detect unknown substances or active metabolites with biological/toxicological activity. There are about 300 permitted active fungicides and the development of an analytical procedure for even one group of substances involves a great deal of time and expense. Nonetheless, the possibility still remains that the polluter will already have released an alternative product into the environment that is not detected by the measurement programme that has been set up.

Procedure:

The organic substances contained in a 500 ml sample of surface water were enriched by solid phase extraction on methanol-conditioned RP18 silica gel cartridges according to DIN V 38407-11:1995-01 (1995), extracted with methanol, evaporated to dryness under $N_2$ and dissolved in 400 μl methanol. For the chromatographic separation 50–100 μl of this extract was applied onto a High-Performance Thin-Layer Chromatography (HPTLC) plate, Silica Gel 60 $F_{254}$. The chromatography was performed by AMD with a 33-step solvent gradient based on dichloromethane with acetonitrile as a polar and n-hexane as a non-polar component described in DIN V 38407-11:1995-01 (1995).

A preliminary identification was then carried out using the spectral library available, where the running distance of the standard substance is considered. Figure 2a/b shows the first results of screening for fungicides in surface waters by spectral analysis after chromatography.

The effects of fungicides were detected using mould spores as the detection organism which were directly applied onto the HPTLC plate (see Figure 3). Strains of penicillium were obtained from isolates of food from Dr. Stempka (State Institute for Health and Environment of the Saarland) and maintained on Sabouraud agar plates (oxoid). Spore suspension of penicillium for the assay was prepared using a glucose-mineral salts medium. The method is described in detail in Chapter 12.

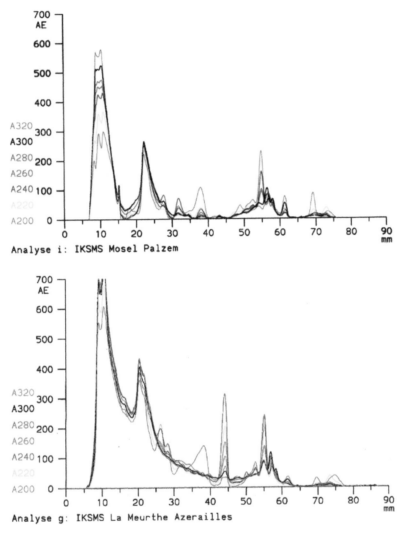

**FIG. 2**  Selection of a chromatogram of organic compounds according to their polarity in water samples of the river basin Mosel at Palzem/Germany (a) and of the river basin La Meurthe at Azerailles/France (b), detection and documentation by a multi-wavelength scan at seven wavelengths from 200 to 320 nm; TLC Scanner 3 from Camag, Muttenz, Schweiz (Weins 2006).

Figure 3 indicates the likely presence of two to three fungicides in samples 1 to 10. When trying to identify the contaminants, it is possible to exclude numerous fungicides (e.g., procymidon, vinclozolin captafol, chlozolinat etc.), since their position on the chromatogram does not correspond to that of the substances detected here.

It was decided to test these samples for carbendazim, imazalil, bitertanol, etc., and in the case of sample 8, for azoles (see Table 1).

## 2.3  WASTEWATER CHARACTERISATION BY EDA WITH *PHOTOBACTERIUM PHOSPHOREUM*

The comparatively quick and low-cost bioassay with the luminescent marine bacterium *Photobacterium phosphoreum*, strain NRRL-B-11177, has gained a considerable popularity for the monitoring of various industrial effluents and for the toxicity of different chemicals. Many toxic

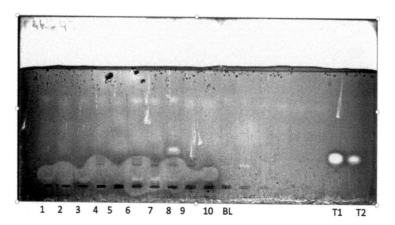

**FIG. 3** Testing surface waters for the possible presence of fungicides; 1–11 surface water samples, T1, T2 standards of Tebuconazol 20 ng, 10 ng, Test organism: *Penicillium* spec. post chromatographically, 1–3 Luxembourg, 4–7 France, 8–10 Germany, BL blank; Incubation time 10 hours, detection at 254 nm (Weins 2006).

**TABLE 1**
**Results of screening for fungicides in surface waters by spectral analysis post chromatography.**

| Country | Fungicide investigated | Comments |
|---|---|---|
| Luxembourg | Imazalil, fenpropidin, benalaxyl | - |
| France | Benalaxyl, isoproteron or metoxuron, carbendazim, fenpropimorph or benalaxyl, iprodion | |
| Saarland/Germany | Carbendazim, imazalil, epoxiconazol | |
| Rhineland-Palatinate/ Germany | Carbendazim, imazalil, benalaxyl | Caffeine, emitted from communicipal wastewater treatment plants at migration distance of 44 mm |

substances (nearly 1350 individual organic compounds) show an inhibition of the bioluminescence of *Photobacterium phosphoreum* and *Vibrio fischeri in vitro* (Kaiser and Palabrica 1991). In HPTLC, these substances have been identified post chromatographically *in situ* on the plate by dipping the plate for 2 seconds into a suspension of bacteria and determinating the difference of photone-emission using a cooled charged coupled device camera in a dark chamber (see Chapter 12).

A linear correlation between the inhibition of bioluminescence of *Photobacterium phosphoreum* and the concentration of an inhibitor could be shown. Under the condition described, the detection limit of pentachlorophenol (PCP) was found to be between 10 and 20 ng on the HPTLC plate while the detection limit of dichlorophenol could be observed in the range of 7.5 ng (Weins 2006).

To detect as many organic pollutants as possible, 40 ml sample of wastewater from a coke plant were extracted at pH 7 and pH 2 with cyclohexane or chloroform (a, b). The extract was evaporated to dryness under $N_2$ and then taken up in 40 µl cyclohexane or chloroform. In order to check for the possible presence of readily volatile compounds, a parallel substance enrichment is carried out on methanol-conditioned RP18[3] silica gel cartridges by solid phase extraction, in accordance with DIN V 38407-11:1995-01 (1995) (c).

---

3   RP18: reversed phase with alkane substitutent attached to silicone of a chain length of 18 hydrocarbon groups.

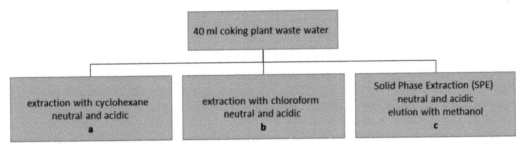

**FIG. 4** Sample of wastewater sample into polar and non-polar substance groups by different methods.

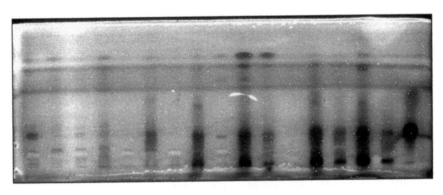

K1n  2n  K3s  4s  K5n  6n  K7s  8s  K9n  10n  11n  K12s  13s  14s  15s  16s  17

**FIG. 5** Luminescent bacteria test on the chromatogram (exposure time 800 sec.): (1–4 cyclohexane extracts, 5–8 chloroform extracts, 9–14 enrichment via RP18 cartridges, 15–16 methanolic extraction of the dried residue; 17 DCP/PCP 30 ng; K = control; n = water sample neutral; s = water sample acid at pH 2) (Weins, 2006).

Additional 40 ml of wastewater from a coke plant were freeze dried and the dry residue was extracted with 400 μl methanol. Figure 4 summarises the sample preparation of 40 ml coking plant wastewater.

For the chromatographic separation, 5–50 μl of these extracts was applied onto an HPTLC plate, Silica Gel 60 $F_{254}$. The chromatography was performed by AMD with a 33-step solvent gradient based on dichloromethane with acetonitrile as a polar and n-hexane as a non-polar component described in DIN V 38407-11:1995-01 (1995).

After removing the mobile phase by a cold air stream, the plate was dipped in a suspension of *Photobacterium phosphoreum*. The method is described in detail in Chapter 12. The biological detection in HPTLC shows that even working standards, such as pentachlorophenol or dichlorophenol, are very often contaminated with toxic impurities, which can only be detected after chromatographic separation. Coking plant wastewater, which was found to inhibit the luminescence of luminescent bacteria in an *in vitro* test (DIN EN ISO 11348-1 2009), has been investigated by various extraction procedures, in order to identify the substances responsible for this inhibition. The inhibition of the bioluminescence is documented in Figure 5.

It was possible to detect varying quantities of bioluminescence-inhibiting substances as a function of the extraction. The recovery rates for bioluminescence inhibitors increases with the polarity of the extraction method. Direct measurement in native water samples was not possible on account of the high salt content. It was possible to identify pentachlorophenol (PCP) unequivocally as one of the bioluminescence inhibitors in these fractions.

## 2.4 Using the HPTLC-bioluminescence bacteria assay for the determination of acute toxicities in marine sediments

For an integrated ecological risk assessment of marine sediment contamination, the determination of target-compound concentrations by for example, mass spectrometric methods, is not sufficient to explain sediment toxicity. Due to the presence of a multitude of environmental contaminants in this complex matrix causing a mixed toxicity, the identification and assessment of main toxicants is a challenge (Logemann et al. 2019).

Marine sediments play an essential role for marine habitats (Snelgrove, 1997). However, they are also threatened by various man-made pressures such as bottom-trawling fishery, excavation measures, offshore wind energy and oil production or by contaminants entering the marine environment through rivers, atmosphere or point source releases.

Logemann et al. investigated fourteen marine sediment samples from the German Bight representing a wide range of contaminant loads and sediment properties. In this study, an EDA approach was developed using HPTLC coupled to bioluminescence bacteria detection with *Aliivibrio fischeri* for the determination of marine sediment acute toxicity. The HPTLC could be optimized with a fast, two-step gradient in order to separate main hydrophobic organic contaminant (HOC) classes found in marine sediments. The results of this study show that the HPTLC-bioluminescence bacteria assay is a promising tool for the toxicity assessment of marine sediments. Differences between the acute toxicity of spatial sediment samples could be detected and the acute toxicity of an inhibition zone could be linked to polycyclic aromatic hydrocarbons (PAH) concentrations.

## 2.5 Detection of antibiotic activity in water samples

It has been demonstrated that large numbers of antibiotic-resistant bacteria are present in the environment. In 1999, Feuerpfeil et al. described that, on the one hand, these bacteria are released directly into the environment during the application of slurry and dung from intensive animal rearing, and, on the other hand, they collect in water treatment plants arriving in wastewaters from clinical and domestic sources and from there, they reach the environment in the treated wastewater (Feuerpfeil et al. 1999).

Measurement of the antibiotically active substances in water acquires increasing importance in this context.

In Figure 6, the profile of a water basin is presented. To detect as many organic pollutants as possible, a 12.5 ml water sample was enriched 1:100 by RP18 solid phase extraction. Besides that, 5

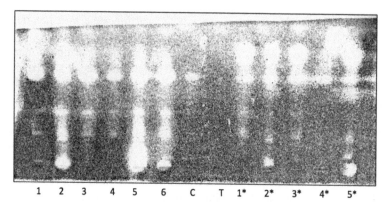

**FIG. 6** Presence of antibiotically active substances, a profile of a water basin; 1–6 RP 18 methanolic extracts; 1*-5*freeze-dried methanolic extracts; 2, 2*, 5, 5* are samples of outlets of a wastewater treatment plant; C: control, T: track without substances (Weins 2006)

ml of the water samples were freeze dried and extracted with 100 µl methanol. 5 to 100 µl of these extracts was applied onto the HPTLC plate silica gel60, $F_{254}$.

After chromatography of these water sample extracts by the AMD technique, *Bacillus subtilis* was used as the indicator organism in the activity analysis that follows. The growth of the test organism on the thin-layer plate was inhibited by antibiotically active inhibitors and was indicated by the production of white zones of inhibition. Detection was carried out by means of a bacterial vitality test, where the bacterial lawn on the thin-layer chromatogram was sprayed with 3-(4,5-dimethylthiazol-2-yl)-2,5-diphenyltetrazoliumbromide (MTT) salt as described in Chapter 12 (Weins 2006).

In Figure 6, it is remarkable that water treatment plant circuits in particular are a source of antibiotically active substances in the environment. Track 1 and 1* show the antibiotic inhibition zones before the outlet of a wastewater treatment plant, track 2 and 2* are samples of the wastewater effluents and track 3 and 3* are just behind the outlet of the wastewater treatment plant where the antibiotic effect is slightly diluted.

## 2.6 DETECTION OF INHIBITORS OF CHOLINESTERASE IN SURFACE WATER

The inhibition of cholinesterase has long been recognised as a biochemical method for the detection of the enzyme-inhibiting effects of organophosphate esters and insecticidal carbamates. The inhibitory effect is determined by the reduction in the enzymatic hydrolysis of 1-naphthyl acetate to 1-naphthol and acetic acid, followed by coupling of the 1-napthol to yield a violet-blue dyestuff (diazonium salt, *Echtblau B*). White zones of inhibition on a coloured background are produced around toxicologically active substances (Geike 1969). The detection limit should be proportional to the inhibition constant of the particular substance and can lie in the lower picogram range.

The sample preparation, extraction, application and chromatographic method were the same as described in section 2.5 above.

When carrying out EDA for cholinesterase inhibitors in the water samples, it was found in particular that the effect of these inhibitors was elevated in the water samples from the outfalls of water treatment plants (Fig. 7).

The size of the zone of inhibition is determined by the amount applied, on the one hand, and by the specific inhibitory activity of the substance, on the other hand.

It becomes obvious that on track 4 (sample taken just behind the outlet of the wastewater treatment plant) the inhibitory effect of cholinesterase is slightly decreased by dilution.

Tacrine, a pharmaceutical drug that has particularly been associated with Alzheimer's disease at high doses (160 mg per day), was measured at detection limit of 8 pg per zone with this method.

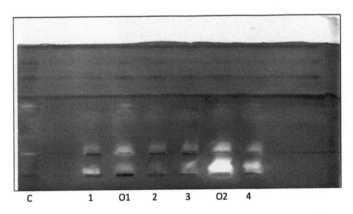

**FIG. 7**  Presence of cholinesterase inhibitors in a river profile, O1, O2 are water sample extracts from treatment plant outflows C: control of an uncontaminated water sample; detection at 254 nm (Weins 2006).

Further investigations have revealed that the outfalls of communal water treatment plants exhibit particularly high degrees of contamination with cholinesterase inhibitors. The analytical procedure can thus be used to document the presence of bio-active substances in communal waste waters.

## 2.7  Detection of estrogenic and androgenic compounds in wastewater

The persistence of endocrine disrupting compounds (EDCs) throughout wastewater treatment processes poses a significant health threat to humans and to the environment. The analysis of EDCs in wastewater remains a challenge for several reasons, including:

- the multitude of bioactive but partially unknown compounds
- the complexity of the wastewater matrix
- the required analytical sensitivity

The estrogen (ER) and the androgen (AR) receptors are prominent members of a hormone receptor super-family that mediates a wide range of significant biological activities. These vary from reproductive development to the regulation of the cardiovascular system, the immune system, the central nervous system and more (Davey and Grossmann 2016; Yaşar et al. 2016). Detecting specific pollutants that exhibit hormonal activity in complex environmental samples (e.g., treated wastewater) is very challenging due to the complexity of the matrix. Such samples may contain a large variety of EDCs, as well as numerous unknown EDC metabolites, which may also exert endocrine-disrupting activity.

In the following analysis, the content and presence of EE2 (17 α-ethinylestradiol), E2 (17β-estradiol and E1 (estrone) is described. Figure 8 shows chromatograms of different concentrations of a mixture of E1, EE2 and E2 (Lv2 – Lv15) as standards in comparison with methanolic extracts of a water sample with different known and unknown estrogenic compounds. The detection was performed with the p-YES test, which is described in detail in Chapter 12.

To evaluate the results shown in Figure 8, one first must identify the compounds with the same migration distance. Here, we find E1, E2 and a small amount of EE2. Furthermore, EDA must determine the strength of the special effect. This is why unknown substances with a specific effect have to be expressed in so-called equivalent values (i.e., in 17β-estradiol (E2) equivalent values EEQ). Figure 8 shows the summation of all the equivalent values of the sample results in 2500 EEQ pg/l.

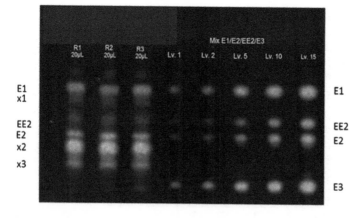

| Subst. | EEQ pg/l | % |
|---|---|---|
| E1 | 490 | 20 |
| x1 | 70 | 3 |
| EE2 | 50 | 2 |
| E2 | 310 | 12 |
| x2 | 1300 | 52 |
| x3 | 280 | 11 |
| Σ | 2500 | |

**FIG. 8**  Detection of methanolic extracts of a water sample with known and different unknown estrogenic compounds (provided by Sebastian Buchinger 2018, German Federal Institute of Hydrology, Koblenz, Germany).

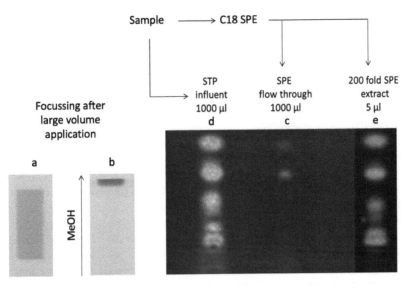

**FIG. 9** Comparison of the chromatogram of 1000 μl of the native water sample with 5 μl of an extract of 1000 μl water sample, enriched 200-fold by SPE (provided by Sebastian Buchinger 2018, German Federal Institute of Hydrology, Koblenz, Germany).

NOTE: Direct measurement of estrogens in a native water sample is possible on account of the composition of the matrix. If the content of salt is too high, for example in samples of wastewater of a coking plant, a direct application of the sample should be avoided.

As the predicted-no-effect concentration (PNEC) is 0.1 ng/L for EE2 (17 α-ethinylestradiol), 2 ng/L for E2 (17β-Estradiol), 6 and 60 ng/L for E1 (estrone), it is a great challenge for the analytical procedure to detect these active agents in a complex matrix. In Chapter 12, it could be demonstrated that the Lower Limit of Quantification[4] (LLOQ) of estradiol (E2) is 0.1 pg/band on the TLC plate and the LLOQ of estrone is 1.0 pg/band. Therefore, the sample preparation, including the enrichment, is of great importance.

To detect as many organic pollutants as possible, EDA allows the native sample to be applied onto the plate for separating the compounds according to their polarity. Figure 9 shows the influence of the sample preparation according to the detection of the effective compounds. In the first track of the chromatogram, 1000 μl of the native sample was applied in a long square (a). During the first step of the chromatography, the components of the sample can be focused to one band by methanol during a migration distance of 20 mm (b). The discharge of the solid phase extraction (SPE) contains some EEQ equivalents, which normally get lost during the procedure of enrichment (c). The detection zones in the chromatogram of the native sample (d) are much more intensive than those of the extract after solid phase extraction with an enrichment factor 200 (e).

In 2020, Moscovici published an article in which he and his co-authors describe the development of yeast-based sensors (*Saccharomyces cerevisiae*) that detect the presence of chemicals exerting androgenic and estrogenic activity by expressing spectrally different fluorescent proteins. This test has been first characterised in a conventional 96-well microtitre plate procedure in liquid medium. Both effects of the estrogenic and androgenic activity can be determined at the same time in the *in vitro* experiment (Moscovici et al. 2020).

---

4 LLOQ: lower limit at which an assay provides quantitative results within the linear range of the assay.

## 3   CONCLUSION

- Regarding environmental risk assessment, EDA assesses the likelihood of the organic pollutants causing harm to the environment. EDA with HPTLC, in combination with biological and biochemical detection modes, is able to
  - identify any hazards (i.e., possible sources of harm)
  - describe the harm they might cause
  - evaluate the risk of occurrence
  - give the result of the present effect in equivalent values with regard to a known standard
- The need to focus on the presence of unknown but bio-active compounds renders traditional detection methodologies, e.g., liquid- or gas-chromatography coupled to mass spectrometry (LC/MS and GC/MS, respectively), less suitable for the detection of EDCs in complex samples (Riegraf et al. 2019).
- Planar bio-tests are a suitable tool to detect emerging contaminants in environmental samples.
- EDA is increasingly used in environmental monitoring to detect and identify key toxicants. HPTLC has proven to be a very suitable fractionation technique for this purpose. However, HPTLC is limited in its separation efficiency. Thus, separated fractions could still contain many different components and identification of the effective substances remains difficult (Stütz et al. 2020)
- The combination of chemical separation by HPTLC and the effect-based assay by yeast-based sensor strains allowed the separation of environmental samples and the discovery of individual sample components exhibiting hormonal activity. These active compounds can then be removed from the HPTLC plate and identified via traditional analytical methods (e.g., LC/MS, GC/MS) (Moscovici et al. 2020).
- An enrichment factor is necessary to perform the EDA for most water samples (e.g., surface water or ground water). The enrichment of organic compounds from water samples is optimised in parallel to the work with biological test systems. Different materials were tested for the solid-phase extraction (SPE) with substances of various polarities for pH values 2 and 7. Application of the native samples can be performed as well due to the LLOQ and the appropriate components of the matrix
- This bio-profiling of a sample is like a fingerprint and enables sources of the pollution to be found out.

**Note:** Policy makers will be enabled to take specific actions against the presence of harmful substances in our environment from results gained by EDA.

## 4   REFERENCES

Burger, K. 1984. "Dünnschichtchromatographie mit Gradienten-Elution im Vergleich zur Säulenflüssigkeits-Chromatographie." *Fresenius´ Zeitschrift für Analytische Chemie*, 318, 228–223

Burger, K. 1995. "Thin-Layer Chromatography with Automated Multiple Development (AMD-TLC)." In: Stan, HJ. (eds) *Analysis of Pesticides in Ground and Surface Water II. Chemistry of Plant Protection*, vol 12. Springer, Berlin, Heidelberg. https://doi.org/10.1007/978-3-662-01063-1_7

Davey, Rachel A., and Mathis Grossmann. 2016. "Androgen Receptor Structure, Function and Biology: From Bench to Bedside." *The Clinical Biochemist Reviews* 37 (1): 3–15. https://pubmed.ncbi.nlm.nih.gov/27057074/.

DIN EN ISO 11348-1. 2009."*Quality – Determination of the Inhibitory Effect of Water Samples on the Light Emission of Vibrio Fischeri (Luminescent Bacteria Test) – Part 1: Method Using Freshly Prepared Bacteria (ISO 11348-1:2007).*" Berlin: Deutsches Institut für Normung. Retrieved from https://www.beuth.de/en/standard/din-en-iso-11348-1/113162789 on 24/07/2022.

DIN V 38407-11:1995-01. 1995. "*German standard methods for the examination of water, waste water and sludge – Jointly determinable substances (group F) – Part 11: Determination of selected organic plant protecting agents by Automated Multiple Development (AMD)-technique (F 11).*" Retrieved from https://www.beuth.de/de/vornorm/din-v-38407-11/2444820 on 24/07/2022

EPA Environmental Protection Agency. 2022. "Unspecified Toxic Chemicals" https://www.epa.gov/caddis-vol2/unspecified-toxic-chemicals; accessed May 16, 2022

Feuerpfeil, I., J. López-Pila, R. Schmidt, E. Schneider, and R. Szewzyk. 1999. "Antibiotikaresistente Bakterien und Antibiotika in der Umwelt." *Bundesgesundheitsblatt – Gesundheitsforschung – Gesundheitsschutz* 42 (1): 37–50. https://doi.org/10.1007/s001030050057. https://www.springermedizin.de/antibiotikaresistente-bakterien-und-antibiotika-in-der-umwelt/8013968.

Geike, F. 1969. "Dünnschichtchromatographisch-enzymatischer Nachweis und zum Wirkungsmechanismus von chlorierten Kohlenwasserstoff-Insektiziden." 1969. *Journal of Chromatography* A 44 (January): 95–102. https://doi.org/10.1016/S0021-9673(01)92502-6.

Kaiser, Klaus L.E., and Virginia S. Palabrica. 1991. "*Photobacterium phosphoreum* Toxicity Data Index." *Water Quality Research Journal* 26 (3): 361–431. https://doi.org/10.2166/wqrj.1991.017.

Kookana, R. S., S. Baskaran, and R. Naidu. 1998. "Pesticide Fate and Behaviour in Australian Soils in Relation to Contamination and Management of Soil and Water: A Review." *Soil Research* 36 (5): 715. https://doi.org/10.1071/s97109.

Logemann, A., M. Schafberg, B. Brockmeyer. 2019. "Using the HPTLC-Bioluminescence Bacteria Assay for the Determination of Acute Toxicities in Marine Sediments and Its Eligibility as a Monitoring Assessment Tool." 2019. *Chemosphere* 233 (October): 936–45. https://doi.org/10.1016/j.chemosphere.2019.05.246.

Mandal S., A. Kunhikrishnan, N.S. Bolan, H. Wijesekara, and R. Naidu. 2016. "Application of Biochar Produced from Biowaste Materials for Environmental Protection and Sustainable Agriculture Production." *Environmental Materials and Waste*, January, 73–89. https://doi.org/10.1016/B978-0-12-803837-6.00004-4.

Migrator. 2011. "What Is an Environmental Risk Assessment?" Nibusinessinfo.co.uk. November 9, 2011. https://www.nibusinessinfo.co.uk/content/what-environmental-risk-assessment.

Moscovici, Liat, Carolin Riegraf, Nidaa Abu-Rmailah, Hadas Atias, Dror Shakibai, Sebastian Buchinger, Georg Reifferscheid, and Shimshon Belkin. 2020. "Yeast-Based Fluorescent Sensors for the Simultaneous Detection of Estrogenic and Androgenic Compounds, Coupled with High-Performance Thin-Layer Chromatography." *Biosensors* 10 (11). https://doi.org/10.3390/bios10110169.

Oberleitner, Daniela, Lena Stütz, Wolfgang Schulz, Axel Bergmann, and Christine Achten. 2020. "Seasonal Performance Assessment of Four Riverbank Filtration Sites by Combined Non-Target and Effect-Directed Analysis." *Chemosphere* 261 (December): 127706. https://doi.org/10.1016/j.chemosphere.2020.127706.

Riegraf, C.; G. Reifferscheid, S. Belkin, L Moscovici, D. Shakibai, H. Hollert and S. Buchinger. 2019. "Combination of Yeast-Based *in vitro* Screens with High-Performance Thin-Layer Chromatography as a Novel Tool for the Detection of Hormonal and Dioxin-like Compounds." *Analytica Chimica Acta* 1081 (November): 218–30. https://doi.org/10.1016/j.aca.2019.07.018.

Sangion, Alessandro and Paola Gramatica. 2016. "Hazard of Pharmaceuticals for Aquatic Environment: Prioritization by Structural Approaches and Prediction of Ecotoxicity." *Environment International* 95 (October): 131–43. https://doi.org/10.1016/j.envint.2016.08.008.

Snelgrove, Paul V. R. 1997. "The Importance of Marine Sediment Biodiversity in Ecosystem Processes." *Ambio* 26 (8): 578–83. https://www.jstor.org/stable/4314672.

Stütz, L., W. Schulz and R. Winzenbacher. 2020. "Identification of Acetylcholinesterase Inhibitors in Water by Combining Two-Dimensional Thin-Layer Chromatography and High-Resolution Mass Spectrometry." *Journal of Chromatography* A 1624 (August): 461239. https://doi.org/10.1016/j.chroma.2020.461239.

Weins, Christel. 2006. "Möglichkeiten und Grenzen der wirkungsbezogenen Analytik mit der Hochleistungs-Dünnschichtchromatographie", Dissertation May 02, University of Basel, Accessed April 04, 2021 https://edoc.unibas.ch/411/1/DissB_7500.pdf

WHO. 2003. "HEALTH RISKS of PERSISTENT ORGANIC POLLUTANTS from LONG-RANGE TRANSBOUNDARY AIR POLLUTION." World Health Organization Accessed June 26, 2021. https://apps.who.int/iris/bitstream/handle/10665/107471/e78963.pdf?sequence=1&isAllowed=y.

Yaşar, Pelin, Gamze Ayaz, Sırma Damla User, Gizem Güpür, and Mesut Muyan. 2016. "Molecular Mechanism of Estrogen-Estrogen Receptor Signaling." *Reproductive Medicine and Biology* 16 (1): 4–20. https://doi.org/10.1002/rmb2.12006.

# Epilogue One

## *Medicinal Agroecology as a path to overcoming Molecular Colonialism and the vicious circle of poisoning*

*Larissa Bombardi and Marina Cristina Campos Peralta*

With the growing world economy, particularly after World War II, agriculture had started to take on a global scale, not only in the sense that a significant part of the agricultural production started to be globally commercialized, becoming a commodity[1] traded on the stock exchange, but also because it began to become dependent on the (synthetic) chemical industries producing patented fertilizers and pesticides and, more recently, dependent on patented seeds.

This process, obviously, did not occur homogeneously on the planet. In the case of pesticides, for instance, there has been a Global North/South divide that separated the countries that concentrated their synthetic pesticide productions (Global North) from the ones that massively and coercively consumed them (Global South). In this epilogue, we present the example of Brazil as one of the major agricultural powers in Latin America and as the fourth largest producing countries in the world.

The industrialization of agriculture – which allowed agricultural production to be carried out on a very large scale and in a simplifying way – creating gigantic monocultures – became known as the "Green Revolution" and had, as a justification for its implementation, the promise of overcoming hunger, through the use of technology (Porto-Gonçalves 2006).

However, more than half a century has passed since the inception of Green Revolution technologies and, in spite of it, hunger remains. Even more so, the number of hungry people in the world has increased. In 2020, from 9.2% to 10.4% of the worldwide population faced hunger (between 720 and 811 million people) (UNICEF 2021). This number is much higher if we consider not only hunger, but the lack of a nutritionally balanced diet, to which everyone should have access as part of human rights.

Whilst hunger has increased, humans and their environment have been intensely contaminated by synthetic chemical substances used in agriculture.

To look at the human and environmental tragedy resulting from this agricultural model, we focus on Brazil as the largest worldwide exporter of soy (Conab 2021), beef, sugar, coffee (Embrapa 2021) and orange juice (FAEG 2021), among other products.

In Brazil, the emblematic expansion of soy – which currently covers an area equivalent to the entire territory of Germany and whose production has grown exponentially (Bombardi, 2021) – shows us how devastating the monoculture expansion scenario for exportation is.

DOI: 10.1201/9781003146902-18

Not only soy, but the expansion of other monocultures destined to become *commodities* has fomented drastic environmental and social impacts, such as the accelerated increase in deforestation and the intensive use of pesticides, which are used on a large scale.

Between 2010 and 2020, the use of pesticides in Brazil substantially increased by 78.3% (IBAMA 2010-2020). As a consequence of this increase, we are witnessing chemical violence, oftentimes indirect, silent and subtle, which arises as an unfolding of the aforementioned Green Revolution.

We may call this form of chemical violence 'Molecular Colonialism' (Mendes 2017; Inhabitants 2018), since substances banned for more than ten or twenty years in the European Union because of risks to human and environmental health continue to be sold cynically – by companies headquartered in the European Union – to countries such as Brazil (Bombardi 2021).

These companies are currently organized in an oligopolistic way, that is, through mergers and acquisitions that made possible a very strong mechanism of capital concentration and with it powerful structures of lobbyism (Bombardi 2019).

Thus, there is an asymmetry in the production and sales of these substances, which are majorly produced and exported by countries of the Global North, a fact that may be noticed in the anamorphosis shown in Figure 1.

This process of 'Molecular Colonialism' becomes even more evident when we consider that the European Union, which leads the worldwide pesticide market, exports substances that are – as already outlined above – banned in its own territory (Public Eye 2020; Bombardi 2021). The reasons for these bans are linked to the fact that, for instance, these substances are associated with the development of various types of cancer, fetal malformations and hormonal dysfunctions amongst other diseases (PAN International 2021).

Thus, the intense use of pesticides can be considered to cause severe impacts on the health of the population, especially in countries like Brazil, where more than 56,000 people were poisoned by pesticides between 2010 and 2019 alone – which corresponds to 15 people per day poisoned by pesticides – as shown in Figure 2 (Bombardi 2021).

Poisoning by pesticides in Brazil goes as far as what can be considered a form of infanticide: about 20% of the intoxicated population are children and adolescents from neonates up to 19 years of age.

Between 2010 and 2019, 3,754 children from 0 to 14 years of age were poisoned by pesticides and, among them, 542 were neonates up to 12 months of age. In other words, babies who do not even move by themselves do get exposed. Maps from Figures 3 and 4 represent this severe detriment caused to children.

In Brazil, poisonings from the use of pesticides may be analyzed thanks to public data provided by the public health services: *Sistema Único de Saúde* (SUS). However, while the numbers are already alarming, in actual fact they are very likely to be even higher due to underreporting.

It is estimated that in recent years, for each reported case, 50 other cases are not reported (Fiocruz 2015), so it is possible that, in Brazil alone, we have had more than 25,000 babies under one year of age poisoned by pesticides.

In addition to acute intoxications, the cases of chronic poisoning by synthetic pesticides, described in Brazilian scientific literature, are alarming (see also Chapter 3 for chronic neurological diseases like Parkinson´s disease).

Just as an example, we may mention the Southeast region of Ceará (a state from the Brazilian Northeast) – with production of tropical fruits for exportation underintensive use of pesticides – in which cases of fetal malformation and precocious puberty were studied and reported. The following synthetic substances have been found in the water from these children's homes: Carbaryl, Procymidone, Carbofuran, Fenitrothion, Tebuconazole, Clethodim, Tepraloxydim, Glyphosate, Abamectin, Difenoconazole, Flumioxazin, Fosetyl-Al, Cyromazine, Imidacloprid, Azoxystrobin and Endosulfan (Aguiar 2017; see also Chapters 12 to 14 for analytical approaches like HPTLC/EDA).

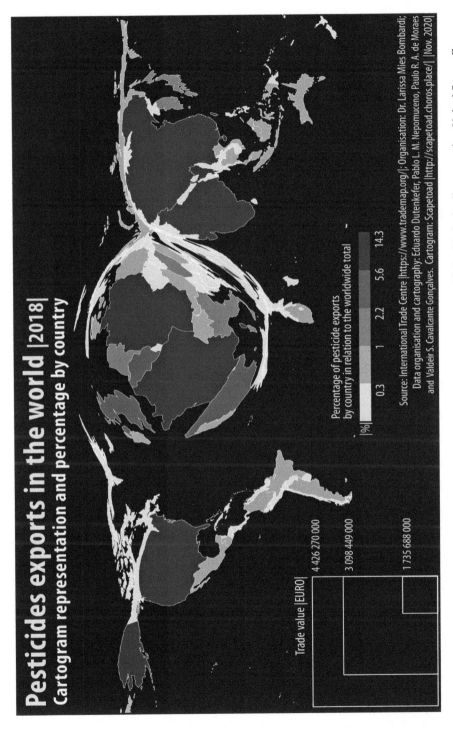

**FIG. 1** Distorted projection mapping (anamorphosis) of worldwide pesticide exports in 2018 with the leading countries: United States, Germany, France, India and China. In 2018, companies belonging to the European Union (mainly Germany), China and the United States, together, were responsible for 83% of the worldwide pesticide sales (AgNews 2021). The eleven largest global agrochemical companies in 2020 represent almost 90% of the worldwide pesticide sales, and the first four represent almost 60% of this total, including: Syngenta (China) with 18.14% of sales, Bayer Crop Science (Germany) with 16.16%, BASF (Germany) with 11.4% and Corteva (USA) with 10.44% of sales (AgNews 2021).

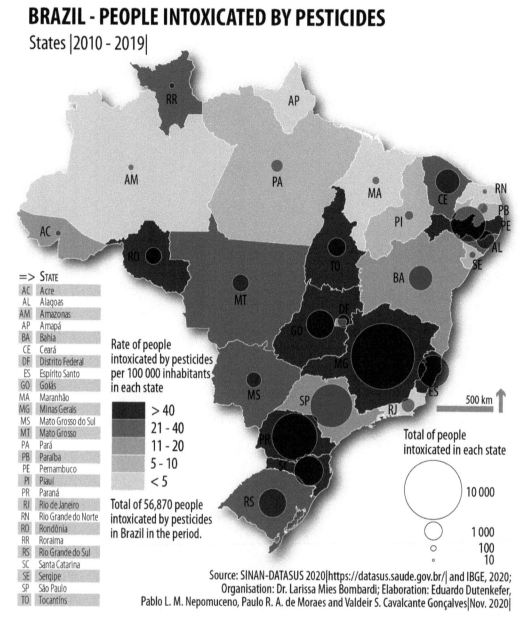

**FIG. 2** Mapping of pesticide intoxications in Brazil between 2010 and 2019; source: Bombardi 2021. Long-term effects from such intoxications, however, are not taken into account, such as chronic neurologic disorders (see also Chapter 3 of this book).

The dissertation mentioned in the previous paragraph demonstrates the existence of real tragedies in this region: from a couple that had three children with fetal malformation (and who died eventually) to two-year-old children with breasts and pubic hair due to exposure to pesticides suffered primarily by their parents (Aguiar 2017). These examples illustrate quite well what we declare as 'infanticide' resulting from approaches of 'Molecular Colonialism'.

Somehow, obviously not with the same impact, these substances return to the countries that produced them, through the foods that contain them. We call this a 'vicious circle of poisoning'

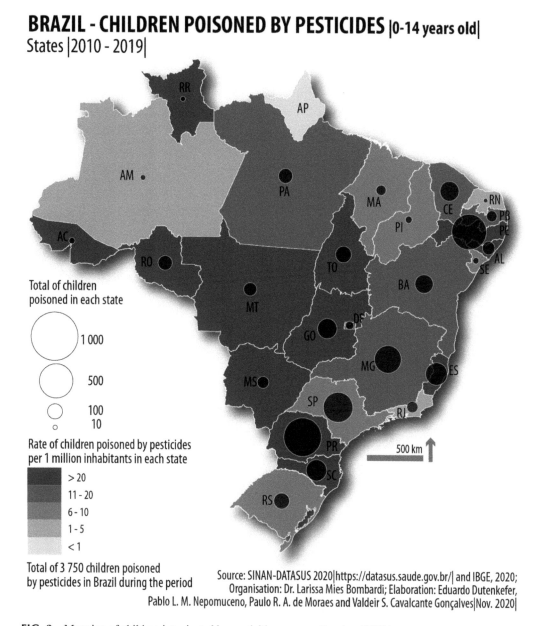

# BRAZIL - CHILDREN POISONED BY PESTICIDES |0-14 years old|
## States |2010 - 2019|

Total of children
poisoned in each state

1 000

500

100
10

Rate of children poisoned by pesticides
per 1 million inhabitants in each state

> 20

11 - 20

6 - 10

1 - 5

< 1

Total of 3 750 children poisoned
by pesticides in Brazil during the period

Source: SINAN-DATASUS 2020|https://datasus.saude.gov.br/| and IBGE, 2020;
Organisation: Dr. Larissa Mies Bombardi; Elaboration: Eduardo Dutenkefer,
Pablo L. M. Nepomuceno, Paulo R. A. de Moraes and Valdeir S. Cavalcante Gonçalves|Nov. 2020|

**FIG. 3**    Mapping of children intoxicated by pesticides; source: Bombardi 2021.

(Galt 2008). Europe is a major food importer, and about 70% of food that Europe imports from
Brazil showed pesticide residues and about 7% of the total showed residues of synthetic substances
banned in the European Union and/or residues beyond the limit allowed by the European Union, as
seen in Figure 5.

In view of this reality of exposure it becomes important to highlight that chemical-dependent
agriculture is primarily a necessity of the industry and its business models (see also Chapter 3) – it
has not been an essential demand coming from agriculture itself, as we postulate.

There is, obviously, an inversion in considering that agriculture depends on such substances. It
is the chemical industries that depend on agriculture to make themselves viable in this sector of

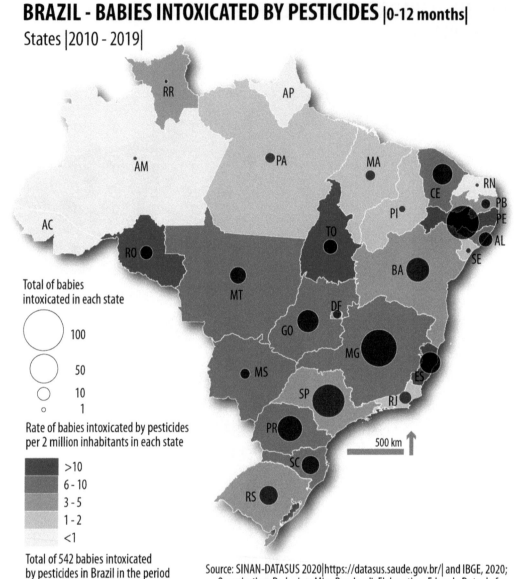

# BRAZIL - BABIES INTOXICATED BY PESTICIDES |0-12 months|
## States |2010 - 2019|

Total of babies intoxicated in each state

- 100
- 50
- 10
- 1

Rate of babies intoxicated by pesticides per 2 million inhabitants in each state

- >10
- 6 - 10
- 3 - 5
- 1 - 2
- <1

Total of 542 babies intoxicated by pesticides in Brazil in the period

500 km

Source: SINAN-DATASUS 2020|https://datasus.saude.gov.br/| and IBGE, 2020;
Organisation: Dr. Larissa Mies Bombardi; Elaboration: Eduardo Dutenkefer,
Pablo L. M. Nepomuceno, Paulo R. A. de Moraes and Valdeir S. Cavalcante Gonçalves|Nov. 2020|

**FIG. 4** Babies intoxicated by pesticides; source: Bombardi 2021.

the economy as well. Conversely, traditional peoples (indigenous peasants) developed agricultural practices that had never required transgenic seeds nor synthetic chemical inputs.

In order to value both traditional/popular knowledge and scientific knowledge, agroecology emerges as a science, political movement and social practice (ABA, n.d.) increasingly needed on a global scale, not least in view of greenhouse gas emmissions and climate change (see e.g. Chapter 8).

Agroecology faces the model imposed by the 'Green Revolution', which reveals itself to be externally imposed: 'monoculturising' dangerously, being unhealthy and colonialist in its approach. Agroecology appears as an alternative, proposing changes to the way economic processes

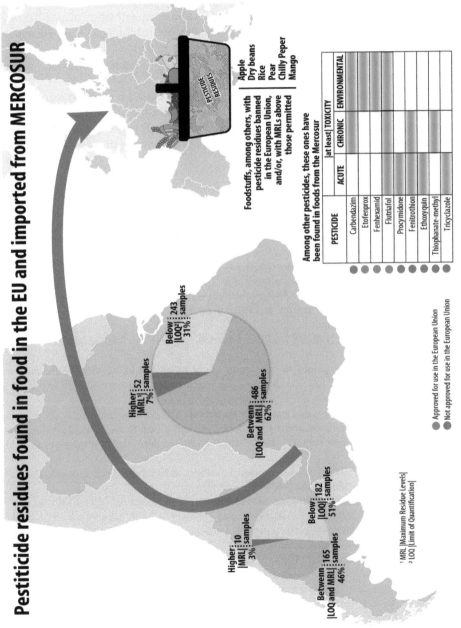

**FIG. 5**   Mapping of 'the vicious circle of poisoning' (Galt 2008); source: Bombardi 2021.

unfold – processes of production, distribution, circulation and consumption of food and other goods from primary production.

For this, it bases studies on popular and traditional knowledge and practices, combining them with scientific knowledge, through the appreciation of a diverse dialogue, in which multiple perspectives and epistemologies – historically excluded – must be valued (Altieri 2012; Caporal 2020; Nicholls et al. 2016; Gliessman and Engles 2015).

Agroecology brings concrete bases that allow healthy food production (Fiebrig et al. 2020) that takes place in a fair way, guaranteeing access to land and quality of life for human beings in a more equal way. This knowledge needs to be considered and more evidenced.

The need for a progressive transition to the path of agroecology is urgent. Otherwise, we will continue on a route that will result in a collision against ourselves.

Hence, the great importance of this work that the reader has in hands! It comes at a good time, with extensive and diverse scientific exmples, brought by researchers from almost every continent, in which we may see agroecology, not just as an 'alternative', which would greatly diminish its potential. But, more than that, as 'the way' for us to overcome the impacts of a sort of agriculture that sickens human beings and the planet.

## REFERENCES

ABA. n.d. "Quem Somos." *Associação Brasileira de Agroecologia.* Accessed February 1, 2022. https://aba-agroecologia.org.br/sobre-a-aba-agroecologia/sobre-a-aba/#:~:text=Por%20fim%2C%20define%2Dse%20a.

Aguiar, Ada Cristina Pontes. 2017. *Más-formações congênitas, puberdade precoce e agrotóxicos: Uma herança maldita do agronegócio para a Chapada do Apodi (CE).* Fortaleza: Universidade Federal do Ceará.

AgNews. 2021. "Ranking List of Global Top 20 Agrochemical Enterprises for FY2020 | 15 Realize Sales Growth, 11 Chinese Enterprises Contribute to 37% of Total Sales." Grainews. https://news.agropages.com/News/NewsDetail---40437.htm.

Altieri, Miguel. 2012. *Agroecologia: Bases Científicas Para Uma Agricultura Sustentável.* 3. ed. Rio de Janeiro: Expressão Popular.

Bombardi, Larissa Mies. 2019. A Geography of Agrotoxins use in Brazil and its Relations to the Eurpean Union. São Paulo: FFLCH-USP. Accessed March 2022. http://www.livrosabertos.sibi.usp.br/portaldelivrosUSP/catalog/book/352

Bombardi, Larissa Mies. 2021. *Geography of Asymmetry: The Vicious Cycle of Pesticides and Colonialism in the Commercial Relationship between Mercosur and the European Union.* Brussels: The Left (European Parliamentary Group). Accessed February 2022. https://left.eu/events/eu-mercosur-the-vicious-circle-of-pesticides/

Caporal, Francisco Roberto. 2020. "Transição Agroecológica e o Papel da Extensão Rural." *Extensão Rural* 27 (3): 7–19. https://doi.org/10.5902/2318179638420.

CONAB. 2021. "Produção de Grãos Tem Previsão de Aumento de 5,7%, Chegando a 271,7 Milhões de T." Companhia Nacional de Abastecimento. 2021 Accessed February 1, 2022. https://www.conab.gov.br/ultimas-noticias/3989-producao-de-graos-tem-previsao-de-aumento-de-5-7-chegando-a-271-7-milhoes-de-toneladas#:~:text=A%20soja%20mant%C3%A9m%20o%20seu.

EMBRAPA. 2021. "Brasil é O Quarto Maior Produtor de Grãos E O Maior Exportador de Carne Bovina Do Mundo, Diz Estudo." Empresa Brasileira de Pesquisa Agropecuária. Accessed February 1, 2022. https://www.embrapa.br/busca-de-noticias/-/noticia/62619259/brasil-e-o-quarto-maior-produtor-de-graos-e-o-maior-exportador-de-carne-bovina-do-mundo-diz-estudo.

FAEG. 2021. "Brasil Se Destaca Como Maior Produtor Mundial de Laranja E Exportador de Suco Da Fruta." Sistema Federação da Agricultura e Pecuária de Goiás. Accessed February 1, 2022. https://sistemafaeg.com.br/faeg/noticias/citrus/brasil-se-destaca-como-maior-produtor-mundial-de-laranja-e-exportador-de-suco-da-fruta.

Fiebrig, Immo, Sabine Zikeli, Sonja Bach, and Sabine Gruber. 2020. "Perspectives on Permaculture for Commercial Farming: Aspirations and Realities." *Organic Agriculture*, 10. https://doi.org/10.1007/s13165-020-00281-8.

FIOCRUZ. 2015. "Estudo Aponta Subnotificação de Mortes Por Agrotóxicos." *Agência Fiocruz de Notícias*. December 8, 2015. Accessed February 1, 2022. https://agencia.fiocruz.br/estudo-aponta-subnotifica cao-de-mortes-por-agrotoxicos#:~:text=O%20pr%C3%B3prio%20Minist%C3%A9rio%20da%20 Sa%C3%BAde.

Galt, Ryan E. 2008. "Beyond the Circle of Poison: Significant Shifts in the Global Pesticide Complex, 1976-2008." *Global Environmental Change* 18 (4): 786–99. https://doi.org/10.1016/j.gloenvcha.2008.07.003.

Gliessman, Stephen R, and Eric Engles. 2015. *Agroecology: The Ecology of Sustainable Food Systems*. Boca Raton: CRC Press.

IBAMA. 2010–2020. "Relatórios de Comercialização de Agrotóxicos." Instituto Brasileiro do Meio Ambiente e dos Recursos Naturais Renováveis. Report series 2010-2020. Accessed February 10, 2022. https:// www.ibama.gov.br/agrotoxicos/relatorios-de-comercializacao-de-agrotoxicos#boletinsanuais.

Inhabitants. 2018. *Colonialismo Molecular: Uma geografia dos agrotóxicos no Brasil/Molecular Colonialism: A Geography of Agrochemicals in Brazil*. Vimeo. October 13, 2018. https://vimeo.com/294971699.

Mendes, Margarida. 2017. *"Molecular Colonialism"*. Anthropocene Curriculum. 2017. Accessed February 1st, 2022. https://www.anthropocene-curriculum.org/contribution/molecular-colonialism.

Nicholls, Clara Ines, Miguel Altieri and L Vazquez. 2016. "Agroecology: Principles for the Conversion and Redesign of Farming Systems." *Journal of Ecosystem & Ecography* 01 (s5). https://doi.org/10.4172/2157-7625.s5-010.

PAN International.. 2021. PAN (Pesticides Action Network) International List of Highly Hazardous Pesticides. Accessed March 22nd 2022. https://pan-international.org/wp-content/uploads/PAN_HHP_List.pdf.

Porto-Gonçalves, Carlos Walter. 2006. *A globalização da natureza a a natureza da globalização*. Rio De Janeiro: Civilização Brasileira.

Public Eye. 2022. Banned in Europe: How the EU exports pesticides too dangerous for use in Europe. Accessed March 22nd, 2022. https://www.publiceye.ch/en/topics/pesticides/banned-in-europe

UNICEF. 2021. "The State of Food Security and Nutrition in the World 2021." United Nations Children´s Fund. Accessed Feb 12, 2022. https://data.unicef.org/resources/sofi-2021/.

# Epilogue Two
## Tibetan Buddhist Medicine: An ecological perspective ending with Medicine Buddha Wishes

*Anna Elisabeth Bach*

*In a world with so many sources of knowledge*
*it is a true enrichment,*
*when practical experience and useful methods meet*
*to create happiness or avoid suffering.*

Lama Ole Nydahl, 2016

## 1. INTRODUCTION

Tibetan Buddhist Medicine (TBM) is a holistic healing system encompassing body, speech and mind (or 'perceiver'). This system is comprised of information and experience transmitted in the Himalayas for more than a thousand years. It also incorporates influences from other ancient cultures originating in India, including ancient territories of Sikkim, Ladakh and Kashmir; further-more Nepal, Bhutan, China, Afghanistan and Persia as well as Greece. TBM is a living system of knowledge that has been transmitted from teacher to student in an unbroken lineage until the present and continues to be taught today. The philosophical foundation of this healing system is Buddhism.

## 2. HISTORICAL PERSPECTIVE OF TBM

The Himalayas and Tibet have long been known as 'Land of Medicinal Herbs' renowned for the abundance of natural herbs that sustain health and prolong life. Thousands of years ago, indigenous peoples there developed a medical tradition of their own, practising shamanic rituals and sacrificing animals to gain well-being and healing for their communities and livestock. As Buddhism spread throughout Tibet, however, use of these ancient traditions and rituals declined and people adopted a new view and different customs.

Buddha Sakyamuni (563–483 BCE) not only spread Buddhism throughout India but also gave medical teachings to his students at several locations. He taught Buddhist Medicine (BM) in the so-called Medicine Forest near present-day Bodhgaya (Bihar, India) in Udhiyana (now Afghanistan) and other places in India. Buddhist medical science (Tib. *so wa rig pa*) was passed down through an unbroken lineage of great Buddhist masters including Manjushri, Jivaka, Saraha, Nagarjuna, Chandrabiananda, Bibyi Gahbyed and Belha Gahdzesma (also known as Bimala Lhatse), Vaghbata I, Vaghbata II, Yuthog Yönten Gönpo the Elder, Shantirakshita and Padmasambhava.

DOI: 10.1201/9781003146902-19

The Himalayas and Tibet have long been known as the 'Land of Medicinal Herbs', renowned for the abundance of natural herbs that sustain health and prolong life. Thousands of years ago, indigenous peoples there developed their own medical tradition, practicing shamanic rituals and sacrificing animals to gain well-being and healing for their communities and livestock. However, as Buddhism spread throughout Tibet and the Himalaya, the use of these ancient traditions and rituals declined and people adopted a new view and new behaviours to the development of TBM.

To Tibet, Buddhism and, along with it, BM, was brought by the great Indian master Padmasambhava (Tib. *Guru Rinpoche*), during the 8th century BCE and taught there by himself whilst closely supporting his Tibetan consort Yeshe Tsogyal, later recognized as female Bodhisattva.

Some sources report that the first BM teachings reached Tibet as early as the 2nd century CE. There is actually mention of four medical transmissions: One early one in the 2nd century by Bibyi Gahbyed and Belha Gahdzesma and in the 8th century by Padmasambhava and Yuthog Yönten Gönpo the Elder (708-833 CE) followed by two later ones in the 11th century by Atisha and in the 13th century by Yuthog Yönten Gönpo the Younger.

## 3.  HOLISTIC PERSPECTIVE

The way sentient beings experience life and things in life is regarded as lying within the mind/perceiver. Underlying ignorance (Tib. *ma rig pa*) is the source and basis for every kind of the suffering of sentient beings – both mental and physical – leading to imbalance and disharmony. From this ignorance arises the illusion of the existence of a 'self' or 'ego', which in turn causes three major disturbing emotions:

*   attachment (tib. rLung energy principle), commonly named Wind;
*   aversion (Tib. mKhris pa energy principle) commonly named Fire; and
*   confusion (Tib.bad kan energy principle) commonly named Earth and Water.

The interplay between body and mind/perceiver causing suffering, disharmony and illness can, conversely, also lead to happiness, harmony and health.

Buddha Sakyamuni recognized this and taught methods for how to overcome and transform suffering, reaching enlightenment eventually. The essence of TBM is contained in the base text and is known as 'The Four Tantras' (Tib. *rGyü bZhi*). It defines the meaning of health, where mind/perceiver and body are in harmony. It explains what measures are necessary to re-establish harmony and how to reach the ultimate goal of enlightenment.

While body, speech and mind are central to TBM, any disturbance here is not the only cause of imbalance or disharmony. Other such aspects of imbalance comprise nutrition, seasons and climate as well as environmental influences including 'disturbing energies'. The main means of diagnosis are through assessment of pulse measurings, tongue appearance and urine inspection, supported by interviewing.

The following methods are commonly used for treatment and healing: Meditation, change in behaviour and nutrition, body exercises, taking herbal infusions and herb mixtures including jewel pills (pills composed of rare herbal, animal and mineral ingredients following large and complex formulas), massage, mugwort heat treatment, hot fire and hot stone treatment, baths, blood-letting, cupping and golden needle treatment, of which some treatments are not accepted or legal in western medicine.

## 4.  ICONOGRAPHY AND AGRO-ECOLOGY

Herbal or plant medicine in TBM is connected with the Medicine Buddha as shown in Figure 1 and as represented in the Medicine Buddha´s Healing Powerfield *Tanadug* (Fig. 2). According to the

**FIG. 1**   *Sangye Menla*, the Medicine Buddha.

underlying teachings, in principle any material in nature can be used as medicine. However, today this has been limited in the sense that many plants with healing properties are already extinct or endangered and much precious knowledge about the healing powers of plants has been lost over the millennia, calling for the importance of biological regeneration and cultural preservation.

One example of the importance of TBM is the unique grain grown in the high Himalayas, tsamp pa (tib.), barley (for the potential role of barley in agroecological systems in the Andean highlands also see Altieri et al. 1998). It forms the basic nutrition of the people in the whole Himalayan region, regardless of the respective country. The task of barley is to nourish development of body and mind with its various systems whilst supporting the different stages of existence, from toddler to old age, thus providing health and a long life.

## 5.   GOOD TBM COLLECTION AND PREPARATION PRACTICES

In TBM medicinal plants should only be collected under certain specific conditions, for example, from clean places, hills or mountains, in the right season and month and at the correct time of day. Flowers have to be collected at a time when their fragrance and colour will be preserved. The individual parts of a medicinal plant should each be collected within a specific window of time; for example, the medical leaves should be gathered before the plant blooms and when the water inside the stem flows upward. If the medicinal plant has already bloomed, then it is no longer medically

**FIG. 2**  *Tanadug*, Healing Powerfield of the Medicine Budhha.

effective. Fruits should be collected after the leaves have fallen off. Moreover, the medicinal plant's fruits should be free from insects and should not be completely ripe. The bark or skin of a medicinal plant should be collected when the plant blooms and its roots should mainly be gathered in winter.

Medicinal plants growing at higher altitudes are generally used to treat 'hot diseases' and those from lower altitudes are used for 'cold diseases'. Also, if one and the same plant is intended for the treatment of different diseases, it may be prepared under different conditions. The treatment of hot diseases requires the plant, flowers, leaves and roots to be dried in a cool place away from the fire or the sun.

For treatment of cold diseases, medicinal plant material should be dried in the sun and not in cold and windy places. Medicinal plants should not be put in contact with smoke, be it from fire, cigarettes or snuff, since this would destroy their potency.

Traditionally, knowledge of the exact recipes for the various herbal mixtures is only disclosed to the best medical school graduates of each year. In the Himalayas, this knowledge is still transmitted within so-called family lineages by the *amchis* or practitioners of TBM – from father to son or from mother to daughter. In rare cases it may also be disclosed to students outside a family lineage.

*Amchis* are bound by an honour code to only collect healing plants in a manner that minimizes environmental impact. Of the 7,000 plant species in the Himalayas, 700 are still used today for medicinal applications. However, due to lack of formal conservation efforts, 220 of these are now considered to be moderately endangered, while 20 species have been listed as endangered including several varieties of orchids (oral explanation by the Principal of the Botanical Garden, Kathmandu, Nepal, in the year 2007).

## 6.  LITERATURE FOR RECOMMENDED FURTHER READING

Clark, Barry. 1995. *The Quintessence Tantras of Tibetan Medicine*. Ithaca, N.Y., USA: Snow Lion Publications.

Clifford, Terry. 1992. *Tibetan Buddhist Medicine and Psychiatry: The Diamond Healing*. York Beach (Maine): S. Weiser.

Dakpa, Tenzin. 2007. *Tibetan Medicinal Plants: An Illustrated Guide to Identification and Practical Use*. New Delhi: Paljor Publ.

Dawa, Doctor. 1999. *A Clear Mirror of Tibetan Medicinal Plants*. Rome, Italy: Tibet Domani.

Dönden, Yeshi. 1995. *The Ambrosia Heart Tantra: The Secret Oral Teaching on the Eight Branches of the Science of Healing*. Dharamsala: Library Of Tibetan Works And Archives.

Dönden, Yeshi and B. Alan Wallace. 2000. *Healing from the Source: The Science and Lore of Tibetan Medicine*. Ithaca, Ny: Snow Lion Publications.

Pa-Saṅs-Yon-TanSman-Rams-Pa, and Yon-Tan-Rgya-Mtsho. 1998. *Dictionary of Tibetan Materia Medica*. Delhi: Motilal Banarsidass Publishers.

Tshe-Ring-Thag-Gcod, and Tshe-Ring-Sgrol-Ma. 2005. *Bod-Lugs Sman-Rtsis Kyi Tshig-Mdzod Bod-Dbyin Shan-Sbyar = Tibetan-English Dictionary of Tibetan Medicine and Astrology*. Dharamsala: Drungtso Publications.

This epilogue is ended with Medicine Buddha Wishes of the 4th Shamar Rinpoche Chökyi Dragpa (1453–1524), Holder of the Karma Kagyu Lineage. It has been adapted and updated from Bach (2016).

*I bow down to the victorious conqueror, Medicine Buddha, the king of aquamarine light.*
*May the power of truth grow.*
*May the knowledge of the healing teachings bloom without degeneration in every universe including this place of meditation.*
*May body, speech and mind of all beings be free from illness and suffering.*
*May all diseases that are arising or are already present be cured.*
*May all healing conditions come together.*
*May there not be a single medical practitioner who rejects his patients.*
*May there not be a single patient who abandons his medical practitioner.*
*May all food, medicine and behaviour be counted among the healing conditions.*
*May they not lead to death.*
*May the exact cause, condition and nature of sickness be fully understood and, after this understanding, may all exist in highest joy due to joyful completely recognized knowledge.*
*May it be known that every disease arises from confusion and karma.*
*May everyone work on removing these (confusion and karma).*
*If sickness arises, may the motivation arise to take the suffering from all beings on oneself.*
*May especially the mental diseases – craving, hate and ignorance – be completely eliminated.*

## 7.  ACKNOWLEDGMENTS

I am grateful for the infinitely precious substantial support that continues to make my work with Buddhist Tibetan Medicine possible.

I expressly wish to thank the 17th Karmapa Thaye Dorje, head of the Karma Kagyu lineage, Löpon Tsechu Rinpoche (departed 2003), Lama Ole Nydahl, and Hannah Nydahl (departed 2007), the Tibetan doctors Kunsang Dorje, Tashi Pedon, and Prof. Dr. Pasang Yonten Arya; my husband Lhakpa Gyalje Sherpa, who knows the plants of his home region Solukhumbu and

Sharkhumbu in Nepal by heart; and my parents Liselotte (departed 2010) and Engelbert Brager (departed 2022).

## 8. REFERENCES

Altieri, Miguel A., P. Rosset, and L. A. Thrupp. 1998. Review of *The Potential of Agroecology to Combat Hunger in the Developing World*. *International Food Policy Research Institut* 2020 Policy Brief (October). http://www.ifpri.org/publication/potential-agroecology-combat-hunger-developing-world. Accessed online 14 May 2021.

Bach, Anna Elisabeth. 2016. *Tibetan Buddhist Medicine. An Introduction*. https://www.academia.edu/36298 860/Tibetan_Buddhist_Medicine_Introduction. Accessed online 11 March 2021.

# Index

*Note*: Page numbers in *italics* refer to figures; Page numbers in **bold** refer to tables. Page numbers followed by 'n' refer to notes.

Index

**317**

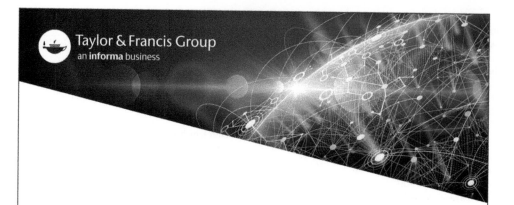